Analytical Raman Spectroscopy

CHEMICAL ANALYSIS

A SERIES OF MONOGRAPHS ON
ANALYTICAL CHEMISTRY AND ITS APPLICATIONS

VOLUME 114

A WILEY-INTERSCIENCE PUBLICATION

JOHN WILEY & SONS, INC.

New York / Chichester / Brisbane / Toronto / Singapore

Analytical Raman Spectroscopy

Edited by

JEANETTE G. GRASSELLI

Department of Chemistry
Ohio University
Athens, Ohio

and

BERNARD J. BULKIN

BP Research Centre
Sunbury-on-Thames
Middlesex, England

A WILEY-INTERSCIENCE PUBLICATION

JOHN WILEY & SONS, INC.

New York / Chichester / Brisbane / Toronto / Singapore

Library of Congress Cataloging in Publication Data:

Analytical Raman spectroscopy / edited by Jeanette G. Grasselli and Bernard J. Bulkin.
p. cm.—(Chemical analysis, ISSN 0069-2883 ; v. 114)
"A Wiley-Interscience publication"
Includes bibliographical references and index.
ISBN 0-471-51955-3
1. Raman spectroscopy. I. Grasselli, Jeanette G. II. Bulkin, Bernard J. III. Series.
QD96.R34A53 1991
543′.08584—dc20 90-25131
CIP

Printed in the United States of America

10 9 8 7 6 5 4 3 2 1

CONTRIBUTORS

Bernard J. Bulkin, BP Research Centre, Sunbury-on-Thames, Middlesex, England

D. Bruce Chase, Central Research & Development Department, E. I. DuPont de Nemours & Co., Inc., Wilmington, Delaware

Donald L. Gerrard, BP Research Centre, Sunbury-on-Thames, Middlesex, England

Jeanette G. Grasselli, Department of Chemistry, Ohio University, Athens, Ohio

Pham V. Huong, Laboratorie de Spectroscopie, Moleculaire et Cristalline, Universite de Bordeaux I, Talence, France

Donald E. Irish, Department of Chemistry, University of Waterloo, Waterloo, Ontario, Canada

Charles K. Mann, Department of Chemistry, Florida State University, Tallahassee, Florida

Mehmed Mehicic, BP Research, Warrensville Research & Environmental Science Center, Cleveland, Ohio

William F. Murphy, Steacie Institute for Molecular Sciences, National Research Council, Ottawa, Canada

Toru Ozeki, Department of Chemistry, University of Waterloo, Waterloo, Ontario, Canada. *Present address*: Hyogo University of Teacher Education, Shimokume, Yashiro-cho, Kato-gun, Hyogo, Japan

Fred H. Pollak, Department of Physics, Brooklyn College of the City University of New York, Brooklyn, New York

John F. Rabolt, IBM Research Division, Almaden Research Center, San Jose, California

James R. Scherer, Department of Chemistry, University of California, Berkeley, California

Thomas J. Vickers, Department of Chemistry, Florida State University, Tallahassee, Florida

PREFACE

The analytical applications of Raman spectroscopy continue to grow although this growth is slower than for Fourier transform infrared spectroscopy—its vibrational partner. Yet there are many situations in which Raman spectroscopy is the preferred method over infrared spectroscopy because of its capability to do *in situ* analyses, to conduct dynamic studies, to work in aqueous solutions, and to obtain information on samples in any physical state of importance to properties.

The most recent review on Raman spectroscopy in *Analytical Chemistry* (June, 1990) notes that 6000 papers on Raman spectroscopy were published in the two-year period 1987–1989. The International Conference on Raman Spectroscopy, held biennially in even years, continues to attract about 500 participants. This is strong evidence that the field is healthy and even vigorous in many areas. The introduction of Fourier transform Raman spectroscopy has reinforced this status.

The chapters in this book cover the major fields of chemistry and are illustrative of the power of Raman spectroscopy for analysis, in both practical and theoretical problems. Some of these areas will benefit from FT-Raman, but they do not depend on it. The emphasis is on interpretation and information available from the Raman spectrum in solving problems of molecular structure or composition. We are grateful to the authors who have contributed chapters to this volume. As with any edited book, the reader will find some difference in style and degree of coverage of the topics. However, each chapter should serve as an excellent guide to the literature in the subject area. The book is intended for the analytical Raman spectroscopist. There is only peripheral mention of resonance, nonlinear, or time-resolved Raman work. This in no way minimizes the importance of these interesting phenomena. We only needed to keep the book to a reasonable length.

We do owe some acknowledgments. We are especially indebted to Mrs. Marcia Dober Schiele for editorial corrections and many helpful suggestions. We also wish to thank Ms. Debra Cook for administrative assistance and for lots of typing. Finally, to our authors, we appreciate your extraordinary patience and cooperation in all phases of the completion of this book. We believe it will provide an important contribution to the field because of the expertise you have shared with our readers. We hope the readers agree.

JEANETTE G. GRASSELLI
BERNARD J. BULKIN

Athens, Ohio
Middlesex, England
July 1991

CONTENTS

CHAPTER 1 THE RAMAN EFFECT: AN INTRODUCTION **1**
Bernard J. Bulkin

1.1. Background 1
1.2. Classical Description of Raman Spectroscopy 4
1.3. Result of Quantum Mechanical Treatment of Raman Scattering 8
1.4. Selection Rules: Contrasting IR and Raman Spectra 9
1.5. Depolarization Ratios 12
1.6. Resonance Raman Effect 13
1.7. Nonlinear Raman Effects 15
1.8. Guide to the Literature of Raman Spectroscopy 17
References 18

CHAPTER 2 MODERN RAMAN INSTRUMENTATION AND TECHNIQUES **21**
D. Bruce Chase

2.1. Introduction 21
2.2. Components of Modern Raman Spectrometers 22
2.2.1. Sources 22
2.2.1.1. Continuous Wave Lasers *22*
2.2.1.2. Pulsed Lasers *24*
2.2.1.3. Lenses and Filters *24*
2.2.2. Collection Optics 26
2.2.3. Dispersing Optical Elements: Monochromators 28
2.2.4. Detection Systems 29
2.2.4.1. Single-Channel Systems *29*

2.2.4.2. Multichannel Systems *29*
2.2.4.3. Computers in Detection Systems *30*
2.3. Sample Handling 30
2.3.1. General 30
2.3.2. Microsampling 30
2.4. Problems: Fluorescence and Thermal Sample Degradation 31
2.5. Fourier Transform Raman Spectroscopy 32
2.5.1. Components 32
2.5.2. Examples of the Advantages of Fourier Transform Raman Measurements 35
2.5.3. Problems 36
2.6. Future Efforts 37
2.7. Addendum 39
2.7.1. Lasers 39
2.7.2. Lenses and Filters 40
2.7.3. Monochromators 40
2.7.4. Detectors 40
2.7.5. Microsampling 41
2.7.6. FT-Raman Spectroscopy 41
References 42

CHAPTER 3 EXPERIMENTAL CONSIDERATIONS FOR ACCURATE POLARIZATION MEASUREMENTS **45**
James R. Scherer

3.1. Introduction 45
3.2. Tools for Tuning 46
3.3. Light Path Geometry 47
3.4. Sample Container 53
3.5. Polarization Scrambler 54
3.6. The Analyzer 56
3.7. Polarization Orientation of the Excitation Laser Beam 56
3.8. Standard Raman Measurement 57
References 57

CHAPTER 4 RAMAN SPECTROSCOPY OF INORGANIC SPECIES IN SOLUTION 59
Donald E. Irish and Toru Ozeki

4.1. Introduction 59
4.2. Identification of Inorganic Species in Solution 60
4.2.1. Presentation of the Raman Spectrum 60
4.2.2. Interpretation 62
4.3. Quantitative Analysis 76
4.3.1. Correlating Raman Intensities and Species Concentrations 76
4.3.2. Spectral Analysis with Bandfitting Programs 80
4.3.3. Factor Analysis 83
4.3.4. The GAUSS-Z Program 87
4.3.5. Raman Difference Spectroscopy 90
4.4. Conclusion 90
Appendix 91
Acknowledgments 103
References 104

CHAPTER 5 QUANTITATIVE ANALYSIS BY RAMAN SPECTROSCOPY 107
Thomas J. Vickers and Charles K. Mann

5.1. Introduction 107
5.2. Historical Development 108
5.3. Variations in Source, Sample, and Optics: Effect on Quantitative Analyses 109
5.3.1. Ordinate Errors 110
5.3.2. Abscissa Errors 115
5.3.3. Data Treatment 116
5.3.4. Multichannel Versus Scanning Measurements 123
5.4. Fluorescence and Raman Measurements 124
5.5. Comparison of Raman and IR Measurements for Quantitative Analysis 127

5.5.1. Solvent Behavior 128
5.5.2. Emission Versus Absorption 128
5.5.3. Experimental Flexibility 129
5.5.4. Experimental Comparison of Sensitivity 130
Acknowledgments 132
References 133

CHAPTER 6 CHARACTERIZATION OF SEMICONDUCTORS BY RAMAN SPECTROSCOPY 137
Fred H. Pollak

6.1. Introduction 137
6.2. General Background of Raman Scattering, Particularly Relating to Crystalline Solids 138
6.2.1. Polarizability Theory of Raman Scattering 138
6.2.2. Second-Order Raman Scattering 141
6.2.3. Lattice Polarizability Including Free Carrier Effects 142
6.2.4. Fano Effect 146
6.2.5. Polarization Selection Rules 146
6.2.5.1. Atomic Displacement Effects 147
6.2.5.2. Linear-q and Surface Electric Field-Induced Effects 150
6.2.6. Dispersion Relations for Lattice Vibrations 152
6.3. Instrumentation 154
6.4. Silicon and Other Group IV Semiconductors 157
6.4.1. Microcrystalline Geometries 157
6.4.2. Structural and Crystallization Effects in a-Si, a-Ge, and a-C 163
6.4.3. Ion Damage and Laser Annealing 167
6.4.4. Reactive Ion Etching 170
6.4.5. Strain Effects 172
6.4.6. The Metal/Si Interface 174
6.4.7. Laser Writing of Si Microstructures 180
6.4.8. Temperature Probe 180
6.5. Binary Zincblende Semiconductors 180

6.5.1. Carrier Concentration and Space/Charge Layer Effects 181
6.5.2. Microcrystalline Effects 185
6.5.3. Symmetry-Forbidden Transverse Optical Mode Scattering 186
6.5.4. Ion Implantation and Annealing 186
6.5.5. Process-Induced Damage 193
6.6. Alloy Semiconductors 200
6.6.1. Alloy Composition 200
6.6.2. Alloy Potential Fluctuations 200
6.6.2.1. Optic Phonons 200
6.6.2.2. Disorder-Activated Zone-Edge Phonons 205
6.6.3. Ion Implantation and Annealing 206
6.6.4. Carrier Concentration 206
6.6.5. $Hg_{1-x}Cd_xTe$ 207
6.6.6. Heterojunctions 208
6.6.7. Two-Phonon Raman Scattering 209
6.7. Surface Films Including Oxides 211
References 214

CHAPTER 7 POLYMER APPLICATIONS 223
Bernard J. Bulkin
7.1. Introduction 223
7.2. Raman Spectrum of Poly(ethylene terephthalate) 224
7.3. Vibrational Analysis of Polymers as a Foundation to Understanding 226
7.4. Sensitivity of the Raman Spectrum to Conformational Change 230
7.5. Low-Frequency ($< 200\ cm^{-1}$) Spectra 231
7.6. Features in the Spectra of Fibers 233
7.7. Considerations in Selecting Spectroscopic Parameters for Comparing Spectra 235
7.8. Orientation, Conformation, and Crystallinity 237
7.9. Amorphous Orientation and Low-Frequency Spectra 242
7.10. Derivation of Polymer Specific Heats 244

7.11. A Guide to Other Raman Studies of Polymers 248
7.12. Conclusion 251
References 251

CHAPTER 8 ANISOTROPIC SCATTERING PROPERTIES OF UNIAXIALLY ORIENTED POLYMERS: RAMAN STUDIES 253
John F. Rabolt

8.1. Introduction 253
8.2. Experimental Scattering Geometry 254
8.3. Geometric Considerations 256
8.4. Symmetry Considerations 257
8.5. Example: Planar Zigzag Poly(vinylidene fluoride) 269
Acknowledgment 273
References 273

CHAPTER 9 ORGANIC AND PETROCHEMICAL APPLICATIONS OF RAMAN SPECTROSCOPY 275
Donald L. Gerrard

9.1. Introduction 275
9.2. Organic and Petrochemical Sample-Handling Techniques 277
9.2.1. Versatility and Simplicity Owing to the Scattering Process 277
9.2.2. Glass Sample Containers and Aqueous Systems 279
9.2.3. Excitation Using Visible Lasers 282
9.2.3.1. Use of Fiber Optics 283
9.3 Characterization of Nonpolar Groups 286
9.4. Quantitative Studies 288
9.4.1. Internal Standard Method 289
9.4.2. Reacting Systems 290
9.5. Spatial Resolution: Raman Microscopy 291
9.6. Methods of Signal Enhancement 295
9.6.1. Resonance-Enhanced Raman

Spectroscopy and Applications in Organic and Petrochemical Systems 295
9.6.2. Surface-Enhanced Raman Spectroscoy and Applications 301
9.6.2.1. Electrochemical Applications *303*
9.6.2.2. Colloids *303*
9.6.2.3. Metal-Island Films *303*
9.6.2.4. General Analytical Applications for Organic Materials *304*
9.7. Time-Resolved Studies: Examples of *In Situ* Studies of Reacting Systems 304
9.7.1. Homopolymerization of Styrene 308
9.7.2. Homopolymerization of Methyl Methacrylate 310
9.7.3. Copolymerizations and Multicomponent Polymerizations: The Use of Mathematical Deconvolution 311
9.7.4. Other Applications 312
9.8. Fluorescence in Organic Systems 312
9.8.1. Fluorescence Reduction 313
9.8.1.1. Sample Purification *313*
9.8.1.2. "Burning Out" with the Laser Beam *313*
9.8.1.3. The Use of UV and Near-IR Lasers *314*
9.8.1.4. Fluorescence-Quenching Agents *315*
9.8.1.5. Anti-Stokes Raman Spectroscopy *316*
9.8.2. Extraction of the Raman Signal from the Fluorescence Background 316
9.8.2.1. Background Subtraction *316*
9.8.2.2. Use of Picosecond Pulsing and Gating *316*
9.8.2.3. Nonlinear Raman Spectroscopy *317*
9.8.3. Conclusions 317
References 317

CHAPTER 10 RAMAN SPECTROSCOPY IN CATALYSTS **325**
Mehmed Mehicic and Jeanette G. Grasselli

10.1. Introduction 325
10.1.1. Advantages of Raman Spectroscopy in Catalytic Studies 325

10.1.2. Miscellaneous Applications 333
10.2. Zeolites 334
10.3. Hydrodesulfurization Catalysts 339
10.3.1. MoO_3-Based Catalysts 342
10.3.1.1. Background and Preparation Variables *342*
10.3.1.2. Promoter (Cobalt-Containing) Structures *345*
10.3.1.3. Promoter (Nickel-Containing) Structures *347*
10.3.1.4. Activated (Sulfided) Molybdenum-Containing Catalysts *349*
10.3.1.5. Conclusion *351*
10.3.2. WO_3-Based Catalysts 352
10.4. Oxidation Catalysis 355
10.4.1. Vanadium Oxide Catalysts 355
10.4.2. Manganese Oxide Catalysts 358
10.4.3. Bismuth Molybdate Catalysts 359
10.4.4. Iron Oxide Catalysts 374
10.5. Other Systems 375
10.5.1. Metathesis Catalysis 375
10.5.2. Olefin Hydrogenation 377
10.5.3. Syngas Processes 377
10.5.3.1. Fischer–Tropsch Catalysis *378*
10.5.3.2. Hydroformylation Catalysis *380*
10.5.3.3. Methanation *381*
10.5.4. Olefin Dehydrogenation 381
References 382

CHAPTER 11 RAMAN SPECTROSCOPY FOR BIOLOGICAL APPLICATIONS **397**
Pham V. Huong

11.1. Introduction 397
11.2. Specific Problems in Biological Studies 398
11.2.1. Identification of Biomolecules 398

11.2.2. Structure in Aqueous Solutions 398
11.2.3. Dynamics and Interactions 398
11.3. Instrumentation and Procedures 398
11.4. Structures 401
11.4.1. Raman Spectra of Living Organisms/Tissues/Cells 401
11.4.1.1. Eye Lens *402*
11.4.1.2. Biological Membranes *402*
11.4.2. Raman Spectra of Isolated Biomolecules 402
11.4.2.1. Structures and Conformations of t-RNA *402*
11.4.2.1.1. Structures in Solutions and Solids *402*
11.4.2.1.2. Endo-melting (Ordered to Disordered State) *406*
11.4.2.1.3. Conformation Transitions Between Ordered States *407*
11.5. Molecular Dynamics 408
11.6. Interactions 409
11.6.1. Selective Enhancement of Raman Bands Corresponding to the Interaction Site of Hemoglobin 409
11.6.2. Changes in the Active Site of Hemoglobin with Interactions 410
11.6.3. α-Helical and Random-Coil Structures in Proteins 410
11.7. Mechanisms 412
11.7.1. Vision 412
11.7.2. Transmission in Nervous Systems 413
11.7.3. Protonation of Biomolecules 414
11.7.4. S—H/S—S Conversion 415
11.7.5. Carcinogenesis 417
11.8. Conclusions 421
Acknowledgments 421
References 421

CHAPTER 12 CHEMICAL APPLICATIONS OF GAS-PHASE RAMAN SPECTROSCOPY **425**
William F. Murphy

12.1. Introduction 425
12.2. Experimental Aspects 426
12.2.1. Instrumentation 426
12.2.1.1. Source *426*
12.2.1.2. Sample Irradiation *427*
12.2.1.3. Sample Cells *427*
12.2.1.4. Transfer Optics *428*
12.2.1.5. Monochromator and Detector *429*
12.2.2. Instrumental Calibration 430
12.2.2.1. Frequency Calibration *430*
12.2.2.2. Intensity Response Calibration *432*
12.3. Applications 434
12.3.1. Theoretical Background 434
12.3.2. Methodology 437
12.3.2.1. Measurement of Species Abundance *437*
12.3.2.2. Measurement of Temperature *438*
12.3.3. Survey of Applications 440
12.3.3.1. Combustion Diagnostics *440*
12.3.3.2. Remote Sensing *442*
12.3.3.3. Analytical Applications *443*
12.4. Conclusions 446
References 447

INDEX **452**

CHAPTER

1

THE RAMAN EFFECT: AN INTRODUCTION

BERNARD J. BULKIN

BP Research Centre
Sunbury-on-Thames
Middlesex, England

1.1. BACKGROUND

The early 1920s was a time of intense interest in the scattering of electromagnetic radiation by charged particles. The Compton effect, the name given to the quantum mechanical explanation of changes in wavelength of X-ray photons when scattered by electrons, was first documented in 1923. Almost immediately Smekal, in Germany, predicted the inelastic scattering of light by molecules in terms of an analog to the Compton effect. There was a related prediction, in terms of classical electromagnetic theory, by Kramers and Heisenberg in 1925. With this background, the experimental observation of what C. V. Raman in India called at first "feeble fluorescence" and later termed the "New Radiation" seems to have been almost inevitable. It is not surprising, therefore, that it was reported almost simultaneously by both Raman (1) and Landsberg and Mandelstam in Moscow (2) in 1928. Moreover, the great American optical scientist R. W. Wood was so close to this discovery himself that he was able to immediately confirm Raman's observations in his own laboratory. The importance of this phenomenon was recognized by the award of the Nobel Prize in Physics to Raman in 1930, and it has been known as the Raman effect, Raman scattering, or Raman spectroscopy ever since.

The Raman effect, as we shall use the term for most of this book, is a light-scattering effect. Raman scattering is most easily seen as the change in frequency for a small percentage of the intensity in a monochromatic beam

Analytical Raman Spectroscopy, Edited by Jeanette G. Grasselli and Bernard J. Bulkin. Chemical Analysis Series, Vol. 114.
ISBN 0-471-51955-3

as the result of interacting with some material. For Raman spectroscopy as it is usually practiced, these frequency changes occur as the result of coupling between the incident radiation and vibrational energy levels of molecules. However, Raman scattering is well established from electronic and rotational energy levels as well. Although for practical reasons a monochromatic incident beam is used to observe and measure Raman spectra, the effect can often be seen inadvertently—for example, in the measurement of fluorescence.

Raman scattering is observed as the appearance of a signal at a frequency where this signal did not exist previously (or as an increase in an existing signal). [Because Raman spectroscopy is often used to probe vibrational energy levels and is thus closely tied to infrared (IR) absorption spectroscopy, one sometimes hears the term "Raman absorption," but this is incorrect.] The spectroscopic region in which the Raman effect is observed will depend on the energy of the incident radiation and on the molecular energy levels that are involved in shifting this radiation. By far the most common case is when both incident and Raman scattered radiation are in the visible region of the spectrum. However, there are many applications of both ultraviolet (UV) and near-IR-excited Raman spectroscopy.

At the time of the discovery of the Raman effect the practice of experimental IR spectroscopy was very difficult, and it remained so for the next two decades. By contrast, instrumentation for Raman spectroscopy was readily available as a consequence of its similarity with that then used for atomic spectroscopy, and there were many scientists trained in the use of such instrumentation. Sources were usually high-pressure mercury arcs filtered to produce monochromatic radiation, and these were coupled to high-quality spectrometers with photographic detection.

During this period, much was being discovered and proven about molecular structure, and Raman spectroscopy flourished. A major work by Hibben (3) in 1939 summarized the theoretical understanding and tabulated virtually all experimental results to that point. This remains a useful reference work on the Raman spectra of many organic molecules.

In the 1950s and 1960s there was considerable development of commercial IR spectrometers with the introduction of ratio recording, double-beam IR instruments. Despite its early success, there were experimental difficulties with Raman spectroscopy that limited its use. Therefore the technique of Raman spectroscopy was practiced by a relatively few specialists.

This situation began to change with the introduction of the first commercial, reliable, continuous wave (CW) gas lasers (He–Ne lasers with output powers from 0.5 to 80.0 mW) that became readily available in the mid-1960s. At about the same time, developments in diffraction gratings, photomultiplier tubes, and photon-counting equipment made a new generation of commercial Raman spectrometers possible. By 1970 such instruments were in widespread use, and the number has continued to grow ever since. The availability of high-quality holographic gratings in the 1970s eliminated many of the artifacts associated with early Raman instruments, particularly in the low-frequency region of the vibrational spectrum. The Ar^+ laser, a source of intense CW radiation in the blue and green regions of the spectrum, became available and eventually became reliable. This increased the intensity of the exciting radiation available for Raman spectroscopy by more than an order of magnitude, while shifting the experiment to a region where photomultiplier detectors were much more sensitive. In the 1980s diode array detection became possible, bringing a multiplex advantage back to Raman spectroscopy that it had given up in the change from photographic to photoelectric detection. The effective coupling of microscopes to Raman spectrometers also expanded the range of applications.

The growth in analytical applications of Raman spectroscopy, despite continued advances in IR instrumentation, has been stimulated by several key features. Even without microscopes, Raman spectra can be obtained on very small samples. Single crystals comparable in size to those used for X-ray diffraction, single grains of powder, individual filaments from synthetic or natural polymers, and liquid sample volumes as small as 1 nl can all be readily examined, usually without additional sample preparation. Samples can be oriented where that is important, e.g., for single polymer fibers. The weak Raman scattering from water makes the study of aqueous solutions much more facile than their study by IR spectroscopy. This type of study is also aided by the fact that most spectra are obtained in the visible region. This has driven much of the application of Raman spectroscopy to biological systems.

The geometry used in experimental Raman spectroscopy also facilitates the use of special cells, such as high-pressure and/or -temperature cells for *in situ* spectroscopy in reactors. Raman spectroscopy also offers the possibility of observing the entire vibrational spectrum, from low-frequency modes such as lattice, torsional, or chain modes, to modes, that appear in the fingerprint region, and on to higher frequency fundamentals, with a single instrument. The use of low-frequency ($< 200\,cm^{-1}$) modes in analytical applications has not been exploited to its potential.

Raman spectroscopy has been limited in its applications by one major point—fluorescence. As a phenomenon, fluorescence is approximately 10^6–10^8 times stronger than Raman scattering. Often when one tries to excite a Raman spectrum, fluorescence is the only phenomenon observed. Trace impurities, coatings on polymers, additives, etc., may fluoresce so strongly that it is impossible to observe the Raman spectrum of a major component. The use of UV or near-IR excitation has proved to be effective in reducing this problem. The difference in time scales between the process of fluorescence and that of Raman scattering has also been exploited to separate them, though this is experimentally complex.

While the use of laser sources simplifies Raman spectroscopy, it can also lead to other problems. Photochemistry in the intense radiation of a focused laser beam may be severe. For some samples, heating and black-body radiation may be so efficient that it is impossible to observe Raman spectra. Even when samples do not absorb in the visible region, multiphoton processes may occur that alter the spectrum. It is necessary to distinguish these problems from fluorescence, as the experimental methods to overcome them differ.

1.2. CLASSICAL DESCRIPTION OF RAMAN SPECTROSCOPY

By far the simplest way of thinking about the mechanism of Raman spectroscopy is via an energy level diagram such as that shown in Fig. 1.1. An incident photon of energy $h\nu_0$ interacts with a molecule having vibrational energy levels $\nu_1, \nu_2, \ldots$ as shown. Most of the incident radiation is unchanged in energy. It is transmitted, refracted, reflected, or even scattered, but at the same energy. A small portion of the energy, however, is lost to the energy levels shown, and appears as $h(\nu_0 - \nu_1)$, $h(\nu_0 - \nu_2), \ldots$. This is the Raman scattered radiation.

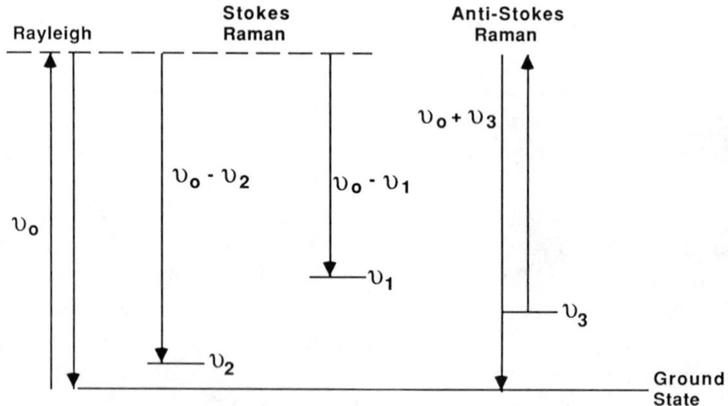

Figure 1.1. Energy level diagram showing the basic transitions involved in Raman scattering from energy levels of $h\nu_i$, when excited by energy of $h\nu_0$.

If $\nu_1, \nu_2, \ldots$ are relatively close to the ground state, at ordinary temperatures these levels will have a significant population determined by the Boltzmann distribution. In this case, molecules in the vibrationally excited states can interact with the incident radiation and return to the ground state. This will result in energies of $h(\nu_0 + \nu_1)$, $h(\nu_0 + \nu_2), \ldots$ being observed. The shifts to lower and higher energy are known as Stokes and anti-Stokes Raman scattering, respectively.

In all spectroscopies there is a mechanism by which the incident radiation interacts with the molecular energy levels. For IR absorption spectroscopy associated with molecular vibrational energy levels, it is the change in dipole moment during the vibration. For Raman spectroscopy, the mechanism has its origins in the general phenomenon of light scattering, in which the electromagnetic radiation interacts with a pulsating, deformable (polarizable) electron cloud. In the specific case of vibrational Raman scattering, this interaction is modulated by the molecular vibrations.

Figure 1.2 illustrates the basic light-scattering experiment. Electromagnetic radiation with electric vectors in the xy plane propagates along the z direction and strikes a (spherical) polarizable sample. This causes the electron cloud

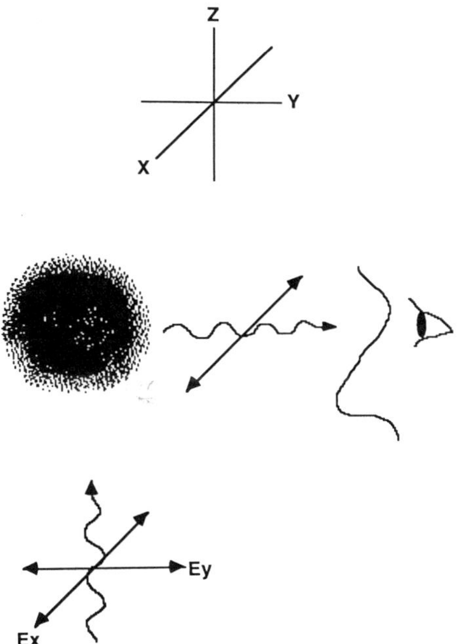

Figure 1.2. Polarization of light by scattering. In the experiment shown, unpolarized radiation propagates along z, and polarized radiation is observed along y.

to oscillate, driven by the frequency of the incident radiation. A pulsating electron cloud radiates in all direction, this radiation being light scattering.

If an observer looks along a direction in the xy plane, the radiation is seen as polarized, as shown in Fig. 1.2. This is because the incident electric vectors force oscillation only in the xy plane and an observer in this plane cannot see electric vectors along the direction of propagation. So it happens that in light scattering we have the possibility of unpolarized incident radiation leading to an observation of polarized scattered radiation. We observe this phenomenon in the sky where polarized solar radiation occurs from light scattering off water droplets and dust in the atmosphere.

Suppose that the incident electric field associated with the light, which is a wave phenomenon, is represented by

$$\mathbf{E} = E_0 \cos 2\pi\nu t,$$

where $\mathbf{E}$ is the time-dependent intensity, E_0 the maximum amplitude, and ν the frequency. This field induces a dipole μ, such that

$$\boldsymbol{\mu} = \alpha\mathbf{E} = \alpha E_0 \cos 2\pi\nu t,$$

where the proportionality constant α is known as the polarizability. Classical theory gives the average rate of total radiation as

$$I = \frac{16\pi^4}{3c^3} \nu^4 \mu_0^2,$$

where μ_0 is the amplitude of μ. For this case the scattered radiation has the same frequency as the incident.

The expression for μ can be rewritten in terms of Cartesian components; in its most general form

$$\begin{aligned}
\mu_x &= \alpha_{xx}E_x + \alpha_{xy}E_y + \alpha_{xz}E_z \\
\mu_y &= \alpha_{yx}E_x + \alpha_{yy}E_y + \alpha_{yz}E_z \\
\mu_z &= \alpha_{zx}E_x + \alpha_{zy}E_y + \alpha_{zz}E_z,
\end{aligned}$$

and this can be rewritten as a matrix equation $\boldsymbol{\mu} = \boldsymbol{\alpha}\mathbf{E}$, that is,

$$\begin{bmatrix} \mu_x \\ \mu_y \\ \mu_z \end{bmatrix} = \begin{bmatrix} \alpha_{xx} & \alpha_{xy} & \alpha_{xz} \\ \alpha_{yx} & \alpha_{yy} & \alpha_{yx} \\ \alpha_{zx} & \alpha_{zy} & \alpha_{zz} \end{bmatrix} \begin{bmatrix} E_x \\ E_y \\ E_z \end{bmatrix}.$$

For almost every case, α is a symmetric matrix ($\alpha_{xy} = \alpha_{yx}$, etc.).

Now suppose that the scattering body is not just a polarizable sphere but has vibrational modes of its own—normal modes Q described by

$$Q_k = Q_k^{\circ} \cos 2\pi \nu_k t.$$

These oscillations can affect the polarizability, and this effect can be written as

$$\alpha = \alpha_0 + \left(\frac{\partial \alpha}{\partial Q_k^{\circ}}\right) Q_k + \text{higher-order terms.}$$

Multiplying by $\mathbf{E}$ gives

$$\alpha \mathbf{E} = \boldsymbol{\mu} = \alpha_0 \mathbf{E} + \left(\frac{\partial \alpha}{\partial Q_k^{\circ}}\right) Q_k \mathbf{E}.$$

The expression for μ now becomes

$$\mu = \alpha_0 E_0 \cos 2\pi \nu t + E_0 Q_k^{\circ} \left(\frac{\partial \alpha}{\partial Q_k}\right) \cos 2\pi \nu t \cos 2\pi \nu_k t.$$

Using a trigonometric identity for the product of two cosines, this can be rewritten as

$$\mu = \alpha_0 E_0 \cos 2\pi \nu t + \tfrac{1}{2} E_0 Q_k^{\circ} \left(\frac{\partial \alpha}{\partial Q_k}\right) [\cos 2\pi(\nu + \nu_k)t + \cos 2\pi(\nu - \nu_k)t].$$

The three terms of this equation represent the three major phenomena observed in a simple Raman spectroscopy experiment: the first term is elastic scattering, known as Rayleigh scattering (see the first diagram in Fig. 1.1); the second term, of frequency $\nu + \nu_k$, is anti-Stokes Raman scattering; and the third term is Stokes Raman scattering.

The classical description gives only a very limited insight into the relative intensities of each of these phenomena. One does expect that $\partial \alpha / \partial Q_k$ will be much smaller than α_0, so that the Raman scattering should be less intense than the Rayleigh scattering. This is in fact the case. Moreover, the classical prediction indicates a simple, linear dependence of Raman scattering on incident beam intensity and sample concentration, again consistent with experiment except for certain special cases to be described later.

The relative intensities of the Stokes and anti-Stokes scattering are only predicted to differ by the ratio of $[(\nu - \nu_k)/(\nu + \nu_k)]^4$, which is not in accord with observation. As indicated earlier, the Boltzmann distribution will be the major factor in determining the relative intensities of these two phenomena. The population of any excited level is always less than that of the ground state, making Stokes Raman scattering always more intense than anti-Stokes.

1.3. RESULT OF QUANTUM MECHANICAL TREATMENT OF RAMAN SCATTERING

A full quantum mechanical treatment of the Raman effect is usually done using time-dependent perturbation theory. The most useful version of this is that found in Long (4), and only certain key results will be given here.

From the classical approach it can be appreciated that the geometry of the sample and that of the experiment (incident, observing directions) will affect the observations. For the purposes of analytical chemistry, the most important samples are liquids and randomly oriented solids. The most commonly used experimental geometry is observation at right angles to excitation, although there are, on occasion, good reasons for observing the scattering in other directions, particularly at 180° to the direction of excitation. A special case of interest, oriented polymers, is discussed in Chapter 8 by Rabolt.

Placzek (5) originally derived the expressions for Raman scattering in different geometries, including the conventional 90° scattering, and put these into a convenient form. In these expressions, the polarizability (α) is divided into two parts:

$$\alpha = \alpha^s + \alpha^a,$$

where α^s is the symmetric or isotropic part and α^a the asymmetric or anisotropic part. These are defined as

$$3\alpha^s = \alpha_{xx} + \alpha_{yy} + \alpha_{zz}$$

$$2(\alpha^a)^2 = [(\alpha_{xx} - \alpha_{yy})^2 + (\alpha_{yy} - \alpha_{zz})^2 + (\alpha_{zz} - \alpha_{xx})^2 + 6(\alpha_{xy}^2 + \alpha_{yz}^2 + \alpha_{zx}^2)].$$

It is possible to make a transformation from Cartesian coordinates to principal axes so that these expressions take the simpler forms

$$\alpha^s = \tfrac{1}{3}(\alpha_1 + \alpha_2 + \alpha_3)$$

and

$$2(\alpha^a)^2 = (\alpha_1 - \alpha_2)^2 + (\alpha_2 - \alpha_3)^2 + (\alpha_3 - \alpha_1)^2.$$

For molecular vibrations, it is not the polarizabilities themselves that we are dealing with but rather the elements of the matrix of polarizability derivatives, $(\partial\alpha/\partial Q)$, usually designated as α'.

Placzek's result for Raman scattering at right angles in terms of these components of the polarizability derivatives connecting a molecule intially in state m and finally in state n is

$$I = \text{constant}\frac{(\nu_0 + \nu_{mn})^4}{\nu_{mn}}\cdot\frac{NI_0}{1 - \exp(-h\nu_{mn}/kt)}[45(\alpha'^s)^2 + 13(\alpha'^a)^2],$$

where N is the number of molecules in state m, and I_0 is the incident intensity. The constants 45 and 13 arise from the orientational averaging process [see Long (4) for details] and are a consequence of the experimental geometry. This yields the ratio of Stokes to anti-Stokes intensity,

$$\frac{I_{\text{Stokes}}}{I_{\text{anti-Stokes}}} = \frac{(\nu_0 - \nu_{mn})^4}{(\nu_0 + \nu_{mn})^4}\exp(h\nu_{mn}/kT),$$

which is verified experimentally at thermal equilibrium. These expressions assume that ν_0 is far from any electronic energy levels of the molecule.

1.4. SELECTION RULES: CONTRASTING IR AND RAMAN SPECTRA

What we have done so far only gives us the terms in the expression for Raman intensity. It does not say whether the key terms in this expression, the α's, are nonzero for a particular vibrational mode. In fact, this is very difficult to predict. But group theory allows us to predict whether these terms *can* be nonzero, using information about the symmetry of a molecule or crystal. In each case, group theory is used to predict whether a transition moment integral can be nonzero. These integrals contain the product of three terms—the wave functions for the ground and excited states, and the operator (in this case the components of the polarizability derivatives) that connects these two states. For a transition to be observed, the product of these three terms must be totally symmetric, that is, it must leave the original molecule unchanged.

One finds that in molecules of high symmetry both IR and Raman spectroscopy are needed to observe the vibrational modes. Even with both techniques, there may still be some vibrations that are totally forbidden. The best known selection rule for IR and Raman spectroscopy is known as the "Rule of Mutual Exclusion," which states that if a molecule has a center of symmetry, vibrations cannot be active in both IR and Raman spectroscopy. This rule has often been applied in molecular structure investigations to determine if a center of symmetry is present. In general, vibrations that do not distort the molecule, symmetric vibrations, are intense in the Raman spectrum while those that maximize the distortion are most intense in the IR spectrum. If the atoms involved in these vibrations are highly polarizable (e.g., sulfur or iodine), then the Raman intensity is high. Some examples of vibrational modes that are of importance in Raman scattering and their frequency ranges are shown in Table 1.1.

Nishimura et al. (6) have summarized the common observations about Raman spectral intensities in four main generalizations:

- Stretching vibrations associated with chemical bonds should be more intense than deformation vibrations.
- Multiple chemical bonds should give rise to intense stretching modes, e.g., a Raman band due to a C=C vibration should be more intense that that due to a C—C vibration.
- Bonds involving atoms of large atomic mass are expected to give rise to stretching vibrations of high Raman intensity. The S—S linkages in proteins are good examples of this.
- Those Raman features arising from normal coordinates involving two in-phase bond stretching motions are more intense than those involving

Table 1.1. Characteristic Wavenumbers and Raman and Infrared Intensities of Groups in Organic Compounds

		Intensity[b]	
Vibration[a]	Region(cm^{-1})	Raman	Infrared
ν(O—H)	3650–3000	w	s
ν(N—H)	3500–3300	m	m
ν(≡C—H)	3300	w	s
ν(=C—H)	3100–3000	s	m
ν(—C—H)	3000–2800	s	s
ν(—S—H)	2600–2550	s	w
ν(C≡N)	2255–2220	m–s	s–0
ν(C≡C)	2250–2100	vs	w–0
ν(C=O)	1820–1680	s–w	vs
ν(C=C)	1900–1500	vs–m	0–w
ν(C=N)	1680–1610	s	m
ν(N=N), aliphatic substituent	1580–1550	m	0

Table 1.1. (*Continued*)

Vibration[a]	Region(cm^{-1})	Intensity[b] Raman	Infrared
ν(N=N), aromatic substituent	1440–1410	m	0
ν_a((C—)NO_2)	1590–1530	m	s
ν_s((C—)NO_2)	1380–1340	vs	m
ν_a((C—)SO_2(–C))	1350–1310	w–0	s
ν_s((C—)SO_2(—C))	1160–1120	s	s
ν((C—)SO(—C))	1070–1020	m	s
ν(C=S)	1250–1000	s	w
$\delta(CH_2)$, $\delta_a(CH_3)$	1470–1400	m	m
$\delta_s(CH_3)$	1380	m–w, s, if at C=C	s–m
ν(CC), aromatics	1600, 1580	s–m	m–s
	1500, 1450	m–w	m–s
	1000	s (in mono-; m-; 1,3,5-derivatives)	0–w
ν(CC), alicyclics, and aliphatic chains	1300–600	s–m	m–w
ν_a(C—O—C)	1150–1060	w	s
ν_s(C—O—C)	970–800	s–m	w–0
ν_a(Si—O—Si)	1110–1000	w–0	vs
ν_s(Si—O—Si)	550–450	vs	w–0
ν(O—O)	900–845	s	0–w
ν(S—S)	550–430	s	0–w
ν(Se—Se)	330–290	s	0–w
ν(C(aromatic)—S)	1100–1080	s	s–m
ν(C(aliphatic)—S)	790–630	s	s–m
ν(C—Cl)	800–550	s	s
ν(C—Br)	700–500	s	s
ν(C—I)	660–480	s	s
δ_s(CC), aliphatic chains	400–250	s–m	w–0
C_n, $n = 3, \ldots, 12$	$2495/n$		
$n > 12$			
Lattice vibrations in molecular crystals (liberations and translational vibrations)	200–20	vs–0	s–0

Source: Reprinted from B. Schrader, *Angew. Chem.* **12**, 882 (1973), with permission of Verlag Chemie, GMBH, Weirheim, Germany.

[a] ν, stretching vibration; δ, bending vibration; ν_s; symmetric vibration; ν_a; antisymmetric vibration.

[b] vs, very strong; s, strong; m, medium; w, weak; 0, very weak or inactive.

a 180° phase difference. Similarly, for cyclic compounds the in-phase "breathing" mode is usually the most intense.

1.5. DEPOLARIZATION RATIOS

An additional piece of information can be derived from Raman spectra of liquids, gases, and clear solids. Refer back to Fig. 1.2. The incident beam is unpolarized in the xy plane, and measurements of the Raman scattering at 90° in the y direction can be made with a polarizer passsing E_x or E_z. The intensity measured in the former case (polarizer passing E_x) is known as $I_{\parallel}$ (parallel to the incident plane) and the latter as $I_{\perp}$. In the usual geometry now used for experiments, a laser is the source of incident radiation and it is completely polarized, along x in this example. Measurement of the depolarization ratio

$$\rho = I_{\perp}/I_{\parallel}$$

is readily done using two positions of an analyzing polarizer in the scattered light. It can be shown (4) that for the experimental geometry just described

$$\rho = \frac{3(\alpha'^{a})^2}{45(\alpha'^{s})^2 + 4(\alpha'^{a})^2}.$$

Again, the actual coefficients (3, 45, and 4) depend on the experimental geometry chosen. Long (4) has summarized results for many different geometries.

Clearly α^s and α^a have different symmetry properties. Only for totally symmetric vibrations, those which maintain all of the symmetry elements of the molecule, can α^s be nonzero. Thus for all other (nontotally symmetric) vibrations, $\rho = 3/4$, whereas for totally symmetric vibrations, $0 \leqslant \rho < 3/4$. If a molecule has nearly spherical symmetry, such as CH_4, then depolarization ratios close to zero are observed. This would be the case for normal vibrations such as the totally symmetric stretching of all four hydrogens. For more complex molecules, however, a full range of values of ρ are generally observed.

Despite the fact that most analytical chemistry is concerned with complex species, depolarization ratios should still be useful in chemical analysis. Freeman (7) has shown that, in complex organic molecules, ρ values often give key structural information in that they vary in a systematic way with substitution, etc. Depolarization ratios have also been used in the study of low-symmetry molecules to determine if the molecules have a plane of symmetry or no symmetry at all (8). The main limitation in the use of ρ values in analytical chemistry has been the poor quality of these data, particularly the lack of interlaboratory reproducibility. While the measurement is inherently simple, it is critically dependent on control of experimental geometry. More attention needs to be given to this problem. An explicit experimental procedure is presented in Chapter 3.

1.6. RESONANCE RAMAN EFFECT

Until now in this chapter we have made a major simplification in the theory of the Raman effect, namely, that no electronic energy levels are close to the energy of the exciting beam. This simplification has led us to omit terms from the intensity expression that contain polarizability derivatives in the numerator and, in the denominator, expressions such as $(\nu_{el} - \nu_0)$, where ν_{el} is the frequency of an electronic transition. When the frequency of the exciting line is far from the frequencies of electronic transitions, then these differences are large and such terms are negligible. As the exciting frequency approaches that of electronic transitions, however, these terms dominate the Raman scattering, and the observed spectrum changes dramatically, showing an enhancement of intensity of certain vibrational modes.

This phenomenon is known as resonance Raman scattering or the resonance Raman effect. Its major features were predicted by Placzek in 1934, and it has been the subject of numerous theoretical treatments in more recent years. The first experimental observation was by Shorygin in 1947. A mild resonance effect known as preresonance Raman scattering is often observed. Figure 1.3 contrasts normal Raman scattering, preresonance Raman, resonance Raman, and fluorescence.

In resonance Raman scattering, the intensity of vibrational modes is affected by the associated electronic transitions. For example, in a complex substituted aromatic species, there might be considerable enhancement of ring modes and very little of vibrational modes associated with aliphatic substituents.

Resonance Raman is extremely important for the study of certain biological molecules, such as hemoglobin, where the resonance spectrum is asssociated only with the heme portion while the major component in terms of molecular weight, the protein globin, is not seen at all when resonance excitation is used. The use of resonance excitation is likely to become more important in the future for all of analytical Raman spectroscopy. It has no analog in IR spectroscopy.

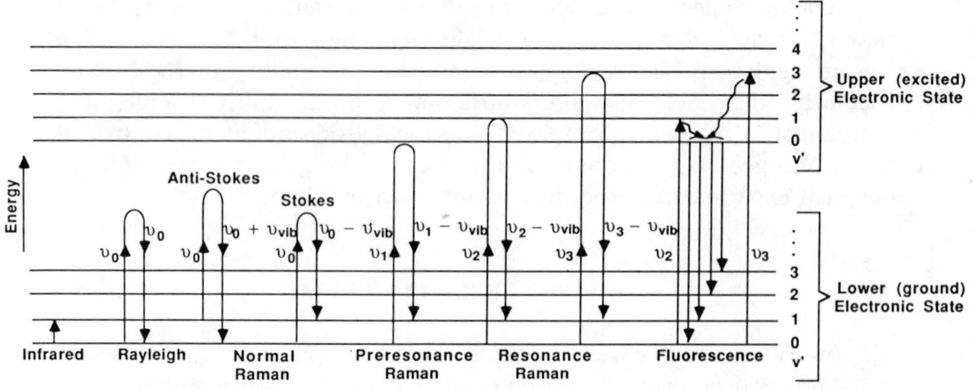

Figure 1.3. Some of the possible consequences of a photon–molecule interaction, showing the relationship between Raman scattering and electronic energy levels. [From Carey (14).]

It is possible to measure resonance Raman scattering of certain species at four or more orders of magnitude lower concentration than that needed for conventional Raman spectra. While visible excitation restricts this to a relatively small class of molecules, UV excitation at wavelengths as low as 200 nm is now readily available, making resonance Raman scattering routinely possible for most molecules.

It is now recognized that there are two distinct mechanisms responsible for resonance Raman intensities. The first (designated as the A term in expressions for this intensity) involves a single excited electronic state. The second (or B term) involves two excited electronic states. Which mechanism is operating can sometimes be determined by studies of the frequency dependence of intensities. When the A-term mechanism operates, one often sees significant enhancement of overtone vibrations in the spectra.

The frequencies of vibrational modes observed in resonance Raman spectra depend only on the ground state of the molecule, whereas the intensities depend upon the electronic excited states. This has been applied in the use of vibrational spectroscopy to elucidate the electronic excited states involved in photochemical processes.

Resonance Raman spectra may show very different depolarization ratios from nonresonance excitation. For certain symmetries Placzek (5) predicted that the polarizability tensor would be unsymmetrical, for example, $\alpha_{xy} \neq \alpha_{yx}$. This has been experimentally verified. If the usual experimental geometry is used for measuring ρ in such situations, depolarization ratios > 1 may be observed.

1.7. NONLINEAR RAMAN EFFECTS

In our treatment of the basic theory of Raman scattering, the induced dipole moment in light scattering was given by the formula

$$\mu = \alpha E,$$

where α is the polarizability. Again, this is simplification. In fact the induced dipole is given by a series of terms of which this is only the first. A more general expression defines the induced dipole as

$$\mu = \alpha \mathrm{E} + \beta \mathrm{EE} + \gamma \mathrm{EEE} + \cdots,$$

where α is the term already considered, β is the hyperpolarizability, and γ is the second hyperpolarizability. Each of these is a tensor quantity, whose components typically drop by 10 orders of magnitude for each higher-order term in the equation. The intensities required to observe effects associated with these terms are obtained with very intense, generally pulsed lasers.

There exists a whole family of "Raman" effects associated with the higher-order terms, and these are collectively known as nonlinear Raman spectroscopy. The β spectroscopies include stimulated Raman gain spectroscopy, inverse Raman spectroscopy, and the Raman-induced Kerr effect. The most developed of the nonlinear Raman effects, and the one that has found broadest chemical application, is a γ effect—coherent anti-Stokes Raman scattering, or CARS. Excellent introductions to both the theory and chemical applications of nonlinear Raman effects are found in the books edited by Harvey (9) and Kiefer and Long (10). Only a brief introduction will be given here. In this introduction ν_e will be the exciting frequency, ν_s will correspond to a Raman-scattered Stokes frequency, and ν_v will correspond to a vibrational energy level.

The hyper-Raman effect is the easiest to understand. It is shown in the energy level diagram of Fig. 1.4 (10). In this effect, two photons from the exciting frequency ν_e are scattered inelastically at frequencies of $2\nu_e - \nu_v$ or $2\nu_e + \nu_v$, corresponding to Stokes and anti-Stokes hyper-Raman scattering. While this effect has not found widespread analytical application, its potential usefulness comes from its very different selection rules from normal Raman scattering. All IR-active modes and all normally IR/Raman-inactive modes are hyper-Raman active.

In stimulated Raman scattering, energy may be supplied to the sample at two or more frequencies ν_e and one or more ν_s's, shown schematically in Fig. 1.5 (10). The effect serves to transfer energy between these two beams. As normally practiced, only ν_e is supplied to the sample, but at very high intensity.

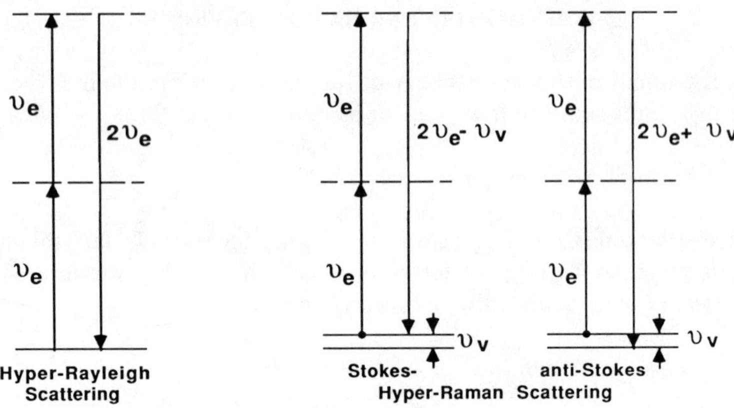

Figure 1.4. Schematic level diagram for hyper-Rayleigh and hyper-Raman scattering. [From Kiefer and Long (10).]

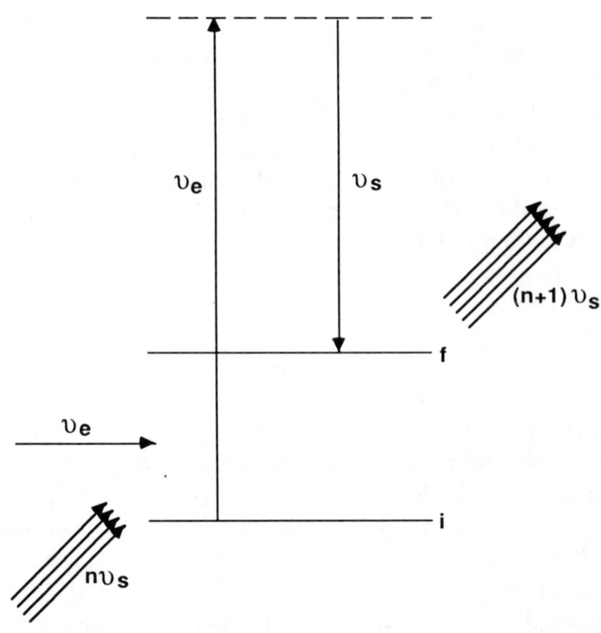

Figure 1.5. Schematic diagram for stimulated Raman scattering as a quantum process. [From Kiefer and Long (10).]

This generates ν_s, and the two photons can then interact to produce the nonlinear process already described. This technique is not of general analytical use because the conversion process is always funneled through one or two vibrational modes and no real "spectrum" is obtained. The conservation of momentum leads to an important difference between stimulated Raman scattering and normal Raman scattering. In the latter, scattering occurs in all directions, whereas in stimulated Raman scattering the emission is confined to one or more cone-shaped areas. This is often observed in two dimensions as a series of rings.

In the modification described next, the technique does become useful. If two lasers, one fixed in frequency and one tunable, or a laser and a continuum, are simultaneously applied to the sample, then energy transfer at a variety of frequencies is seen. This can be observed either as a gain or as a loss in energy, and these techniques are known as stimulated Raman gain spectroscopy and inverse Raman spectroscopy, respectively. The key experimental detail is that the two beams must be spatially and temporally coincident in the sample for the effect to occur. If the sample scatters appreciably, this is almost impossible to realize, so the techniques are restricted to gases, liquids, or clear solids.

CARS is a three-wave version of these techniques. In CARS, energy at ν_e and ν_s is supplied to the sample, and we observe emission at $2\nu_e - \nu_s$. The Stokes counterpart, CSRS, is observed at $2\nu_e + \nu_s$. In CARS and CSRS, the conservation of momentum requirement leads to emission along a line rather than a cone, so that the Raman measurement is a beam in a fixed direction depending upon the vibrational frequency. This is both an experimental difficulty and an analytical advantage. CARS has found its major analytical application in gas-phase spectroscopy, particularly for the study of combustion.

1.8. GUIDE TO THE LITERATURE OF RAMAN SPECTROSCOPY

In this book, a number of experts describe the analytical applications of Raman spectroscopy in diverse fields. With thousands of papers being published each year, it is impossible to be exhaustive in the treatment. Some topics are judged too limited in interest to be included, whereas others are so vast that it has not been possible to summarize them in a chapter.

The most thorough introduction to the theory and experimental practice of Raman spectroscopy, but with emphasis on the theoretical details, is the book by Long (4). An introduction to diverse analytical applications is given by Grasselli, Snavely, and Bulkin (11, 12). Although somewhat dated, Dollish,

Fateley, and Bentley's book (13) on Raman group frequencies remains very useful, as do the books by Hibben (3) and Freeman (7) already cited.

Biological and biochemical applications are a vast subject, treated in several books and numerous review articles. Carey (14) and Tu (15) deal with the Raman literature in detail, with some background on theory and practice, while the most recent edition of Parker's book (16) provides a wealth of information and guidance on biological applications of both IR and Raman spectroscopy. For polymers, the book by Painter, Coleman, and Koenig (17) is essential as an introduction to the methodology required for analyzing spectra, complete with many worked-out examples and literature references.

The biennial reviews of Raman spectroscopy in *Analytical Chemistry* provide the best review of the current literature on analytical applications. Useful review articles also appear in many other places, but the series *Advances in Infrared and Raman Spectra*, edited by R. Hester and R. J. H. Clark, and *Vibrational Spectra and Structure*, edited by J. R. Durig, consistently contain articles of interest. Past volumes of these series are an excellent starting place for anyone looking for a review of the application of Raman spectroscopy to a particular area. The International Conference on Raman Spectroscopy (ICORS) is held biennially, and most of the papers from these conferences have been collected in published proceedings since the early 1970s.

REFERENCES

1. C. V. Raman and K. S. Krishnan, *Nature (London)* **121**, 501 (1928).
2. G. Landsberg and L. Mandelstam, *Naturwissenschaften* **16**, 557 (1928).
3. J. H. Hibben, *The Raman Effect and Its Chemical Applications.* Reinhold, New York, 1939.
4. D. A. Long, *Raman Spectroscopy*, McGraw-Hill, New York, 1977.
5. G. Placzek, in *Handbuch der Radiologie* (E. Marx, ed.), Vol. 6, p. 205. Akademie-Verlag, Leipzig, 1934.
6. Y. Nishimura, A. Hirakawa, and M. Tsuboi, *Adv. Infrared Raman Spectrosc.* **5**, 217 (1978).
7. S. K. Freeman, *Applications of Laser Raman Spectroscopy.* Wiley, New York, 1974.
8. B. J. Bulkin, F. T. Prochaska, and D. Beveridge, *J. Chem. Phys.* **55**, 5828 (1971).
9. A. B. Harvey, ed., *Chemical Applications of Nonlinear Raman Spectroscopy.* Academic Press, New York, 1981.
10. W. Kiefer and D. A. Long, eds., *Non-Linear Raman Spectroscopy and Its Chemical Applications.* Reidel, Dordrecht, The Netherlands, 1982.
11. J. G. Grasselli, M. Snavely, and B. J. Bulkin, *Phys. Rep.* **65**, 231 (1980).

12. J. G. Grasselli, M. Snavely, and B. J. Bulkin, *Chemical Applications of Raman Spectroscopy*. Wiley, New York, 1981.
13. F. R. Dollish, W. G. Fateley, and F. F. Bentley, *Characteristic Raman Frequencies of Organic Compounds*. Wiley, New York, 1974.
14. P. R. Carey, *Biochemical Applications of Raman and Resonance Raman Spectroscopies*. Academic Press, New York, 1982.
15. A. Tu, *Raman Spectroscopy in Biology: Principles and Applications*. Wiley, New York, 1981.
16. F. S. Parker, *Applications of Infrared, Raman, and Resonance Raman Spectroscopy in Biochemistry*. Plenum, New York, 1983.
17. P. C. Painter, M. M. Coleman, and J. L. Koenig, *The Theory of Vibrational Spectroscopy and Its Application to Polymeric Materials*. Wiley, New York, 1982.

CHAPTER

2

MODERN RAMAN INSTRUMENTATION AND TECHNIQUES

D. BRUCE CHASE

Central Research & Development Department
Experimental Station
E.I. DuPont de Nemours & Co., Inc.
Wilmington, Delaware

2.1. INTRODUCTION

Broken down into its basic components, a Raman spectrometer consists of a light source, a collection optic or optics, a dispersing optical element, and a detection system. This chapter will briefly review the current capabilities in each of these areas, as well as discuss relevant data-processing techniques. Finally, some recent developments in an alternate approach to Raman spectroscopy, FT(Fourier transform)-Raman measurements, will be described. Only the processes and instrumentation involved in spontaneous Raman measurements will be covered. The reader interested in nonlinear Raman techniques is referred to the books by Levenson (1) and by Harvey (2), as well as several review articles (3). In addition, several recent books (4, 5) and articles (6) have reviewed the basic experimental procedures involved in Raman spectroscopy.

The driving force behind most changes and improvements in the instrumentation for Raman spectroscopy is the inherent weakness of the scattering phenomenon. For example, benzene has a cross section for Raman scattering of 1.6×10^{-29}. If one assumes standard collection optics, one can calculate a scattering efficiency of approximately 1 in 10^{10}. In order to produce a detectable number of Raman scattered photons an extremely intense light source is needed. This in turn requires a spectrometer with a very high degree of discrimination against internal scattered or stray light.

Analytical Raman Spectroscopy, Edited by Jeanette G. Grasselli and Bernard J. Bulkin. Chemical Analysis Series, Vol. 114.
ISBN 0-471-51955-3

Finally, since very few photons are scattered inelastically, the detection system employed must be extremely sensitive and able to detect small numbers of photons over the dark noise background. Most attempts to improve the sensitivity of nonresonant linear Raman spectroscopy involve increasing the initial laser intensity, increasing the effective collection of scattered photons, decreasing the transmission losses of the spectrometer, or increasing the sensitivity of the detection system.

2.2. COMPONENTS OF MODERN RAMAN SPECTROMETERS

2.2.1. Sources

2.2.1.1. Continuous Wave Lasers

The available sources for Raman spectroscopy are conveniently divided into pulsed and continuous wave (CW) lasers. Most of the work in analytical Raman spectroscopy today is done using a CW laser. The pulsed lasers have their primary applications in ultraviolet (UV) resonance Raman spectroscopy and time-resolved Raman measurements. In the last several years, most applications of Raman spectroscopy have involved the use of CW gas ion lasers, either argon or krypton. These lasers offer sufficient power levels for analytical applications, and the available lines are in the visible region of the spectrum. This is important since the Raman scattering cross section varies as the fourth power of the frequency. Raman spectra obtained at 514.5 nm with an argon laser will be anywhere from 2.3 to 4.5 times more intense than spectra obtained at 632.8 nm with a He–Ne laser operating at the same power. The early stability problems with gas lasers seem to have disappeared, and even the krypton units are relatively reliable. As always, operation at higher power results in shorter tube life. The available lines from both argon and krypton lasers are given in Table 2.1. Note that some of these lines require high-power lasers and specialized optics but the major lines are available even from lasers that are air cooled, thus avoiding the need for the plumbing associated with water cooling. For additional capability in the red region of the visible spectrum, dye lasers (pumped by a high-power CW argon laser) complement the basic gas lasers nicely. Currently available dyes when used with sufficient power from the pumping laser can provide Raman excitation from 450 to 800 nm.

In order to eliminate source fluctuation noise from the spectrum, the output power of the laser must be very stable. All of these gas and dye lasers can be operated in a light-stabilized mode to reduce output fluctuations to less than 0.2%. In most Raman spectroscopy labs an argon ion laser system

Table 2.1. Results from Argon and Krypton CW Lasers[a]

Argon[b]		Krypton[b]	
Wavelength	Typical Output	Wavelength	Typical Output
		413	0.50
4545	1.0		
4579	1.2		
4658	0.60		
		4680	0.30
4727	1.0		
		4762	0.40
4765	2.5		
		4825	0.40
4880	6.0		
4965	2.25		
5017	1.4		
5145	6.4		
		5208	0.70
5287	0.90		
		5309	1.4
		5682	1.0
		6471	3.50
		6764	0.90
		7525	1.10
		7993	0.25

[a]Some lines require special optics.
[b]Outputs were obtained with Spectra Physics 171–18.

is preferred, followed by either a krypton system or a dye laser (if an argon system is available with high enough power to effectively pump the dye laser). Since the available power levels of all these lasers currently exceed the damage threshold for most samples, further increases in power will not directly benefit Raman spectroscopists.

Another source that has seen limited use in analytical Raman spectroscopy is the frequency-doubled Nd/YAG laser (532 nm). It is a solid-state device that offers reliability and ease of operation. The Nd/YAG laser must be frequency doubled to reach the visible range. The doubling units that use the new potassium trihydrogen phosphate (KTP) or potassium dihydrogen phosphate (KDP) crystals are still somewhat temperature sensitive and tend to drift a bit when used at high powers.

2.2.1.2. *Pulsed Lasers*

The situation for pulsed sources is nowhere near as clear. The Nd/YAG laser with a fundamental at 1.06 μm is a powerful, stable laser that is highly reliable. Although it can be operated in a CW mode, most applications involve using it in a pulsed mode with pulse widths of 10 ns and a peak power output of 500 kW. The mode-locked repetition rate is around 80 MHz, and the laser can be Q-switched to produce 10–30 pulses per second. This laser can be used with KTP or KDP crystals to produce output at 532.0, 354.6, or 266.0 nm. An alternate source of pulsed UV radiation is the excimer laser. These two laser systems have been critically compared by Asher and Johnson (7) with respect to UV Raman spectroscopy.

A third choice for a pulsed output system is a synchronously pumped dye laser. The pump laser is mode locked to produce nanosecond pulses at a repetition rate determined by the cavity length. This pulse train is then used to pump the dye laser. By matching cavity lengths and using a cavity dumper, one can obtain very short pulses (of picosecond duration) at a variable repetition rate.

The major applications for pulsed excitation are time-resolved spectroscopy (8) and UV Raman measurements (9). If the primary goal is simply to obtain Raman spectra, the CW approach is favorable, assuming one can work in the visible region. There have been reports (9) that UV-excited Raman spectra offer a large degree of fluorescence rejection and excellent discrimination among conjugated aromatics. If UV operation is required, then pulsed excitation is needed to produce reasonable output powers in the UV range. Resonance Raman spectroscopy with UV excitation may prove to be extremely useful for analytical applications. Figure. 2.1 (7) shows the potential for differentiation of polycyclic aromatics when this technique is used.

2.2.1.3. *Lenses and Filters*

The laser is usually focused onto the sample by using a long focal length lens that may be antireflection coated for the laser line of interest. By varying the focal length, one can change the spot size at the sample. Usually one

Figure 2.1. Ultraviolet resonance Raman spectra of polycyclic aromatic hydrocarbons dissolved in acetonitrile and excited at 255 nm: (a) $5 \times 10^{-3}\,\mu$ triphenylene; (b) $5 \times 10^{-3}\,\mu$ phenanthrene; (c) $5 \times 10^{-3}\,\mu$ naphthalene; (d) $5 \times 10^{-3}\,\mu$ fluorene; (e) $5 \times 10^{-3}\,\mu$ pyrene. The peaks from the solvent have been numerically subtracted; average power, 2.5 mW; number of laser pulses averaged, 36,000 (30-min scan); spectrometer bandpass, $\sim 12\,cm^{-1}$. Reprinted from Asher (7) with permission. Copyright 1984 by the AAAS.

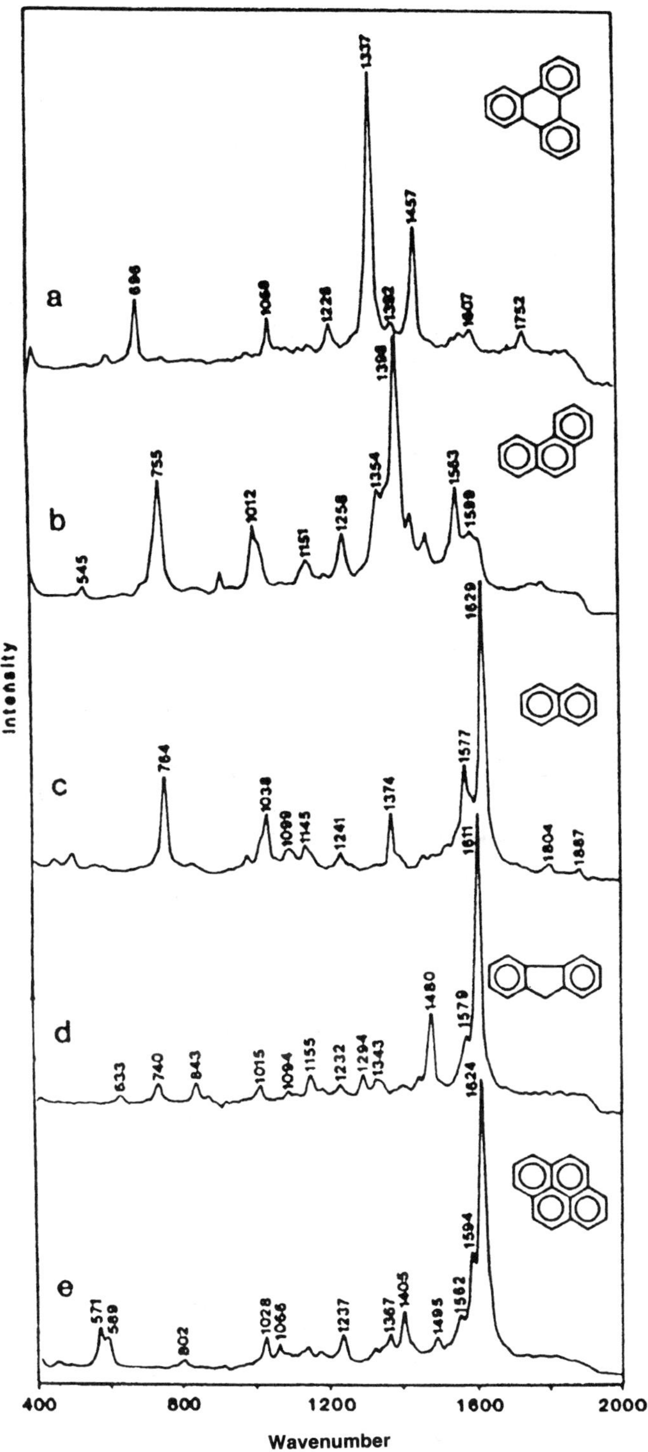
Intensity
Wavenumber
400
800
1200
1600
2000
a
696
1068
1228
1337
1392
1457
1607
1752
b
545
755
1012
1151
1258
1354
1396
1563
1599
c
764
1038
1099
1145
1241
1374
1577
1629
1804
1887
d
633
740
843
1015
1094
1155
1232
1294
1343
1480
1579
1611
e
571
589
802
1028
1066
1237
1367
1405
1495
1582
1594
1624

strives for small spot size at the sample in order to make the collection process simple. In addition to the focusing lens, a polarization rotator and a laser line filter are usually incorporated. The laser line filter is either a narrow band pass interference filter or a prism monochromator. Requirements for an effective filter are a high degree of transmission at the laser line and adequate rejection (more than 99.9%) at other wavelengths. This rejection filter eliminates nonlasing emission lines that are present and can be elastically scattered by the sample producing artifacts in the Raman spectrum. In cases where sample degradation occurs, the focusing lens can be changed to a cylindrical element. This results in a line focus at the sample, reducing the power density and thus reducing the rate of thermal decomposition. The illuminated area on the sample can then be imaged onto the spectrometer slits with very little loss in collection efficiency.

2.2.2. Collection Optics

The collection system is usually composed of two lenses. The first lens, a very short focal length element with a low f number (see Fig. 2.2), will collect the largest possible solid angle. The second lens is used to refocus the collected light onto the entrance slit of the monochromator, matching the f number of the monochromator. This ensures that the monochromator is neither overfilled nor underutilized. Such a two-lens system will result in a magnification of the spot size at the slit. It is important to keep the focused laser beam at the sample as small as possible, at least as small as the slit width/magnification factor.

Almost any collection geometry (i.e., setup of the lenses or other optics) can be utilized in Raman spectroscopy, but specific experiments may require certain optimized collection angles. For example, in the study of absorbed species on metallic substrates, a collection angle of 60° optimizes the observed signal-to-noise ratio (S/N) (10). In practice, most instruments are operated in a 90° or 180° scattering geometry. The 90° system is certainly the easiest to set up, since the illumination axis and the collection axis are separate in space. One simply has to ensure that the two axes intersect at their focal points. The major advantage of the 90° system is the ease of introduction of sampling accessories such as hot or cold cells, or waveguide assemblies.

The two-lens system described above can be easily converted to a 180° system by including a small mirror or prism on the front side of the first lens (Fig. 2.2). There is a slight loss in collection efficiency owing to the central obscuration caused by the prism, but this usually amounts to less than 10%. There are advantages to 180° backscattering especially in resonance Raman studies (11), since sample reabsorption of the Raman scattered radiation is

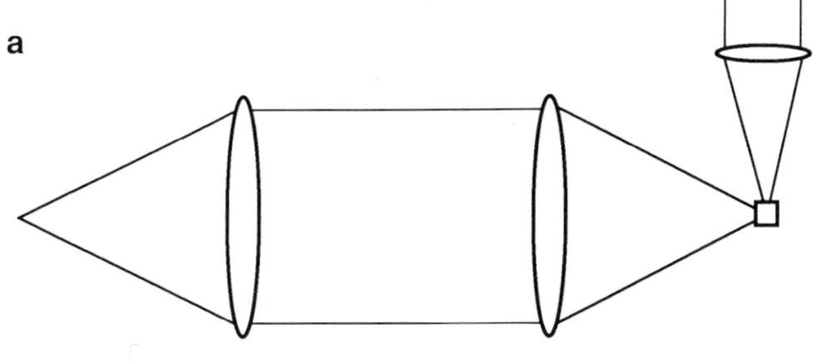

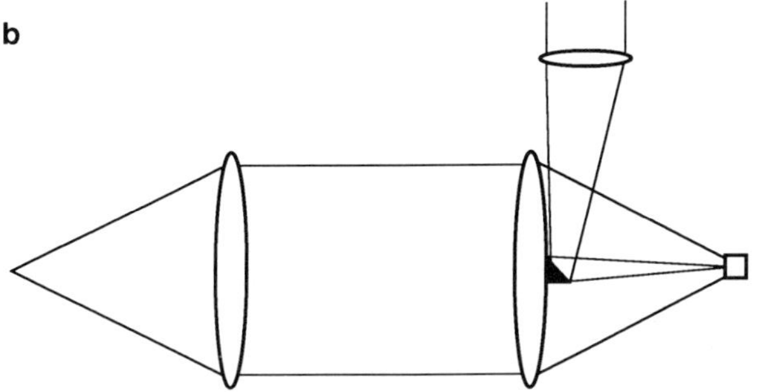

Figure 2.2. (a) A 90° scattering lens collection system. (b) A 180° backscattering lens collection system.

minimized. If suitable care is taken to choose the proper lenses, very good collection systems can be designed; often a good camera lens is a suitable collection element. However, if mirrors are employed, losses due to absorptions in the lenses are eliminated and a larger collection angle can be achieved. A deep dish ellipsoidal mirror can collect nearly π steradians, but because of the difficulty in alignment and because the sample itself masks the collected beam, it is not normally employed. It has, however, been effectively utilized in the FT-Raman system described in Ref. 41. The most notable improvements in collection systems have arisen from work in Raman circular dichroism. This is an extremely weak effect, so much effort has been devoted to maximizing collection efficiency (12, 13).

One other interesting modification to collection optics is the use of fiber optics. Properly designed, a fiber-optic illumination and collection system can incorporate both efficient illumination and a high collection efficiency (14). An additional benefit is the ease of remote sampling.

2.2.3. Dispersing Optical Elements: Monochromators

A monochromator is required to disperse collected light across its exit slits for sequential presentation to a detector or to disperse it across an array detector. Since the elastically scattered component is also collected, the monochromator must be able to discriminate effectively against this large signal. For analytical Raman spectroscopy, the standard dispersion system today is still the double monochromator, or two-grating system. There are several such systems manufactured and they have been described in other chapters on Raman spectroscopy (4, 5). The major problem associated with these systems is transmission loss. Since the first monochromator serves essentially to reduce the stray light level associated with the very strong Rayleigh line, an alternate approach using a single monochromator might be feasible if this intensity could be minimized. Recent developments in filter technology may provide the means of accomplishing this. The colloidal filter developed by Asher and co-workers (15) exhibits very high extinction at the laser line and good transmission characteristics over the rest of the spectrum. The use of such a filter with a single monochromator could increase Raman sensitivity by a significant factor. In his review of Raman spectroscopy at surfaces, Campion (16) notes that much of his work done with a single monochromator suitably filtered for removal of the Rayleigh line. For any monochromator, a critical aspect involves correctly matching the f number with the optical system. An overfilled grating will result in a large increase in stray light.

The increasing use of array detectors has forced the development of efficient, low stray light level spectrographs. These are usually composed of a first-stage double monochromator (used as a bandpass filter and operated in subtractive dispersion), followed by a third grating that acts as a "normal" monochromator and disperses the bandpass of the first stage onto the face of the array detector. This setup is used in state-of-the-art optical multichannel array detection instruments (OMAs). The sole purpose of the first-stage double monochromator is to reject the stray light and select the bandpass for presentation to the spectrograph stage. The major problem with these instruments is again transmission losses. Presumably a single monochromator with sufficient filtering of the Rayleigh line could offer improved performance with an array detector, assuming that the region of interest in the Raman spectrum does not go below several hundred cm^{-1}. For low-frequency Raman spectroscopy (less than 200 cm^{-1}), a double or triple monochromator is required.

2.2.4. Detection Systems

2.2.4.1. Single-Channel Systems

Detection systems conveniently divide into single-channel and multichannel schemes. The standard single-channel detection method now involves photon counting by photomultipliers. Except in very specialized cases, DC (direct current) amplitude detection has nearly vanished, since photon-counting systems now have large dynamic ranges and can deal with a wide range of intensities of scattered light. Photomultiplier tubes generally must exhibit extended red response and low dark count. Both the GaAs photocathode and the standard S-20 response photocathode meet these two criteria when the tubes are cooled. Commonly utilized photomultipliers are the RCA C31000 series and the Hamamatsu R928. The output from the photomultiplier is fed to a discriminator or discriminator amplifier that detects only electron bursts originating on the photocathode. With careful electronic shielding such systems can produce counting statistics that are governed by shot noise. The standard photon-counting detection system is relatively mature, with no major improvements expected. Photomultipliers might be developed with somewhat higher quantum efficiencies, but curent performance levels for single-channel detection will probably not change appreciably.

2.2.4.2. Multichannel Systems

The multichannel detection system includes an array detector which usually has an intensifier coupled to it. Comparisons of available array detectors are given in Campion and Woodruff (17) and Talmi and Busch (18). This type of system should exhibit a full multichannel advantage, since all of the spectral elements of interest are observed at the same time. The drawback of such a system involves a resolution/bandwidth trade-off. If high bandwidth or spectral coverage is required, then resolution must be sacrificed. The array detectors are usually comprised of 1024 elements, of which 700–800 are active. In addition, there is cross talk between adjacent elements, so effective resolution is reduced. In spite of these problems, the gain in sensitivity is significant, as demonstrated by Campion (19) in the detection of monolayers on metal surfaces. Continuing advances are being made in the sensitivity of these array detectors. Techniques for reduction of readout noise and background noise are being developed. The recent introduction of charge coupled device (CCD) and charge injection device (CID) detectors (20) will further improve the results being obtained in multichannel detection.

2.2.4.3. Computers in Detection Systems

A personal computer or microcomputer is now considered an integral part of the detection system in a Raman spectrometer using either single-channel or multichannel detectors. The computer both controls the monochromator and stores in digital form the output from the photon counter or the diode array system. This allows various forms of spectral processing along with signal averaging (21, 22). It also makes the application of frequency calibrations easier. The sinusoidal error function of a standard cosecant drive in a grating spectrometer can be readily corrected for and a frequency offset, determined by the use of discrete line calibration sources, can be applied. Once these frequency calibrations have been applied, spectral subtractions are possible. Signal averaging produces low-noise spectra amenable to resolution enhancement or Fourier self-deconvolution (23, 24).

2.3. SAMPLE HANDLING

2.3.1. General

One of the advantages Raman spectroscopy has in comparison to infrared measurements is the relative ease of sample preparation. Sample handling in Raman spectroscopy can range from ridiculously simple to complex. Particulate solids or liquids can be run in ordinary glass capillaries or cuvettes. The containers should be checked for background fluorescence or impurity bands, but this is not usually a problem with visible excitation. A UV Raman experiment will usually require the use of quartz cuvettes. Gas sampling methods are discussed later in this book in Chapter 12 by Murphy. When samples undergo thermal decomposition, spinning cell techniques can often alleviate the problem (25). In severe cases, dilution with a thermally conductive matrix can help (26). Variable temperature and evacuable cells are readily adapted to most 90° or 180° collection systems. The major problem is locating the collection lens close enough to the sample to maintain a large collection angle. A unique method of sample handling that also dramatically increases sensitivity for measurements of thin polymer films involves waveguide techniques. This approach has been described in detail by Rabolt and Swalen (27).

2.3.2. Microsampling

The area of microsampling in Raman spectroscopy has benefited greatly from the coupling of a good optical microscope with a Raman spectrometer

(28, 29). The objective of the microscope serves both to focus the laser to a small point on the sample, often near the diffraction limit, and then to collect the scattered light from the illuminated region to present to the entrance slit of the monochromator. For small samples (less than 100 μm), this is the optimal illumination/collection scheme. The tight focus required for microscopic illumination results in very-high-power densities, and often sample degradation is a problem. The various aspects of Raman microprobe techniques have been covered in a review by Rosasco (30).

2.4. PROBLEMS: FLUORESCENCE AND THERMAL SAMPLE DEGRADATION

The inherent sensitivity of a good Raman system operating in the visible region is now quite good. When multichannel detectors are used, very weak scattering can be observed. This high sensitivity is, however, drastically reduced when weak scattering occurs in the presence of background fluorescence, a situation often encountered. Extensive efforts have been made on several fronts to circumvent this problem. Often the most successful approach is the oldest: "drench and quench." Irradiation of the sample with the laser for an extended period of time often results in some photobleaching of the fluorescent impurity. Of course, when the sample itself is fluorescent, this approach does not work. Since the peak of the fluorescence usually occurs in the visible region, UV resonance Raman techniques (31) and techniques using red or near-IR lasers (see Section 2.5, next) often can be used to avoid fluorescent interference. Surface-enhanced Raman spectroscopy (32) often is accompanied by a quenching of fluorescence. Finally, time-resolved measurements, which discriminate against the fluorescence based on the differing temporal behaviors of the fluorescent and sample signals, have been used with some success (33, 34).

An associated problem is thermal decomposition of samples under the intense focused laser beam. The standard approach to this problem has been to spin the sample rapidly, allowing the thermal energy to dissipate for any given part of the sample between exposures.

Another technique has involved embedding the sample in a finely divided metallic powder.

The current state of Raman spectroscopy in the visible region is that these two problems limit the use of the technique. In the absence of fluorescence or thermal degradation, the instrumentation, however, is more than adequate for the detection of Raman scattering.

2.5. FOURIER TRANSFORM RAMAN SPECTROSCOPY

2.5.1 Components

As mentioned in the previous section, the two major problems associated with the routine application of Raman spectroscopy are fluorescence and thermal decomposition. The use of FT-Raman spectroscopy often solves these problems, as well as providing other advantages. The FT measurement is done in the near-IR and is a multiplex experiment that uses interferometric optics.

Both problems of fluorescence and thermal decomposition are related to absorption of the incident photons by some species in the sample. Clearly, the way to avoid both of these problems is to use an incident laser frequency that avoids all absorption processes. Since electronic absorptions have an energy threshold below which no absorption takes place, a suitable laser would probably emit in the red or near IR. The Nd/YAG lasers used in FT-Raman measurements emit at 1.064 μm (9395 cm^{-1}) and, for almost all organic systems, this is well below any absorption energy threshold. Unfortunately, photomultipliers and diode array detectors do not currently operate in this region. The available single-channel detectors in the near IR have noise figures significantly above the shot noise limit for the photon fluxes encountered in Raman spectroscopy. In addition, owing to the reduction in cross section for Raman scattering (a factor of 16–20 when going from 5145 A, to 1.064 μm due to the ν^4 dependence), the sensitivity of the experiment is severely reduced. In order to compensate for these losses, a multiplex measurement is required. FT-Raman uses interferometric optics (rather than monochromators) to achieve a multiplex measurement. (Regular Raman measurements can be multiplexed by detection with OMA or CCD devices.)

A Michelson interferometer used in FT-Raman measurements, equipped with a quartz beam-splitter, operates quite well in the red or near-IR region of the spectrum. After initial studies by Chase and Hirschfeld (35) demonstrated the feasibility of this approach, further efforts at optimization of the technique have resulted in dramatic improvements in sensitivity. It is now comparable to conventional single-channel Raman spectroscopy in the visible (36, 37).

As with the conventional measurements, there are several components required for FT-Raman measurements: a source, collection optics, an interferometer, and a detector. The source employed in our lab is a Nd/YAG laser with single-mode output powers of 10 mW to 10 W. The output of the laser is filtered with a laser line interference filter to remove plasma and other nonlasing emission lines. The radiation is focused onto the sample using a 101-mm focal length lens (see Fig. 2.3). The collection element

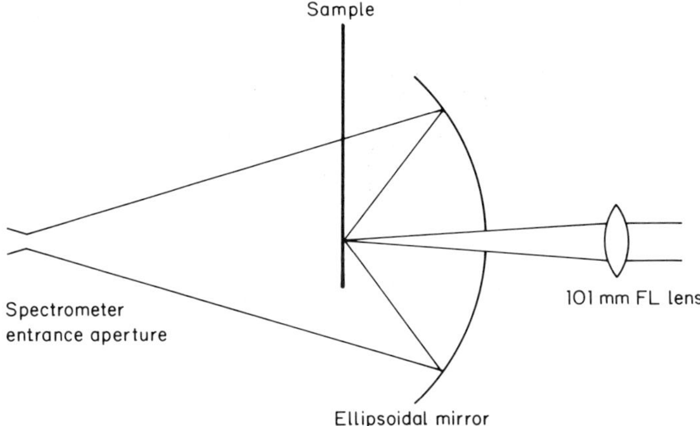

Figure 2.3. Ellipsoidal collection system.

currently employed is a 75-mm-diameter ellipsoidal mirror, with a short focal length of 25 mm and a long focal length of 250 mm. The solid angle of collection approaches π steradians. This is an in-line 180° backscattering arrangement that offers a high degree of collection efficiency but suffers from limited accessibility to the sample, thus making addition of accessories difficult. The advantage in using a 90° collection system and refractive optics have been discussed by Rabolt and co-workers (38). The long focus of the ellipse is held at the entrance aperture of the interferometer. From this point on the experiment is identical to a near-IR emission experiment, and the reader interested in the interferometry aspects is referred to standard texts on FT-IR (39).

Because of the nature of a multiplex measurement, all of the collected light strikes the detector simultaneously. This includes the elastically scattered component, which is more intense than the Raman components by many orders of magnitude. Therefore, if this spectral element is allowed to strike the detector, the detection system will become saturated. There are several approaches to optically filtering the Rayleigh line, and these have been discussed in Chase and Hirschfeld (35). Currently the easiest approach is to use long-pass dielectric multilayer filters. The instrumental layout employed in our lab is shown in Fig. 2.4. The Rayleigh line filters are located at the spectrometer entrance aperture and before the detector. Several filters are needed to supply the very high extinctions required to reduce the intensity of the Rayleigh scattering to a level comparable to the Raman scattering. Owing to the design of the Bomen DA3.02 Michelson interferometer used here, additional filtering is needed to remove unwanted near-IR radiation

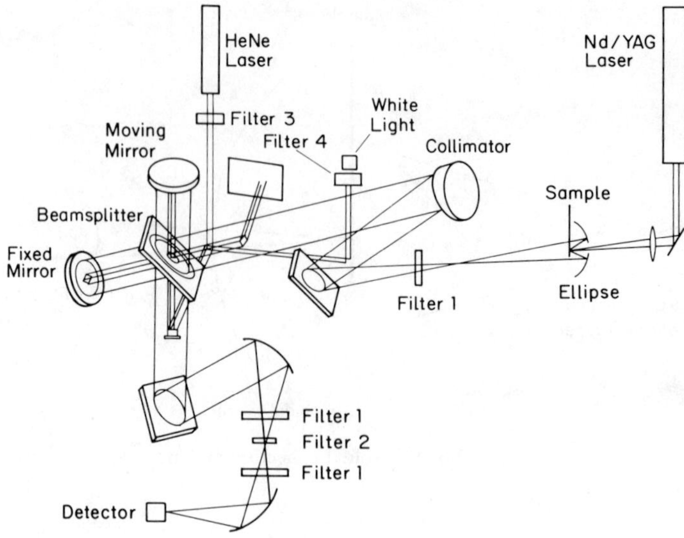

Figure 2.4. Instrumental layout for FT-Raman spectroscopy.

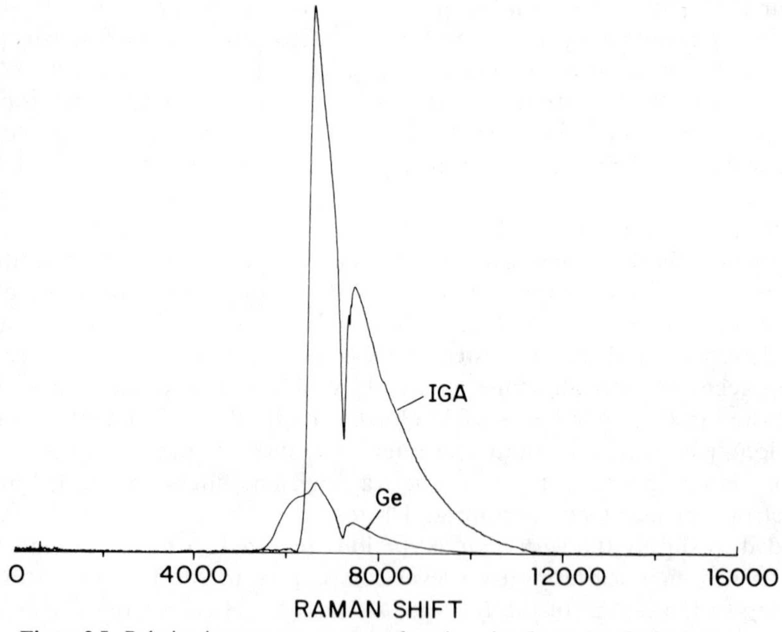

Figure 2.5. Relative instrument response functions for Ge and InGaAs detectors.

from the He–Ne referencing laser and the white light source referencing system.

Several different detectors are available for operation near 1.06 μm. Hendra and co-workers (40) used an InAs element, while initial studies by Rabolt and Chase (36, 37) utilized Ge detectors. Recent developments in the use of of InGaAs (41) have shown it to be a significantly better detector than Ge, as shown in Fig. 2.5. These curves are the instrument response functions for the two detectors scaled for equal noise figures. The InGaAs shows an increase of 7 in its effective sensitivity and appears to be close to a background-limited detector in its effectiveness. Minor improvements in detectors might increase the sensitivity of this type of measurement slightly, but as soon as a shot noise limit is reached, the measurement should be done with a dispersive instrument. In this vein, there have been reports of IR photomultipliers (42). At present, these devices lack sufficient dynamic range, but with further improvements, near IR Raman spectroscopy could be done without resorting to an interferometer.

2.5.2. Examples of the Advantages of Fourier Transform Raman Measurements

Figure 2.6 gives an example of the performance of a typical FT-Raman instrument. The spectra of indene were obtained in 2- and 30-s measurement time, respectively, with a high power of 1.0 W incident on the sample. The instrumental resolution was 4 cm^{-1}. The similarity in the spectra indicate that no sample degradation occurred. Most samples can withstand a high power level in the near-IR since the mechanisms for light absorption are primarily through overtones and combinations of vibrational modes. These are quite weak, and therefore very little of the energy is absorbed by the sample.

The real benefit of working in this region is shown in Fig. 2.7. The upper trace is the Raman spectrum of anthracene taken in the visible region with a conventional scanning monochromator. The high fluorescent background is due to photodecomposition products of anthracene. The lower trace is the FT-Raman spectrum of the same sample, taken at 1.06 μm. The fluorescent background is completely eliminated. Similarly, Fig. 2.8 shows the FT-Raman spectrum of a sample of a fluorescein dye. Most of the work done in the near IR using FT techniques indicates that the problem of background fluorescence is reduced to a negligible amount.

A further advantage offered by FT techniques is the ease of spectral subtraction. The optical referencing of a Michelson interferometer ensures a high degree of precision in the frequency. This allows one to subtract two spectra and be confident that there are no anomalous features due to a lack

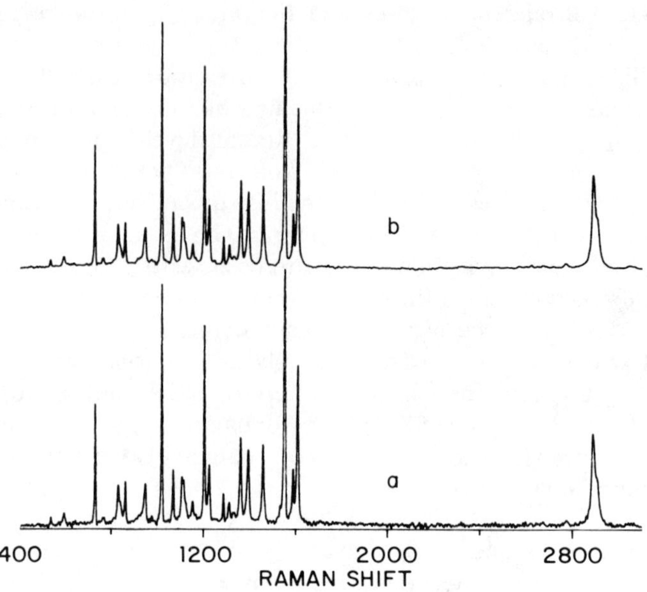

Figure 2.6. FT-Raman spectrum of indene at 1.0 W and (a) 2-s and (b) 30-s measurement time.

of registration between the two spectra. Figure 2.9a shows the FT-Raman spectra of a 30/70 mixture of benzene and toluene. Figure 2.9b presents a spectrum of pure toluene. Figure 2.10 shows the subtraction result (a) and the reference spectrum of pure benzene (b).

2.5.3. Problems

While the FT-Raman experiment using interferometric techniques coupled with near-IR lasers certainly offers improvements to conventional Raman experiments, it is not without problems. The low-frequency region of the spectrum is still not available owing to the elastic scattered component and the required filtering. Current efforts by Rabolt (43) and co-workers have extended the cutoff down to $200\,cm^{-1}$, but for low-frequency studies a dispersive instrument is needed (44). Thermal degradation can still be a problem, especially for inorganic systems that may have electronic absorptions in the near IR. The optimal lab will be equipped with both interferometric and dispersive systems and lasers in both the vsible and near IR. This will provide the full complement of techniques to address most problems in Raman spectroscopy.

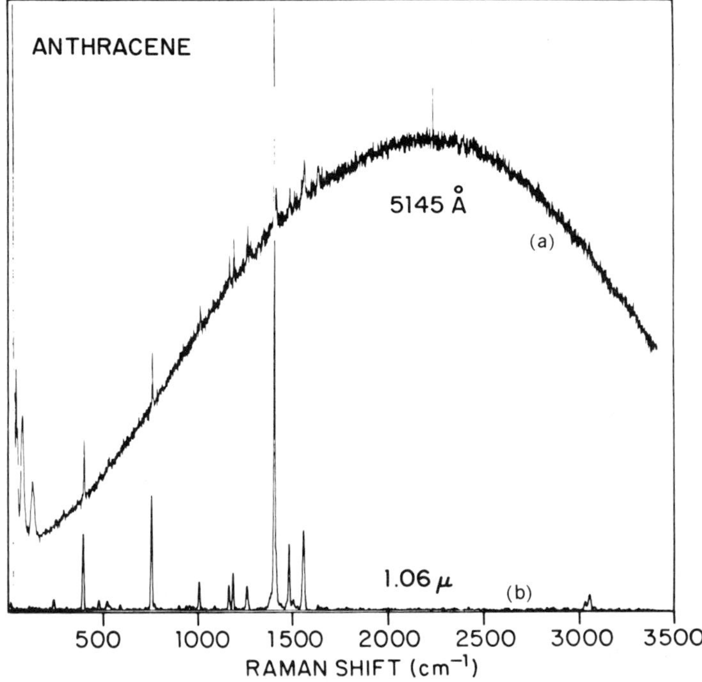

Figure 2.7. (a) Raman spectrum of anthracene at 514.5-nm excitation. (b) FT-Raman spectrum of anthracene at 1.06-μm excitation.

2.6. FUTURE EFFORTS

The analytical use of Raman spectroscopy is dependent upon the removal of the background fluorescence interference so often encountered. I would expect that future efforts will concentrate in two areas. The first will involve UV Raman spectroscopy. If laser systems are improved and become reliable and easily used, then the resulting strong increase in Raman scattering cross section and the minimization of background will combine to make this approach a sensitive tool in trace analysis. For routine Raman spectroscopy, the near IR approach is the other alternative. Whether the experiment is done using interferometers or dispersive instruments, as the sensitivity is improved, there is a great possibility for widespread use of Raman spectroscopy in conjunction with IR for routine structural analysis. Further developments in array detectors may increase their spectral range to include the near IR. In addition, new implementations of Hadamard transform

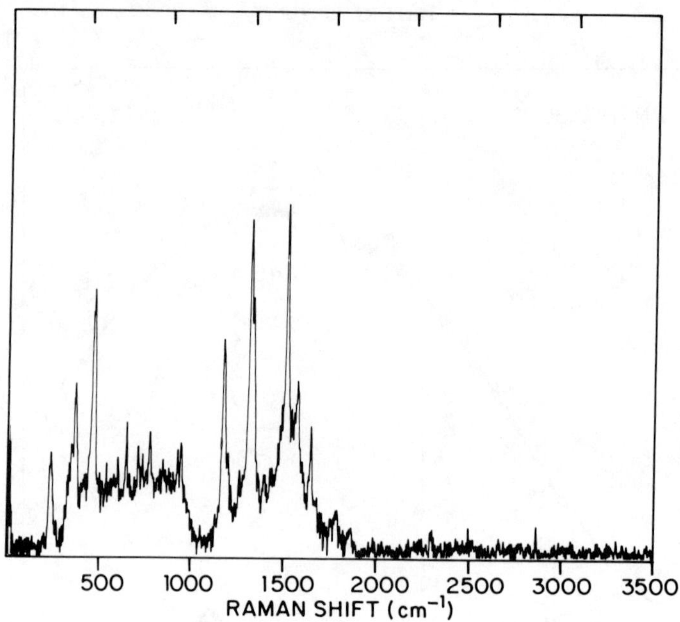

Figure 2.8. FT-Raman spectrum of Rhodamine 6G (solid), a fluorescein dye, at 106-μm excitation, 500 mW, 60-s measurement time.

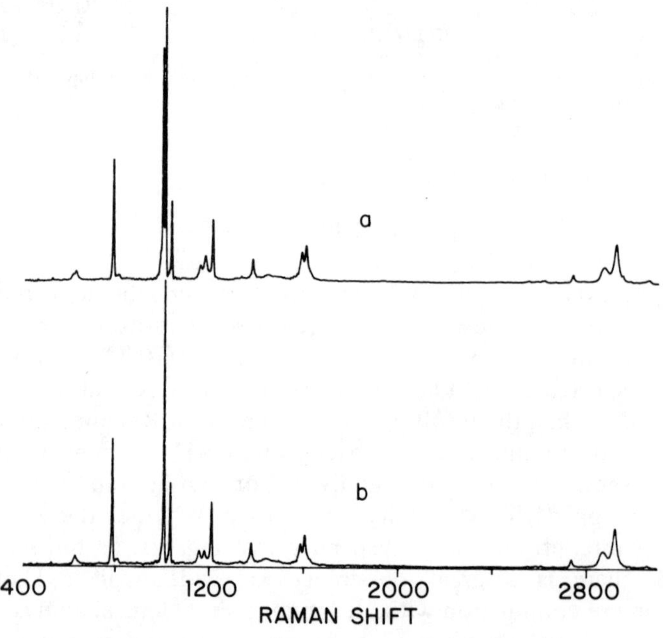

Figure 2.9. (a) FT-Raman spectrum of 70/30 mixture toluene/benzene. (b) FT-Raman spectrum of toluene.

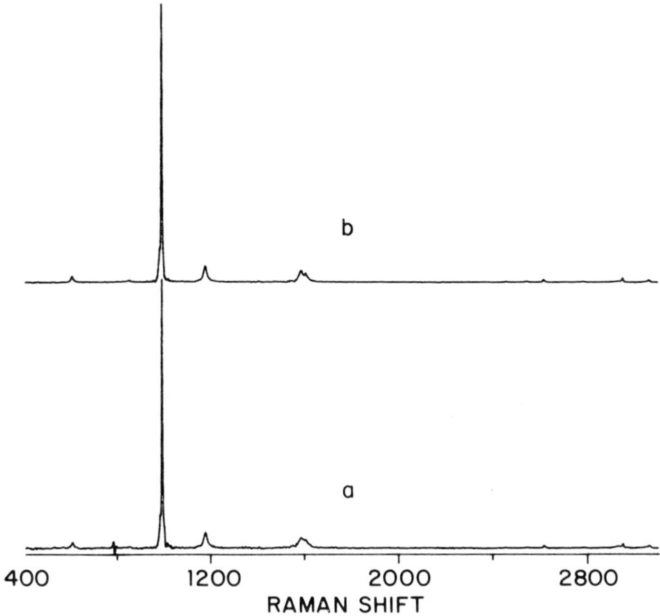

Figure 2.10. (a) Subtraction result from Fig. 2.9. (b) FT-Raman spectrum of benzene.

techniques have demonstrated a potential for Raman spectroscopy (45) that will be evaluated with further experiments.

2.7. ADDENDUM

Since the initial preparation of this chapter, several advances have occurred in many aspects of Raman spectroscopy. I will attempt to summarize some of them in this addenum.

2.7.1. Lasers

The success of FT-Raman spectroscopy has resulted in increased interest in the red excitation region. Diode lasers have made dramatic strides in the last three years and are now available with reasonable power levels (30–300 mW) at 785 nm and 830 nm. These lasers can serve as excellent sources for Raman scattering in the near infrared. Typical results are shown by McCreery and co-workers (46). Tunable excitation in the region 670 nm to 1100 nm can be obtained with a Ti sapphire laser. This unit is essentially a solid state analog to a dye laser. Pumped by a CW argon ion laser, output power conversions of 10–15% are readily obtained. A Ti sapphire system is an excellent replacement for a krypton ion laser. The parallel development of red-extended CCD detectors, coupled with these diode lasers, has allowed the production of Raman spectrometers with excellent fluorescence rejection and reasonable sensitivity.

2.7.2. Lenses and Filters

The major new development in this area is the introduction of holographic filters. These devices have the sharp cut-off characteristics of the multilayer dielectrics, but do not exhibit the harmonic structure in the transmission curve. The transmission is high (80–90%) and featureless, as opposed to the dielectrics which have numerous features in the transmission curve because of interferences among the multiple layers. The holographic filters are angle tunable and as a result angle sensitive. They operate best in a collimated beam, although they can be used in convergent beams if the angular spread is not severe. The extinction at the laser line can be as high as OD 5.

2.7.3. Monochromators

The major change in this area is the rapid development of single monochromators for Raman spectroscopy. Good cut-off filters such as the holographic units remove most of the Rayleigh line, allowing the single monochromators to function as dispersing units with high efficiency. Both scanning and fixed focal plane units are available from several manufacturers. An especially promising monochromator is from J-Y. This unit has aberration-corrected gratings which appear to offer superior imaging characteristics. The only drawback is the lack of a scanning capability. To change wavelength range, a new grating must be installed.

2.7.4. Detectors

In the past several years, CCD detectors have made tremendous strides. Denton has reviewed these developments in a clear, concise article (47). The use of red-extended CCD detectors for Raman spectroscopy has been discussed by McCreery (46) and Pemberton (48). In addition, the use of full imaging capability has been developed by Pelletier in conjunction with an echelle spectrometer for Raman studies (49). The major weakness of a CCD detection system is the limited spectral bandwidth obtainable with reasonable gratings. If a full spectral scan is required (50–4000 cm^{-1}), several intermediate spectra must be obtained and then pasted together. This presents some severe challenges for preservation of frequency precision. However, if only a limited spectral region is of interest, the sensitivity of these detectors can not be matched. The most recent entry in the field is a device from Tektronix which claims quantum efficiencies in excess of 75% from 400 to 750 nm and a quantum efficiency of 15% at 1000 nm. This unit achieves these properties through backside illumination and an effective anti-reflection coating. The one drawback of most CCD units is a cosmic ray sensitivity. This results in spikes in the spectrum which limit integration times to values less than imposed by dark count considerations.

2.7.5. Microsampling

Raman microscopes continue to be highly utilized, especially in industrial laboratories. The only major change in this area is the recent introduction of a confocal microscope for Raman spectroscopy by DILORS (50). This approach should help to minimize the spectral contamination caused by depth of focus in ordinary microscopes.

2.7.6. FT-Raman Spectroscopy

There are now five major instrument companies offering turnkey FT-Raman systems. The performance in terms of sensitivity appears to have leveled with no major increases expected. There is a great deal of interest in converting from main frame Nd/YAG lasers to the diode-pumped versions coming onto the market. The noise performance of these units is significantly better. They run with little or no water cooling and at 110 V. Their small size makes incorporation into a lab spectrometer easier. The only limitation is low power (less than 1 W), and this limitation is changing almost daily. Several companies promise several watts by the end of 1991.

The holographic filters discussed above are impacting FT-Raman spectroscopy as well. Two filters are sufficient to give Rayleigh line rejection, and spectral information down to 150 cm^{-1} can be obtained. Another filter type is the Chevron unit (35). This system has been shown by Nicolet, and possibly Bruker, to give Raman data down to 60 cm^{-1}. Additionally, they allow data to be obtained on both Stokes and anti-Stokes bands (51).

The choice in detectors now appears to have swung back to Ge. The use of a PIN Ge detector operated at 77K and biased at 250 V gives slightly better performance than InGaAs and increases the spectral range out to 3500 cm^{-1} Raman shift. This detector is, however, sensitive to cosmic rays, and efforts must be made to ensure these spikes do not contaminate the interferogram.

I believe FT-Raman and the parallel development of multichannel detection coupled with diode lasers will see tremendous gains in the next few years. Raman spectroscopy, through these two developments, may actually become a routine tool for the industrial analytical spectroscopist.

REFERENCES

1. M. O. Levenson, *Introduction to Nonlinear Laser Spectroscopy*. Academic Press, New York, 1982.
2. A. Harvey, ed., *Chemical Applications of Nonlinear Raman Spectroscopy*. Academic Press, New York, 1981.
3. W. Keifer and A. Beckman, *J. Mol. Struct.* **13**, 83 (1984)
4. J. G. Grasselli, B. Bulkin, and M. Snavely, *Chemical Applications of Raman Spectroscopy*. Wiley, New York, 1981.
5. D. P. Strommen and K. Nakamoto, *Laboratory Raman Spectroscopy*. Wiley, New York, 1984.
6. D. L. Gerrard, *Anal. Proc.* (*London*) **57**(2), 538 (1985).
7. S. A. Asher and C. R. Johnson, *Science* **225**, 311 (1984).
8. A. Laberau and M. Stockburger, eds., *Time-Resolved Vibrational Spectroscopy*. Springer-Verlag, Berlin, 1985.
9. B. Hudson, *Spectroscopy* **1**(1) 22 (1987).
10. A. Campion et al., *Surf. Sci.* **115**, L153-L158 (1982).
11. F. Parker, *Applications of Infrared, Raman and Resonance Raman Spectroscopy in Biochemistry*. Plenum, New York, 1983.
12. P. L. Polavarpu, *Appl. Spectrosc.* **37**(5), 447 (1983)
13. M. R. Obodi et al., *J. Raman Spectrosc.* **16**(6) 366 (1985).
14. S. D. Schwab and R. L. McCreery, *Anal. Chem.* **56**, 2199 (1984).
15. P. L. Flaugh, S. E. O'Donnell, and S. A. Asher, *Appl. Spectrosc.* **38**(6), 847 (1984).
16. A. Campion in *Vibrational Spectroscopy of Molecules at Surfaces* (J. Yates and T. Madey, eds.), Chapter 8. Plenum, New York, 1987.
17. A. Campion and W. Woodruff, *Anal. Chem.* **59**(22), 1299A (1987).
18. Y. Talmi and K. Busch, *MultiChannel Image Detectors* (Y. Talmi, ed.), Vol. 2, Am. Chem. Soc., Washington, D.C., 1983.
19. D. Harradene and A. Campion, *Chem. Phys. Lett.* **135**(6), 501 (1987).
20. W. Wang, L. R. Hudson, and H. Tseng, *Opt. Eng.* **26**(9), 844 (1987).
21. R. W. Snyder and T. M. Hooker, *Appl. Spectrosc.* **38**(1), 58 (1984).
22. S. A. Dyer and D. S. Hardin, *Appl. Scpectrosc.* **39**(4), 655 (1985).
23. H. Bowley et al., *Appl. Spectrosc.* **36**(6), 1004 (1985).
24. F. Ni and H. A, Scheraga, *J. Raman Spectrosc.* **16**(5), 337 (1985).
25. J. Eng, R. Czenuszewicz, and T. Spiro, *J. Raman Spectrosc.* **16**(6), 432 (1985).
26. J. Nimmo, A. Bouill, A. McConnell, and W. Smith, *J. Raman Spectrosc.* **16**(4), 245 (1985).

27. J. Rabolt and J. Swalen, in *Spectroscopy of Surfaces* (R. J. H. Clark and R. Hester, eds.), Chapter 1. Wiley, New York, 1988.
28. J. B. Hopkins and L. A. Farrow, *J. Appl. Phys.* **59**(4), 1103 (1986).
29. B. Wopenka and J. D. Pastens, *Appl. Spectrosc.* **40**(2), 144 (1986).
30. G. J. Rosasco, *Adv. Infrared Raman Spectros.* **7**, 223 (1980).
31. S. and Asher C. Johnson, *Science* **225**, 311 (1984).
32. H. Metieu, *Prog. Surf. Sci.* **17**, 162 (1984).
33. J. Watanabe, S. Kinoshita, and T. Kusheda, *Rev. Sci. Instrum.* **56**(6), 1195 (1985).
34. A. VanHoek and A. Visser, *Anal. Instrum.* **14**(2), 143 (1985).
35. B. Chase and T. Hirschfeld, *Appl. Spectrosc.* **40**(2), 133 (1986).
36. B. Chase, J. Am. Chem. Soc. **108**(24), 7485 (1986).
37. C. Zimba et al., *Appl. Spectrosc.* **41**(5), 721 (1987).
38. V. Hallmark et al., *Spectroscopy* **2**(6), 40 (1987).
39. P. R. Griffiths and J. DeHaseth, *Fourier Transform Infrared Spectroscopy.* Wiley, New York, 1986.
40. P. Hendra et al., *Appl. Spectrosc.* **42**(5), 796 (1988).
41. B. Chase, *Microchim. Acta, Special Issue III* **1987**, 81 (1988).
42. M. D. Petroff, M. G. Stapelbroek, W. and A. Kleinhans, *Appl. Phys. Lett.* **51**(6), 406 (1987)
43. J. Rabolt, private communication.
44. D. R. Porterfield and A. Campion, *J. Am. Chem. Soc.* **110**(2), 408 (1988).
45. D. C. Tilotta, R. Freeman, W. and Fateley, *Appl. Spectrosc.* **41**(8), 1280 (1987).
46. Y. Wang and R. L. McCreey, *Anal. Chem.* **61**, 2647 (1989).
47. P. M. Epperson et al., *Anal Chem.* **60**, 327A (1988).
48. J. E. Pemberton and R. L. Sobocinski, *J. Am. Chem. Soc.* **111**, 432 (1989).
49. M. Pelletier, *Appl. Spectrosc.* **44**(10), 1699 (1990).
50. Pittsburgh Conference 1991.
51. J. G. Radziszewski and J. Michel, *Appl. Spectrosc.* **44**(3), 414 (1990).

CHAPTER

3

EXPERIMENTAL CONSIDERATIONS FOR ACCURATE POLARIZATION MEASUREMENTS

JAMES R. SCHERER

Department of Chemistry
University of California
Berkeley, California

3.1. INTRODUCTION

Intensity measurements of polarized Raman scattering depend on a number of factors that, if understood and carefully adjusted, can routinely lead to accurate values of depolarization ratios or isotropic and anisotropic spectra. Polarized scattering from liquids (or gases) is conveniently measured in a Cartesian coordinate system X, where the directions of incident radiation X_i, incident polarization X_j, scattered polarization X_k, and direction of observation X_l are indicated in a format devised by Damen, Porto, and Tell (1) as $X_i(X_jX_k)X_l$. This notation completely describes any spontaneous Raman scattering experiment. As might be guessed from the foregoing, our ability to experimentally define these directions will greatly affect the outcome of the measurements. Liquid samples are usually confined in cells that permit measurement from a volume element that falls on the intersecting incident beam axis and the observation axis. With single crystals having angled faces, it may be necessary to physically displace these two axes so that the monochromator can "see" the incident beam inside the crystal. However, the incident beam axis and observation axis are maintained orthogonal in most measurements. Even though single crystals will often have unique requirements regarding directions of incident and scattered light, many of the tricks discussed here will still be useful in setting up the appropriate experimental geometry.

If one takes the measures described here to correctly align the experimental geometry and the directions of polarization for the incident and scattered

---

Analytical Raman Spectroscopy, Edited by Jeanette G. Grasselli and Bernard J. Bulkin. Chemical Analysis Series, Vol. 114.
ISBN 0-471-51955-3

beams, it will be hard *not* to obtain highly accurate measurements of band polarization.

3.2. TOOLS FOR TUNING

(1) *Professional Carpenter's Square.* This tool will be used for many operations, i.e., establishing a parallel laser beam above a plane, or orthogonality of a slit relative to a plane, etc. The inside corner should be compared with a machinist's right-angle standard, and each of the two arms should have parallel edges.

(2) *A High-Quality Front-Surface Mirror* ($\sim$50 mm $\times$ 50 mm). The front and back surfaces should be parallel. This can be checked with a micrometer ($<$0.25 μm deviation) or with a small He–Ne laser. The mirror is placed on a flat surface and the laser beam reflected from its surface at a convenient angle to a distant wall ($\sim$4 m). The spot on the wall should not deviate as the mirror is turned on the surface.

(3) *Penta Prism.* This device provides a very convenient way of turning a beam by exactly 90°. While its cost prohibits its use as a 45° diagonal mirror, it is quite helpful for correctly positioning a 45° diagonal mirror. The penta prism is inserted into the path of a laser beam (parallel to a surface), and the new beam is directed to the target in the sample plane. The prism is rotated till the new beam is the same distance from the surface as the old beam. Two variable apertures are put into the new beam: one near the penta prism, and the other near the target. The penta prism is replaced with an orthogonally mounted mirror, and the mirror is adjusted to pass the beam through the two apertures. The mirror is secured and the apertures are removed.

(4) *The Absolute Horizontal Mirror.* Many monochromators are placed on optical tables that are flat to within a few thousands of a centimeter over several meters. In these cases the table top may be used as the "reference plane" for all subsequent definitions of orientation. The flat table surface is particularly useful if the table is floating on an air cushion and has no permanent horizontal position. For monochromators that are not on an optical table but are resting on a solid earth support, one can define a horizontal surface using a penta prism in conjunction with a true vertical beam. A true vertical laser beam can be established by deflecting a horizontal beam downward on to the surface of a pool of mercury in a glass dish ($\sim$10 cm in diameter and 5–10 mm deep) covered with silicone oil about the consistency of honey ($\sim$5 mm deep). This viscously damped mirror always returns a true vertical beam back on itself and is easily seen on the back

side of any aperture that is placed in the beam path. This device is the beam analog of the mechanical plumb bob in much the same way as the penta prism is the beam analog of the mechanical square. The true vertical beam can be adjusted to a higher precision than a horizontal beam using a level. However, a beam can be leveled quite accurately by using the penta prism to deflect the horizontal beam downward onto the mercury mirror and adjusting the horizontal beam up or down until the reflected beam from the mercury mirror returns on itself. When not in use, the mercury mirror should be covered to prevent dust from settling on the surface and stored in a ventilated environment.

(5) *The Rochon Polarizer.* This device can separate a beam of unpolarized light into two linearly polarized components: one beam is undeflected and polarized in one orientation, and the second beam is deflected a few degrees (depending on construction) and polarized 90° to the undeflected beam. This device is used to establish the orthogonality of our experimental beam polarization directions. A rectangular metal holder should be machined with flat edges and at least one very good right angle. The Rochon should be mounted loosely in the center of the rectangular holder. One of the good edges of the assembly is placed on a flat support in the path of a polarized alignment laser ($\sim$2 mW He–Ne laser polarized with a Polaroid HN32 or HN22 sheet), and the beam is observed as a spot on a white piece of paper some distance from the Rochon. The polaroid is adjusted to have its pass polarization approximately horizontal (or vertical). The Rochon is rotated about the beam axis in the rectangular holder until one of the two passed beams is extinguished completely. The holder is then rotated by 180° (about a vertical axis) on the flat support. If the same beam is not completely extinguished, rotate the polaroid to partially extinguish the beam and repeat adjustment of the Rochon. At some position of the polaroid, the Rochon will extinguish exactly in both (front/back) orientations. Fix the Rochon plate in the holder. A repeat of this measurement with the holder resting on the side adjacent to the good right angle should exactly extinguish the other beam previously passed by the Rochon in the first adjustments. These two edges of the Rochon assembly should be marked as the standard edges and will be used in subsequent definitions of polarization direction.

3.3. LIGHT PATH GEOMETRY

We will begin our spatial beam definitions by considering first the geometry of the monochromator and then making all beam adjustments to fit this geometry. The entrance slit of the monochromator is like the proverbial "eye of the needle." It defines the point through which our scattered light must pass. But, in order for our monochromator to function properly, the light entering the entrance slit must fill the first collimating mirror.

The line formed by the center of the entrance slit and the center of the first collimator defines the monochromator optic axis. Some instrument manufacturers will provide a mask with a cross hair or central mark that can be placed over the first collimator and a small hole or aperture that defines the central part of the entrance slit (required for extended entrance slits). The first requirement for beam alignment is that the optic axis of the monochromator be parallel to either the plane of the floating optical table, Case I, or perpendicular (parallel) to an established true vertical (horizontal), Case II. If the optical table has an orthogonal grid of screw holes in its surface, the monochromator should be moved so that the optic axis is parallel to holes running the length or width of the table. If no holes are provided, the reference line should be established in some other way. It is also convenient to orient an optical rail parallel to the optic axis.

Having established the optic axis on the table and having fixed (permanently, if possible) the lateral position of the monochromator, we now determine the slit orientation. The slits of the monochromator should be either parallel or perpendicular to the reference plane (Case I) or reference vertical (Case II). This can be checked by back lighting the whole entrance slit with a white light source, putting a lens close to the front of the entrance slit, and imaging the slit on a nearby wall (in a darkened room). If the slits are perpendicular to the reference plane, place one arm of the square on the reference plane and the other upright, and see if the image of the slit falls along the upright edge of the square. If the slits are parallel to the reference plane (Case II), support a straight edge by two equidistant spacing blocks and intercept the slit image. In either case, tilt the monochromator about the optic axis until the slit image is parallel to the square edge.

A monochromator can function quite well with its entrance slits internally compatible but skew to an outside framework. However, it will not be effective in maximizing the signal from an orthogonally aligned sampling area. The above adjustment can make for a rather odd-looking instrument, but it will guarantee either maximum measured scattered intensity or a visit from the monochromator representative to straighten up the slits. If the primary reference is a true vertical (Case II), a plumb bob may be positioned to bisect the slit image.

We have now established the optic axis of the monochromator relative to the reference plane, the orientation of the slit relative to the reference plane, and the position of the optical rail parallel to the optic axis (on which sampling apparatus can rest). The next step is to ensure that the optic axis passes through the center of the light-collection optic (which is usually positioned for some degree of image magnification). Make sure that the light-collection lens has enough aperture to match the f number of the monochromator and the length of extended image for the image magnification desired (i.e., that the projection of the first collimator through the extended entrance slit falls within the lens aperture nearest the slit). With the alignment laser centered on the optic axis, center a pinhole on the axis so that it can be slid along the axis on the parallel optical rail to change position of the point source. With the light-collection lens in place and positioned for low magnification, the light-collection lens should be moved laterally until the image of the pinhole falls on the center of the entrance slit *and* the center of the first collimator. The light-collection lens should be firmly locked into position so that the only motion permitted is movement along the optical rail. Move the pinhole and lens position for higher image magnification. The spot on the first collimator should remain centered. Repeat adjustments until this condition is satisfied.

Some instrument manufacturers allow the user to move the light-collection lens laterally to accommodate positioning of the sample image onto the entrance slit. This practice can lead to serious intensity loss, since the collected light must necessarily fall off the first collinator if it does not travel along the optic axis. Also, vignetting of the useful light collected by the lens at the lens aperture can occur if the lens aperture is closely matched to the f number of the monochromator. Furthermore, if the direction of light collection is not centered on the optic axis, the ill-defined experimental geometry can give rise to off-axis polarization errors. The best recommended practice is to make imaging adjustments by leaving the light-collection optics stationary and moving the sample (since *it* is not properly located on the optic axis).

For convenience we will label the optic axis of the monochromator the Y axis and the reference plane the XY plane (see Fig. 3.1). In Table 3.1 are ten different schemes for excitation and polarized (X or Z) or total ($X + Z$) light collection. The scattering intensities are listed in terms of mean polarizability $\bar{\alpha}$ and anisotropy β. The intensity errors caused by off-axis light collection by the lens are qualitatively indicated ($+, -$) for the cases of a completely polarized line ($\beta = 0$) and a depolarized line ($\bar{\alpha} = 0$). The depolarization ratio is defined as the second intensity value divided by the first intensity value for each pair of experimental configurations. Note that geometries 3 and 4 use a different definition for the depolarization ratio than geometries 1, 2, and 5. Geometry 5 involves 0° forward scattering or 180° backscattering. Note that the errors due to off-axis light collection are small. However, these collection geometries either introduce too much exciting light into the monochromator (0° forward), which can increase background from stray light, or—in the case of 180° backscattering—block a significant amount of Raman scattering from reaching the monochromator because of

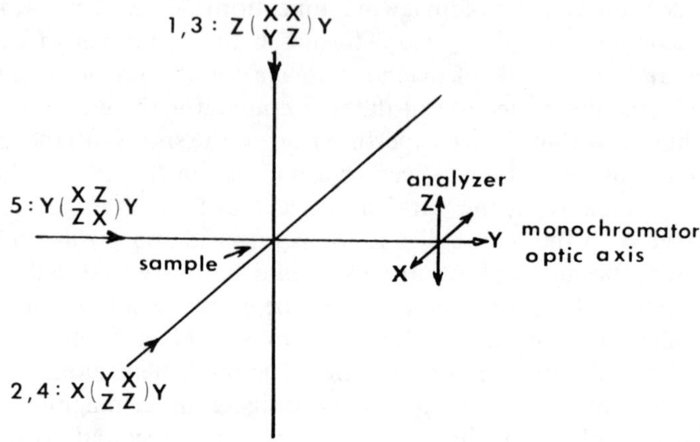

Figure 3.1. Excitation and light-collection geometries.

the small diagonal mirror or prism that must be placed on the optic axis to direct the beam to the sample (2). In addition, since the laser beam path is on the viewing axis, Raman scattering from the sample container is not spatially separated from the sample. This scattering cannot be removed by masking, and it adds to the spectrum of the sample. These complications make these collection geometries unattractive for obtaining high-quality spectra.

In Table 3.1 we note that the most accurate depolarization ratios are obtained from geometries 1 and 2 for both highly polarized and depolarized bands. Geometries 3 and 4 give rise to moderate positive errors in the depolarization ratio of highly polarized bands (i.e., typically 0.03 to 0.06 depending on the light-collection angle), but depolarized bands show small errors that usually make the ratio slightly greater than 6/7. Considering the almost double-scattered light intensity for this experiment, the small increase in error might be tolerated.

We have now reached the point where we must define the position of the exciting beam. When the monochromator has vertical entrance slits, the beam usually enters the sampling area from either above the sampling (monochromator) plane or from below (geometry 1 or 3). For instruments having horizontal entrance slits, the beam enters from the side (along the X axis, geometry 2 or 4). The only remaining task is to ensure that the laser beam is normal to the reference plane (XY) (Case I) and intersects the optic axis (Y); or that it is true vertical (Case II) and intersects the optic axis (Y). For instruments with horizontal slits (geometries 2 and 4), the laser beam

Table 3.1. Experimental Geometries for Polarization Measurements

Geometry	Monochromator Slit Direction	Scattering Geometry	Intensity	Light-Collection Errors: Polarized Bands	Light-Collection Errors: Depolarized Bands
1	Vertical (Z)	$Z(XX)Y$	$45\bar{\alpha}^2 + 4\beta^2$	Small $--$	Very small $-$
		$Z(XZ)Y$	$3\beta^2$	None	None
2	Horizontal (X)	$X(ZZ)Y$	$45\bar{\alpha}^2 + 4\beta^2$	Small $--$	Very small $-$
		$X(ZX)Y$	$3\beta^2$	None	None
3	Vertical (Z)	$Z(X, X+Z)Y$	$45\bar{\alpha}^2 + 7\beta^2$	Small $--$	Very small $-$
		$Z(Y, X+Z)Y$	$6\beta^2$	Moderate $++++$	Very small $++$
4	Horizontal (X)	$X(Z, X+Z)Y$	$45\bar{\alpha}^2 + 7\beta^2$	Small $--$	Very small $-$
		$X(Y, X+Z)Y$	$6\beta^2$	Moderate $++++$	Very small $++$
5	Horizontal or vertical (X or Z)	$Y(X, X)Y$	$45\bar{\alpha}^2 + 4\beta^2$	Small $--$	Very small $-$
		$Y(X, Z)Y$	$3\beta^2$	None	None

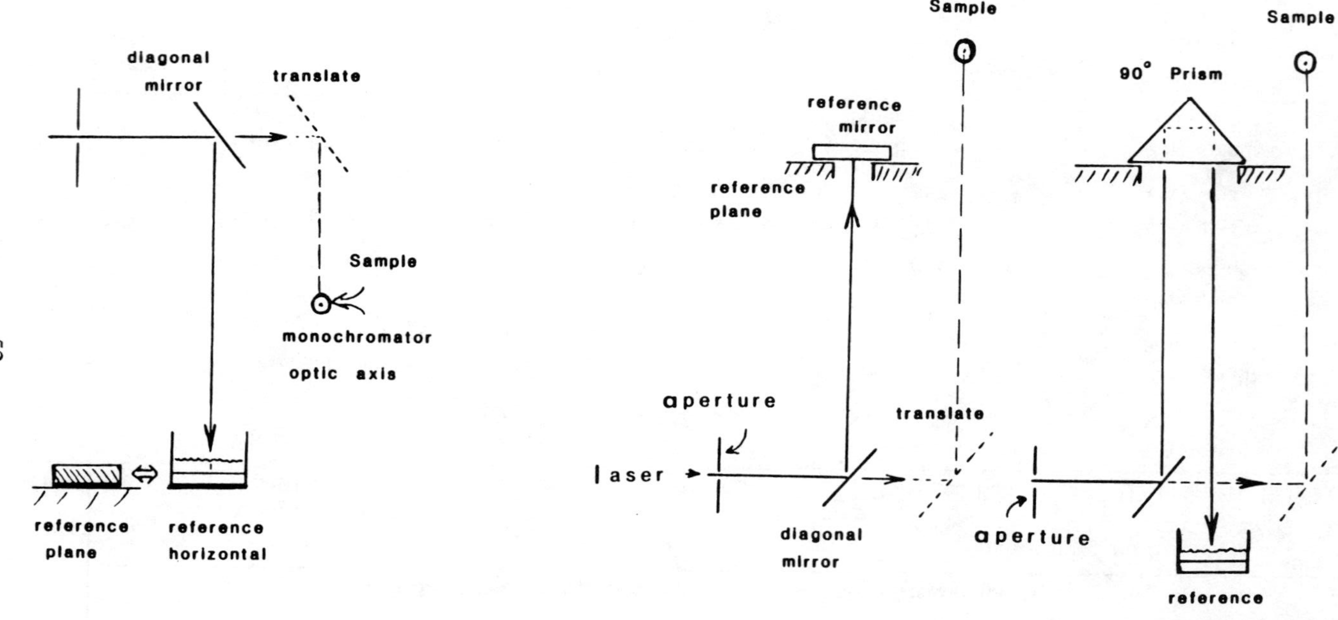

Figure 3.2. Establishing a perpendicular or vertical beam relative to a reference plane or reference horizontal.

must be parallel to the reference plane (Case I) or to true horizontal (Case II) and intersect the optic axis (Y).

If the beam approaches the sample from above, it can be adjusted to be normal to the front-face mirror resting on the reference plane and then translated to the point along the monochromator or optic axis where the sample will be placed (see Fig. 3.2). If the reference is true vertical, the beam orientation can be adjusted with the mercury mirror and subsequently translated to the correct position. If the beam is coming from below the sample, the mirror can be placed front surface down on the reference plane over the beam, and the return beam made to pass back through an aperture placed near the laser. Once perpendicular to the reference plane, it can be translated to intersect the sample. The direction of the beam can be made parallel to true vertical by using a good 90° prism to direct the beam to the horizontal mercury mirror and adjusting the diagonal mirror to send the beam back through an aperture. The intersection of the optic axis and the incident laser beam can be seen by positioning the alignment He–Ne laser to pass along the optic axis and observing the intersection of beams on a piece of paper or marking the intersection of the vertical (ZY) plane with the reference plane (XY) and observing the incident laser beam on this intersecting line.

3.4. SAMPLE CONTAINER

Having established the direction of incident and scattered radiation, we can place a liquid sample in a suitable container and move it to the correct position by means of translators. Two configurations of observing Raman scattering from liquids in capillary tubes are illustrated in Fig. 3.3. The magnified image of the illuminated sample at the intersection with the optic axis is projected onto the length of the entrance slit. The configuration shown in Fig. 3.3a uses a lens ($\sim$6 cm focal length) to produce a collimated beam inside the capillary. The capillary length is oriented parallel to the direction of the entrance slit (vertical or horizontal), and the beam inside the tube is imaged onto the slit. Spurious Raman scattering from the lens at the tip of the capillary is masked by an aperture at the entrance slit. A second geometry is shown in Fig. 3.3b. In this case, the beam is focused inside a capillary tube through the walls and Raman scattering is observed at 90°. The slit orientation is parallel to the beam direction. Consequently, a larger diameter tube must be used to obtain as much scattering as can be obtained from the geometry in Fig. 3.3a (assuming equivalent image magnification and slit length). In this configuration the scattering from glass at the points of laser entry and exit should be masked at the entrance slit. The collimated beam

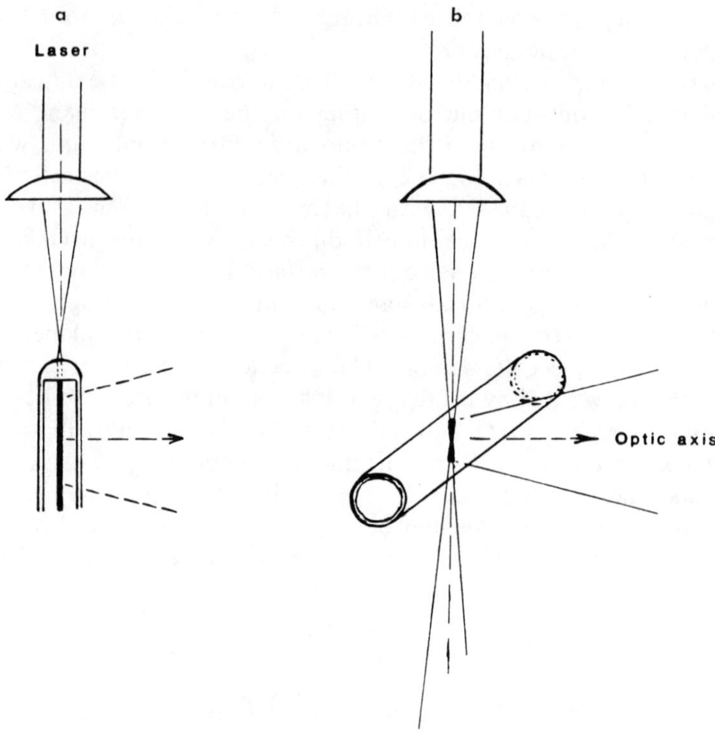

Figure 3.3. Two common geometries for observing Raman scattering from samples in capillary tubes. In configuration (a) the beam remains collimated inside the tube if the focusing lens and the capillary are translated together relative to the laser beam. This facilitates matching the beam inside the tube to the entrance slit. In configuration (b) translation of the lens off the center of the laser-beam axis produces an angular beam inside the tube and possible misalignment with the slit orientation. Also, the focused beam must be normal to the tube surface where the beam enters the tube or beam deviations (and consequent polarization errors) in the plane of the slit and optic axis can be appreciable. This can be minimized by moving the capillary so that the beam position, with and without the capillary, is unchanged.

in geometry 3a provides better image matching (and higher intensity) with the straight slits of the monochromator than the focused beam in 3b.

3.5. POLARIZATION SCRAMBLER

Grating and prism monochromators have different transmission efficiencies for light polarized parallel ($\|$) and perpendicular ($\perp$) to the entrance slit. A convenient method of measuring light polarized both $\|$ and $\perp$ to the entrance slit is to scramble the light with a "scrambler plate" that is positioned after the analyzer and before and entrance slit. A commonly used scrambler is a thin birefringent wedge whose optic axis is in the plane ⊥ to the monochromator optic axis and 45° to the slit. The wedge angle is in a plane that passes through the slit. A 15-mm-long wedge of calcite with a thickness difference of 0.5 mm provides 129.5 cycles retardation at 0.6563 μm or 182 cycles retardation at 0.4861 μm over the length of wedge (3). A quartz scrambler with similar thickness difference provides only 6.8 and 9.6 cycles retardation at 0.6563 and 0.4861 μm, respectively. An even number of cycles retardation produces a transmitted beam having all orientations of polarization. However, addition of a half cycle of retardation will produce a degree of polarization ellipticity in the transmitted beam that depends on the relative amounts of fully scrambled and elliptically polarized light. Consequently, there must be several cycles of retardation along the working aperture of the scrambler plate so that the transmitted light of an incomplete cycle is not a significant fraction of the total. Increasing the wedge angle increases the retardation difference, which improves the scrambling. But it also increases the beam deviation, which puts the transmitted light off the center of the first collimator, which, in turn, causes an intensity loss. The beam deviation of a scrambling wedge can be conveniently measured with an alignment laser. Compensation of the wedge with a matching wedge of glass whose index of refraction is between that of the two refractive indices of the birefringent wedge will minimize the deviation of the two emerging beams produced by the scrambler from the monochromator optic axis. A calcite scrambler with the dimensions given above can be compensated with Bausch and Lomb's Light Barium Crown No. 573574 so that the two beams deviate by ± 3.0 mrad about the optic axis over the wavelength range of 0.4–1.0 μm. The two beams produced by a quartz scrambler of similar dimensions only deviate by 0.3 mrad (at 0.589-μm wavelength) from each other, but they both deviate by 18.6 mrad from the optic axis and should be compensated to avoid intensity loss at the first collimator.

Bailey and Scherer (3) estimate that where the path difference leads to 16.5 cycles of retardation and the monochromator transmittance ratio $T_{\parallel}/T_{\perp}$ (parallel and perpendicular to the slit) is 5/1, the maximum deviation of $T_{\parallel}/T_{\perp}$ from 1.0 (the perfect case) for the scrambler-equipped monochromator is ± 0.013. This means that path differences smaller than 16 cycles can lead to depolarization ratio errors greater than 1.3%. The calcite scrambler described above meets these criteria with a 2.4-mm working aperture height, but the quartz scrambler needs about double the wedge angle (and certainly compensation) and a full 15-mm working aperture. Restriction of the slit height for this scrambler will almost certainly lead to errors in polarization intensity measurements, particularly in spectral regions where the grating efficiencies for the ∥ and ⊥ cases are not equal. These errors can become serious even in the case of the calcite scrambler, when the device is placed within 1 or 2 cm from the entrance slit and a small sample area is mistakenly imaged on the scrambler plate instead of the entrance slit (the slit being apertured to a small height).

The operation of the scrambler plate can be checked by recording the spectra with the two analyzer orientations of an unpolarized white light source that is imaged onto a white (MgO smoked glass or chalk) surface. Care should be taken to ensure that the position of the surface is at the right spot to produce a focused image at the entrance slit and not the scrambler plate, and that the slit height is the same as that which will be used during measurements of Raman spectra. An inadequate scrambler plate will cause an out-of-phase sinisoidal variation in the intensity recorded with the two analyzer orientations.

3.6. THE ANALYZER

The analyzer should be capable of having an extinction ratio of between 10^{-5} and 10^{-4} (i.e., pass less then 0.010% of the light of one polarization when the analyzer is set to reject it). Polaroid sheet HN32 has an extinction ratio between 10^{-4} and 10^{-5} from 488 to 740 nm with a transmittance of better than 68% over this range and HN22 has an extinction ratio between 1^{-5} and 10^{-6} over the same range with a transmittance of $\sim$45%. A good extinction ratio is necessary to cleanly separate overlapping bands of different polarization and intensity (i.e., separate a strong and highly polarized band from a weak depolarized band). The analyzer should be checked and adjusted to give good extinction in the position in which it will be used during a spectral scan. This check is done with the Rochon assembly described earlier. The alignment laser is aimed at the entrance slit with the analyzer in position. The Rochon is placed on one of its standard edges on the reference plane so that the straight-through beam passes through the analyzer and the entrance slit. One of the orientations of the analyzer should extinguish this beam, or at least bring it to minimum intensity. If it does not, the analyzer should be rotated or adjusted to give minimum intensity. The other orientation of the analyzer should extinguish the beam when the Rochon is placed on its other edge. If it does not, the analyzer positions are not orthogonal and should be modified.

3.7. POLARIZATION ORIENTATION OF THE EXCITATION LASER BEAM

A Glan-laser prism polarizer (with an extinction ratio of $\sim 2 \times 10^{-5}$) is usually placed in the excitation laser beam path ahead of the sample to clean up incident polarization. This device may have a slight beam divergence and should be in place in the earlier stages of beam alignment. It should be put in a mount capable of being reproducibly rotated (for easy change of incident polarization direction). The standard edge of the Rochon prism is positioned $\parallel$ (or $\perp$) to the optic axis (use the previously defined line), and the Glan-laser prism is rotated to extinguish one of the beams emerging from the Rochon. A polarization rotator positioned ahead of the Glan-laser prism provides a convenient way of changing power levels emerging from the Glan-laser prism. The extinction of the alternate beam from the Rochon can be checked by rotating the Glan-laser prism (and the polarization rotator) by 90°.

3.8. STANDARD RAMAN MEASUREMENT

The depolarization ratios of Raman bands can be affected by resolution. Therefore it is important to use comparable resolution in comparing results with the published literature. The depolarization ratios of the 459 cm^{-1} band of CCl_4 have been accurately reported by Kiefer and Topp (4) and plotted as a function of cm^{-1} by Kint, Elsken, and Scherer (5). The depolarized bands at 218 and 314 cm^{-1} are reasonably insensitive to resolution and should have depolarization ratios of 3/4 (or 6/7 depending on experimental geometry). The Raman spectra of liquid water exhibit broad bands in the 2800 to 3900 cm^{-1} regions, and polarization measurements may be compared with those obtained by Scherer, Go, and Kint (6).

REFERENCES

1. T. C. Damen, S. P. S. Porto, and B. Tell, *Phys. Rev.* [*2*] **142**, 570 (1966).
2. J. R. Scherer, G. F. Bailey, and S. Kint, *Anal. Chem.* **43**, 1917 (1971).
3. G. F. Bailey and J. R. Scherer, *Spectrosc. Lett.* **2(9)**, 261 (1969).
4. W. Kiefer and J. A. Topp, *Appl. Spectrosc.* **28**, 26 (1974).
5. S. Kint, R. H. Elsken, and J. R. Scherer, *Appl. Spectrosc.* **30**, 281 (1976).
6. J. R. Scherer, M. K. Go, and S. Kint, *J. Phys. Chem.* **78**, 1304 (1974).

CHAPTER

4

RAMAN SPECTROSCOPY OF INORGANIC SPECIES IN SOLUTION

DONALD E. IRISH AND TORU OZEKI*

Department of Chemistry
University of Waterloo
Waterloo, Ontario, Canada

4.1. INTRODUCTION

Raman spectroscopy is a powerful tool for the study of inorganic species in solution. The vibrational spectrum provides frequencies, intensities, and other band properties that often allow one to identify the species present and thus the chemical processes taking place, to measure the species populations and hence equilibrium constants and rate constants, and to determine the effects of temperature and pressure. The ability to probe systems conveniently at high temperatures and pressures by using Raman spectroscopy has resulted in considerable recent interest by geochemists and geologists (1, 2). The interface and interphase associated with electrodes is being actively studied (3), and solution chemistry and liquid dynamics continue to be investigated by chemists and physicists using this technique.

Raman spectroscopy, which probes the microscopic properties of solutions, is a powerful complement to methods such as conductance, solubility, calorimetry, and potentiometry, providing information on macroscopic properties. However, both the accuracy and precision with which populations can be measured using Raman spectroscopy, and hence derived quantities such as degrees of dissociation, equilibrium constants, rate constants, or thermodynamic parameters can be obtained, are considerably

* *Present address*: Hyogo University of Teacher Education, Shimokume, Yashiro-cho, Kato-gun, Hyogo, Japan, 673-14.

Analytical Raman Spectroscopy, Edited by Jeanette G. Grasselli and Bernard J. Bulkin. Chemical Analysis Series, Vol. 114.
ISBN 0-471-51955-3

less than are obtainable from potentiometric measurements, for example. The lack of knowledge of activity coefficients makes a rigorous transition to thermodynamic variables almost impossible. Nevertheless, the detailed information derived from Raman spectra about species, equilibrium states, and the degree of dissociation of an electrolyte makes a unique and useful contribution to the subject.

A number of good reviews have appeared over the years (4–12). Thus our objective in this chapter is not to provide an overview of the literature, but rather to present in some detail case studies and examples of how Raman spectroscopy has been used to elucidate the chemistry of selected systems. We shall first discuss identification of species. This is followed by the procedure for measuring species concentrations and extracting equilibrium constants. In this section we shall describe a number of the computer programs that are enabling investigators to extract considerably more information from the spectra than ever before.

4.2. IDENTIFICATION OF INORGANIC SPECIES IN SOLUTION

4.2.1. Presentation of the Raman Spectrum

Spontaneous Raman scattering is generally measured by collecting scattered radiation contained within a solid collection angle that depends on the collection geometry and optics, at each frequency of interest, $\tilde{\nu}$, and plotting the intensity of the radiation, $I(\tilde{\nu})$, versus $\tilde{\nu}$. The measured Raman intensity, $I(\tilde{\nu})$, is given by

$$I(\tilde{\nu}) = C(\tilde{\nu}_0 - \tilde{\nu}_i)^4 \tilde{\nu}_i^{-1} B^{-1} S_i, \tag{1}$$

where C is a constant that depends on the instrument response, slitwidth, solid collection angle, and attenuation due to absorptivity (color) of the sample; $\tilde{\nu}_0$ is the absolute frequency (in wavenumber units) of the laser excitation line; $\tilde{\nu}_i$ is the frequency difference of the scattered radiation (i.e., the Raman shift); B is a temperature factor given by

$$B = 1 - \exp(-h\tilde{\nu}_i c/kT) \tag{2}$$

when the Boltzmann distribution is applied; and S_i is the intrinsic molar scattering coefficient at frequency $\tilde{\nu}_i$ (4).

Most of the published spectra of species in solution are presented in this $I(\tilde{\nu})$ versus $\tilde{\nu}$ format. Important additional information can be obtained by separating and measuring the light of different polarizations. If we consider light incident on the sample along the x axis and collected at 90° along the y axis, the $x(zz)y$ configuration provides the intensity of light scattered parallel ($I_{\parallel}$) to the polarization of the incoming laser beam, and the $x(zx)y$ configuration provides the intensity of light scattered perpendicular ($I_{\perp}$) to the polarization of the incoming laser beam. The ratio

$$\rho = I_{\perp}/I_{\parallel}, \tag{3}$$

called the *degree of depolarization* or the *depolarization ratio*, is a useful parameter because it allows one to designate whether a Raman band originates from a totally symmetric mode of vibration ($\rho < 0.75$) or not ($\rho = 0.75$). However, there are many cases where the conclusion is ambiguous: $I_{\parallel}$ contains information about both the isotropic and anisotropic polarizability tensor components; $I_{\perp}$ contains only contributions from the anisotropic polarizability tensor components. Another way of portraying the spectrum is to plot I_{iso} (intensity of isotropic components) versus $\tilde{\nu}$, where

$$I_{\text{iso}} = I_{\parallel} - \tfrac{4}{3} I_{\perp}. \tag{4}$$

This spectrum contains only the isotropic polarizability elements and thus only contributions from symmetric vibrational modes. The anisotropic spectrum, I_{aniso}, is given by

$$I_{\text{aniso}} = \tfrac{4}{3} I_{\perp}. \tag{5}$$

We show examples of these various formats below.

For spectra with $\tilde{\nu} < 300\,\text{cm}^{-1}$ a reduced function, $R(\tilde{\nu})$, is recommended rather than Eq. (1) [see Brooker et al. (13)]:

$$\begin{aligned} R(\tilde{\nu}) &= I(\tilde{\nu})(\tilde{\nu}_0 - \tilde{\nu}_i)^{-4} \tilde{\nu}_i B \\ &= C S_i. \end{aligned} \tag{6}$$

This corrected intensity is directly proportional to the intrinsic molar scattering coefficient, S_i, as can be seen by comparing Eqs. (6) and (1). This quantity removes the temperature- and frequency-dependent terms from the directly measured $I(\tilde{\nu})$. It essentially removes the contribution from the Rayleigh scattering and thus may reveal Raman bands at low wavenumbers, which might otherwise be masked. The isotropic and anisotropic spectra can also be presented in this form; for example,

$$R_{\text{iso}} = R_{\parallel} - \tfrac{4}{3} R_{\perp}. \tag{7}$$

4.2.2. Interpretation

The internal motions of a vibrating molecule, although apparently complex, can be resolved into a number of relatively simple vibratory motions known as the normal modes of vibration. Each of these modes has a characteristic frequency, dependent on the masses of the atoms and the forces between them. The number of modes is $3N-6$ for a nonlinear molecule and $3N-5$ for a linear molecule containing N atoms. The activity of the normal modes in Raman or infrared (IR) spectroscopy is dependent on the geometry of the molecule. The shape of a molecule is specified by the symmetry elements that can be ascribed to it (planes of symmetry, axes of rotation, center of symmetry), which leads to a designation in terms of the molecular point group. The reader is referred to standard works for the method of classifying molecules by point groups (14–16). These works also address the subject of group theory—the mathematics that allows one to correlate the symmetry properties of the normal modes of vibration to the symmetry of the molecule. Such a group-theoretical analysis does not require knowledge of the frequencies or the detailed forms of the normal modes. For an isolated molecule the analysis reveals the number of normal modes and their symmetry type, the numbers of Raman active modes and IR-active modes, the number of polarized Raman lines, and the activity of overtones and combination bands. Thus the Raman selection rules are revealed. It is worth emphasizing that these rules rigorously apply to molecules in isolation, i.e., molecules unperturbed by their environment. To quote Halford (17), "There will be no selection rules operating in the liquid phase," and again, "The relaxation of selection rules is associated with the development of intermolecular forces" (18). In a solution there exist solute–solute and solute–solvent interactions that are large and time dependent. Thus we should not be surprised if the selection rules inferred from group theory fail on occasion.

As a first example let us consider the common molecular symmetry based on the tetrahedron. The perchlorate ion, ClO_4^-, is a good example of such a species. We will compare the results of group theory with data from an actual Raman spectrum and use group theory (selection rules) to interpret the spectrum. Group theory leads to a reduced representation, $\Gamma(T_d)$, for the normal modes given by

$$\Gamma(T_d) = a_1(\mathrm{R}) + e(\mathrm{R}) + 2f_2(\mathrm{R}, \mathrm{IR}),$$

where R indicates Raman-active modes, and IR, infrared-active modes. The $(3N-6)$ rule indicates nine normal modes of vibration. Because of the symmetry some of these are degenerate, i.e., several modes have exactly the same frequency and energy. The representation tells us that one mode is totally symmetric with respect to all symmetry operations; it is the a_1 mode—the symmetric stretching and contracting of the Cl—O bonds. This mode is Raman active and appears in the isotropic spectrum but not in the anisotropic spectrum. The depolarization ratio ρ is less than 0.75 and in fact should be zero (if unperturbed by the surroundings). It is IR forbidden (there is no change of dipole accompanying the motion), but it is the strongest band in the Raman spectrum. The representation tells us that a second band is Raman active only; it is designated e, indicating that it is doubly degenerate; it is a deformation mode. Two bands are triply degenerate and are labeled f_2. These are both Raman and IR active. One is the antisymmetric Cl—O stretching mode; this is intense in the IR, as a large dipole moment change

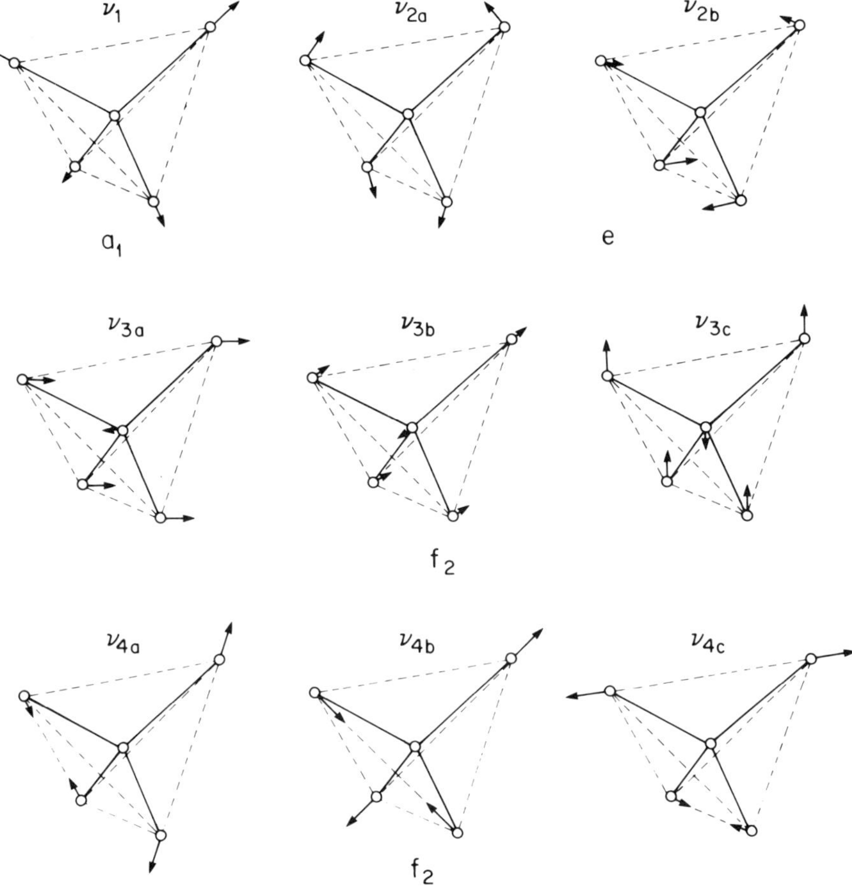

Figure 4.1. Normal vibrations of a tetrahedral MX_4 molecule.

accompanies the motion, but weak in the Raman. The other is a deformation mode. The forms of these normal modes are presented in Fig. 4.1. From this example the reader can correctly conclude that Raman and IR spectroscopies are complementary; because of the difference in selection rules, the correct identification and geometry may often require the measurement of both types of spectra. However, using known frequencies and selection rules, identification is frequently possible using only one or the other spectroscopy.

The Raman spectrum of the ClO_4^- anion is shown in Fig. 4.2. In agreement with the reduced representation, there are four bands. The low-frequency bands are attributed to the deformations; the intense line is the symmetric stretching mode a_1; and the weak, higher-frequency line is the antisymmetric stretch f_2. The numbering scheme follows the Herzberg notation, viz., from high frequency to low frequency in the order of decreasing symmetry. Thus $\nu_1(a_1) = 936\ \text{cm}^{-1}$; $\nu_2(e) = 462\ \text{cm}^{-1}$; $\nu_3(f_2) = 1113\ \text{cm}^{-1}$; and $\nu_4(f_2) = 629\ \text{cm}^{-1}$. The known IR activity is consistent with this assignment. If this species had been square planar (D_{4h} point group) instead of tetrahedral (T_d point group) the reduced representation would be

$$\Gamma(D_{4h}) = a_{1g}(\text{R}) + a_{2u}(\text{IR}) + b_{1g}(\text{R}) + b_{2g}(\text{R}) + b_{2u}(\text{IR}) + 2e_u(\text{IR}).$$

The observation of four Raman bands in the spectrum of ClO_4^-, rather than three, suggests T_d symmetry. The measurement of the IR spectrum provides conclusive evidence. The D_{4h} geometry requires three IR bands, all of which

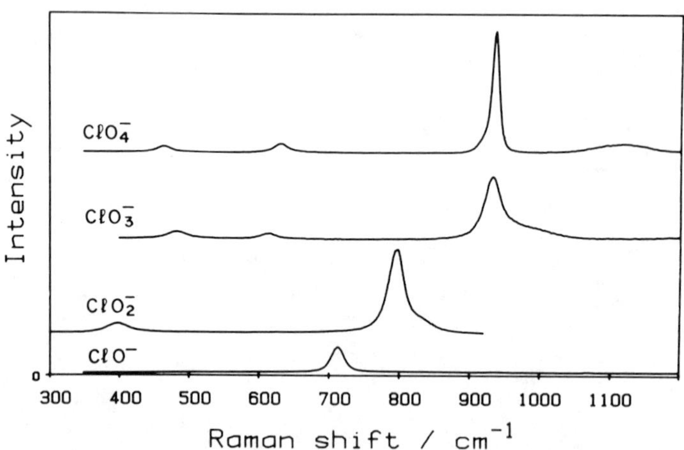

Figure 4.2. Total Raman intensity of ClO_4^-, ClO_3^-, ClO_2^-, and ClO^-. Data are courtesy of Professor H. H. Eysel, University of Heidelberg.

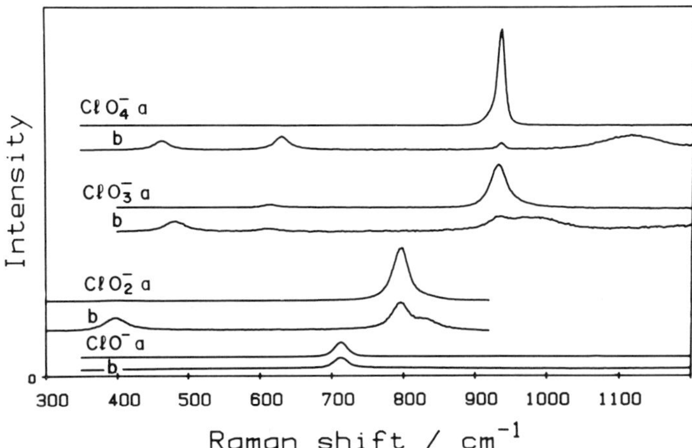

Figure 4.3. Isotropic (a) and anisotropic (b) intensity of ClO_4^-, ClO_3^-, ClO_2^-, and ClO^-. The anisotropic intensity, *b*, is scaled four times the isotropic intensity, *a*. Data are courtesy of Professor H. H. Eysel, University of Heidelberg.

are noncoincident with Raman bands, by virtue of the center of symmetry and the mutual exclusion principle. We thus see how the two spectroscopies, with the guidance of selection rules, work together to lead one to the correct assignment.

In Figs. 4.2 and 4.3 we show a comparison of the spectra of perchlorate ion ClO_4^-, chlorate ion ClO_3^-, chlorite ion ClO_2^-, and hypochlorite ion ClO^-. The total intensity is given in Fig. 4.2, the isotropic and anisotropic intensities in Fig. 4.3. The symmetry has been lowered from T_d for ClO_4^-, to C_{3v} for ClO_3^-, to C_{2v} for ClO_2^-, and to $C_{\infty v}$ for ClO^-. Both the force constants and the geometry change. The species symbols are given in Fig. 4.4. For C_{3v}, C_{2v}, and $C_{\infty v}$ symmetries, all normal modes are both IR and Raman coincident so the IR spectrum does not provide substantial new information. The diatomic ClO^- ion has only one vibration. If we examine the isotropic intensity we see that ClO_4^- has only one band, and ClO_3^- and ClO_2^- have two each, although on the scale presented the $\sim 400\,cm^{-1}$ contribution for the latter is almost unobservable. The anisotropic spectra clearly identify the nonsymmetric modes.

Even for a species as common and apparently simple as the perchlorate anion the spectrum is not entirely clear-cut. A shoulder can be seen on the low-frequency side of the $\nu_1(a_1)$ band in both Figs. 4.2 and 4.3. The origin of this extra band is not entirely clear (19). It may arise from an overtone of ν_2 in Fermi resonance with $\nu_1(a_1)$ or from a hot band, but the arguments

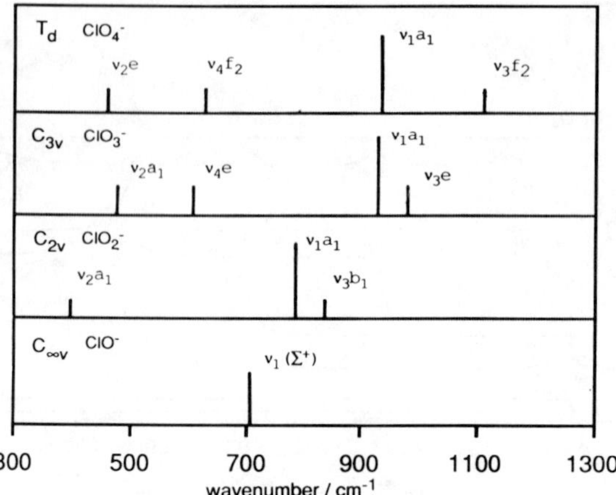

Figure 4.4. Observed fundamental vibration spectra of the oxychloride anions with the reduced representation species symbol of each mode given.

are not totally conclusive. Even for this simple species in solution the Raman spectrum has many intriguing subtleties and changes with temperature, pressure, and composition to whet the curiosity of investigators and challenge their interpretation of the results.

The oxyanions described above are molecular species that retain their integrity in solution. In contrast, metal halide complexes such as $ZnBr_4^{2-}$ and $FeCl_4^-$ are formed from the aquated simple cations and anions, probably in steps, and thus are more labile. We consider first $ZnBr_4^{2-}$. The species clearly gives a four-line spectrum consistent with T_d symmetry (Fig. 4.5). The bromide ligands are exchanging with bromide ions in the solution, but on a time scale that is slow compared to the time for vibration. One can convince oneself that the species is formed by the sequential addition of bromide ion to aquated zinc ion by replacement of water:

$$Zn^{2+} + xBr^- = ZnBr_x^{2-x}$$

Thus, in Fig. 4.6, one can see that as the ratio (R) of bromide ion to zinc ion is increased the spectrum in the region of the symmetric stretch changes from three bands to two bands to one band, corresponding to the change in the relative populations of the species $ZnBr_2$, $ZnBr_3^-$, and $ZnBr_4^{2-}$. For the former two species the rigorous deduction of the molecular geometry is also not possible from the spectra. The number of bands and their polarization states cannot be clearly inferred from the spectra because of their intrinsic

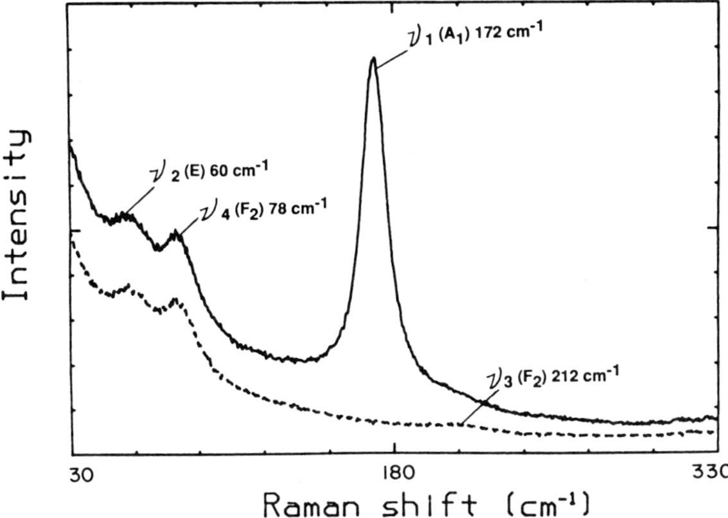

Figure 4.5. The Raman spectrum of $ZnBr_4^{2-}$. The solid line is parallel polarization. The dashed line is perpendicular polarization. Reprinted from Yang et al. (20), with permission of the Journal of Solution Chemistry.

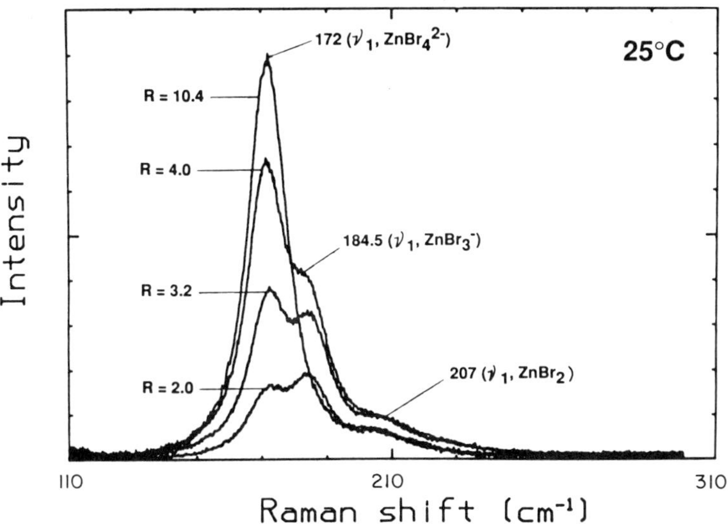

Figure 4.6. The Raman spectra of aqueous solutions of $ZnBr_2$/NaBr mixtures at 25°C. The ratio of bromide ion to zinc ion, R(Br/Zn), is specified on the figure. The totally symmetric stretching vibrations of the complexes $ZnBr_4^{2-}$, $ZnBr_3^-$, and $ZnBr_2$ are shown. Reprinted from Yang et al. (20), with permission of the Journal of Solution Chemistry.

weakness and the proximity of the exciting line. Even recasting the spectra in reduced form did not apparently help (20). Another relatively common but still not completely understood phenomenon that complicates such spectra is the role of the solvent. The zinc ion is aquated by six water molecules; the final halo-complex has only four ligands. Thus the geometry has changed from octahedral (O_h) to tetrahedral (T_d) during the sequential replacement of water molecules by bromide ions. It is believed, in this case, that the change occurs at the second step, during conversion of $ZnBr(H_2O)_5$ to $ZnBr_2(H_2O)_2$ (21, 22). Such changes should affect the spectra, but spectral changes resulting from changes in geometry are not easily detected. Challenges abound.

The spectra of halometal complexes are often difficult to interpret for yet another reason: band overlap. The Raman spectra of, for example, aqueous zinc chloride solutions (23), indium chloride solutions (24), and aqueous cadmium bromide solutions (25) do not show distinct bands of the several contributing species, as shown in Fig. 4.6, but rather an envelope of intensity that encompasses the several bands (Fig. 4.7). Even on forcing the equilibria by addition of excess halide to favor the highest complex, $ZnCl_4^{2-}$ or $CdBr_4^{2-}$, the well-resolved four-line spectrum characteristic of a tetrahedral species is not obtained (Fig. 4.8). Thus the investigator must here be cautioned: reliable conclusions can only be reached by examination of many spectra recorded

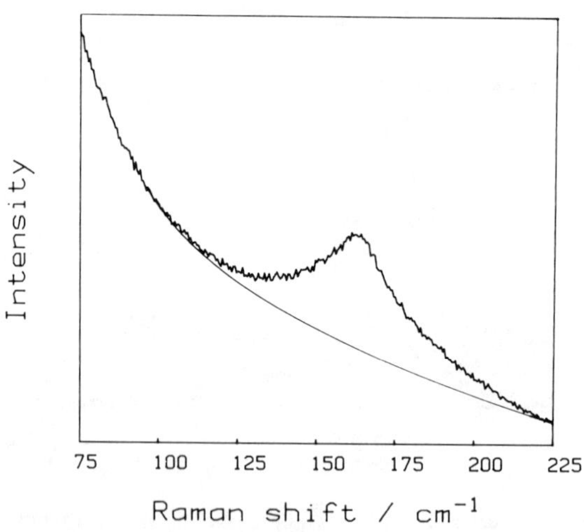

Figure 4.7. The Raman spectrum of 1.0 *m* $CdBr_2$ solution at 25°C. The envelope of intensity at 159 cm^{-1} is generated by three species. Compare with Fig. 4.6. Reprinted from Anderson and Irish (25), with permission of the Journal of Solution Chemistry.

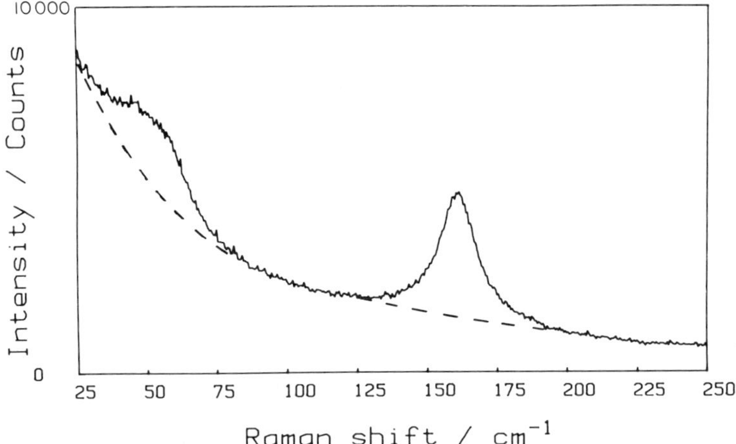

Figure 4.8. The Raman spectrum of the tetrahedral species $CdBr_4^{2-}$ ($R = 9.74$, 25°C). Compare this spectrum with Fig. 4.5 and note the lack of the characteristic four lines. Reprinted from Anderson and Irish (25), with permission of the Journal of Solution Chemistry.

of solutions of differing concentration, composition, and temperature. We shall see later that the use of the computer facilitates such interpretation.

The symmetry of the tetrahedral species will be lowered by addition of a sixth atom to one of the outer atoms. An example of this is the addition of a hydrogen atom to sulfate to give bisulfate anion. The latter can be described in terms of C_{3v} symmetry if the OH group is considered as a point mass (a distorted five-atom tetrahedron); otherwise it will have C_s or C_1 symmetry. The normal modes of vibration will span the following representations for the two cases, C_{3v} and C_s, respectively:

$$\Gamma(C_{3v}) = 3a_1(\mathrm{R, IR}) + 3e(\mathrm{R, IR})$$

$$\Gamma(C_s) = 8a'(\mathrm{R, IR}) + 4a''(\mathrm{R, IR}).$$

The spectrum is shown in Fig. 4.9. The spectrum has been isolated from that of sulfate by raising the temperature of the solution to 300°C, under a pressure of 10 MPa (26). In the upper panel a very weak feature at 977 cm^{-1} is the most intense line of the residual sulfate ion. The band positions and their assignments are presented in Table 4.1. Similarly the addition of a proton to the perchlorate anion converts it to perchloric acid as shown in Fig. 4.10, but because this acid is very strong the bands of $HClO_4$ are only detected when the water/acid ratio is less than 1.8 (19). Data for these species also are given in Table 4.1.

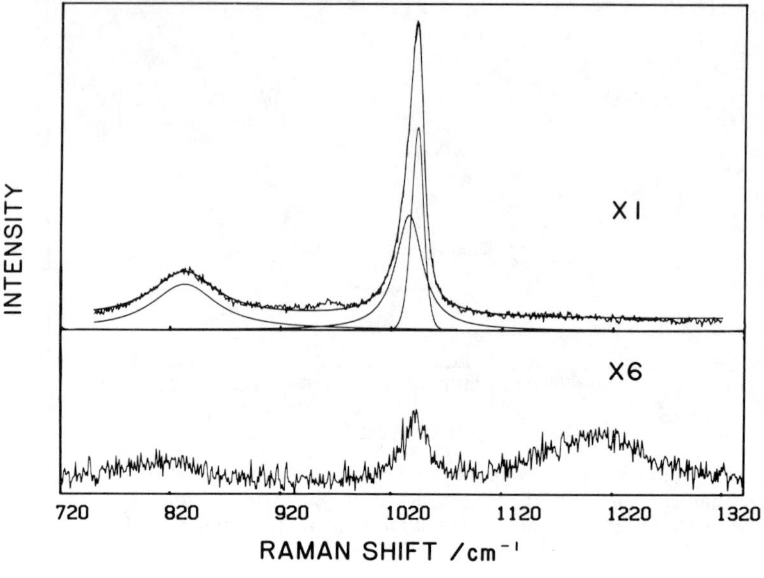

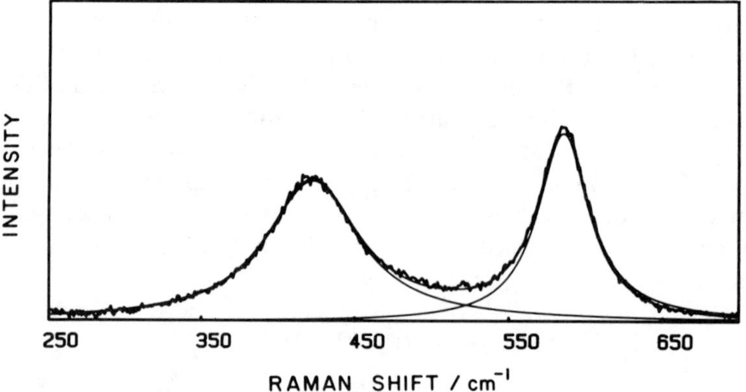

Figure 4.9. The Raman spectrum of the bisulfate anion in 3.8 *m* NH_4HSO_4 aqueous solution at 300°C. Upper panels: the 720–1320 cm^{-1} region; isotropic, × 1; anisotropic, × 6. Lower panel: the 250–700 cm^{-1} region showing the fit to two symmetrical bands. Reprinted from Dawson et al. (26), with permission. Copyright 1986, American Chemical Society.

Another common oxyanion in aqueous solution is the nitrate ion, NO_3^-, with D_{3h} symmetry. We will discuss it and the related species HNO_3 and NO_2^- briefly. The hypothetical "free" nitrate ion should exhibit a four-band spectrum. The representation is

$$\Gamma(D_{3h}) = a_1'(\mathrm{R}) + a_2''(\mathrm{IR}) + 2e'(\mathrm{R, IR}).$$

Table 4.1. Raman Bands (cm^{-1}) of the Tetrahedral Species Sulfate and Perchlorate, and Their Protonated Forms Bisulfate and Perchloric Acid

	SO_4^{2-} (25°C)	HSO_4^- [a] (300°C)	Assignments under C_{3v} Symmetry	ClO_4^- [b] (25°C)	$HClO_4$ (25°C)	Assignments under C_{3v} Symmetry
		422 p	δ_{as} SO_3 def., e		432 dp	δ_{as} ClO_3 def., e
$\nu_2(e)$	451 dp[c]			462dp		
		585 p	δ_s SO_3 def., a_1		578 p	δ_s ClO_3 def., a_1
$\nu_4(f_2)$	613 dp			629 dp	~585	δ_{as} ClO_3 def.
		831 p	ν_s S-(OH) str., a_1	928 sh	757 p	ν_s Cl-(OH) st., a_1
$\nu_1(a_1)$	980 p			935 p		
		1035 p			1039 p	ν_s ClO_3 str., a_1
		1043 p	ν_s SO_3 str., a_1			
$\nu_3(f_2)$	1104 dp	1195 dp	ν_{as} SO_3 str., e	1113 dp	1220 dp	ν_{as} ClO_3 str., e

[a] 3.8 m NH_4HSO_4 at 300°C and 10-MPa pressure.
[b] In the system $HClO_4 \cdot H_2O$ at 25°C.
[c] Key: p, polarized; dp, depolarized; sh, shoulder.

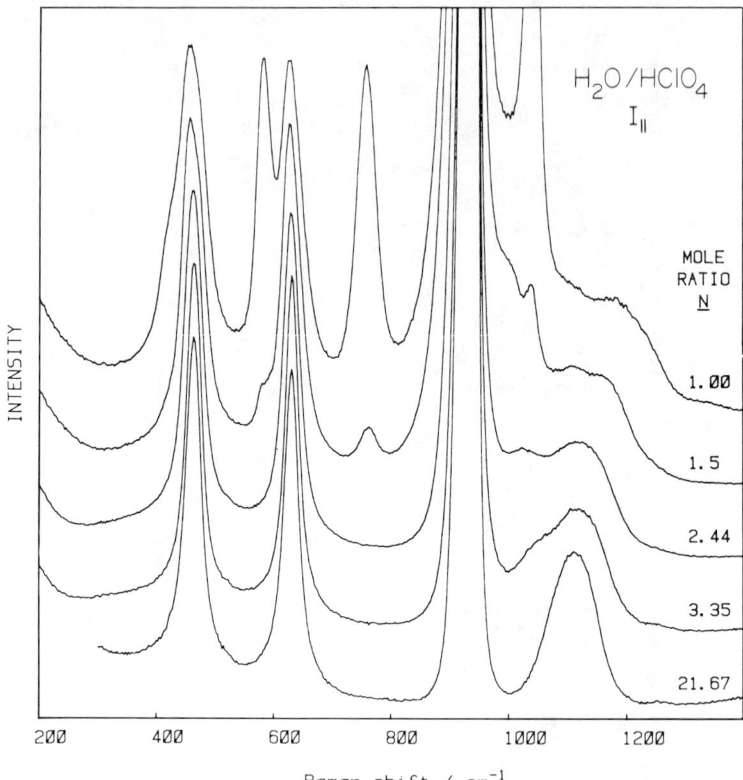

Figure 4.10. Raman spectra ($I_{\parallel}$) of aqueous solutions of perchloric acid in the 200–1400 cm^{-1} region at 25°C. The mole ratio, N, of water to $HClO_4$ is given along the right-hand side. Note that for $N = 1.5$ and 1.0 there are new bands in the spectrum characteristic of molecular $HClO_4$. Reprinted from Ratcliffe and Irish (19), with permission of the National Research Council of Canada from the Canadian Journal of Chemistry, 1984.

The Raman bands occur at 1048 cm^{-1} (a_1'), 1380 cm^{-1} (e'), and 717 cm^{-1} (e'). However in aqueous solutions the 1380 cm^{-1} band appears as a doublet, 1348 and 1406 cm^{-1}; this lifting of the degeneracy is attributed to interaction with water. The IR-active mode, a_2'', occurs at 830 cm^{-1}. The nitrate ion can bind to many different cations, thus forming an ion pair or complex ion. When this occurs a new set of bands arises, somewhat displaced from those of solvated nitrate ion; thus, since both bound and free forms exist in equilibrium, all lines of nitrate are essentially doubled, although for some systems the separation of some of the components is almost undetectable. Computer analysis helps in resolving such spectra. The spectra of nitrate complexes have been reviewed in depth (7).

On addition of a proton to NO_3^-, nitric acid, HNO_3, is formed. Although the HNO_3 molecule strictly has C_s symmetry, the bands of the (HO)—NO_2 unit in solution behave as though the molecule is a C_{2v} species. Under this symmetry the bands and mode descriptions are as follows: 640 cm^{-1} for δ[(OH)—N—O rock] of b_1; 688 cm^{-1}, δ(O—N—O scissors), a_1; undetected, δ(out-of-plane); 955 cm^{-1}, ν[N—(OH) stretch], a_1; 1304 cm^{-1}, ν(NO_2 symmetric stretch), a_1; 1558 cm^{-1}, 2δ(out-of-plane), a_1; 1673 cm^{-1}, ν(NO_2 antisymmetric stretch), b_1. The spectrum is shown in Fig. 4.11. The bands of HNO_3 increase relative to those of NO_3^- when the solution is heated (27) as shown in Fig. 4.12; similar results are obtained for the acids HSO_4^- and $HClO_4$ on heating.

The nitrite ion, NO_2^-, is nonlinear and thus of C_{2v} symmetry. Three Raman bands are expected and observed (Fig. 4.13). These are 1331 cm^{-1} for ν_1 (NO_2 symmetric stretch) of a_1; 817 cm^{-1}, ν_2 (symmetric deformation), a_1; 1242 cm^{-1}, ν_3 (NO_2 antisymmetric stretch), b_1. This ion also binds to metal ions to form complexes in solution. An example is the formation of cadmium nitrite complexes; spectra are illustrated in Fig. 4.13. A characteristic feature here is the new band at 861 cm^{-1}; the bands from the stretching vibrations of the complexed nitrite occur at almost the same positions as those of the free nitrite ion and differ largely in molar intensity. Note that although the spectrum suggests only one set of lines for the complex, detailed

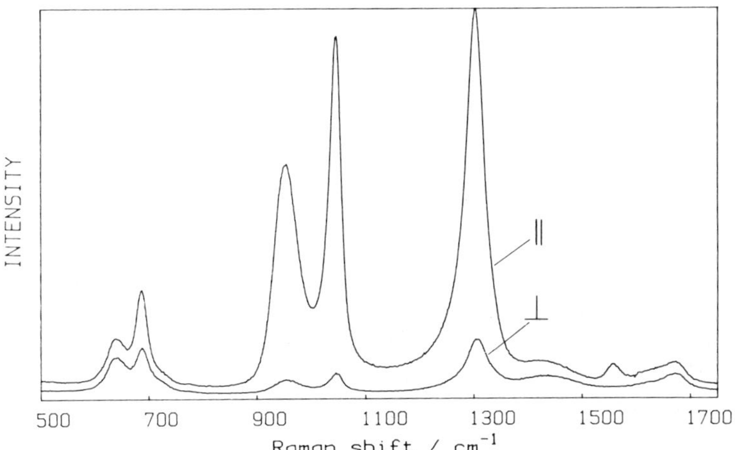

Figure 4.11. The Raman spectrum of a solution of nitric acid with 1.5 moles water per mole HNO_3, at 25°C. Bands at ~720 cm^{-1}, weak shoulder; 1046 cm^{-1}, strong; ~1430 cm^{-1}, weak; ~1648 cm^{-1}, shoulder arise from the nitrate ion. The other bands arise from the HNO_3 molecule. Reprinted from Ratcliffe and Irish (27), with permission of the National Research Council of Canada from the Canadian Journal of Chemistry, 1985.

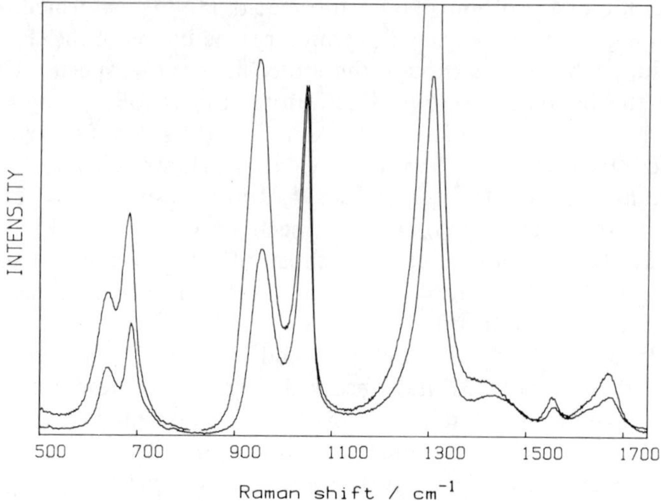

Figure 4.12. Superimposed spectra of a solution of nitric acid with 1.5 moles water per mole HNO_3 at 3°C (lower) and 100°C (upper), normalized on the symmetric stretching band of NO_3^-. Note the increase in intensity of the bands of molecular nitric acid with increasing temperature. Reprinted from Ratcliffe and Irish (27), with permission of the National Research Council of Canada from the Canadian Journal of Chemistry, 1985.

intensity studies indicate that there are four different complexes with superimposed bands (28): $Cd(NO_2)^+$, $Cd(NO_2)_2$, $Cd(NO_2)_3^-$, and $Cd(NO_2)_4^{2-}$. Again it is important to note that the investigator can be misled if conclusions are based on only a single spectrum of a single composition. Many spectra of solutions with different concentrations and metal/ligand ratios must be analyzed quantitatively. We will discuss this aspect in Section 4.3, next.

It is beyond the scope of this chapter to provide spectra of the many inorganic species that can occur in solution. Good compilations of such exist (29, 30). Our objective has been to illustrate spectra of selected species, show how these species change in solution by formation of acids or complex ions, and explain that, even in the spectra of "simple" species there exist subtle features that are not easily rationalized. We now turn our attention to quantitative analysis.

Figure 4.13. Comparison of Raman spectra of $NaNO_2$ and Cd^{2+}/NO_2^- solutions, from 700 to 1500 cm^{-1}: (a) 1.475 M $NaNO_2$; (b) 0.400 M Cd^{2+}, 3.62 M NO_2^-; (c) 0.896 M Cd^{2+}, 1.852 M NO_2^-. Reprinted from Irish and Thorpe (28), with permission of the National Research Council of Canada from the Canadian Journal of Chemistry, 1975.

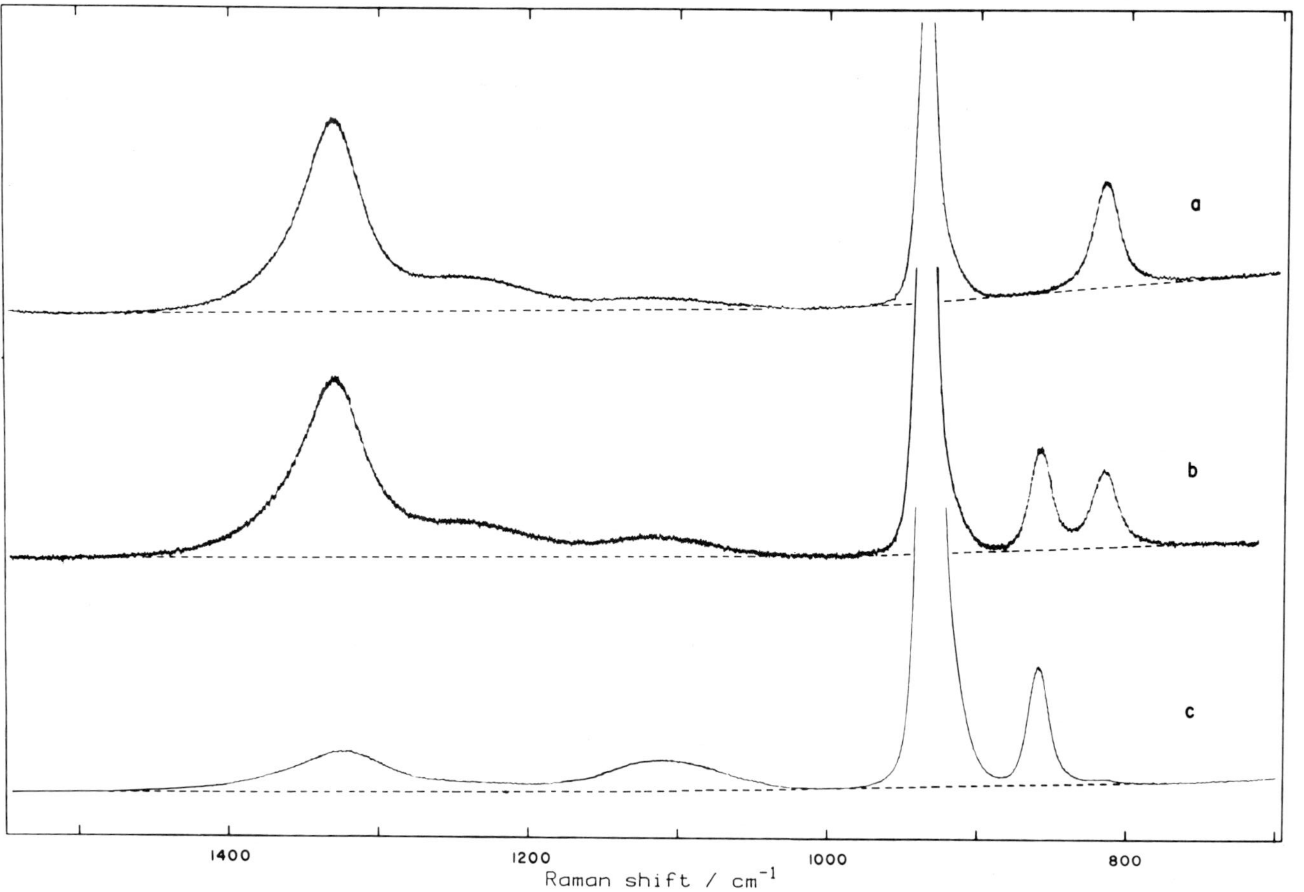
a
b
c
INTENSITY
1400
1200
1000
800
Raman shift / cm⁻¹

4.3. QUANTITATIVE ANALYSIS

4.3.1. Correlating Raman Intensities and Species Concentrations

Raman spectroscopy is largely a single-beam technique. This has placed limitations on its use for measurement of species concentrations. Variations in the intensity of the exciting light from one scan to the next, variations in the sample, including changes in refractive index and optical attenuation, and variations in the orientation of different samples in the optical path all cause a lack of reproducibility. To overcome these problems it has become routine to introduce into the sample a small quantity of a presumably inert material with an easily detected Raman signal to be used as an internal intensity reference. Such a substance must not interact chemically with the species of interest, it must be stable, and it must have Raman signals distinct from those of the species of interest. The essence of the method is illustrated in Fig. 4.14. Two lines of the spectrum, one from the standard and one from the sample, are shown in the upper-right side of the figure. The areas of these are measured. With today's digitized, computer-stored spectra, this is a relatively easy task. The definition of a correct baseline is frequently the largest source of error. The relative integrated intensity, I_R, is defined as

$$I_R = \left(\frac{\text{area of band from sample}}{\text{area of band from standard}}\right)\cdot\text{concentration of standard.} \qquad (8)$$

This quantity is essentially the integrated intensity from the species per integrated intensity per mole per liter of standard. This quantity is frequently

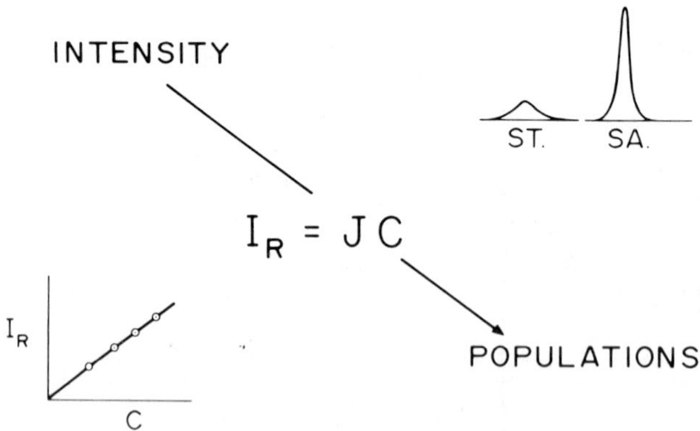

Figure 4.14. Schematic diagram to illustrate the method of measuring relative Raman intensity.

found to be directly proportional to the concentration of the species of interest, as illustrated in the lower left side of Fig. 4.14. Thus

$$I_R = J \cdot C \tag{9}$$

The slope of the graph is designated as the molar intensity, or the molar scattering coefficient, J. Knowledge of J for a particular species allows one to calculate the concentration C from the measured I_R. A more detailed exposition of quantitative analysis by Raman spectroscopy is contained in Chapter 5 by Vickers and Mann.

We present an example taken from the work of Dawson et al. (26). In Fig. 4.15, the Raman signals from the 935 cm^{-1} line of perchlorate anion from a 0.309 m $NaClO_4$ solution and the 980 cm^{-1} line of sulfate anion from a 0.619 m Na_2SO_4 solution are shown. The molality, m, (mole solute per kilogram solvent water) was used because it is a temperature-independent concentration unit. As discussed above, the shape of the 935 cm^{-1} line is asymmetrical and the band has been resolved into two closely spaced bands. The sum of these two components (i.e., the total band area) has been measured for the standard. A linear baseline has been adopted. From these raw data a value of I_R is obtained. A set of values is plotted against the molality of Na_2SO_4 in Fig. 4.16. These data were obtained with the sample held at 100°C and 10-MPa pressure in a furnace assembly (31). The slope of the plot is J for which a value of 0.644 ± 0.006 was obtained. This value was used to calculate the concentration of SO_4^{2-} in acid solutions of ammonium hydrogen

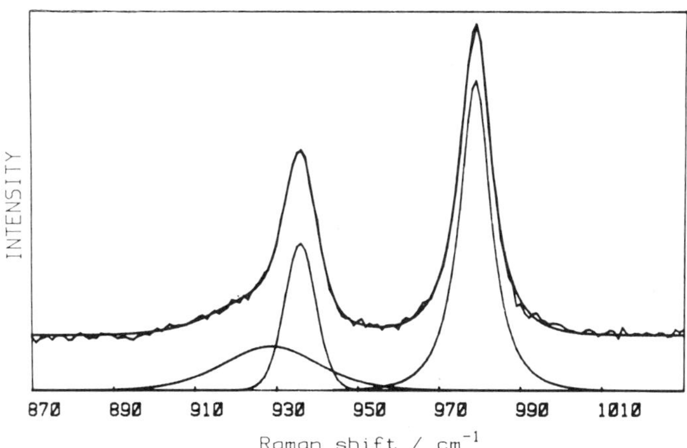

Figure 4.15. The Raman bands from 0.309 m $NaClO_4$ and 0.619 m Na_2SO_4.

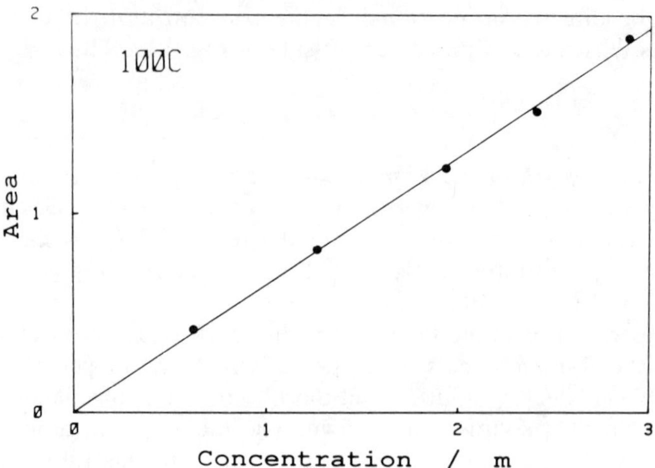

Figure 4.16. A plot of the relative integrated band intensity of the 980 cm^{-1} band of sulfate anion versus the concentration of Na_2SO_4, measured at 100°C.

sulfate (26), also measured at 100°C. The objective of that study was to measure the degree of dissociation of HSO_4^- in solutions with temperatures ranging from 25° to 300°C. The J value was measured independently at each temperature of interest.

In order to compare the J values obtained at different temperatures, it is necessary to correct the intensity for the variation in temperature. It is clear from examination of Eq. (1) that the correction lies in the B factor (temperature). One way of avoiding this correction at this stage is to employ the reduced intensities $R(\tilde{\nu})$ [Eq. (6)] instead of $I(\tilde{\nu})$, which was used by Dawson et al. (26). Let us see the magnitude of this correction. We will compare the J value obtained at 100°C with that obtained at 25°C, viz., 0.636 ± 0.022. The uncorrected ratio is $J_{100}J_{25} = 1.01$ (2). Four band intensities have been employed to obtain this ratio. Thus the correction factor can be expressed as

$$(B_{980}^{100}/B_{935}^{100})(B_{935}^{25}/B_{980}^{25}),$$

where the subscript designates the wavenumber of the band and the superscript designates the temperature. This factor equals 1.002. Thus within the accuracy of the measured J values the correction makes no difference. Even for 300°C the correction is only 1.008. This is a consequence of the relatively large values of the wavenumbers of the bands being considered. For low-frequency bands the correction can be large. Thus a factor of 0.74 was applied when comparing the J values for the 160 cm^{-1} line of $CdBr_4^{2-}$ measured at

200° and 25°C; the 1032 cm^{-1} band of the trifluoromethanesulfonate anion ($TFMS^-$) was used as the internal reference in that case (25). For the sulfate/perchlorate systems the molal intensity values were found to be essentially independent of temperature over the range 25–250°C; a plot of J versus temperature had a slope of only $(\partial J/\partial T)p = -1.34(10^{-4}) \pm 0.57(10^{-4})$ over a 225 C° range (26). In contrast, the J value for $CdBr_4^{2-}$ decreased by about 10% for a 175 C° range (25).

In this section we have demonstrated the method of correlating integrated Raman intensities with species concentrations. Several points require emphasis. The selection of a substance to be used as an internal reference is often difficult. In aqueous solutions the perchlorate anion has frequently been used because it is believed that, in most cases, it does not form complexes with the cations being studied. This may not always be true. Also, as the temperature of the solution is raised, perchlorate may undergo decomposition or may oxidize the sample. An example of this was recently encountered in the study of iron(III) chloride complexes at elevated temperature (32). To get around this problem another noncomplexing anion, $TFMS^-$, has been used. It has a number of useful, intense bands, as shown in Fig. 4.17, and is noncomplexing. It was used in high-temperature studies of $ZnBr_2$ (20) and $CdBr_2$ (25). However $TFMS^-$ is also susceptible to metal-ion-catalyzed decomposition at elevated temperature (33), so its suitability in a particular

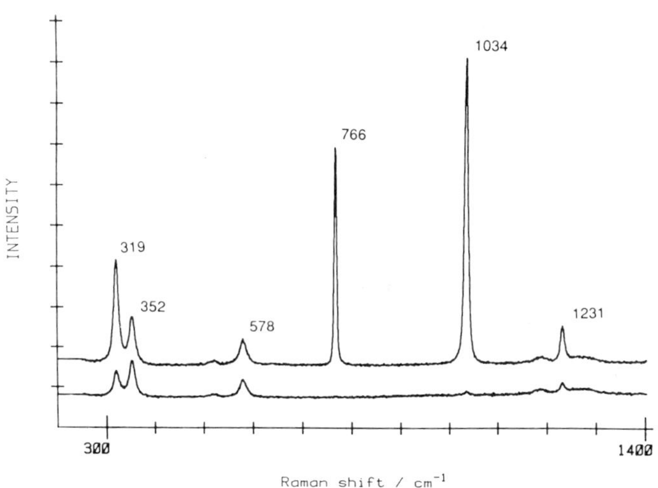

Figure 4.17. The Raman spectrum of a 3 *m* solution of trifluoromethanesulfonic acid. Upper, $I_{\parallel}$; lower $I_{\perp}$. (Reproduced by permission of M. M. Yang, Ph.D. Thesis, Princeton University, 1988.)

experiment must be considered carefully. Both ClO_4^- and TFMS$^-$ can be introduced as the strong acid, thus lowering pH and counteracting hydrolysis. In some cases a Raman line of the solvent can be used as a standard; thus for the study of nitrate salts in *N*-methlylacetamide (NMA) the 833 cm^{-1} band of the solvent was used (34). However, care must be taken to ensure that the band of the reference is not affected by the solute. For aqueous electrolyte solutions the bands of water (the 1640 cm^{-1} bending vibration and the broad envelope of intensity at 3400 cm^{-1} resulting from O—H stretching) are not considered to be good choices. Water molecules bind to cations strongly; frequently the bending mode may be a doublet arising from bound water and bulk solvent water. The hydrogen-bonded structure of liquid water is altered by both temperature and solute; this is reflected in changes in the shape of the ~3400 cm^{-1} band envelope (19, 35). Thus water can be used as an intensity reference only in some special cases.

Also the uncertainty of the baseline is a significant source of error. In some cases a linear baseline is justified (e.g., see Fig. 4.15). In some cases, when one uses digitized spectra, a background spectrum from a blank can be successfully subtracted from the spectrum of interest. However, the nature of the solvent structure and hence the extent of the Rayleigh scattering may be different for blank and sample, thus leading to an inadequate fit. Near the exciting line a background shape can be synthesized from baseline points on either side of the bands of interest and subtracted from the spectrum (Fig. 4.18); exponential-quadratic functions of the form

$$Y = A_1 \exp[P_1 X] + A_2 + A_3(P_2 - X) + A_4(P_2 - X)^2 \tag{10}$$

have been found to yield suitable empirical representations of the baseline in the region close to the exciting line (36). In this expression Y denotes intensity, and X, the wavenumber shift; the constants A_1 to A_4 are obtained by nonlinear least square regression, and P_1 and P_2 are specified constants. Another approach in this region of the spectrum is to use the reduced intensity $R(\tilde{\nu})$.

Reference was made above to the application of a bandfitting program for the analysis of overlapping Raman bands. We now address this theme in detail.

4.3.2. Spectral Analysis with Bandfitting Programs

Frequently the measured spectrum consists of a number of bands from different species that overlap each other to some degree. The overlap may be moderate, as shown in Fig. 4.15, or severe, as shown in Fig. 4.7. To estimate the contribution of particular bands, computer programs are used. These

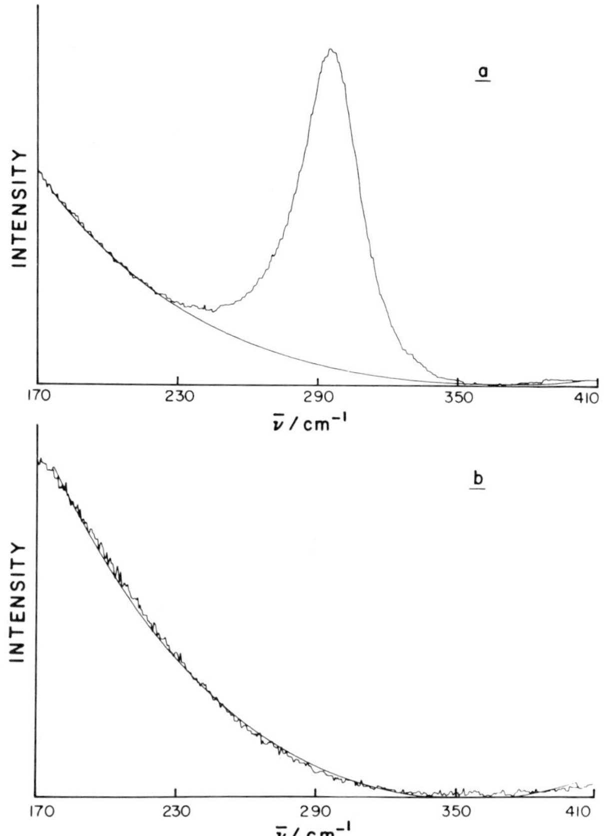

Figure 4.18. (a) A spectrum of a solution containing In^{3+} and Cl^-, with an exponential-quadratic baseline; (b) the fit of the exponential-quadratic function to a blank for the In^{3+}/Cl^- study. Reprinted from Bulmer et al. (36), with permission.

techniques are almost routine now that data are stored in digital format on computers.

One of the widely used programs is that developed by Pitha and Jones (37) for IR spectroscopy and subsequently modified by Murphy and Bernstein (38) for Raman spectroscopy. We shall call this program BNDFIT (for bandfit). The fitting is dependent on a prior assumption of the shape of a band. Pitha and Jones found that a product of a Gaussian $[Y = a\exp(-X^2)]$ and a Lorentzian $[Y = b(1 + X^2)^{-1}]$ function could be used to model IR band shapes. Thus, the expression

$$I(\tilde{\nu}) = B_2 \exp[-B_3(\tilde{\nu} - B_1)^2]/[1 + B_4(\tilde{\nu} - B_1)^2] \qquad (11)$$

can be used. Here, $I(\tilde{\nu})$ is the intensity at each wavenumber, B_{2-4} are the three fitted regression parameters, and $(\tilde{\nu} - B_1)$ is the distance from the peak maximum. The program is constructed in an interactive mode so that the number of component peaks, peak positions, and shapes can be either fixed or varied as necessary to obtain a good fit via a damped least squares method. The program output includes the four parameters B_{1-4}, the full width at half height (FWHH), the area of each band, and the contribution of the Lorentz or Gauss functions to the shape.

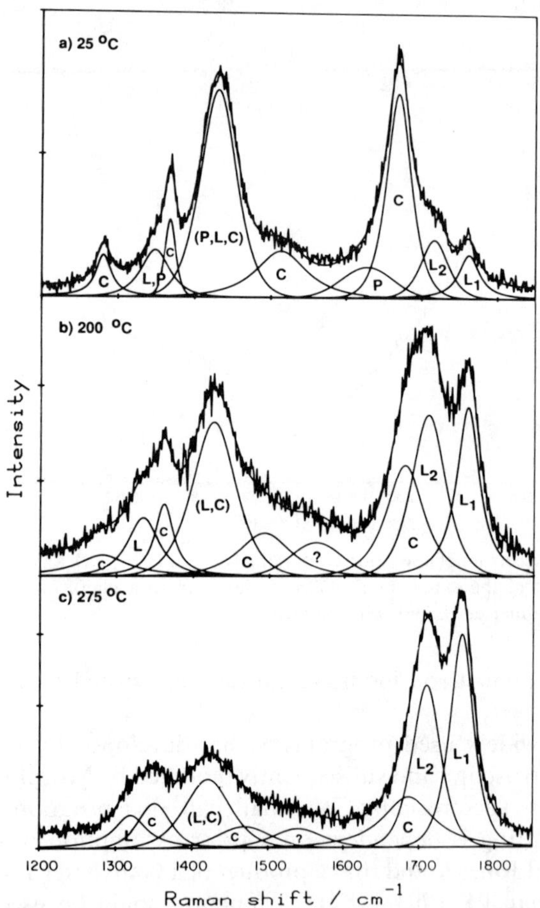

Figure 4.19. The spectra and the results after analysis with the program BNDFIT of glacial acetic acid in the range 1200–1850 cm^{-1} at (a) 25°C, (b) 200°C, and (c) 275°C. Reprinted from Semmler and Irish (39), with permission of the Journal of Solution Chemistry.

Bandfitting of the complex spectrum of acetic acid is shown in Fig. 4.19 (39). Generally one must subject a series of spectra of samples with different compositions or temperatures to bandfitting in order to be satisfied that the result for any one spectrum is reasonable. In deciding the number of bands required to fit a given spectral envelope, the "principle of parsimony," or—as it is more generally called—"Ockham's razor," has been used (that is, simplicity is preferable to unnecessary complexity) (40). With the development of factor analysis (or principal component analysis) another numerical technique has become available to indicate the minimum number of component bands contributing to the spectrum. This method will next be described.

4.3.3. Factor Analysis

Factor analysis, also called principal component analysis (PCA), is a computer-based procedure that can provide the number of species (i.e., the number of components, NC) contributing Raman intensity to a complicated spectrum. The procedure does not assume a particular band shape; rather, it is based on standard theorems of linear algebra. Shurvell and Bulmer (41) and Malinowski and Howery (42) have written detailed accounts of the method. Here we will present an overview of the method. In the Appendix to this chapter we shall present examples that will make the procedure more comprehensible for those readers who are new to the subject.

For Raman spectral measurements an expression analogous to Beer's Law [see Eq. (9)] is written

$$I_{ij} = \sum_{k=1}^{NC} J_{ik} C_{kj}, \tag{12}$$

where I_{ij} is the Raman intensity at the ith wavenumber in the jth solution; J_{ik} is the molar intensity of the kth species at the ith wavenumber; and C_{kj} is the concentration of the kth species in the jth solution. In matrix notation Eq. (12) becomes

$$\mathbf{I} = \mathbf{JC}, \tag{13}$$

where the dimensions are as follows:

I is the NW × NS Raman-intensity matrix (data matrix);
J is the NW × NC molar-intensity matrix (spectral matrix);
C is the NC × NS concentration matrix.

Here NW is the number of wavenumbers digitized, NS is the number of solutions studied, and NC is the number of scattering species. Usually experiments are such that NW > NS > NC. If we know NW and NS, the problem is to deduce NC. If the spectrum of one of the individual components is not a linear combination of the spectra of the other components and if the concentration of one or more species cannot be expressed as a linear combination of the concentrations of the other components for all experiments performed, then the number of scattering components is given by the rank of the matrix **I**:

$$\text{rank}\,\mathbf{I} = \text{NC}. \tag{14}$$

Matrix I_{ij} is generally a large matrix. For example, spectra may have been recorded at $0.5\,\text{cm}^{-1}$ intervals over $500\,\text{cm}^{-1}$ for ten different solutions. Thus there would be 10 columns and 1001 rows. When it is possible to obtain the isolated spectrum of a single species, the rank of **I** can be reduced by 1 using rank annihilation factor analysis.

The rank of **I** is determined by examining the eigenvalues and eigenvectors of the *second moment* (*square*) *matrix* of dimensions NS × NS:

$$\mathbf{Z} = {}^t\mathbf{I}\mathbf{I}, \tag{15}$$

where ${}^t\mathbf{I}$ is the transpose of **I** and the matrix **Z** has the same rank as **I**. These eigenvectors are "abstract," as they have no physical significance. The number of statistically nonzero eigenvalues equals NC. If the spectra were perfect and free of noise, PCA would lead to an unambiguous value of NC. However the blemishes in real spectra and experimental error contribute as independent eigenvectors. Each eigenvalue measures the relative importance of the eigenvector. A large eigenvalue indicates an important factor, whereas a very small eigenvalue indicates an unimportant factor such as instrumental noise. The set of eigenvectors associated entirely with experimental error are identified by subjecting the results to several statistical tests.

Once NC has been measured it may be possible to extend the analysis and estimate the spectrum of each component and its concentration in the mixture. Spectral isolation factor analysis (SPI) can be used. As an example consider the conversion of DABCO (1,4-diazabicyclo[2.2.2]octane) to the monoprotonated and diprotonated forms by titration:

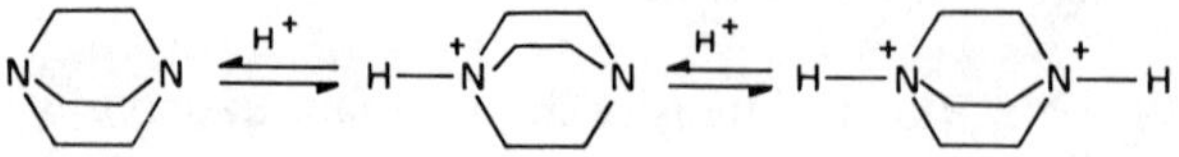

Each of these species has a distinct, characteristic Raman spectrum in the

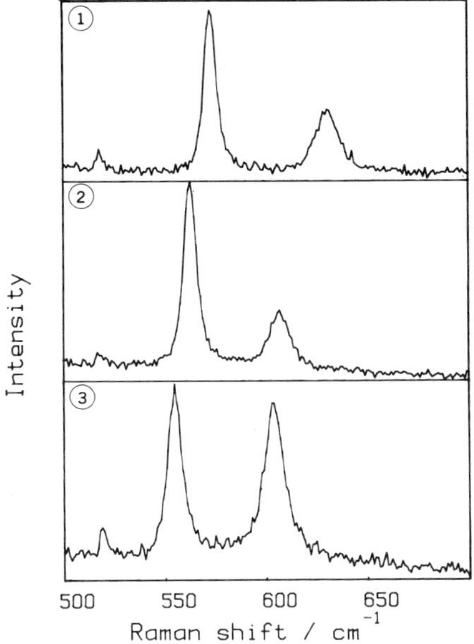

Figure 4.20. The Raman spectra of (1) DABCO, (2) $DABCOH^+$, and (3) $DABCOH_2^{2+}$ in the spectral region 500–700 cm^{-1} at pH 10, 6, and 1 respectively.

spectral region 500–650 cm^{-1}. This is apparent in Fig. 4.20. Application of SPI to the overlapped spectra generates the three-component spectra illustrated in Fig. 4.21. The concentrations of the components were also extracted, but the resulting plot of relative concentrations versus pH has only roughly approximated the composition versus pH curve determined from the acidity constants or from the BNDFIT approach (43). The resultant information is useful, although the error in the concentrations is relatively large. The method has been applied for the first time to surface-enhanced Raman spectra to identify and quantify surface species at an electrode (44). For SPI to be rigorously applicable there should be at least one wavenumber position at which only each single species contributes (45). If this condition is not met, SPI will fail to converge (46). In such a case one can use the knowledge of NC obtained by PCA to analyze the spectra with BNDFIT. This latter approach was taken in the analysis of the spectra shown in Fig. 4.19 (39).

Gampp et al. proposed a new approach that takes knowledge of the chemical equilibria into account in order to obtain correct concentrations and spectra (47, 48). A method based on their approach, which we call factor analysis with equilibrium constraints (FAEC), has been developed by Ozeki

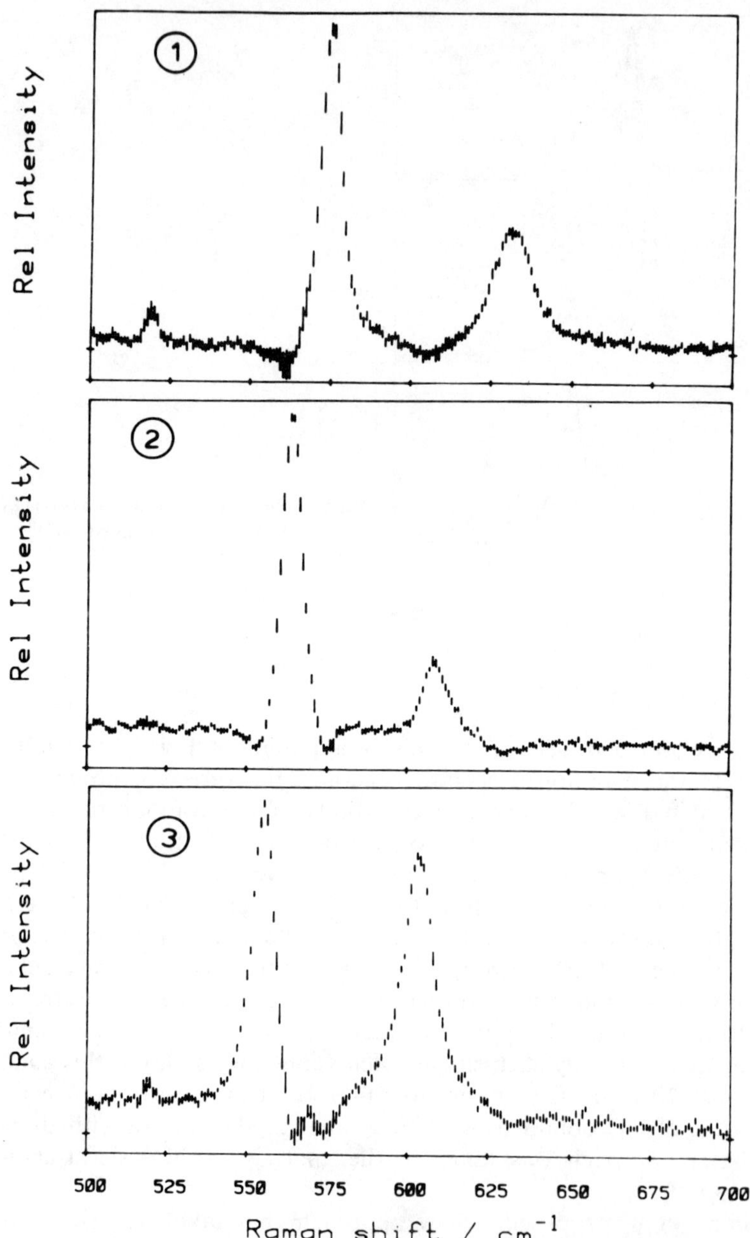

Figure 4.21. Spectra of (1) DABCO, (2) $DABCOH^+$, and (3) $DABCOH_2^{2+}$, isolated by SPI. Reprinted from Guzonas and Irish (43), with permission of the National Research Council of Canada from the Canadian Journal of Chemistry, 1988.

et al. and applied to the analysis of UV spectra (49). This method has now been applied to the Raman data for the stepwise formation of zinc bromide complexes in dimethyl sulfoxide solutions (50). In this method an equilibrium or equilibria are identified and the concentrations of species are expressed in terms of the equilibrium constants. By assuming a value(s) for the equilibrium constant(s), one can write an expression for the concentration matrix $\mathbf{C}^{\circ}$. This matrix can be compared with the concentration matrix $\mathbf{C}^*$ obtained from factor analysis, after the bases of the matrix are suitably altered by matrix multiplication (to rotate the factor axis) so as to give a minimum sum of the squares of the differences. The former matrix $\mathbf{C}^{\circ}$ is altered by refinement of the K values until the two matrices $\mathbf{C}^{\circ}$ and $\mathbf{C}^*$ virtually agree, i.e., until the measured difference function is a minimum. At this stage one has a good estimate for the concentrations of species and the equilibrium constants. By matrix multiplication the true spectra of the components can then be found. A worked-out example is provided in the Appendix. This combination of equilibrium and spectral information appears to provide a very powerful method for the study of processes in solution.

4.3.4. The GAUSS-*Z* Program

Frequently in solutions, anions and cations combine to give a series of complexes. We will consider a series of mononuclear complexes of the form $MA, MA_2, MA_3, \ldots, MA_i, \ldots, MA_n$, where M = metal cation and A = ligand. Charges will depend on the particular cation and ligand and have been omitted for simplicity. Equilibria are established at each temperature and composition:

$$\begin{aligned} M + A &= MA, \\ MA + A &= MA_2, \qquad \ldots, \\ MA_{i-1} + A &= MA_i, \qquad \ldots, \\ MA_{n-1} + A &= MA_n. \end{aligned} \tag{16}$$

The populations of each constituent can be described by defining the degree of formation of each species α_i, the cumulative formation constants β_i, and the stepwise formation constants K_i:

$$\alpha_i = \frac{[MA_i]}{[M]_t}; \qquad \beta_i = \frac{[MA_i]}{[M][A]^i}; \qquad K_i = \frac{[MA_i]}{[MA_{i-1}][A]}. \tag{17}$$

Here β_n is related to K_i by

$$\beta_n = \prod_{i=1}^{n} K_i. \tag{18}$$

The average ligand number, $\bar{n}$, is given by the ratio of the concentration of total bound ligand $[A]_b$, to the total metal concentration $[M]_t$:

$$\bar{n} = [A]_b/[M]_t. \tag{19}$$

If polynuclear species are absent, $\bar{n}$ is given by

$$\bar{n} = \frac{\beta_1[A] + 2\beta_2[A]^2 + \cdots}{1 + \beta_1[A] + \beta_2[A]^2 + \cdots}, \tag{20}$$

where [A] is concentration of free, unbound ligand. Note that this expression is independent of the concentration of M. Thus the formation curve, which is a plot of $\bar{n}$ versus [A], is a monotonically increasing curve independent of [M]. All data points ($\bar{n}$, [A]) from samples with different concentrations of M should lie on the same curve. (If the data lie on a family of curves where each curve corresponds to a different value of $[M]_t$, the formation curves indicate the existence of polynuclear complexes.) By statistical procedures the problem is to obtain a unique set of β_n that best fits the $\bar{n}$ versus [A] data. The GAUSS-*Z* least squares program does this (51, 52). An average ligand number, $\bar{n}_{calc}$, is calculated from estimated β's, $[M]_t$, and [A] data. The program then performs a nonlinear least squares refinement of the β's to minimize the sum of the squares of the differences between $\bar{n}_{calc}$ and $\bar{n}_{obs}$. From the number and values of the β's one can infer what equilibria exist and their equilibrium constants.

This method can be applied to Raman spectra as follows: A series of solutions is prepared with a wide range of total metal concentration $[M]_t$ and total ligand concentration $[A]_t$. For a given temperature, [A] the concentration of unbound ligand (free ion) is obtained by measuring the relative integrated intensity of one of its Raman bands, $I_{R,f}$. Thus $[A] = I_{R,f}/J_f$, where J_f has been obtained in a separate experiment as described earlier. From [A] one can compute $[A]_b$, the concentration of bound A, thus:

$$[A]_b = [A]_t - [A]. \tag{21}$$

The average ligand number $\bar{n}$ [Eq. (19)] is then obtained. Thus the primary data are measured.

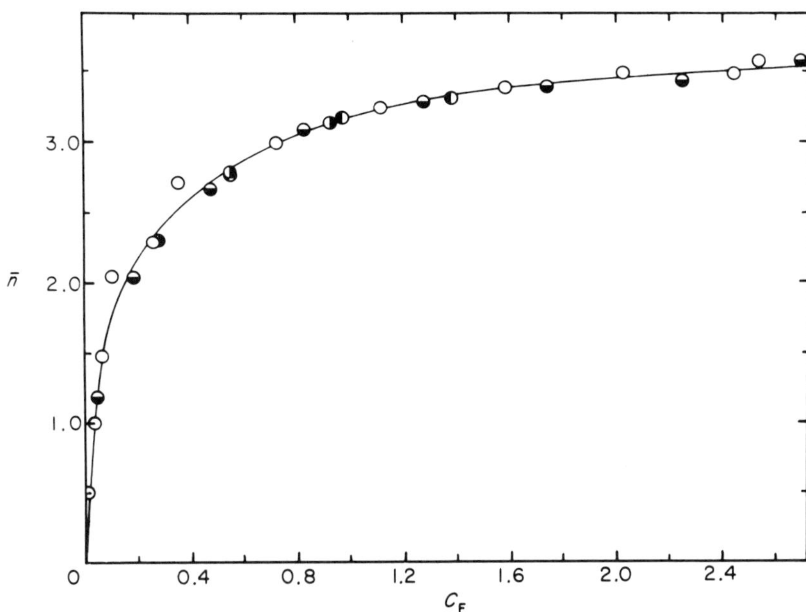

Figure 4.22. The formation curve, $\bar{n}$ versus free nitrite concentration: ◐, 1.00 M Cd^{2+}; ○, 0.90 M Cd^{2+}; ◒, 0.40 M Cd^{2+}: ◑, 0.30 M Cd^{2+}. Reprinted from Irish and Thorpe (28), with permission of the National Research Council of Canada from the Canadian Journal of Chemistry, 1975.

This procedure has been applied to the system Cd^{2+}/NO_2^- in water at 25°C (28). Raman spectra of these species are illustrated in Fig. 4.13. The formation curve is shown in Fig. 4.22. The curve is reproduced by four β's: $\beta_1 = 61.6$, $\beta_2 = 993$, $\beta_3 = 2390$, and $\beta_4 = 1899$, corresponding to the formation of $Cd(NO_2)^+$, $Cd(NO_2)_2$, $Cd(NO_2)_3^-$, and $Cd(NO_2)_4^{2-}$, respectively. The equilibrium constants are in reasonable agreement with values obtained from dilute solutions by potentiometry (53) and polarography (54). If the results are to be meaningful, it is necessary to obtain many pairs of ($\bar{n}$, [A]) data of high reliability. The method has been applied in Raman studies of Sr^{2+}/NO_3^- (55) and Gd^{3+}/NO_3^- (56).

The GAUSS-*Z* method can be applied when FAEC is not applicable. For the Cd^{2+}/NO_2^- system all of the complexes contribute to the single 861 cm^{-1} band intensity; each complex does not have a distinct spectrum. The FAEC mentioned in the previous section cannot be applied to this system because it requires that all of the species participating in the reactions must have unique Raman peaks. Even for such a system, some modification of FAEC is possible. An example illustrating the comparison of GAUSS-*Z* and the modified FAEC has been reported for magnesium acetate complex formation (57).

4.3.5. Raman Difference Spectroscopy

Today most Raman instruments are controlled by computers. Data in digitized form are stored for subsequent analysis. It is thus straightforward to subtract one spectrum from another and thus remove solvent bands, or to calculate isotropic and anisotropic spectra or reduced spectra $[R(\tilde{\nu})]$. To some extent this overcomes one disadvantage of Raman spectroscopy—that it is inherently a single-channel technique.

When solvent spectra are subtracted, it is frequently observed that derivative-like features remain at the positions where solvent bands occurred. These features are indicative of either frequency shifts or bandwidth changes. Small shifts or changes, of magnitude $\sim 0.1\ cm^{-1}$, can be reproducibly measured with a precision of about 5% by using a rotating sample system. The concept of a rotating sample system applied to Raman difference spectroscopy (RDS) was introduced by Kiefer (58, 59) and further developed by Laane and Kiefer (60) and others (61–63). Laane (64) has provided a detailed review of the instrumentation, theory, and examples. Briefly stated, as a divided sample cell rotates, data are collected alternately from one sample and then from the other and stored in separate channels ready for subtraction, addition, or other computation. Associated electronics are activated by a rotating trigger wheel through which light from a luminescent diode passes to generate a switching signal. The wheel also blocks the laser beam so that the cell divider is not illuminated. The assembly has been adapted for polarization measurements (60, 65) and for four-channel detection (60). The latter permits polarization measurements to be made on two different samples simultaneously. A rotating-lens system has been developed for samples that cannot be rotated (66).

The technique has been tested with various mixtures of organic liquids. Small frequency shifts and bandwidth changes have been detected that permit the study of solvent effects (67, 68), isotopic shifts (69, 70), ion interactions (61, 71), and interactions in biochemical systems (64). The technique thus has potential for the study of inorganic species in solution.

4.4. CONCLUSION

In this chapter we have outlined the approach taken by the chemist to elucidate the processes that occur in solutions containing inorganic species. The same procedure can, of course, be applied to mixtures of organic liquids.

Raman spectroscopy provides specific information about the identity of species and their populations and thus provides a powerful complement to methods such as ultrasonic relaxation that are not species specific (72). Raman spectroscopy is being extended to an ever greater range of temperatures and pressures. As technology changes, one can expect the measurement of data at ever lower concentrations, approaching "infinite dilution."

APPENDIX

Understanding Factor Analysis

Here we present an example to explain the basic principles underlying factor analysis. The first objective, as explained earlier, is to determine the number of species, NC, contributing to the spectrum. Factor analysis will give the least number of statistically significant factors (optically independent species) in the system needed to reproduce the observed spectra. We will analyze four hypothetical Raman spectra consisting of seven intensity values "recorded" at seven different wavenumbers, for each of four different solutions, a, b, c, and d. These are plotted in Fig. 4.23.

Step 1. Formation of Data Matrix I. A data matrix **I** is formed from the Raman intensities; if the intensities of NS sample solutions were measured at NW different wavenumber points, the data matrix **I** of NW rows and NS columns is formed by aligning those intensities as each column of the matrix

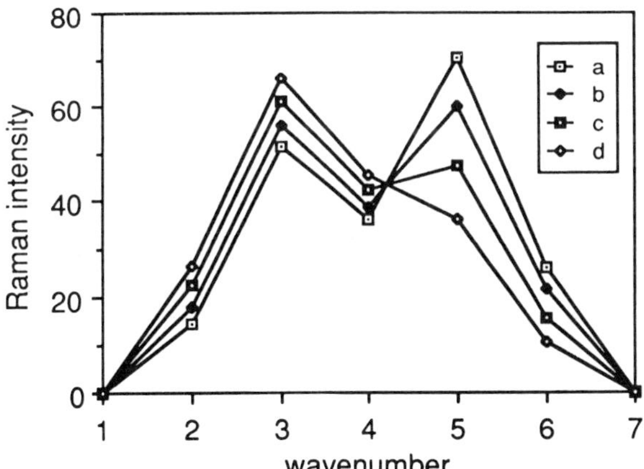

Figure 4.23. Hypothetical Raman spectra of four different solutions designated a, b, c, and d.

(A-1). The matrix will thus be of dimension NW × NS. For the hypothetical set of data it will be 7 × 4. Each column (x_1 to x_7) corresponds to the spectrum of each sample x (a to d).

$$\mathbf{I}=\begin{bmatrix} a_1 & b_1 & c_1 & d_1 \\ a_2 & b_2 & c_2 & d_2 \\ a_3 & b_3 & c_3 & d_3 \\ a_4 & b_4 & c_4 & d_4 \\ a_5 & b_5 & c_5 & d_5 \\ a_6 & b_6 & c_6 & d_6 \\ a_7 & b_7 & c_7 & d_7 \end{bmatrix}=\begin{bmatrix} 0.0 & 0.0 & 0.0 & 0.0 \\ 14.70 & 18.26 & 22.62 & 26.53 \\ 51.30 & 55.66 & 60.98 & 65.75 \\ 36.10 & 38.87 & 42.26 & 45.30 \\ 70.20 & 59.90 & 47.32 & 36.04 \\ 26.40 & 21.65 & 15.84 & 10.63 \\ 0.0 & 0.0 & 0.0 & 0.0 \end{bmatrix} \tag{A-1}$$

Step 2. Second Moment Matrix and Eigenvalue. A *second moment matrix* **Z** is obtained by the multiplication of the data matrix **I** with its transposed matrix ${}^t\mathbf{I}$. Sometimes, the matrix **Z** is called the *covariance matrix* [Eqs. (A-2) and (A-3)].

$$\mathbf{Z}={}^t\mathbf{I}\mathbf{I} \tag{A-2}$$

$$\mathbf{Z}=\begin{bmatrix} a_1 & a_2 & a_3 & a_4 & a_5 & a_6 & a_7 \\ b_1 & b_2 & b_3 & b_4 & b_5 & b_6 & b_7 \\ c_1 & c_2 & c_3 & c_4 & c_5 & c_6 & c_7 \\ d_1 & d_2 & d_3 & d_4 & d_5 & d_6 & d_7 \end{bmatrix}\begin{bmatrix} a_1 & b_1 & c_1 & d_1 \\ a_2 & b_2 & c_2 & d_2 \\ a_3 & b_3 & c_3 & d_3 \\ a_4 & b_4 & c_4 & d_4 \\ a_5 & b_5 & c_5 & d_5 \\ a_6 & b_6 & c_6 & d_6 \\ a_7 & b_7 & c_7 & d_7 \end{bmatrix}$$

$$=\begin{bmatrix} z_{11} & z_{12} & z_{13} & z_{14} \\ z_{21} & z_{22} & z_{23} & z_{24} \\ z_{31} & z_{32} & z_{33} & z_{34} \\ z_{41} & z_{42} & z_{43} & z_{44} \end{bmatrix}$$

$$=\begin{bmatrix} 9775.99 & 9303.53 & 8726.41 & 8208.94 \\ 9303.53 & 8999.07 & 8627.24 & 8293.83 \\ 8726.41 & 8627.24 & 8506.22 & 8397.71 \\ 8208.94 & 8293.83 & 8397.71 & 8490.87 \end{bmatrix}. \tag{A-3}$$

The second moment matrix is a square matrix of NS rows and NS columns; the element z_{ij} of the ith row and the jth column is the same as that of the jth row and the ith column: $z_{ij}=z_{ji}$. Each row or each column of the matrix

Z can be regarded as a vector. This matrix, as a whole, reveals a space containing such vectors.

Step 3. Statistical Analysis to Select the Number of Factors—NC. Now we search for the minimum dimension of the space (rank of the matrix), that is, the minimum number of base vectors. For this purpose, several methods have been proposed: the *centroid method*, the *minimum residual method*, the *principal factor method*, etc. The details of these methods are given by Malinowski and Howery (42). We apply the *Jacobi method* to the second moment matrix **Z** [see Eq. (A-3)]. The Jacobi method is the one developed for the calculation using the computer, and its algorithm is explained in books of computer mathematics (73). An *eigenvalue matrix* **E** and an *eigenvector matrix* **Q** are obtained [Eqs. (A-5) and (A-6)].

$$\mathbf{Z} = \mathbf{Q}\mathbf{E}{}^t\mathbf{Q}$$

$$= \begin{bmatrix} q_{11} & q_{21} & q_{31} & q_{41} \\ q_{12} & q_{22} & q_{32} & q_{42} \\ q_{13} & q_{23} & q_{33} & q_{43} \\ q_{14} & q_{24} & q_{34} & q_{44} \end{bmatrix} \begin{bmatrix} \lambda_1 & 0 & 0 & 0 \\ 0 & \lambda_2 & 0 & 0 \\ 0 & 0 & \lambda_3 & 0 \\ 0 & 0 & 0 & \lambda_4 \end{bmatrix} \begin{bmatrix} q_{11} & q_{12} & q_{13} & q_{14} \\ q_{21} & q_{22} & q_{23} & q_{24} \\ q_{31} & q_{32} & q_{33} & q_{34} \\ q_{41} & q_{42} & q_{43} & q_{44} \end{bmatrix}, \quad \text{(A-4)}$$

where

$$\mathbf{E} = \begin{bmatrix} 34750.86 & 0.0 & 0.0 & 0.0 \\ 0.0 & 0.00192 & 0.0 & 0.0 \\ 0.0 & 0.0 & 0.00048 & 0.0 \\ 0.0 & 0.0 & 0.0 & 1021.28 \end{bmatrix} \quad \text{(A-5)}$$

and

$$\mathbf{Q} = \begin{bmatrix} 0.51896 & -0.49176 & -0.28398 & -0.63891 \\ 0.50723 & 0.82734 & 0.03225 & -0.23913 \\ 0.49290 & -0.26098 & 0.79169 & 0.24934 \\ 0.48005 & -0.07460 & -0.53995 & 0.68734 \end{bmatrix}. \quad \text{(A-6)}$$

Both the **E** and the **Q** have the same number of rows as the **Z**: in this case, NS. Now for the example, four eigenvalues are obtained (λ_1 to λ_4); but two eigenvalues are very large (34, 750.86 and 1021.28) and the others are very small (0.00192 and 0.00048). The number of the independent factors in the spectra, NC, should be equal to the number of eigenvalues having large values. For the present example, NC = 2 is assumed. But experimental error may contribute large eigenvalues, and the selection of the number of factors becomes difficult if only the magnitude of the eigenvalues is checked. In such cases, the factor indicator function, IND, which was introduced by Malinowski, is useful (42, p. 82; 74):

$$\mathrm{IND} = \left(\sum_{j=\mathrm{NC}+1}^{j=\mathrm{NS}} \lambda_j \right)^{1/2} \mathrm{NW}^{-1/2}(\mathrm{NS} - \mathrm{NC})^{-5/2}. \tag{A-7}$$

Here, NW and NS are the number of rows and of columns in the data matrix **I** respectively; and λ_j is the jth eigenvalue. The sum is taken over all eigenvalues from the (NC + 1)th to the NSth. In this expression, the condition NW > NS is assumed; for the case NW < NS, NW and NS must be interchanged.

For the present example, the IND function gives the values shown in the accompanying tabulation.

NC	IND	
1	0.774857	
2	0.003273	(a minimum value)
3	0.008281	
4	—	(not calculated because 0 is divided by 0)

The IND function reaches a minimum when the correct number of factors is introduced and the error is random and fairly uniform throughout the entire data matrix. From this result, NC = 2 is assumed.

Another way to get the number of factors, NC, is to check the reproducibility between the original matrix **I** and the reproduced data matrix **I**′, assuming a correct number for NC (46, 49). Consider the case of NC factors. In this case, the eigenvectors corresponding to the NC largest eigenvalues are taken out and are aligned in rows to construct a new matrix **C**. This matrix **C** has NC rows and NS columns and is known as the first expression for the concentration matrix. For the example, we have Eq. (A-8):

$$\mathbf{C} = \begin{bmatrix} q_{k1} & q_{k2} & q_{k3} & q_{k4} \\ q_{l1} & q_{l2} & q_{l3} & q_{l4} \end{bmatrix} = \begin{bmatrix} 0.51896 & 0.50723 & 0.49290 & 0.48005 \\ -0.63891 & -0.23913 & 0.24934 & 0.68734 \end{bmatrix}, \tag{A-8}$$

where k is the column number whose eigenvalue λ_k has the largest value: $k = 1$. Similarly, l is the number whose eigenvalue λ_l has the second largest value; $l = 4$. By multiplying the transpose of the concentration matrix **C** with the data matrix **I**, a new matrix **R** is obtained. This matrix **R** has NW rows and NC columns and is known as the first expression of the spectral matrix.

In the text, the spectral matrix is expressed by using a symbol **J**, but here a symbol **R** is used to avoid confusion between the element of the matrix j_{ij} and subscript j [see Eq. (A-9)].

$$\mathbf{R} = \mathbf{I}^t\mathbf{C}$$

$$= \begin{bmatrix} r_{11} & r_{12} \\ r_{21} & r_{22} \\ r_{31} & r_{32} \\ r_{41} & r_{42} \\ r_{51} & r_{52} \\ r_{61} & r_{62} \\ r_{71} & r_{72} \end{bmatrix} = \begin{bmatrix} 0.0 & 0.0 \\ 40.776 & 10.117 \\ 116.475 & 14.312 \\ 81.027 & 9.314 \\ 107.439 & -22.605 \\ 37.593 & -10.788 \\ 0.0 & 0.0 \end{bmatrix}. \tag{A-9}$$

When the spectral matrix **R** is multiplied by the concentration matrix **C**, a hypothetical data matrix **I**′ is produced (A-10).

$$\mathbf{I}' = \mathbf{RC}$$

$$= \begin{bmatrix} a'_1 & b'_1 & c'_1 & d'_1 \\ a'_2 & b'_2 & c'_2 & d'_2 \\ a'_3 & b'_3 & c'_3 & d'_3 \\ a'_4 & b'_4 & c'_4 & d'_4 \\ a'_5 & b'_5 & c'_5 & d'_5 \\ a'_6 & b'_6 & c'_6 & d'_6 \\ a'_7 & b'_7 & c'_7 & d'_7 \end{bmatrix} = \begin{bmatrix} 0.0 & 0.0 & 0.0 & 0.0 \\ 14.697 & 18.263 & 22.621 & 26.528 \\ 51.302 & 55.657 & 60.979 & 65.751 \\ 36.099 & 38.872 & 42.260 & 45.299 \\ 70.199 & 59.901 & 47.320 & 36.039 \\ 26.402 & 21.648 & 15.839 & 10.631 \\ 0.0 & 0.0 & 0.0 & 0.0 \end{bmatrix}. \tag{A-10}$$

The elements of this matrix **I**′ are very close to those of the original data matrix **I**. This suggests that the assumption NC = 2 is correct. The reproducibility of the hypothetical data matrix, however, can be examined by calculating an error function ER from the sum of the squares of the differences between the elements of the two matrices **I** and **I**′ in the following manner:

$$\mathrm{ER} = [\textstyle\sum_i \sum_j (x_{ij} - x'_{ij})^2]^{1/2} / [\sum_i \sum_j (x_{ij})^2]^{1/2}, \tag{A-11}$$

where $x_{1j} = a_j$, $x_{2j} = b_j$, $x_{3j} = c_j$, and $x_{4j} = d_j$, respectively, from Eq. (A-1), and similarly for x'_{1j} to x'_{4j} from Eq. (A-10).

For the present example, the ER function gives the values shown in the accompanying tabulation. The optimum number of factors, NC = 2, is assumed to be the number where the convergence is observed.

NC	ER
1	0.16897
2	0.00004
3	0.00003
4	0.0

According to the treatment discussed so far, the reliable number of species, NC, which are optically active and are participating in the measured spectra, has been obtained. Note that the obtained number of factors NC is the least number sufficient to reproduce the observed spectra; the procedure does not suggest that this number is exactly the number of species in the sample solutions. Some species might be optically inactive; two or more species might have exactly the same Raman spectrum. In such cases, the correct number of species participating in the reaction cannot be obtained from the treatment of the aforementioned factor analysis. Sometimes, some additional chemical intuition is required.

Step 4. To Obtain the Spectra and Concentrations of the Components. Now, suppose the correct number of factors is NC. The next step is to get the most reliable representation of the pure spectrum of each species and the most reliable representation of its concentration distribution among the sample solutions. Expressions for the spectral matrix **R** and the concentration matrix **C** have already been obtained. If they are correct, each column of **R** should correspond to a pure spectrum and each row of the concentration matrix **C** should correspond to a concentration distribution of each pure component. However, as seen in **R** (A-9) and **C** (A-8), some elements are negative; and such negative quantities do not make sense either as the Raman intensity or as the concentration. Even if there are no negative elements, the spectra obtained by chance are not necessarily the correct spectra. The treatment carried out so far gives only the least number of optically independent species in the sample solutions.

In factor analysis, many combinations of spectra and concentration distributions—namely, combinations of **R** and **C** matrices—can reproduce the same spectral data matrix **I**. Now we introduce a set of two regular matrices, $\mathbf{T_R}$ and $\mathbf{T_C}$, which fulfill the relation shown in Eq. (A-12),

$$\mathbf{T_R}\mathbf{T_C} = \mathbf{U}$$

$$= \begin{bmatrix} tr_{11} & tr_{12} & tr_{13} & tr_{14} \\ tr_{21} & tr_{22} & tr_{23} & tr_{24} \\ tr_{31} & tr_{32} & tr_{33} & tr_{34} \\ tr_{41} & tr_{42} & tr_{43} & tr_{44} \end{bmatrix} \begin{bmatrix} tc_{11} & tc_{12} & tc_{13} & tc_{14} \\ tc_{21} & tc_{22} & tc_{23} & tc_{24} \\ tc_{31} & tc_{32} & tc_{33} & tc_{34} \\ tc_{41} & tc_{42} & tc_{43} & tc_{44} \end{bmatrix}$$

$$= \begin{bmatrix} 1 & 0 & 0 & 0 \\ 0 & 1 & 0 & 0 \\ 0 & 0 & 1 & 0 \\ 0 & 0 & 0 & 1 \end{bmatrix}. \quad \text{(A-12)}$$

where **U** is the unit matrix. With these matrices we can rotate the bases of **R** and **C** while keeping their product the same as the original **I** $(= \mathbf{RC} = \mathbf{RT_R T_C C})$.

Examples of the rotation matrices that fulfill the condition of (A-12) are shown in Eqs. (A-13) and (A-14):

$$\mathbf{T_R} = \begin{bmatrix} 0.4824 & 0.5526 \\ -1.9443 & 1.9255 \end{bmatrix}; \quad \text{(A-13)}$$

$$\mathbf{T_C} = \begin{bmatrix} 0.9612 & -0.2759 \\ 0.9706 & 0.2408 \end{bmatrix}. \quad \text{(A-14)}$$

Each of these rotates the **R** matrix or the **C** matrix as shown in Eqs. (A-15) and (A-16).

$$\mathbf{R^*} = \mathbf{RT_R}$$
$$= \begin{bmatrix} r^*_{11} & r^*_{12} \\ r^*_{21} & r^*_{22} \\ r^*_{31} & r^*_{32} \\ r^*_{41} & r^*_{42} \\ r^*_{51} & r^*_{52} \\ r^*_{61} & r^*_{62} \\ r^*_{71} & r^*_{72} \end{bmatrix} = \begin{bmatrix} 0.000 & 0.000 \\ 0.000 & 45.705 \\ 29.611 & 100.000 \\ 21.902 & 68.221 \\ 100.000 & 17.237 \\ 40.834 & 0.000 \\ 0.000 & 0.000 \end{bmatrix}; \quad \text{(A-15)}$$

$$\mathbf{C^*} = \mathbf{T_C C}$$
$$= \begin{bmatrix} c^*_{1a} & c^*_{1b} & c^*_{1c} & c^*_{1d} \\ c^*_{2a} & c^*_{2b} & c^*_{2c} & c^*_{2d} \end{bmatrix} = \begin{bmatrix} 0.64656 & 0.53014 & 0.38790 & 0.26035 \\ 0.32157 & 0.39959 & 0.49493 & 0.58042 \end{bmatrix}. \quad \text{(A-16)}$$

The resultant spectra of these two factor species and their concentration distributions are shown in Fig. 4.24. They seem to be reasonable, but there is no guarantee that they are unique and correct because $\mathbf{T_R}$ and $\mathbf{T_C}$ were arbitrarily chosen.

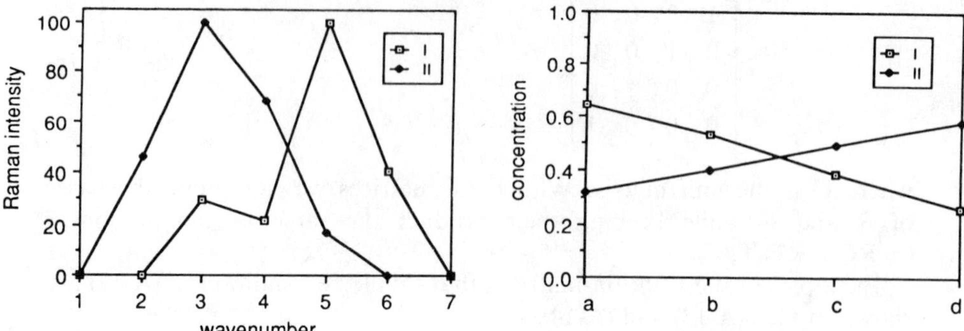

Figure 4.24. The resultant spectra of species I and II and their concentration distribution in the four solutions designated a, b, c, and d.

Any $\mathbf{T_R}$ and $\mathbf{T_C}$, whose product is the unit matrix, can give **R** and **C** matrices. But most of them thus obtained are not reasonable. One criterion enabling us to remove less reliable **R** and **C** matrices is to avoid the negative elements for both of the **R** and **C**. This criterion still does not give a unique set of **R** and **C**; however at least this decreases the number of wrong candidates for the **R** and **C** matrices. The criterion to avoid negative intensity for **R** is sometimes called the "pure wavenumber constraint" because it assumes the existence of a unique wavenumber point at which only one species has finite intensity and the other species have zero intensity, and because it assumes that this kind of unique wavenumber point exists for each species. The rotation matrices, $\mathbf{T_R}$ and $\mathbf{T_C}$, obtained based on this constraint along with another assumption that each species has the same maximum intensity, are the ones given in Eqs. (A-13) and (A-14). Note in Eq. (A-15) that in column 1 there is a maximum intensity of 100 units for r^*_{51} and in column 2 for r^*_{32} (following the "same maximum constraint" assumption). Also at r^*_{22} there is finite intensity, 45.705, compared to zero for r^*_{21}, and vice versa for r^*_{61} and r^*_{62} (the "pure wavenumber constraint"). This is illustrated in Fig. 4.24, where these spectral points (matrix elements) are plotted.

The criterion to avoid negative concentrations for **C** is called the "pure component constraint" because it assumes the existence of a sample containing only one species I and therefore having a spectrum of pure species I, and similarly another sample containing only one species II that provides the spectrum of pure species II. For the present example, the reasonable regions of the spectra and those of the concentration distributions obtained based on these two extreme constraints are shown in Fig. 4.25. When the pure component constraint was applied, another assumption that the sum of the amounts of the two species was always kept constant was also being

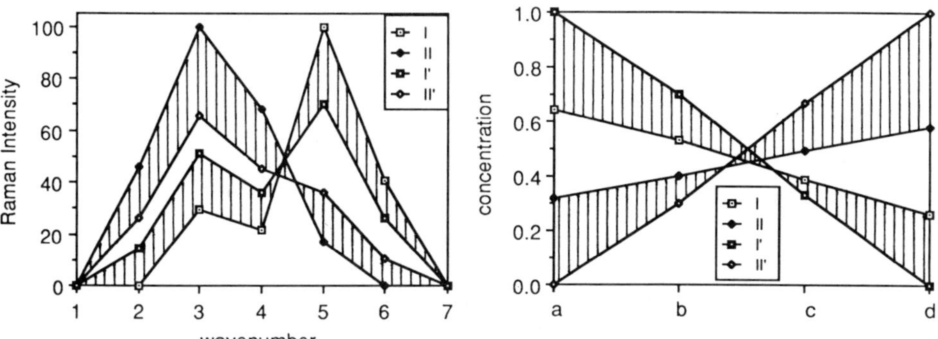

Figure 4.25. The true spectrum of species I should lie between the lines designated I and I′. Similarly the true spectrum of species II should lie between the lines designated II and II′. The limits of the concentrations of species I and II are similarly given in the right-hand panel.

invoked. The details of these kinds of treatments are given in several references (see 42, 45, 46, 75, 76).

Factor Analysis Taking Equilibria into Account

In the previous section, we obtained one possibility for the spectral matrix **R** and the concentration matrix **C**. Unfortunately, we could not conclude that they are correct because the rotation matrices were not determined uniquely. If we can get one of **R** or **C** uniquely, we can get the remainder automatically by using the inverse matrix of the rotation matrix ($\mathbf{T_R} = \mathbf{T_C^{-1}}$ and vice versa).

Now let us consider the concentration matrix **C**. It was mentioned earlier that multiplying an adequate matrix by **C** may give a correct concentration matrix $\mathbf{C}^\circ$. Consider the system of the four sample solutions, with two species P and PX that are optically active and related to each other by the following reaction:

$$\mathrm{P + X = PX}. \tag{A-17}$$

The equilibrium constant of this reaction is K:

$$K = [\mathrm{PX}]/[\mathrm{P}][\mathrm{X}]. \tag{A-18}$$

Now suppose $\mathbf{T_C}$ is the correct rotation matrix to give the correct concentration matrix $\mathbf{C}^\circ$. Then

$$\mathbf{C}^\circ = \mathbf{T_C C}. \tag{A-19}$$

This equation can be rewritten by introducing the elements of the matrix (A-20):

$$\begin{bmatrix} [\mathrm{P}]_a & [\mathrm{P}]_b & [\mathrm{P}]_c & [\mathrm{P}]_d \\ [\mathrm{PX}]_a & [\mathrm{PX}]_b & [\mathrm{PX}]_c & [\mathrm{PX}]_d \end{bmatrix} = \begin{bmatrix} t_{11} & t_{12} \\ t_{21} & t_{22} \end{bmatrix} \begin{bmatrix} c_{1a} & c_{1b} & c_{1c} & c_{1d} \\ c_{2a} & c_{2b} & c_{2c} & c_{2d} \end{bmatrix}. \tag{A-20}$$

Each element in the left-hand matrix is a true concentration (hence in square brackets) of the species P or PX, as specified, in each of the samples *a* to *d*. The material balance equations for X and P are

$$\begin{aligned} C_{\mathrm{X}} &= (\text{total X}) = [\mathrm{X}] + [\mathrm{PX}] \\ C_{\mathrm{P}} &= (\text{total P}) = [\mathrm{P}] + [\mathrm{PX}]. \end{aligned} \tag{A-21}$$

Then, the equilibrium equation becomes

$$K = [\mathrm{PX}]/(C_{\mathrm{P}} - [\mathrm{PX}])(C_{\mathrm{X}} - [\mathrm{PX}]). \tag{A-22}$$

We obtain a quadratic equation for [PX]

$$K[\mathrm{PX}]^2 - (KC_{\mathrm{P}} + KC_{\mathrm{X}} + 1)[\mathrm{PX}] + KC_{\mathrm{P}}C_{\mathrm{X}} = 0. \tag{A-23}$$

Thus

$$[\mathrm{PX}] = \{b - (b^2 - 4K^2C_{\mathrm{P}}C_{\mathrm{X}})^{1/2}\}/2K, \tag{A-24}$$

where

$$b = KC_{\mathrm{P}} + KC_{\mathrm{X}} + 1. \tag{A-25}$$

Also

$$[\mathrm{P}] = C_{\mathrm{P}} - [\mathrm{PX}]. \tag{A-26}$$

From these equations, we can evaluate the values of [P] and [PX]. As an example, suppose $C_{\mathrm{P}} = 1.0$ M, and four values of C_{X} are 0.2, 0.5, 1.0, and 2.0 M in samples *a* to *d*, respectively. Now let us make the assumption that the equilibrium constant K is 5.0 (remember that this value is not correct). Then the $\mathbf{C}^{\mathrm{o}}$ matrix is obtained as shown in Eq. (A-27), from the above equations for [P] and [PX]:

$$\mathbf{C}^{\mathrm{o}} = \begin{bmatrix} 0.83852 & 0.62170 & 0.35826 & 0.14833 \\ 0.16148 & 0.37830 & 0.64174 & 0.85167 \end{bmatrix}. \tag{A-27}$$

Equation (A-19) can be expressed in another way, as shown in Eq. (A-28):

$$[\mathrm{P}]_j = t_{11}c_{1j} + t_{12}c_{2j} \qquad (j = a \text{ to } d) \qquad \text{(A-28)}$$
$$[\mathrm{PX}]_j = t_{21}c_{1j} + t_{22}c_{2j}.$$

Here, the values of [P] and [PX] are related to the elements of the concentration matrix **C**, which is a candidate for the concentration distributions but is not correct. For the values of c_{ij}, let us use those of the **C** matrix obtained for the example given in the previous section, expressed in (A-16):

$$\mathbf{C} = \begin{bmatrix} 0.64656 & 0.53014 & 0.38790 & 0.26035 \\ 0.32157 & 0.39959 & 0.49493 & 0.58042 \end{bmatrix}. \qquad \text{(A-16)}$$

Then a set of t_{ij}, which gives the closest agreement with the elements of the matrix $\mathbf{C}^{\mathrm{o}}$ in (A-27), can be obtained by applying the least squares method. Essentially, a set of t_{ij} is obtained such that the sum of the squares of the differences between the new values $t_{1i}c_{ij}$ (or $t_{2i}c_{ij}$), and the corresponding values in the $\mathbf{C}^{\mathrm{o}}$ trial matrix is minimized. The result is shown in Eq. (A-29):

$$\mathbf{T}_{\mathrm{C}} = \begin{bmatrix} 1.504683 & -0.433707 \\ -0.616848 & 1.758360 \end{bmatrix}. \qquad \text{(A-29)}$$

The rotated matrix becomes (A-30):

$$\mathbf{T}_{\mathrm{C}}\mathbf{C} = \begin{bmatrix} 0.83340 & 0.62439 & 0.36900 & 0.14001 \\ 0.16660 & 0.37561 & 0.63099 & 0.85999 \end{bmatrix}. \qquad \text{(A-30)}$$

$\mathbf{C}^{\mathrm{o}}$ for $K = 5.0$ was as shown in Eq. (A-27):

$$\mathbf{C}^{\mathrm{o}} = \begin{bmatrix} 0.83852 & 0.62170 & 0.35826 & 0.14833 \\ 0.16148 & 0.37830 & 0.64174 & 0.85167 \end{bmatrix}. \qquad \text{(A-27)}$$

These two matrices, $\mathbf{T}_{\mathrm{C}}\mathbf{C}$ and $\mathbf{C}^{\mathrm{o}}$, are not equal because the value of K assumed here ($K = 5.0$) is not the correct value. The sum of the error is calculated as follows:

$$\mathrm{LSM} = \{\textstyle\sum_i \sum_j [(\mathbf{C}^{\mathrm{o}})_{i,j} - (\mathbf{T}_{\mathrm{C}}\mathbf{C})_{i,j}]^2/(\mathrm{NS}\cdot\mathrm{NC})\}^{1/2}, \qquad \text{(A-31)}$$

where the product of NS and NC (NS·NC) corresponds to the number of the elements in the matrix **C**. With $K = 5.0$, the value 0.0035 was obtained for LSM. Our goal is to look for the K value that gives the lowest value for the LSM. The iterative calculation using a computer is carried out. One K value gives one LSM value. By the iterative refinement of K values a minimum LSM value is obtained, with an optimum K and $\mathbf{T_C}$. As a result, the following $\mathbf{T_C}$ was obtained at $K = 3.515$:

$$\mathbf{T_C} = \begin{bmatrix} 1.479991 & -0.332043 \\ -0.592155 & 1.656696 \end{bmatrix}. \tag{A-32}$$

The rotated matrix becomes:

$$\mathbf{T_C C} = \begin{bmatrix} 0.85013 & 0.65192 & 0.40974 & 0.19259 \\ 0.14987 & 0.34807 & 0.59026 & 0.80741 \end{bmatrix}. \tag{A-33}$$

The concentrations of [P] and [PX] calculated using $K = 3.515$ are:

$$\mathbf{C^o} = \begin{bmatrix} 0.85014 & 0.65190 & 0.40976 & 0.19259 \\ 0.14986 & 0.34810 & 0.59024 & 0.80741 \end{bmatrix}. \tag{A-34}$$

The value of LSM becomes 0.794×10^{-5}. Virtually perfect agreement exists between $\mathbf{T_C C}$ and $\mathbf{C^o}$ for $K = 3.515$; thus this value is considered the best estimate for the example equilibrium.

From the inverse of $\mathbf{T_C}$, we can get $\mathbf{T_R}$ [Eq. (A-35)]; thus the spectral matrix $\mathbf{R}$ is also rotated to the optimum matrix $\mathbf{R^o}$, as shown in Eq. (A-36).

$$\mathbf{T_R} = \begin{bmatrix} 0.734588 & 0.147230 \\ 0.262565 & 0.656236 \end{bmatrix} \tag{A-35}$$

$$\mathbf{R^o} = \mathbf{R T_R} = \begin{bmatrix} 0.000 & 0.000 \\ 12.000 & 29.993 \\ 48.009 & 69.983 \\ 34.001 & 47.994 \\ 77.984 & 26.034 \\ 29.996 & 6.012 \\ 0.000 & 0.000 \end{bmatrix}. \tag{A-36}$$

The result is shown in Fig. 4.26. It is clear that the obtained spectra $\mathbf{R^o}$ are among the expected candidates shown in Fig. 4.25. The dependence of the value of LSM upon the K is shown in Fig. 4.27. Obviously LSM has a

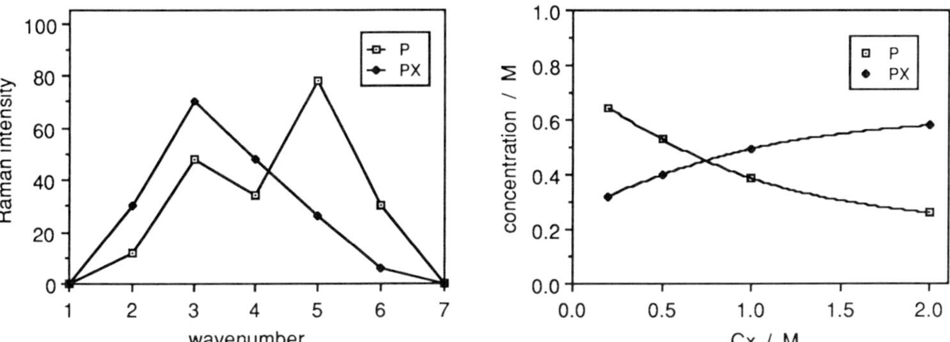

Figure 4.26. The final calculated spectra of the two species I and II, now designated P and PX, and the dependence of their concentrations on the total ligand concentration.

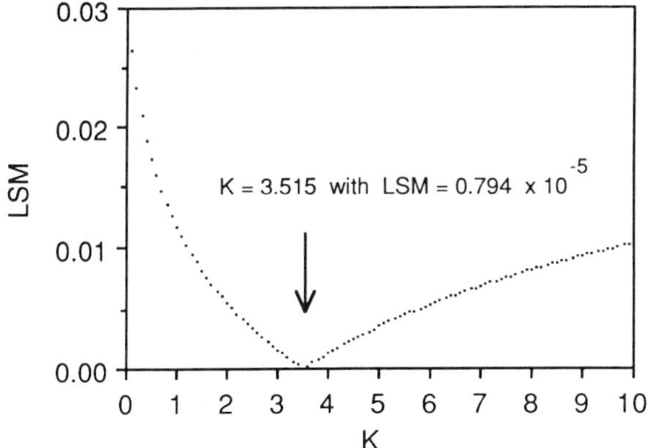

Figure 4.27. An illustration of how the value of the error function LSM depends on the selected value of the equilibrium constant, K. The optimum value of K is that value for which LSM is minimized.

minimum value at $K = 3.515$. Two successful applications of this method have been reported for UV spectra (49) and Raman spectra (50).

ACKNOWLEDGMENTS

This work was supported in part by the Office of Naval Research. The authors express their thanks to their co-workers for their assistance and contributions: Mr. B. Anderson and Drs. D. Guzonas, M. Krell, J. Semmler, and M. Yang.

REFERENCES

1. G. H. Brimhall and D. A. Crerar, in *Thermodynamic Modeling of Geological Materials: Minerals, Fluids and Melts* (I.S.E. Carmichael and H. Eugster, eds.), Vol. 17, Chapter 8, p. 235. Rev. Mineral, Mineral. Soc. Am., Washington, D.C., 1987.
2. M. Buback, D. A. Crerar, and L. M. Vogel, in *Hydrothermal Experimental Techniques* (G. C. Ulmer and H. L. Barnes, eds.), Chapter 14, p. 333. Wiley (Interscience), New York, 1987.
3. R. L. Birke and J. R. Lombardi, in *Spectroelectrochemistry Theory and Practice* (R. J. Gale, ed.), Chapter 6, p. 263. Plenum, New York, 1988.
4. M. H. Brooker, in *The Chemical Physics of Solvation. Part B. Spectroscopy of Solvation* (R. R. Dogonadze, E. Kálmán, A. A. Kornyshev, and J. Ulstrup, eds.), Chapter 4, p. 119. Elsevier, Netherlands, 1986.
5. M. H. Brooker and G. N. Papatheodorou, *Adv. Molten Salt Chem.* **5**, 27 (1983).
6. C. I. Ratcliffe and D. E. Irish, in *Water Science Reviews 3* (F. Franks, ed.), Chapter 1, p. 1. Cambridge University Press, Cambridge, 1988.
7. D. E. Irish and M. H. Brooker, *Adv. Infrared Raman Spectrosc.* **2**, 212 (1976).
8. D. J. Gardiner, *Adv. Infrared Raman Spectrosc.* **3**, 167 (1977).
9. D. E. Irish, in *Physical Chemistry of Organic Solvent Systems* (A. K. Covington and T. Dickinson, eds.), Chapter 4, p. 433. Plenum, New York, 1973.
10. R. E. Verrall, in *Water: A Comprehensive Treatise* (F. Franks, ed.), Vol. 3, Chapter 5, p. 211. Plenum, New York, 1973.
11. T. H. Lilley, in *Water: A Comprehensive Treatise* (F. Franks, ed.), Vol 3, Chapter 6, p. 265. Plenum, New York, 1973.
12. D. E. Irish, in *Ionic Interactions* (S. Petrucci, ed.), Vol. 2, Chapter 9, p. 188. Academic Press, New York, 1971.
13. M. H. Brooker, O. Faurskov Nielsen, and E. Praestgaard, *J. Raman Spectrosc.* **19**, 71 (1988).
14. S. F. A. Kettle, *Symmetry and Structure.* Wiley, Chichester, 1985.
15. J. R. Ferraro and J. S. Ziomek, *Introductory Group Theory.* Plenum, New York, 1975.
16. G. Davidson, *Introductory Group Theory for Chemists.* Elsevier, London, 1971.
17. R. S. Halford, *J. Chem. Phys.* **14**, 8 (1946).
18. R. S. Halford and O. A. Schaeffer, *J. Chem. Phys.* **14**, 141 (1946).
19. C. I. Ratcliffe and D. E. Irish, *Can. J. Chem.* **62**, 1134 (1984).
20. M. M. Yang, D. A. Crerar, and D. E. Irish, *J. Solution Chem.* **17**, 751 (1988).
21. S. Ahrland and N. O. Björk, *Acta Chem. Scand.* **A30**, 249 (1976).
22. P. L. Goggin, G. Johansson, M. Maeda, and H. Wakita, *Acta Chem. Scand.* **38**, 625 (1984).
23. D. E. Irish, B. McCarroll, and T. F. Young, *J. Chem. Phys.* **39**, 3436 (1963).
24. T. Jarv, J. T. Bulmer, and D. E. Irish, *J. Phys. Chem.* **81**, 649 (1977).

25. B. G. Anderson and D. E. Irish, *J. Solution Chem.* **17**, 763 (1988).
26. B. S. W. Dawson, D. E. Irish, and G. E. Toogood, *J. Phys. Chem.* **90**, 334 (1986).
27. C. I. Ratcliffe and D. E. Irish, *Can. J. Chem.* **63**, 3521 (1985).
28. D. E. Irish and R. V. Thorpe, *Can. J. Chem.* **53**, 1414 (1975).
29. K. Nakamoto, *Infrared and Raman Spectra of Inorganic and Coordination Compounds*. Wiley, New York, 1986.
30. S. D. Ross, *Inorganic Infrared & Raman Spectroscopy*. McGraw-Hill, London, 1972.
31. D. E. Irish, T. Jarv, and C. I. Ratcliffe, *Appl. Spectrosc.* **36**, 137 (1982).
32. K. Murata, D. E. Irish, and G. E. Toogood, *Can. J. Chem.* **67**, 517 (1989).
33. L. Fabes and T. W. Swaddle, *Can. J. Chem.* **53**, 3053 (1975).
34. D. E. Irish, T. G. Chang, S. Y. Tang, and S. Petrucci, *J. Phys. Chem.* **85**, 1686 (1981).
35. C. I. Ratcliffe and D. E. Irish, *J. Phys. Chem.* **86**, 4897 (1982).
36. J. T. Bulmer, D. E. Irish, F. W. Grossman, G. Herriot, M. Tseng, and A. J. Weerheim, *Appl. Spectrosc.* **29**, 506 (1975).
37. J. Pitha and R. N. Jones, *Can. J. Chem.* **44**, 3031 (1966); **45**, 2347 (1967); *Optimization Methods of Fitting Curves to Infrared Band Envelopes*, NRC Bull. No. 12. National Research Council of Canada, Ottawa, 1968.
38. W. F. Murphy and H. J. Bernstein, *J. Phys. Chem.* **76**, 1147 (1972).
39. J. Semmler and D. E. Irish, *J. Solution Chem.* **17**, 805 (1988).
40. E. A. Moody, in *Dictionary of Scientific Biography* (C.C. Gillispie, ed.), Vol. 10, p 171. Charles Scribner's Sons, New York, 1974; S. F. Mason, *Main Currents of Scientific Thought*, p. 91. Henry Schuman, New York, 1953.
41. H. F. Shurvell and J. T. Bulmer, in *Vibrational Spectra and Structure* (J. R. Durig, ed.), Vol. 6, Chapter 2, p. 91. Elsevier, Amsterdam, 1977.
42. E. R. Malinowski and D. G. Howery, *Factor Analysis in Chemistry*. Wiley, New York, 1980.
43. D. A. Guzonas and D. E. Irish, *Can. J. Chem.* **66**, 1249 (1988).
44. D. A. Guzonas, D. E. Irish, and G. F. Atkinson, *Langmuir* **5**, 787 (1989).
45. F. J. Knorr and J. H. Futrell, *Anal. Chem.* **51**, 1236 (1979).
46. T. Ozeki, H. Kihara, S. Hikime, and S. Ikeda, *Anal. Sci.* **3**, 285 (1987).
47. H. Gampp, M. Maeder, C. J. Meyer, and A. D. Zuberbühler, *Talanta* **32**, 95 (1985).
48. H. Gampp, M. Maeder, C. J. Meyer, and A. D. Zuberbühler, *Talanta* **32**, 257 (1985).
49. T. Ozeki, H. Kihara, and S. Ikeda, *Anal. Chem.* **60**, 2055 (1988).
50. J. van Heumen, T. Ozeki, and D. E. Irish, *Can. J. Chem.* **67**, 2030 (1989).
51. F. J. C. Rossotti, H. S. Rossotti, and R. J. Whewell, *J. Inorg. Nucl. Chem.* **33**, 2051 (1971).
52. R. S. Tobias and M. Yasuda, *Inorg. Chem.* **2**, 1307 (1963).

53. I. Leden, Ph.D. Dissertation, University of Lund, Lund, Sweden (1943).
54. A. Swinarski and A. Grodyicki, *Rocz. Chem.* **39**, 1155 (1965).
55. J. T. Bulmer, T. G. Chang, P. J. Gleeson, and D. E. Irish, *J. Solution Chem.* **4**, 969 (1975).
56 A. S. C. Cheung and D. E. Irish, *J. Inorg. Nucl. Chem.* **43**, 1383 (1981).
57 J. Semmler, D. E. Irish, and T. Ozeki, *Geochim. Cosmochim. Acta* **54**, 947 (1990).
58. W. Kiefer, *Appl. Spectrosc.* **27**, 253 (1973).
59. W. Kiefer, W. J. Schmid, and J. A. Topp, *Appl. Spectrosc.* **29**, 434 (1975).
60. J. Laane and W. Kiefer, *Appl. Spectrosc.* **35**, 428 (1981); *J. Chem. Phys.* **72**, 5305 (1980).
61. D. J. Gardiner, R. B. Girling, and R. E. Hester, *J. Chem. Soc., Faraday Trans. 2* **71**, 709 (1975).
62. H. van den Boom and R. E. Breemer, *Rev. Sci. Instrum.* **46**, 1664 (1975).
63. D. L. Rousseau, *J. Raman Spectrosc.* **10**, 94 (1981).
64. J. Laane, in *Vibrational Spectra and Structure* (J. R. Durig, ed.), Vol. 12, Chapter 6, p. 405. Elsevier, Amsterdam, 1983.
65. W. Kiefer and J. A. Topp, *Appl. Spectrosc.* **28**, 26 (1974).
66. N. Zimmerer and W. Kiefer, *Appl. Spectrosc.* **28**, 279 (1974).
67. J. Laane, *J. Chem. Phys.* **75**, 2539 (1981).
68. J. Laane and W. Kiefer, *Appl. Spectrosc.* **35**, 267 (1981).
69. J. Laane, H. Eichele, H. P. Hohenberger, and W. Kiefer, *J. Mol. Spectrosc.* **86** 262 (1981).
70. J. Laane and W. Kiefer, *J. Chem. Phys.* **73**, 4971 (1980).
71. M. M. Strube and J. Laane, *36th Symp. Mol. Spectrosc.*, Ohio State University, June, *1981* Paper TG3 (1981); J. Laane and M. M. Strube, *183rd Am. Chem. Soc. Meet.*, Las Vegas, Nevada, March, *1982* Paper PHYS 060 (1982).
72. D. E. Irish, S. Y. Tang, H. Talts, and S. Petrucci, *J. Phys. Chem.* **83**, 3268 (1979).
73 R. J. Wherry, Sr., *Contributions to Correlational Analysis*, p. 178. Academic Press, London, 1984.
74. E. R. Malinowski, *Anal. Chem.* **49** 612 (1977).
75. W. H. Lawton and E. A. Sylvestre, *Technometrics* **13**, 617 (1971).
76. E. A. Sylvestre, W. H. Lawton, and M. S. Maggio, *Technometrics* **16**, 353 (1974).

CHAPTER

5

QUANTITATIVE ANALYSIS BY RAMAN SPECTROSCOPY

THOMAS J. VICKERS AND CHARLES K. MANN

Department of Chemistry
Florida State University
Tallahassee, Florida

5.1. INTRODUCTION

Quantitative analysis by Raman spectroscopy has not kept pace with recent rapid growth in the use of Raman spectroscopy for structural and qualitative analysis. We take as the starting point for this chapter that this failure to make full use of the Raman technique is a major omission that calls for rapid correction. It is our contention that quantitative Raman spectroscopy can be made as routine and reliable as absorption spectroscopy and will provide the analyst with vastly more information. Raman spectroscopy should be employed in a variety of applications requiring rapid, direct analyses, such as process-stream monitoring and environmental screening. This chapter examines the obstacles, real and imagined, that must be overcome to achieve this goal, describes the current state of the field, and points to directions for further development.

Raman spectroscopy shares with other emission techniques the handicap of historical development as a single-beam technique and the limitations which that implies for quantitative analysis owing to uncorrected variations in source, sample, and optics. Further, the Raman-scattering phenomenon is inherently weak compared to a process such as absorption of radiation, and this has led to unwarranted assumptions about the sensitivity of Raman spectroscopy relative to other approaches. Finally, it is readily apparent that the conditions required for Raman spectroscopy will also give rise to strong fluorescence from many samples, and there is a widespread assumption that

Analytical Raman Spectroscopy, Edited by Jeanette G. Grasselli and Bernard J. Bulkin. Chemical Analysis Series, Vol. 114.
ISBN 0-471-51955-3

the occurrence of fluorescence will make the observation of Raman spectra impossible or, at least, exceedingly difficult. Each of these difficulties will be addressed below.

There are, of course, a number of recognized advantages to Raman spectroscopy. It provides vibrational spectra that are rich in highly reproducible detailed features, providing the possibility of highly selective determinations. In comparing Raman and infrared (IR) spectroscopies, the Raman approach is advantageous in that aqueous solutions pose no special problems, the low-frequency spectral region is easily obtained, optical components such as windows, sample containers, and optical fibers can be made out of relatively inexpensive and readily available materials, and superior single-channel and multichannel detectors are available. Compared to absorption techniques in general, the Raman approach permits considerable flexibility in the physical state of the sample and in the range of concentrations that can be examined without alteration in the system. Finally, resonance and surface-enhancement phenomena offer possibilities for extending the sensitivity and selectivity of Raman spectroscopy.

5.2. HISTORICAL DEVELOPMENT

The 1971 review article by Irish and Chen (1) provides a convenient starting point for consideration of the development of quantitative analysis by Raman spectroscopy. By 1971, lasers had completely replaced the balky and bulky mercury-vapor lamps as the excitation source, and Raman spectroscopy seemed for the first time to hold some promise for the analyst. The aforementioned review provides an extensive bibliography of quantitative Raman measurements prior to 1970. The referenced studies are primarily directed toward the physical chemistry of electrolyte solutions. The basis of quantitative Raman analysis is succinctly described, including the use of internal and external standards to compensate for instrumental and sample variations. However, the availability of laboratory computers that would make these corrections rapid and routine was still some years in the future, and the promise of Raman spectroscopy remained unfulfilled for most analytical chemists.

The book *Chemical Applications of Raman Spectroscopy* by Grasselli, Snaveley, and Bulkin (2) cites a dozen miscellaneous applications of quantitative Raman spectroscopy for the period from 1970 to 1979. Of greater interest in the context of this chapter are a number of reports during this period of efforts to apply Raman spectroscopy for routine analyses, such as the determination of pollutants in natural water samples. Bradley and Frenzel (3, 4) reported the detection of the herbicide 2,4-dichlorophenoxyacetic acid (2,4-D) in aqueous solutions at the 500 ppm level. Braunlich et al. (5), from theoretical considerations, projected a 1–10 ppm detection limit for organic pollutants in water. Brown and co-workers showed that inorganic anions could be detected in the 25–75 ppm range with conventional laboratory equipment (6) and in the 100–200 ppm range when using remote detection equipment (7). Cunningham et al. (8) reported detection limits in the 4–40 ppm range for various anionic species in water. Miller (9) applied Raman spectroscopy to the determination of oxyanions in nuclear waste material. Furuya et al. (10) examined the determination of nitrate ion in waste and treated water and demonstrated a detection limit of 2 ppm under ideal conditions and about an order of magnitude higher for real samples.

During this period initial efforts to employ resonance enhancement to improve the sensitivity of Raman measurements for trace components were also reported. Van Haverbeke, Brown, and co-workers applied resonance Raman to the determination of pesticides and fungicides (11) and to dyes (12) with absorption bands in the near-UV (ultraviolet) and visible range. Van Haverbeke and Herman described the use of resonance Raman for several phenolic compounds (13) and pesticides (14) following derivatization of the analyte to provide near-UV absorption. Furuya et al. (15) used a similar derivatization approach in the determination of nitrite ion in waste and treated waters, reporting a detection limit of about 0.0005 ppm. Morris (16, 17) determined catecholamines by resonance Raman spectroscopy of their aminochromes. Reviews appeared on the application of resonance Raman to the determination of pollutants in water (18) and on the general utility of resonance Raman spectroscopy for chemical analyses (19).

The current decade has seen further evolution in the use of Raman spectroscopy for quantitative analysis. Applications have included the determination of phenols (20, 21), sulfur oxyanions (22, 23), pharmaceuticals (24–26), antitumor agents (27), aromatic amines (28), azo dyes (29), and ethanol in fermentation broth (30). The analysis of fluid inclusions in geologic samples (31) and of the headspace in sealed drug vials (32) has further demonstrated the flexibility of the technique for rapid, direct analysis. Of particular significance has been the extension of measurements into the deep-UV range, permitting determination of many more molecular species by resonance Raman spectroscopy without prior derivatization (33–39) and the development of surface enhancement on silver as an analytical technique (40–47).

5.3. VARIATIONS IN SOURCE, SAMPLE, AND OPTICS: EFFECT ON QUANTITATIVE ANALYSES

A technique for routine quantitative analysis must be "robust," meaning that correct results are produced even when there are significant variations in the measurement environment. The technique should also indicate the presence of conditions that will produce incorrect results. The success of absorption spectroscopy for routine analysis owes much to the use of double-beam systems to achieve the required measurement robustness. Absorption instruments benefit from more than three decades of commercial development, and modern computer-controlled instruments often include various kinds of automated testing to indicate error conditions.

5.3.1. Ordinate Errors

The use of an internal standard in Raman measurements provides a condition comparable to the double-beam approach for absorption measurements and can confer a comparable degree of measurement robustness. In solution measurements, the solvent is often chosen as the internal standard species. Water is, of course, a uniquely important solvent, and happily it also is very convenient as an internal standard for aqueous samples. In the 1000–2000 cm^{-1} range, where so many useful vibrations occur for organic compounds, water has only one band, the rather weak bending mode at 1640 cm^{-1}. This water feature is also unusual in that it is much broader than the typical Raman line. It is therefore easy to discriminate between it and analyte peaks, making water a convenient internal standard. Figure 5.1 (30) illustrates this point.

Figure 5.1(a) shows a portion of the Raman spectrum of a 5% solution of acetone in water in the vicinity of the 1640 cm^{-1} water band. Figure 5.1(b) shows the spectrum of water over the same range. Figure 5.1(c) is the spectrum of acetone, in water, obtained by subtracting a suitably scaled version of Fig. 5.1(b) from Fig. 5.1(a). In subsequent determinations of acetone in water, the spectra in Fig. 5.1(b, c) were taken as references to be fitted by a least squares procedure, and the response to the water peak in the fitting calculation was used as the internal standard for acetone. A similar procedure was used to carry out the simultaneous determination of ethanol, methanol, and acetone in aqueous mixtures, with the results shown in Table 5.1. In a further test (30) of the efficacy of the internal standard approach, ethanol production in a fermentation mixture was monitored over a 6-day period using the water peak as an internal standard.

Raman measurements carried out on several strongly absorbing azo dyes in methanol (29) provided a somewhat different test of the robustness conferred by the internal standard approach. The 1040 cm^{-1} peak of methanol was used as the internal standard. Figure 5.2 shows the effect of the internal standard correction on the analytical curve data for the worst case of the three dyes examined, Eriochrome Blue Black B (EBBB); in the figure, results for three series of measurements are shown. The solid boxes and circles and

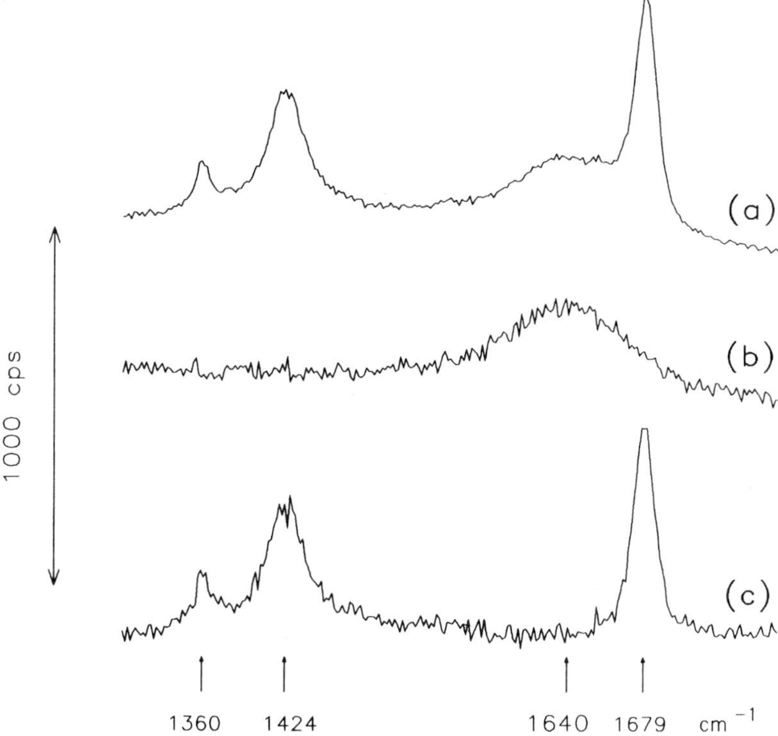

Figure 5.1. Raman peaks of acetone and water. (a) Experimental spectrum of acetone in water. (b) Water spectrum. (c) Acetone spectrum obtained by subtracting scaled (b) from (a). [From Shope et al. (30).]

Table 5.1. Multicomponent Analyses

	Ethanol (%)		Methanol (%)		Acetone (%)	
Sample	Taken	Found	Taken	Found	Taken	Found
A	0.5	0.51 ± 0.01	0.1	0.103 ± 0.002	0.5	0.503 ± 0.002
B	0.1	0.104 ± 0.002	0.5	0.503 ± 0.005	0.5	0.499 ± 0.004
C	0.5	0.506 ± 0.002	0.5	0.501 ± 0.002	0.1	0.102 ± 0.002

Source: From Shope et al. (30).

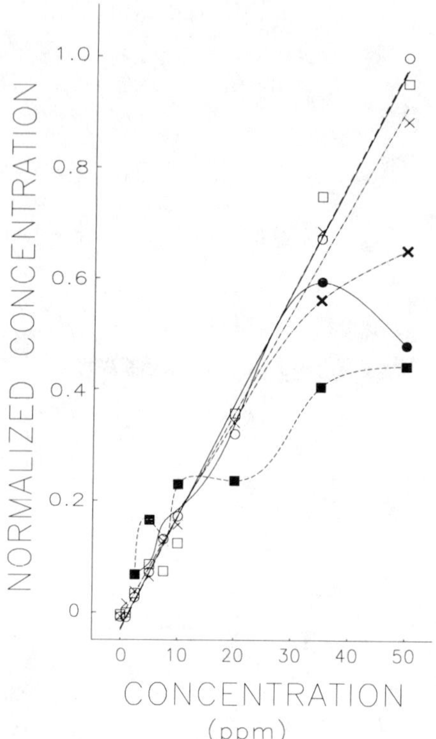

Figure 5.2. Demonstration of the use of an internal standard to correct for variables in analytical curve measurement for Eriochrome Blue Black B in methanol (three runs). [From Womack et al. (29).]

boldface × symbols are for the three sets of uncorrected results. Sample absorbance, variations in exciting beam intensity, and variations in optical alignment from one sample to the next give rise to large deviations from the expected proportionality between response and concentration. The uncorrected points in a given run were connected with the aid of a spline. To avoid confusion, uncorrected points were omitted when they were superimposed on corrected points. The most extreme variations in intensity were caused by the need to attenuate the laser beam as dye concentration increased to avoid deleterious heating effects in the sample. The points corrected by the internal standard are shown in Fig. 5.2 by the open boxes and circles and lightface × symbols. The efficacy of the internal standard approach in even these rather extreme conditions is best seen in the statistical summary in Table 5.2.

Table 5.2. Analytical Curve Statistics for EBBB

Run	Slope (ppm^{-1})	Intercept	Correlation Coefficient	Limit of Detection (ppm)[a]
1	0.0201	−0.0290	0.998	1.6
2	0.0202	−0.0317	0.994	2.7
3	0.0184	−0.0125	0.997	1.8

Source: From Womack et al. (29).

[a]The limit of detection is taken as three times the standard deviation of the intercept of the analytical curve divided by its slope.

Implementation of the internal standard approach with solid samples is analogous to that for solution samples. In most cases in which quantitative analysis is sought, it is possible to select a major component as the internal standard species. This approach has been illustrated in the determination of impurities in a pharmaceutical material (26). Figure 5.3(b) shows the Raman spectra of the pharmaceutical, sulfamethoxazole, and Fig. 5.3(a) the impurity compound, sulfanilamide. The impurity feature is superimposed on the

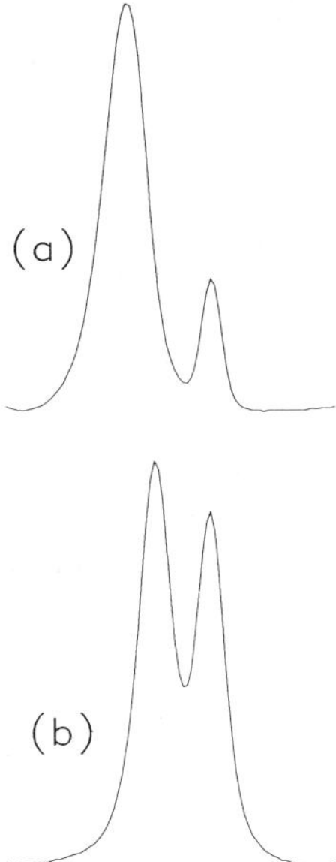

Figure 5.3. Segment of Raman spectrum illustrating spectral overlap of strongest features for (a) sulfanilamide and (b) sulfamethoxazole. [From Mann et al. (26).]

principal peak of the bulk component used as the internal standard. Quantitation was carried out by both least squares fitting and cross correlation, with essentially identical results. Analytical curves prepared with a series of samples with 0, 0.5, 1.0, 2.5, and 5.0% sulfanilamide in sulfamethoxazole gave a correlation coefficient of 0.9992 and limit of detection of 0.12% by the least squares procedure and 0.9991 and 0.11% by the correlation procedure.

It is possible to imagine conditions under which the complete correction of ordinate errors by an internal standard might fail. Among these are changes in shape or position of the relevant lines with changes in sample composition, nonlinearity in the instrumental response function, and differences in sample absorbance at the analyte and standard lines. The first of these seems to be rare in Raman measurements, but an effect of this type has been noted in the measurement of ethanol in water and a method of compensating for this error has been suggested (30). The instrumental components typically used for UV and visible spectroscopy exhibit wide linear ranges, and Raman instruments exhibit good linearity for a wide range of signal values. Some nonlinearity has been noted in the response of multichannel detectors, but even in this case the detector-response function seems to be reproducible for a given detector and is readily corrected in a computerized system.

Difference in sample absorbance is the most likely cause for failure of the internal standard approach in applying quantitative Raman spectroscopy to "real-world" samples, particularly when using resonance enhancement. This difficulty has received considerable attention, and methods to correct for the effect of absorption have been described for various scattering geometries (37, 48–52a). The results of these studies can be summarized as follows: When sample absorbance is the same at the wavelengths of the analyte and standard Raman lines, internal standardization corrects for absorption of the incident and Raman scattered radiation, and behavior such as that described above for the azo dye measurements is obtained. When sample absorbance is different at the wavelengths of the analyte and standard Raman lines, internal standardization does not fully correct for the effects of absorption. The extent to which the internal standardization process fails increases with increasing difference in the sample absorption at the wavelengths of the analyte and standard Raman lines and with increasing sample absorbance. Use of a backscattering geometry and thin samples will reduce the effect of sample absorbance. To carry out routine quantitative measurements on strongly absorbing samples it will be necessary to provide some means of obtaining the absorption spectrum of the sample over the wavelength range covered by the relevant Raman lines as the Raman measurements are made. The absorption data provide the information required to fully correct the Raman results for sample absorbance.

5.3.2. Abscissa Errors

Wavenumber errors as small as a few tenths of a wavenumber can routinely be achieved with scanning spectrometers. With multichannel measurements it should be possible to obtain spectra with much smaller abscissa errors, although close attention must be paid to mechanical and thermal stability to do so. For many purposes such small abscissa errors may be of no consequence, but for some applications of quantitative Raman spectroscopy abscissa errors of this magnitude limit utility. Figure 5.4 illustrates such a case. The portion of the Raman spectrum of sulfamethoxazole shown in Fig. 5.4(a) has been subtracted from versions of itself that have been shifted on the wavenumber axis by ± 0.05 and $\pm 0.02\ \mathrm{cm}^{-1}$ as indicated. Subtraction

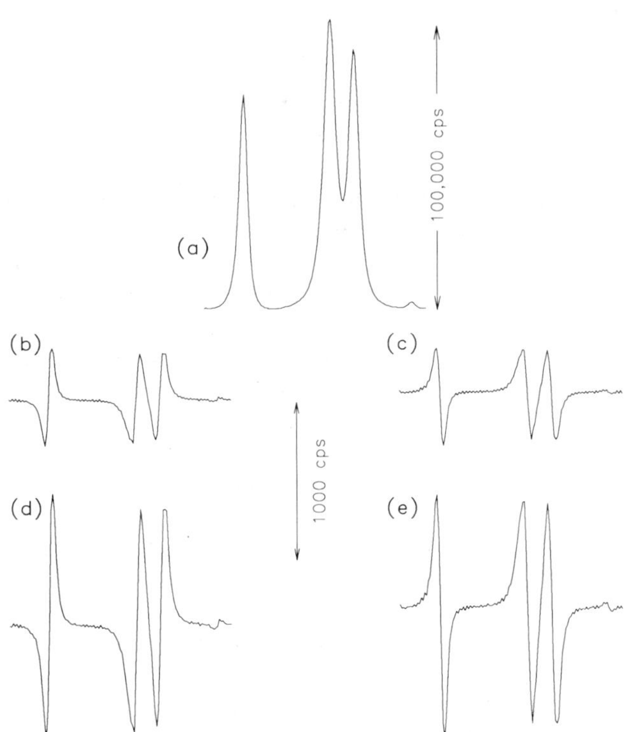

Figure 5.4. Effect of abscissa error on subtraction of major spectral features. (a) Portion of the Raman spectrum of sulfamethaxazole. (b, c) Residuals produced by subtraction for abscissa shifts of $0.02\ \mathrm{cm}^{-1}$. (d, e) Residuals produced by subtraction for abscissa shifts of $\pm 0.05\ \mathrm{cm}^{-1}$.

of spectra with abscissa offset produces signals that resemble derivatives. A shift as little as $0.02\,\text{cm}^{-1}$ produces these artifacts. This is much less than the error encountered in setting a spectrometer to predetermined position. The size of residuals is determined by the amount of abscissa error. For a shift of $0.02\,\text{cm}^{-1}$, the residuals [depicted in Fig. 5.4(b, c)] are about the size that would be expected for the main peak of a 0.5% impurity. If the goal of measurement is the detection and/or determination of impurities in spectral regions in which the bulk material exhibits strong spectral features, then abscissa correction is imperative.

Note that the spectra in Fig. 5.4 correspond to the usual condition that the abscissa error is smaller than a single sampling interval. Thus it is not possible to correct the abscissa error by simply shifting the data set, employing the disappearance of the derivative features after subtraction of the reference as an indicator of success. This limitation could be circumvented, of course, by decreasing the sampling interval, but this is an unsatisfactory solution because it sacrifices the considerable efficiency associated with collecting, transferring, and storing small data sets. A technique is available that makes use of the fast Fourier transform algorithm to permit rapid abscissa correction of any set that meets the Nyquist criterion for adequate sampling of the spectrum.

If a function is compared with one that is identical except for having been displaced along the abscissa, it is found that the magnitudes of corresponding spatial frequency components are identical for both functions and that the effect of the abscissa shift is to change the phases. If a 128-point set has been sampled at $1\,\text{cm}^{-1}$ intervals, a $1\,\text{cm}^{-1}$ abscissa shift produces a phase shift of $2\pi/128$ radians in the first spatial frequency component, twice that for the second, and so forth. To detect abscissa error in a spectrum, corresponding regions of the sample and a reference spectrum are taken, the phase shift for each of the more important spatial frequencies is determined and expressed as abscissa offset, and the average value is computed. The average shift value, after scaling by the frequency as already described, is then used to correct each component of the sample spectrum, and the set is then retransformed to produce the corrected spectrum.

5.3.3. Data Treatment

In order to consider data-treatment possibilities, let us ignore, for a moment, the chemical and physical implications of Raman spectra and think of them as abstract data sets of ordinate values plotted on a uniformly spaced X axis. The data set can be transformed to the Fourier domain without altering its information content. In this form it consists of a group of terms each of which describes a sinusoid. A set initially made up of n data points contains

$n/2$ terms in the Fourier domain. The first term describes a signal with zero frequency, a constant DC (direct current) level; the second term describes a sinusoid that has exactly one period in the space occupied by the data; the third term describes a sinusoid with two periods in the space; and so on. The data can be returned to their original form by summing these terms.

The first Fourier domain term, the DC level, is the average value of the original data set. Consider the example in Fig. 5.5. The useful information is contained in the size and shape of the curve. It does not matter whether the DC level is zero, as in Fig. 5.5(a), or positive, as in Fig. 5.5(b), or negative, as in Fig. 5.5(c). The DC level can be arbitrarily selected without altering the shape of the spectrum. Accordingly, in manipulating the data, the DC level can be set to any convenient value without degrading the information content of the signal. This is very useful, since the major component responsible for the DC level in Raman spectra is usually fluorescence.

Altering any other term of the Fourier domain does change the size and shape of the data in ways that may or may not be desirable. For example, Fig. 5.6(b) was formed by removing the second term from the set in Fig. 5.6(a). The shape is grossly distorted. This evidently is bad if the purpose of the measurement is to determine the shape of the spectrum. It is not necessarily undesirable if the purpose is to identify or quantify the compound responsible for the spectrum. The curve in Fig. 5.6(b) retains much of the original information. If the comparison is made between a sample spectrum, like that shown in Fig. 5.6(b), and a reference spectrum, like that of Fig. 5.5(a), which has been similarly treated by selective frequency exclusion, the sets will be found to be identical, except for experimental noise. Exclusion of a low-frequency term occasionally is useful if the spectral feature of interest is superimposed upon a strongly curved background.

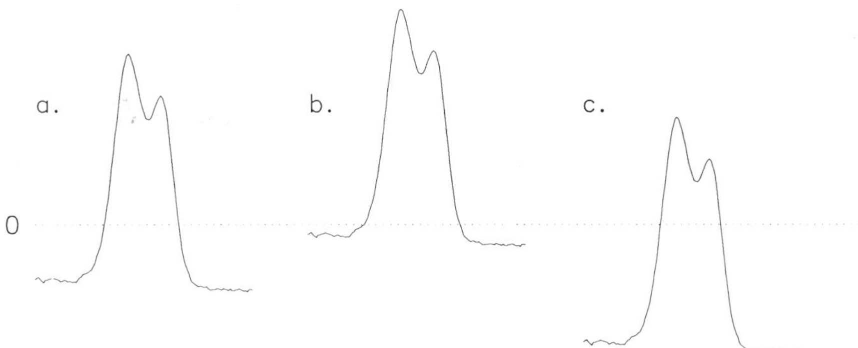

Figure 5.5. Identical Raman peaks with different DC levels. (a) Zero DC level. (b) Positive DC level. (c) Negative DC level.

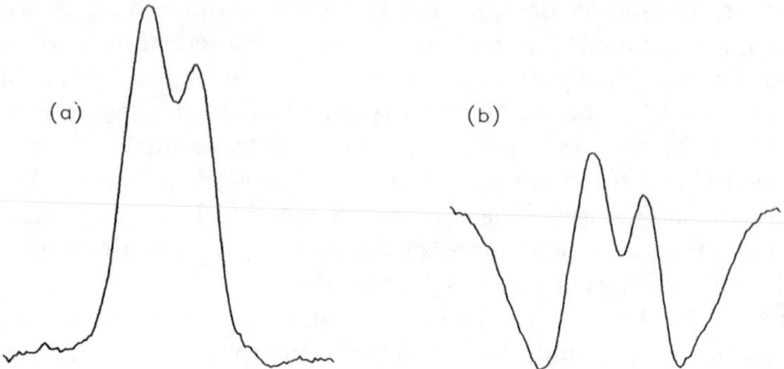

Figure 5.6. Effect of deletion of a single-frequency component in the Fourier domain. (a) Original spectral feature. (b) Same feature after removal of second-frequency term in the Fourier domain.

There is, of course, an additional factor that must be considered in using frequency exclusion. The process is useful only if the signal-to-noise ratio (S/N) is improved as a result. Figure 5.7(a) shows the power density distribution of Fig. 5.6 before removal of the second Fourier domain term. For each term, the crosshatched area represents signal power contributed by noise at the frequency in question; the area of the remainder of the bar corresponds to the power contributed by the spectral features of the analyte. It is quite apparent that exclusion of any of the terms from the second through the seventh causes a decrease in S/N since signal power is larger than noise power. Exclusion of the eighth and all higher terms improves the S/N.

A variety of techniques is available for quantifying spectroscopic peaks; we discuss area, peak height, cross correlation, and least squares fitting. In selecting one, it is useful to note that the largest source of error is likely to be the establishment of the position of the baseline upon which the peaks are situated. Accordingly, data processing procedures that do not require an explicit baseline set give better results when the S/N is small.

In quantifying spectral data, it is of course the area of peaks that is desired. However, it generally is undesirable to make a direct area measurement unless the S/N is very high and the baseline is well defined. When this is not true, small errors have a disproportionate effect on the final result because all points in the spectrum are given equal weight in the calculation.

Peak height measurements usually give better results than area measurements unless there are significant changes in peak shape with concentration. As with area measurement, a peak height measurement requires an explicit baseline location. However, if the measurement is made at the peak maximum, the point of optimum S/N is used. A value based upon a single point in a set is especially vulnerable to high-frequency noise. If peak heights are to be measured, it is desirable to filter the signals beforehand.

Quantification by correlation of the sample with a reference spectrum usually gives better results than either of the aforementioned two methods. Correlation is defined by Eq. (1) (54); here $r(t)$ and $s(t)$ are the reference and sample spectra, $x(t)$ is the correlation result, and τ is a variable that describes the position of $r(t)$ along the abscissa:

$$x(t) = \int_{-\infty}^{\infty} r(t+\tau)s(\tau)\,d\tau. \tag{1}$$

The process can be visualized as one of displacement of one data set relative to the other, by changing τ, with multiplication and summation as indicated by the equation. It can be carried out efficiently by using the Fourier transform. Both sets are transformed. The complex conjugate of one is taken, and the two Fourier domain sets are multiplied together. The correlation value is obtained as the first data point in the set that is produced by inverse transform of the product set.

As a technique of quantification, the correlation value has the same properties as an absorbance in a peak measurement (55, 56). Results are generally better because the whole data set can be used, rather than just one point. The multiplication indicated in Eq. (1) causes the data points of the sample set to be weighted in proportion to the intensity of the corresponding point in the reference set. This is ordinarily proportional to S/N. As indicated in Eq. (1), high-frequency noise is rejected by integration. In those cases in which the background features remain constant from one measurement to the next, a calibration curve can be established by correlation that is unaffected by interference from the background. Thus if a determination is to be made in which a major sample constituent produces a spectral peak interfering with that of the analyte, use of correlation as a quantifying procedure will eliminate the effects of the interference.

A correlation can be conveniently combined with digital filtering by excluding points from the Fourier domain of one set. As illustrated in Fig. 5.7, this would be done by nulling the ninth and higher terms. The major effects of baseline variations are removed by discarding the DC component. Correlation of a reference with any of the sets of Fig. 5.5 would give the same result if the DC level were omitted.

Correlation does not correct for sloping or curved baselines. The most common situation in Raman spectroscopy is a sloping and very slightly curved baseline, with major baseline components contributed by the tail of the Rayleigh line and by fluorescence.

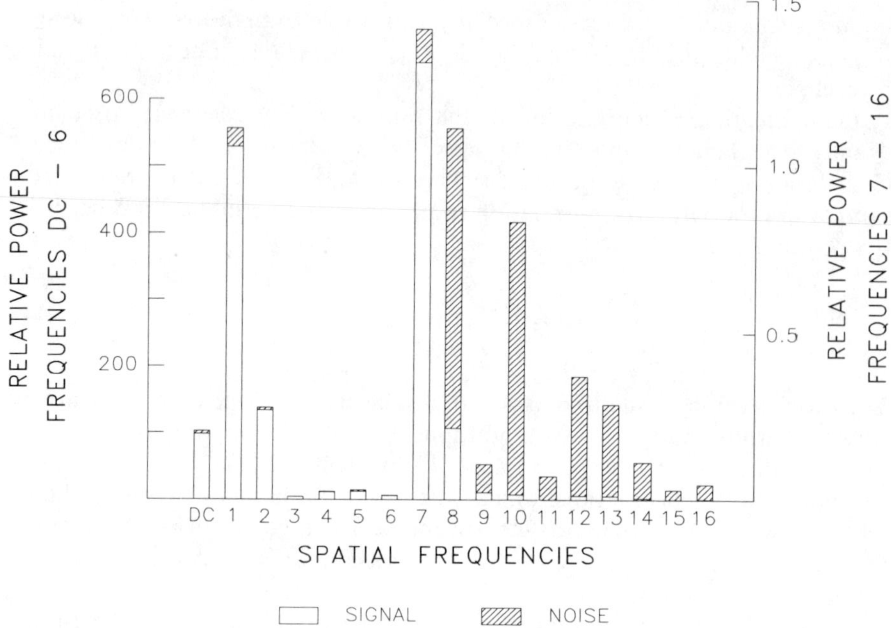

Figure 5.7. Power density distribution of spectrum shown in Fig. 5.6(a). For each term, the crosshatched area represents signal power contributed by noise at the frequency in question and the area of the remainder of the bar corresponds to the power contributed by the spectral feature of the analyte.

If a peak is situated on a linear baseline of any slope and if the signals arise as a result of linear combination of components, an accurate baseline correction can be made by projecting a straight line across the base of the peak and subtracting it point by point from the original data.

Baselines encountered in Raman spectroscopy usually vary quite a lot from one run to the next even when the samples are nominally the same. Both the intensity and direction of slope can be expected to change. Peaks are usually superimposed on slightly curved baselines. The curves in Fig. 5.8. are typical of what might be expected with analyte concentrations near the limit of detection.

Figure 5.8(a) represents a reference spectrum showing high S/N. Figure 5.8(b, c) represents samples of the same size but superimposed on variable, sloping, curved baselines with normally distributed random noise added to give an uncontaminated S/N of 10 (where peak height is taken as signal). These are typical of samples that are fairly near the limit of detection.

The signals in Fig. 5.8 can be quantified by truncating the sets in both

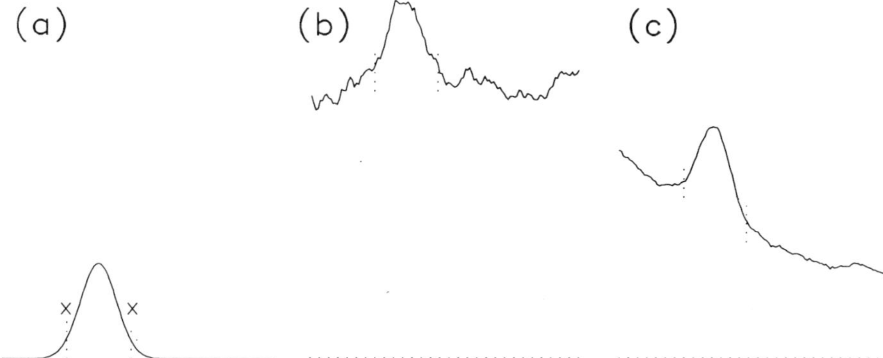

Figure 5.8. Effect of baseline variations on a spectral feature. (a) A high S/N reference spectrum. (b, c) Spectral features the same size as that shown in (a) but contaminated with noise and superimposed on variable, sloping, curved baselines.

the data and Fourier domains. Best results are normally obtained by retaining only the region that contains the most intense section of the reference. In this case the range × – × is retained in each set. A straight line is projected between the reference points and values below it are subtracted. All points outside the segment are set to zero. The DC level is rejected, together with high-frequency components that contain no signal information. In this case the second through tenth terms in the Fourier domain are retained. After this pre-treatment, correlation of the reference with the samples gives 0.948 for Fig. 5.8(b) and 1.057 for Fig. 5.8(c). The correct value would be 1. Here the results are presented as the ratio between the cross correlation of the reference with the sample and the autocorrelation of the reference.

The least squares fit is an alternative method that is preferable under certain circumstances (57, 58). The operation is based upon the idea that a spectrum can be considered to be a vector that can efficiently be manipulated by the techniques of matrix algebra. If a 128-point spectrum is taken as a vector in 128 space, scaling the spectrum to reflect a change in concentration is a manipulation on 128 simultaneous equations. If the data are experimental, evidently there is error in each measurement and the equations are inconsistent. A least squares operation allows resolution of these inconsistencies and makes it possible to use matrix operations.

To illustrate the point, consider the two-dimensional vectors **OA** and **OB** in Fig. 5.9. Suppose that it is desired to fit **OA** to **OB**. Since they are not colinear, **OA** cannot be scaled to equal **OB** and there is no solution. The best that can be had is the scaled version of **OA** that best matches **OB**. Using the least squares fitting routine described below, this is found to be the vector

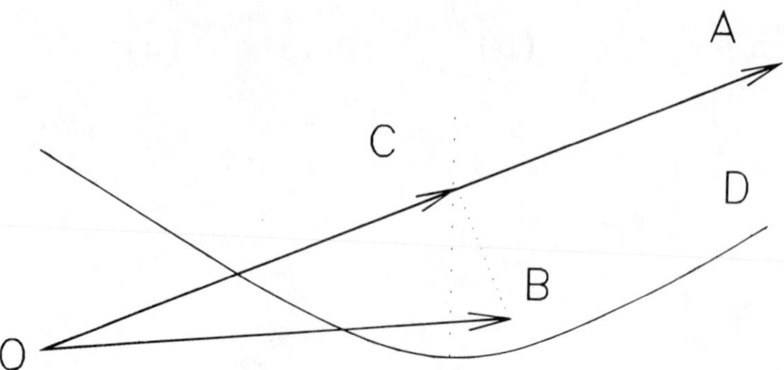

Figure 5.9. Illustration of least squares fitting for vectors in two-dimensional space.

OC. To check the effectiveness of this solution, suppose that **OA** is gradually decreased in length and the distance between it and **OB** is continually measured. The curved line *D* in Fig. 5.9 is a plot of this distance versus the length of **OA**. It passes through a minimum when **OA** = **OC**. In this case, application of a least squares solution is computational overkill since we know in advance that the closest approach is the perpendicular distance from point *B* to **OA**, which is consistent with both of the criteria just described. However, spectra contain larger numbers of points and a matrix solution provides a method for dealing with them in one operation.

Let us consider Eq. (2):

$$\mathbf{X} = \mathbf{RC} + \mathbf{E}. \tag{2}$$

Here **X** represents a spectrum (in matrix notation it would be an $n \times 1$ matrix, with n the number of data points); **R** is a set of reference spectra, one for each of the m components in the sample (**R** is therefore an $n \times m$ matrix); **C** is the concentration matrix, the scale factors by which **R** is fitted to **X** and is an $m \times 1$ matrix; **E** represents the error associated with each measurement. The least squares solution is shown in Eq. (3):

$$\mathbf{C} = (\mathbf{R}^T\mathbf{R})^{-1}\mathbf{R}^T\mathbf{X}. \tag{3}$$

Here the T superscript denotes a transpose, and -1 an inverse. This is a practical computation, since $\mathbf{R}^T\mathbf{R}$ ensures that the matrix to be inverted has dimensions $m \times m$ rather than $n \times n$.

The major restriction on use of the least squares procedure is that there must be a component in the reference matrix for each significant component in the sample set. This restriction can to some extent be circumvented by including simulations of components in the reference matrix. For example, a DC offset can be accounted for by including one column that contains only a constant. A sloping baseline can be accounted for by including a ramp as a component. Taking the example of Fig. 5.8, a simple application of the set in Fig. 5.8(a) would give very bad results because of the large offset. Including a ramp and a DC term in the reference matrix gives 1.037 and 0.882 for Fig. 5.8(b) and 5.8(c), respectively.

The pretreatment described in the discussion of correlation may give a greater degree of freedom from the necessity to account for each sample constituent. In the least squares treatment, if the data set is truncated as indicated by $\times - \times$ in Fig. 5.8(a), the values obtained are 0.949 and 1.053. If, in addition, the digital filtering step is applied, the results are 0.948 and 1.057. These are the same values that are obtained by correlation. This will be true if the two operations are carried out in exactly the same way. In this last operation, each is measuring the extent of similarity between the same two sets.

The least squares fit routine generally runs somewhat faster than correlation, providing it does not include a filtering step. It allows the user to test the effectiveness of the calculation, since a model based upon the references and the calculated fits can be constructed.

5.3.4. Multichannel Versus Scanning Measurements

Multichannel operation is desirable in Raman for the same reasons that apply to other types of spectroscopy. In addition, it offers important advantages that pertain especially to Raman measurements. The ability to see an entire spectrum being updated at 1-s intervals is a substantial advantage. There is a significant advantage in allowing simultaneous measurements of the analyte and internal standard signals. The vital importance of the use of internal standards was discussed in Section 5.3.1. When the measurement is made with a scanning instrument, it is impossible to have simultaneous measurement for any part of the spectrum except that which is superimposed upon the internal standard peak. The stability of scanning instruments is usually good enough to allow effective use of the internal standard, but a truly simultaneous measurement has evident advantages.

The problems associated with abscissa errors in Raman measurements were discussed in Section 5.3.2.

Because Raman is an emission method, the set up of Raman instruments requires optimization of the physical positions of the various optical components. To a considerable extent, the problems with multichannel machines are much the same as those with scanning, single-channel ones, since neither type of instrument can be set to a predetermined position with sufficient accuracy for quantitative measurements. A multichannel detector is clearly advantageous in routine operations, since it can be set one time and then left alone.

At the present time, most multichannel Raman measurements are made with intensified silicon diode array detectors which are offered by several manufacturers. They are furnished with coolers and can be arranged for gated operation. Achievable sensitivities are approximately comparable to those obtained with cooled photon-counting detectors.

The basic dynamic range of array detectors is around 10,000, about an order of magnitude smaller than that of a photon-counting system. However, the difference is mostly made up by providing flexible offset and gain adjustments. Gain can be set both in front of and behind the diodes. The intensifier gain provides control of the light that gets to the diodes and an analog amplifier is provided in front of the analog-to-digital (A/D) converter. Coupled with this is an analog summing function that can be used to provide offset to spread a small peak that rides on top of a larger signal over the range of the A/D converter.

The most important disadvantages of intensified array detectors are their relatively high cost (about $20,000, including controller) and their nonlinear ordinate response. We have examined one detector and found nonlinearities on the order of 1%. These appear to be properties, either of the detector or the intensifier, that remain constant over time. They can therefore be compensated for by calibrating the detector.

Charge coupled devices (CCD) are coming into general use. They appear to have a substantial advantage over diode arrays in that dark current is much smaller, making it possible to use extended integration times to measure small signals without intensification. With an intensified silicon diode array, operated at about −20°C, the maximum useful integration time is perhaps 2 min. Long integration times, obviously valuable in certain applications, are unattractive for chemical analysis. Generally it is more satisfactory to use an intensifier and shorten the analysis turnaround time.

5.4. FLUORESCENCE AND RAMAN MEASUREMENTS

The negative effect of fluorescence on Raman measurements has received considerable attention (59–71). One school of thought favors reducing fluorescence by chemical or physical manipulation of the sample—for example, by addition of quenchers, cleanup of the sample, or burnout of the fluoresence by preliminary exposure to high-intensity radiation. Such approaches are incompatible with the goal of rapid, routine analysis described earlier. A second class of approaches calls for various instrumental modifications, such as those required for time resolution of signals, to reduce the fluorescence contribution to the signal. The complexity of these approaches is again at odds with the requirements for a robust method for routine quantitative analysis.

While the effect of fluorescence is often described as "swamping" the Raman signal, measurements can be made under conditions that ensure a linear combination of the Raman signal and the background. The instrumental modification approaches referenced above, in fact, make this assumption. Although each of the instrumental modification approaches provides a measure of discrimination against the signal due to fluorescence, the fluorescence is not prevented from reaching the detector and the recovered Raman signal is contaminated with the noise due to the fluorescence. An approach that operates under the same conditions but is compatible with routine quantitative Raman spectroscopy has been described in Mann and Vickers (72). This approach takes advantage of the fact that fluorescence peaks are reliably broader than Raman peaks and can therefore be discriminated against in the Fourier domain after the measurement has been made. No additional instrumentation other than the now ubiquitous small computer for data treatment is needed for this approach.

Differences in bandwidth between the Raman and fluoresence signals are sufficiently great that quite good results can be obtained by assuming a linear baseline over the limited range of a single Raman peak and discarding the DC level in the quantification process. This amounts to making the assumption that background contributions to signal power occur only at DC, which is not literally correct, but the procedure is effective because a large fraction of the power contributed by the background is at DC. Over a useful range of sample concentrations, it is possible to arrange that the contributions to total signal power in the first through fourth spatial frequencies from the analyte are enough larger than the corresponding background components to allow the latter to be ignored.

Figures 5.10 and 5.11 (72) illustrate the approach. Figure 5.10 shows a set of uncorrected Raman spectra, all plotted on the same scale, for a concentration series of an analyte that exhibits considerable fluorescence. These spectra show an offset due primarily to the analyte fluorescence. The Raman features apparent for the higher concentration samples are obviously much sharper than the fluorescence. The data set has been truncated in the data domain to retain only the section of the spectrum that contains the strongest target features. The linear background approximation is applied to this set. A straight line is projected between the endpoints, and all values are adjusted to fix the projected line at zero. The result is shown in Fig. 5.11, in which

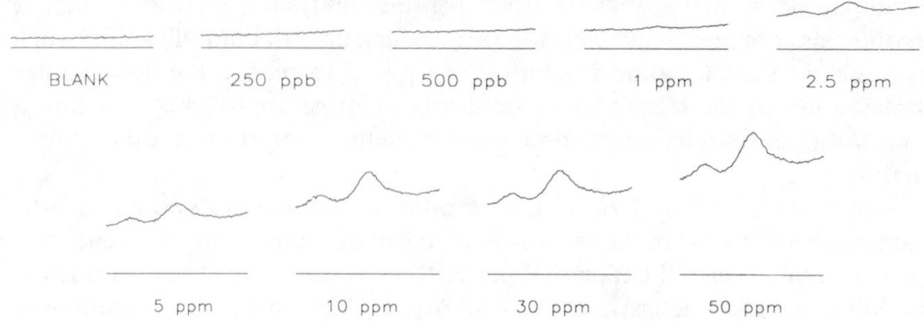

Figure 5.10. Uncorrected Raman spectra for various concentrations of an analyte exhibiting considerable fluorescence. All spectra are to the same scale. [From Mann and Vickers (72).]

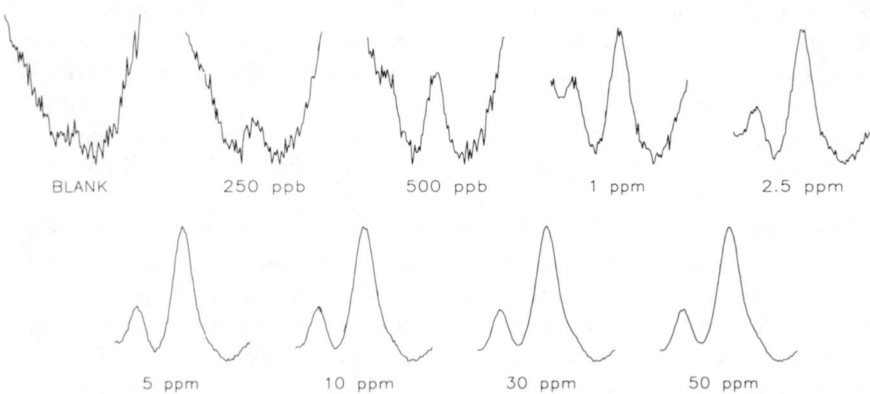

Figure 5.11. Demonstration of effect on fluorescence contribution to spectra by straight-line projection described in text. Spectra are scaled to produce features of the same size. [From Mann and Vicker (72).]

all the spectra have been scaled to produce the same size curves to show details. Quantification is then carried out by any technique that does not require retention of the DC level. For the data shown, this was done by cross correlation, with the DC level omitted from the calculation and the 50-ppm set as reference. Expressing response as a ratio to that of the 50-ppm autocorrelation, the analytical curve has a slope of 1.9988×10^{-2}, intercept of 1.012×10^{-3}, correlation coefficient of 0.999991, and limit of detection of 95 ppb. Considering the procedure followed, the slope should be 0.02 and the intercept zero. The limit of detection is taken as the concentration producing a response equal to three times the standard deviation of the intercept of the analytical curve.

It is apparent in Fig. 5.11 that for concentrations of 1 ppm and less the quantified responses will be systematically low because of the background curvature. From simulations (72) it can shown that the error for the lowest concentration sample, 0.25 ppm, is about 25%. The simulation process provides a means of correcting for the systematic error, if that is desirable.

5.5. COMPARISON OF RAMAN AND IR MEASUREMENTS FOR QUANTITATIVE ANALYSIS

Infrared spectroscopy is widely used for quantitative analysis. Although the use of Raman spectroscopy for this purpose is increasing, as evidenced by the citations in this report, its use is not commonplace. As noted above, this difference is partly a consequence of historical development. Implementation of double-beam design in the 1950s made it possible to use then-available analog-signal-handling techniques to produce IR spectra from which a range of artifacts, such as wavelength-dependent source and detector responses and solvent absorbance, were removed before the spectrum was presented to the user. Instrument users became conditioned to expect this level of performance, and the designers of contemporary instruments now meet this expectation by incorporating dedicated computers into instrument design.

By contrast, the operations that are required to produce corrected emission-mode spectra could not be carried out by analog devices, and users of Raman equipment accordingly have been accustomed to working with uncorrected spectra. Although modern Raman instruments are furnished with dedicated computers that operate the spectrometer and manipulate the data, the design tradition is largely unchanged. The user is responsible for making corrections, a situation that meets the needs of spectroscopy specialists but is unsatisfactory for those who wish to do routine analysis.

This tradition apparently still influences the direction of research on Raman instrumentation. Emphasis is placed on attempts to produce raw signals that do not contain objectionable artifacts, rather than to remove or compensate for them, as is done in absorption spectroscopy. While this approach is obviously desirable in principle, the instruments that have so far been manufactured are not well suited for chemical analysis. The various aforementioned instrumental approaches enabling one to discriminate against fluorescence in Raman measurements and the recently developed Fourier transform (FT) Raman spectrometer are striking examples (73–76). By shifting to near-IR excitation, the FT-Raman instrument produces signals in which fluorescence has been reduced but at the expense of Raman intensity that is greatly reduced. For certain kinds of studies this is a desirable trade-off, but for chemical analysis it is almost certainly not.

Given the lead in instrumental development for IR measurements, there is no incentive for development of convenient Raman instrumentation unless the Raman approach is superior in some way. There are several respects in which the Raman approach is superior for routine analysis.

5.5.1. Solvent Behavior

Because Raman measurements can be made in the visible and UV ranges, most of the common solvents are well suited to Raman measurements, and water is perhaps the best. By contrast, virtually all solvents absorb strongly in the IR range and thus are not well suited for IR measurements. This difficulty has been minimized by the development of very thin cells and high-performance electronics that can utilize the small signals produced. However, even with modern equipment, IR measurements in aqueous solutions are sharply circumscribed. This point is illustrated in the recent application of an attenuated total reflection (ATR) attachment in an FT-IR instrument to determine acetone, ethanol, and methanol in a model fermentation broth (77). In Raman measurements of the same system the strongest analyte features, blocked by solvent absorption in the IR measurements, can be used to give better quantitative sensitivity (30).

5.5.2. Emission Versus Absorption

When comparisons are made between methods of chemical analysis, emphasis is sometimes given to sensitivity, defined as the ability to discriminate between signals produced by low-concentration samples and those produced by unresponsive blank samples. An emission technique offers a fundamental advantage in this comparison because it involves detection of a small signal superimposed upon dark noise. The comparable absorbance measurement requires discrimination between two large signals. However, for practical analyses, a more relevant criterion is the ability to use signals produced by low-concentration-sample components superimposed on those produced by high-concentration components. In this context, absorbance measurements are at a considerable disadvantage, and emission measurements enjoy a fundamental advantage because of the wider linear dynamic range that can be obtained.

Both the lower and upper limits in an absorbance measurement are circumscribed by fundamental factors. The lower limit is set by the ability to discriminate between two large signals, and the upper by cutoff of the transmitted radiation. With dispersive spectrometers, response nonlinearity at the high absorbance end is attributable to stray light in the monochromator. The best commercial systems for the mid-IR range exhibit about 0.2% stray light, which implies more than 1% deviation from linearity for absorbances in excess of 1.1. For FT-IR systems stray light is not a problem, but there are other factors that contribute to system nonlinearity at the high absorbance end. One manufacturer expresses this as 0.02% equivalent stray light, which implies an upper limit of 2.5 absorbance for 1% deviation from linearity. Ordinate repeatability is reported by various manufacturers to be 0.1% transmittance, implying an absorbance uncertainty of about 4.3×10^{-4} absorbance units at the low absorbance end of the scale. With the usual criterion that the limit of detection is reached when the signal is three times the uncertainty, the limiting detectable absorbance is about 0.0013 absorbance units.

The lower limit in an emission measurement is reached when the signal becomes so weak that it cannot be distinguished from the dark noise. With currently available multichannel systems the limiting detectable signal is less than 10 counts. However, this point is rarely relevant in Raman measurements on real samples. When mixtures are examined, various components contribute to the observed signal. This process leads to increasing signal intensity in an emission measurement, but signal cutoff in an absorbance measurement. In emission measurements there are, in practice, two limits on the size of a small component that can be detected when it is superimposed upon a larger one. One occurs when the noise associated with the larger exceeds one-third the size of the smaller. The other is imposed by the dynamic range of the measuring system.

In the visible and UV ranges, detector noise power is proportional to the square root of the signal. On this basis, if the smaller component is to be detected, the major component cannot exceed the square of one-third the size of the smaller. However, the noise is fairly uniformly distributed over all spatial frequencies while the information is not. In a typical Raman measurement of 512 data points, approximately 90% of the useful signal information occurs in four spatial frequencies. Therefore, after bandwidth restriction the effective noise will be proportional to the square root of the major component signal divided by 64. When the major component signal is on the order of 10,000 counts, approximately the point at which diode saturation becomes evident in a diode array multichannel detector, the limiting detectable signal for the smaller component is still less than 40 counts.

5.5.3. Experimental Flexibility

The advantage of emission measurements is especially apparent when one is dealing with solid samples. The same experimental conditions can be used for both solution and solid samples. Spectroscopic features do reflect changes in molecular environment in going from liquid to solid state, but the quality

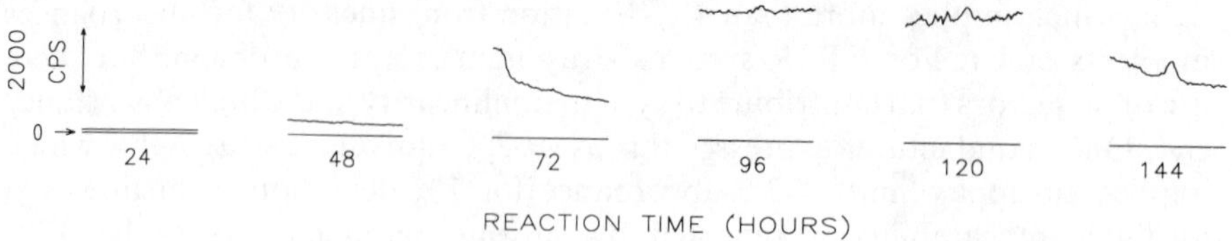

Figure 5.12. Raman measurements during the course of a fermentation showing growth of ethanol peak and changes in background features. [From Shope et al. (30).]

of the spectra obtained from solids or slurries is just as good as from liquids. Absorption measurements on solid samples are, by contrast, much more troublesome. Reflectance is often used for IR measurements, but IR reflectance spectra, unlike Raman spectra, are affected by sample properties such as particle size.

The flexibility of the Raman-scattering experiment was illustrated in monitoring ethanol formation during a fermentation (30). The fermentation mixture, including suspended solids, was pumped through a short segment of glass tubing for measurement. During the course of the fermentation, various components, some of which fluoresced and some of which were opaque, were formed and consumed, producing a widely varying baseline, as illustrated in Fig. 5.12. These variations did not interfere with measurement of either the ethanol signal or the internal standard water peak.

5.5.4. Experimental Comparison of Sensitivity

There appears to be a widely held view that, all other considerations aside, Raman signals are too weak to provide practical levels of sensitivity for routine analysis. To provide an objective basis for comparison of the analytical capabilities of Raman and IR measurements, a direct comparison was made. Sulfamethoxazole and sulfanilamide in acetonitrile were taken as the test materials. Sulfamethoxazole and sulfanilamide both show strong IR bands, which occur in a favorable position in the absorbance spectrum of acetonitrile, and large Raman scattering cross sections. IR measurements were made with a research-grade instrument (Perkin–Elmer 983); Raman measurements were made with a laboratory-constructed system employing an intensified diode array detector. Comparable spectral regions were obtained using each instrument, with total measurement time kept constant. Data processing was the same for each method.

Table 5.3 summarizes results for two runs by each method for sulfamethoxazole in acetonitrile. In each case five standards covered the range from 0.2 to 1.5 wt.% and a 5% standard was employed as the reference.

Table 5.3. Comparison of Raman and IR Measurements on Sulfamethoxazole in Acetonitrile

	Raman		IR	
	Run 1	Run 2	Run 1	Run 2
Slope	0.200	0.198	0.240	0.236
Intercept	−0.0006	−0.0007	−0.0007	0.0017
Correlation coefficient	0.99977	0.99999	0.99973	0.99994
LOD[a]	0.034	0.008	0.037	0.018

[a] LOD = limit of detection, defined as three times the standard deviation of the intercept divided by the slope.

It is evident that the IR and unenhanced Raman methods provide essentially equivalent results in terms of limit of detection. With the procedure used for quantitation, the expected slope and intercept are 0.2 and 0. The higher slope for the IR measurements reflects the fact that the 5% reference falls beyond the linear response range of the IR instrument.

A somewhat different view of the relative sensitivity of IR and Raman approaches was provided by measuring the response for sulfanilamide in acetonitrile with various levels of sulfamethoxazole as a concomitant. Sulfanilamide and sulfamethoxazole show overlapping strong bands due to the S—O stretch. Thus the sulfamethoxazole provides a background signal component on which the sulfanilamide signal must be measured, but (as noted earlier) the consequences of this are quite different for absorption and emission modes of measurement. Figure 5.13 shows the results, plotting response versus sulfanilamide concentration for five levels of sulfamethoxazole concentration ranging from 0 to 3%.

For the Raman measurements shown in Fig. 5.13(a), it is apparent that the results are independent of sulfamethoxazole concentration; all the points fall on essentially the same line with the expected slope and intercept. The same is not true for the IR measurements shown in Fig. 5.13(b). Even with no added sulfamethoxazole, nonlinearity in the response is evident for the highest sulfanilamide concentration. The addition of sulfamethoxazole increases total sample absorbance at the measurement position. Owing to the limited linear range of the absorbance measurement, as total absorbance increases, the signal observed for the sulfanilamide decreases, producing the observed downward trend in position and slope of the sulfanilamide analytical curves. For routine analysis the situation modeled by the sulfanilamide-plus-sulfamethoxazole samples is a common one, and the Raman approach offers a clear advantage.

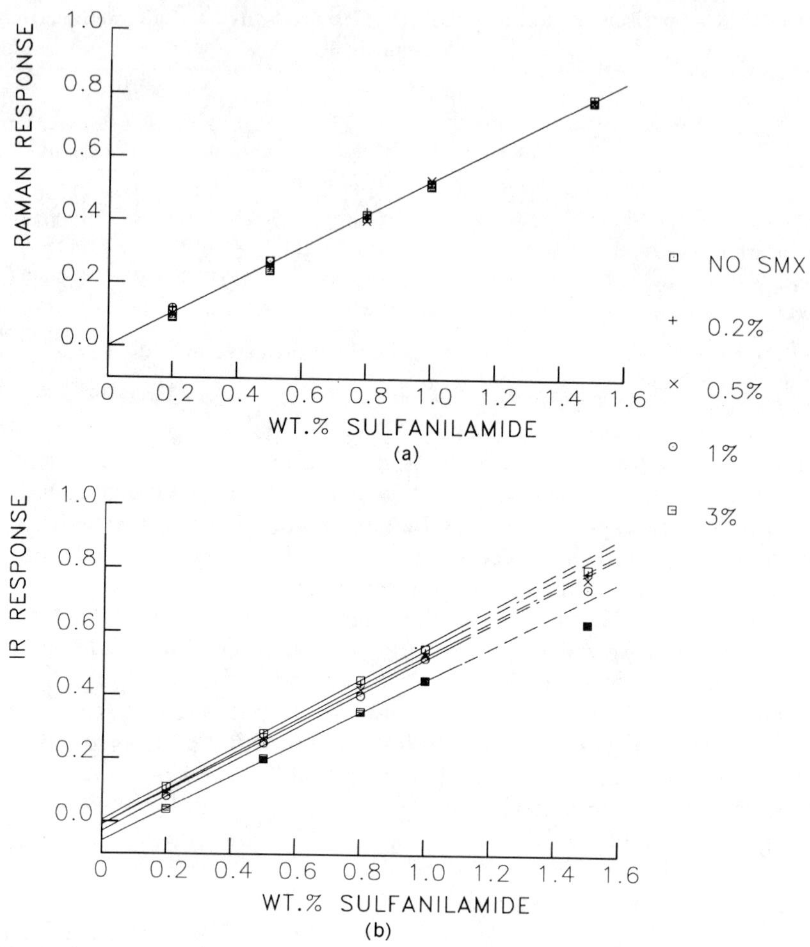

Figure 5.13. Analytical curves for sulfanilamide in acetonitrile with various amounts of sulfamethaxazole as concomitant. (a) Raman measurement. (b) IR measurement.

ACKNOWLEDGMENTS

We acknowledge with gratitude the support and encouragement of our work on quantitative Raman spectroscopy by Mort Smutz and John Tokar of the National Oceanic and Atmospheric Administration and Warren McAllister of the Burroughs Wellcome Company, and the contributions of the students who have participated in this work: Nancy Marley, Yong-Chien Ling, Tom King, Liang-Tsair Lin, David Womack, Tom Shope, Mei-Fang Shyu, Chan Kong Chong, and Jianxiong Zhu.

REFERENCES

1. D. E. Irish and H. Chen, *Appl. Spectrosc.* **25**, 1–6 (1971).
2. J. G. Grasselli, M. K. Snaveley, and B. J. Bulkin, *Chemical Applications of Raman Spectroscopy*, pp. 24–28. Wiley (Interscience), New York, (1981).
3. E. B. Bradley and C. A. Frenzel, *Water Res.* **4**, 125–128 (1970).
4. J. M. Reeves, E. B. Bradley, and C. A. Frenzel, *Water Res.* **7**, 1417–1429 (1973).
5. G. Braunlich, G. Gamer, and M. S. Petty, *Water Res.* **7**, 1643–1647 (1973).
6. S. F. Baldwin and C. W. Brown, *Water Res.* **6**, 1601–1604 (1972).
7. M. Ahmadjian and C. W. Brown, *Environ. Sci. Technol.* **7**, 452 (1973).
8. K. M. Cunningham, M. C. Goldberg, and E. R. Weiner, *Anal. Chem.* **49**, 70–75 (1977).
9. A. G. Miller, *Anal. Chem.* **49**, 2044–2048 (1977).
10. N. Furuya, A. Matsuyuki, S. Higuchi, and S. Tanaka, *Water Res.* **13**, 371–374 (1979).
11. R. J. Thibeau, L. Van Haverbeke, and C. W. Brown, *Appl. Spectrosc.* **32**, 98–100 (1978).
12. L. Van Haverbeke, P. F. Lynch, and C. W. Brown, *Anal. Chem.* **50**, 315–317 (1978).
13. L. Van Haverbeke and M. A. Herman, *Anal. Chem.* **51**, 932–936 (1979).
14. L. Van Haverbeke and M. A. Herman, *Pergamon Ser. Environ. Sci.* **7**, 127–131 (1982).
15. N. Furuya, A. Matsuyuki, S. Higuchi, and S. Tanaka, *Water Res.* **14**, 747–752 (1980).
16. M. D. Morris, *Anal. Chem.* **47**, 2453–2454 (1975).
17. M. D. Morris, *Anal. Lett.* **9**, 469–478 (1976).
18. L. Van Haverbeke, J. F. Janssens, and M. A. Herman, *Int. J. Environ. Anal. Chem.* **10**, 205–215 (1981).
19. M. D. Morris and D. J. Wallan, *Anal. Chem.* **51**, 182A–192A (1979).
20. N. A. Marley, C. K. Mann, and T. J. Vickers, *Appl. Spectrosc.* **38**, 540–543 (1984).
21. N. A. Marley, C. K. Mann, and T. J. Vickers, *Appl. Spectrosc.* **39**, 628–633 (1985).
22. Y.-C. Ling, T. J. Vickers, and C. K. Mann, *Appl. Spectrosc.* **39**, 463–470 (1985).
23. S. Sato, S. Higuchi, and S. Tanaka, *Appl. Spectrosc.* **39**, 822–827 (1985).
24. S. Sato, S. Higuchi, and S. Tanaka, *Anal. Chim. Acta* **120**, 209–215 (1980).
25. T. H. King, C. K. Mann, and T. J. Vickers, *J. Pharm. Sci.* **74**, 443–447 (1985).
26. C. K. Mann, T. J. Vickers, and J. D. Womack, *Appl. Spectrosc.* **41**, 1324–1329 (1987).
27. L. Tosi and A. Garnier-Suillerot, *J. Chem. Soc., Dalton Trans.* pp. 103–108 (1982).
28. K. Ikeda, S. Higuchi, and S. Tanaka, *Bunseki Kagaku* **30**, 701–705 (1981); *Chem. Abstr.* **96**, 45625 (1982).

29. J. D. Womack, T. J. Vickers, and C. K. Mann, *Appl. Spectrosc.* **41**, 117–119 (1987).
30. T. B. Shope, T. J. Vickers, and C. K. Mann, *Appl. Spectrosc.* **41**, 908–912 (1987).
31. B. Wopenka and J. D. Pasteris, *Appl. Spectrosc.* **40**, 144–151 (1986).
32. L. P. Powell and A. Campion, *Anal. Chem.* **58**, 2350–2352 (1986).
33. S. A. Asher, *Anal. Chem.* **56**, 720–724 (1984).
34. S. A. Asher, C. R. Johnson, and J. Murtaugh, *Rev. Sci. Instrum.* **54**, 1657–1662 (1983).
35. C. R. Johnson and S. A. Asher, *Anal. Chem.* **56**, 2258–2261 (1984).
36. M. Ludwig and S. A. Asher, *Appl. Spectrosc.* **41**, 331–332 (1987).
37. L.-T. Lin, C. K. Mann, and T. J. Vickers, *Appl. Spectrosc.* **41**, 422–426 (1987).
38. C. M. Jones, V. L. De Vito, P. A. Harmon, and S. A. Asher, *Appl. Spectrosc.* **41**, 1268–1275 (1987).
39. R. Rumelfanger, S. A. Asher, and M. B. Perry, *Appl. Spectrosc.* **42**, 267–272 (1988).
40. C. D. Tran, *Anal. Chem.* **56**, 824–826 (1984).
41. T. Vo-Dinh, M. Y. K. Hiromoto, G. M. Begun, and R. L. Moody, *Anal. Chem.* **56**, 1667–1670 (1984).
42. T. M. Meier, A. Wokaun, and T. Vo-Dinh, *J. Phys. Chem.* **89**, 1843–1846 (1985).
43. P. D. Enlow, M. Buncick, R. J. Warmack, and T. Vo-Dinh, *Anal. Chem.* **58**, 1119–1123 (1986).
44. C. Jennings, R. Aroca, A.-M. Hor, and R. O. Loutfy, *Anal. Chem.* **56**, 2033–2035 (1984).
45. E. Gautner, D. Steinert, and J. Reinhardt, *Anal. Chem.* **57**, 1658–1662 (1985).
46. R.-S. Sheng, L. Zhu, and M. D. Morris, *Anal. Chem.* **58**, 1116–1119 (1986).
47. E. L. Torres and J. D. Winefordner, *Anal. Chem.* **59**, 1626–1632 (1987).
48. W. Rauch and H. Bettermann, *Appl. Spectrosc.* **42**, 520–521 (1988).
49. W. Kiefer, *Adv. Infrared Raman Spectrosc.* **3**, 18–20 (1977).
50. I. Tsukamoto, H. Nagai, and K. Machida, *J. Raman Spectrosc.* **17**, 313 (1986).
51. D. F. Shriver and J. B. R. Dunn, *Appl. Spectrosc.* **28**, 319–323 (1974).
52. T. C. Strekas, D. H. Adams, A. Packer, and T. G. Spiro, *Appl. Spectrosc.* **28**, 324–327 (1974).

52a. J. D. Womack, C. K. Mann, and T. J. Vickers, *Appl. Spectrosc.* **43**, 527 (1989).

53. C. K. Mann and T. J. Vickers, *Appl. Spectrosc.* **41**, 317–319 (1987).
54. E. O. Brigham, *The Fast Fourier Transform*, p. 64. Prentice-Hall, Englewood Cliffs, New Jersey, 1974.
55. C. K. Mann, J. R. Goleniewski, and C. A. Sismanidis, *Appl. Spectrosc.* **36**, 223–227 (1982).
56. L. L. Tyson, Y.-C. Ling, and C. K. Mann, *Appl. Spectrosc.* **38**, 663–667 (1984).
57. D. C. Haaland and R. G. Easterling, *Appl. Spectrosc.* **34**, 539 (1980).

58. G. Strang, *Linear Algebra and Its Applications*, pp. 112–115. Academic Press, New York, 1980.
59. R. P. Van Duyne, D. L. Jeanmaire, and D. F. Shriver, *Anal. Chem.* **46**, 213–222 (1974).
60. F. L. Galeener, *Chem. Phys. Lett.* **48**, 7 (1977).
61. A. Lowenschuss and A. Moss, *Appl. Spectrosc.* **36**, 183–184 (1982).
62. P. P. Yaney, *J. Opt. Soc. Am.* **62**, 1297 (1972).
63. R. J. Nemanish, S. A. Solin, and J. Doehler, *Rev. Sci. Instrum.* **47**, 741 (1976).
64. T. L. Gustafson and F. E. Lytle, *Anal. Chem.* **54**, 634–637 (1982).
65. S. M. Angel, M. K. DeArmand, K. W. Hanck, and D. W. Wertz, *Anal. Chem.* **56**, 3000–3001 (1984).
66. A. Z. Genack, *Anal. Chem.* **56**, 2957–2960 (1984).
67. J. N. Demas and R. A. Keller, *Anal. Chem.* **57**, 538–545 (1985).
68. F. V. Bright and G. M. Hieftje, *Appl. Spectrosc.* **40**, 583–587 (1986).
69. S. Higuchi, E. J. Yu, and S. Tanaka, *Appl. Spectrosc.* **41**, 413–416 (1987).
70. L. A. Spino, D. W. Armstrong, A. M. Alak, and T. Vo-Dinh, *Appl. Spectrosc.* **41**, 771–773 (1987).
71. K. Kamogawa, T. Fujii, and T. Kitagawa, *Appl. Spectrosc.* **42**, 248–255 (1988).
72. C. K. Mann and T. J. Vickers, *Appl. Spectrosc.* **41**, 427–430 (1987).
73. T. Hirschfeld and B. Chase, *Appl. Spectrosc.* **40**, 133–137 (1986).
74. D. J. Moffatt, H. Buijs, and W. F. Murphy, *Appl. Spectrosc.* **40**, 1079–1081 (1986).
75. C. G. Zimba, V. M. Hallmark, J. D. Swalen, and J. F. Rabolt, *Appl. Spectrosc.* **41**, 721–726 (1987).
76. B. Chase, *Anal. Chem.* **59**, 881A–889A. (1987).
77. D. Kuehl and R. Crocombe, *Appl. Spectrosc.* **38**, 907 (1984).

CHAPTER

6

CHARACTERIZATION OF SEMICONDUCTORS BY RAMAN SPECTROSCOPY

FRED H. POLLAK

Department of Physics
Brooklyn College of the City University of New York
Brooklyn, New York

6.1 INTRODUCTION

Rapid advances in semiconductor technology, including thin-film deposition, interface preparation, structure fabrication, as well as microelectronic processing, have increased the need for characterization techniques that can provide greater information. Experimental probes must provide insight into properties on a smaller dimensional scale with greater sensitivity and at higher precision than ever before. Spectroscopic techniques offer powerful tools for probing relevant materials and device parameters.

Raman scattering is an extremely useful method for the characterization of semiconductors, either in bulk, thin-film or device form, as well as semiconductor interfaces such as semiconductor/vacuum, Schottky barriers, metal-insulator-semiconductor (MIS), heterojunctions, etc. Raman scattering (RS) is a function of the electron–phonon (lattice vibration) interaction and therefore is an important complement to other spectroscopic techniques [scanning ellipsometry, reflectivity and modulated reflectivity, luminescence, X-ray photoelectron spectroscopy (XPS), Auger electron spectroscopy (AES), secondary ion mass spectroscopy (SIMS), transmission electron microscopy (TEM), etc.], many of which are functions of the electronic states of the solid. This spectroscopic technique is probably the most useful method for the study of lattice vibrations and their interactions with other excitations such as plasmons, etc. Since lattice vibrations are very sensitive to local environments, RS can give information about material (or device) structure and/or quality

Analytical Raman Spectroscopy, Edited by Jeanette G. Grasselli and Bernard J. Bulkin. Chemical Analysis Series, Vol. 114.
ISBN 0-471-51955-3

on the scale of a few lattice constants. For example, this spectroscopic method can be used to study the microscopic nature of structural and/or compositional disorder, clustering, etc. Also, since Raman scattering is a second-order optical process, it contains important symmetry information not available from first-order interactions. For example, in cubic materials such as diamond- and zincblende-type semiconductors first-order optical processes are not sensitive to the polarization of the radiation (relative to the crystal axes), whereas in RS the scattering intensity for a given phonon mode is a function of both the incident and scattered light polarizations. This dependence leads to symmetry information that can be utilized to characterize many materials parameters. In RS, peak positions as well as polarization selection rules are sensitive to perturbations, both external and internal, such as electric fields, strain, or temperature. Furthermore, by using various excitation lines with different wavelengths (and hence penetration depths) it is possible to (a) perform nondestructive depth-profiling measurements and (b) excite resonance interactions.

In this paper we review the general background of RS including polarization selection rules (and the influence of symmetry breaking on these selection rules). Instrumentation is discussed, including some recent developments such as optical multichannel analyzers (OMAs) and the Raman microprobe. In Section 6.4 the use of RS to study elemental semiconductors such as Si is reviewed, including amorphous and microcrystalline phases, interactions of various metals such as Pt, Pd, Ti, and W with Si, laser-annealed material, determination of strains at the interface of Si on various substrates, etc. Section 6.5 is devoted to binary materials, particularly GaAs. Among the areas covered are carrier-concentration determination (n-type material, with $N \geqslant 10^{17}\,\mathrm{cm}^{-3}$), crystal quality, ion-damaged and laser-annealed effects, process-induced damage, and various interfaces such as semiconductor/vacuum, Schottky barrier, or MIS. Various aspects of alloy semiconductors and their structure are discussed in the following section. Topics covered include determination of alloy composition, deviations from crystallinity in HgCdTe, microscopic alloy potential fluctuations, single heterostructures, etc. In the final section we review the use of RS to study the nature of surface films and oxides on compound semiconductors.

6.2. GENERAL BACKGROUND OF RAMAN SCATTERING, PARTICULARLY RELATING TO CRYSTALLINE SOLIDS

6.2.1. Polarizability Theory of Raman Scattering

In RS the incident electromagnetic field couples with the phonon (lattice vibration) field through the induced dipole moment (1–4). This is accomplished by the variation of the polarization tensor χ_{lm}. In the section below we shall discuss in some detail RS in terms of a semiclassical theory (1–4). The quantum mechanical results are given in Hayes and Loudon (1).

The polarizability tensor $\tilde{\chi}$ is defined by

$$\mathbf{P} = \tilde{\chi}\mathbf{E}, \tag{1}$$

where $\mathbf{P}$ is the dielectric polarization (dipole moment per unit volume) and $\mathbf{E}$ is the electric field. Of course, we also have the relations

$$\mathbf{D} = \mathbf{E} + 4\pi\mathbf{P}, \tag{2a}$$

$$\mathbf{D} = \mathbf{E} + 4\pi\tilde{\chi}\mathbf{E}, \tag{2b}$$

and

$$\mathbf{D} = \tilde{\varepsilon}\mathbf{E}, \tag{2c}$$

where $\mathbf{D}$ is the displacement field and $\tilde{\varepsilon}$ is the dielectric constant tensor. From Eqs. (2) it is clear that

$$\tilde{\varepsilon} = 1 + 4\pi\tilde{\chi}, \tag{3}$$

so that the dielectric constant and polarization are intimately related.

The polarization tensor can be expanded as

$$\chi_{lm} = \chi_{lm}^{(0)} + \chi_{lm}^{(1)}u + \chi_{lm}^{(2)}u^2 + \cdots, \tag{4}$$

where u is a nuclear displacement during a normal vibration. Thus, for the jth normal mode

$$u_j = u_{j0} \exp(i\omega_j t), \tag{5}$$

where ω_j is the frequency of the jth normal mode and

$$\chi_{lm}^{(1)} = \chi_{lm,j}^{(1)} = \left(\frac{\partial \chi_{lm}}{\partial u_j}\right)_{u_j=0}. \tag{6}$$

Since

$$\mathbf{E} = \mathbf{E}_0 \exp(i\omega_i t) \tag{7}$$

denotes the electric field of frequency ω_i we can write from Eqs. (1), (4), (5), and (7)

$$\mathbf{P} = \chi^{(0)} E_0 \exp(i\omega_i t) + \chi^{(1)} u_{j0} E_0 \exp[i(\omega_i \pm \omega_j)t] + \chi^{(2)} u_{j0}^2 E_0 \exp[i(\omega_i \pm 2\omega_j)t] + \cdots. \tag{8}$$

The first term of Eq. (8) gives rise to Rayleigh scattering in which the frequency of the radiation remains unchanged during the scattering process (i.e., elastic scattering). The second term, involving the derivative of the polarizability, constitutes *first-order* RS, which is an inelastic scattering process. The incident photon has (ω_i and $\mathbf{k}_i$), where ω_i and $\mathbf{k}_i$ are the frequency and wavevector of the photon, respectively, and the system makes a transition from an initial state (n, v), where n and v represent the electronic and vibrational quantum numbers, to an intermediate state (n', v'). The system eventually decays to the final state $(n, v \pm 1)$. In the process a phonon $(\omega_j, \mathbf{q}_j)$ is created or annihilated and a photon of different frequency and wavevector (ω_s, $\mathbf{k}_s$) is emitted so that energy and momentum are conserved:

$$\hbar\omega_s = \hbar\omega_i \pm \hbar\omega_j \tag{9a}$$

$$\hbar\mathbf{k}_s = \hbar\mathbf{k}_i \pm \hbar\mathbf{q}_j. \tag{9b}$$

In Eqs. (9) plus the (+) sign denotes anti-Stokes scattering while the minus (−) sign denotes Stokes scattering.

The frequency of the exciting radiation is usually in a range where here is little dispersion, i.e., in a region where $\omega_i \gg \omega_j$. Thus the relative difference between ω_i and ω_s is small. Therefore, to a good approximation $|\mathbf{k}_s| = |\mathbf{k}_i| = k$. The conservation of momentum thus reduces to the Bragg condition

$$q_j = 2k \sin\tfrac{1}{2}\phi = [2\omega n(\omega)/c] \sin\tfrac{1}{2}\phi, \tag{10}$$

where ϕ is the angle between the incident and scattered directions, and $n(\omega)$ is the index of refraction of the medium at ω. Thus the wavevector of the phonon involved in the scattering process depends on the scattering angle. For visible (or near-visible) light, which is usually employed for exciting the Raman effect, k is between 10^4 and 10^5 cm^{-1}, which is about 10^{-4} to 10^{-3} of q_{max} for phonons [edge of the Brillouin zone (BZ)]. Thus, in a perfect crystal in first-order RS only phonons from near the center of the BZ are involved:

$$q_j \approx 0. \tag{11}$$

This is a very important "selection" rule with considerable implicati will be discussed later.

In the cases where the incident and scattered waves are damped inside the scattering volume, $\mathbf{k}_i$ and $\mathbf{k}_s$ are complex. This situation occurs when the energy of the incident light is near an interband electronic transition and hence there is strong absorption. The inelastic scattering is then due to excitations having a range of wavevector

$$\Delta q = |\operatorname{Im} \mathbf{k}_i| + |\operatorname{Im} \mathbf{k}_s| \tag{12}$$

about $q = \operatorname{Re}(\mathbf{k}_i - \mathbf{k}_s)$, that is, there is a *relaxation of the* $q \approx 0$ *selection rule* stated above in Eq. (11).

The differential scattering cross section for RS, i.e., the scattering energy per unit time and solid angle $d\Omega$, is given by

$$\frac{\mathrm{d}\sigma}{\mathrm{d}\Omega} = V^2\left(\frac{\omega_s}{c}\right)^4 |\hat{e}_i \cdot \chi_{lm}^{(1)} \cdot \hat{e}_s|^2 \begin{cases} n_j & \text{Stokes} \\ (n_j + 1) & \text{Anti-Stokes} \end{cases} \tag{13}$$

where $\hat{e}_i$ and $\hat{e}_s$ are the unit polarization vectors of the incident and scattered light, respectively, and V is the scattering volume. The symmetry property of $\chi_{lm}^{(1)}$ is called the Raman tensor. The quantity n_j is the phonon occupation number

$$n_j = [\exp(\hbar\omega_j/kT) - 1]^{-1}. \tag{14}$$

6.2.2. Second-Order Raman Scattering

The second-order Raman effect is a scattering process in which two phonons participate, as shown in the third term on the right-hand side of Eq. (8) (1–4). They may both be created (giving a Stokes component in the scattered light), or one may be created and the other destroyed (giving a Stokes and anti-Stokes component), or both may be destroyed (giving an anti-Stokes component).

If the two created phonons have branch indices σ and σ' and their wavevectors are q and q', then energy and momentum conservation give

$$\hbar\omega_{\sigma q} + \hbar\omega_{\sigma' q'} = \hbar\omega_i - \hbar\omega_s \tag{15}$$

and

$$\hbar\mathbf{q}_\sigma + \hbar\mathbf{q}'_{\sigma'} = \hbar\mathbf{k}_i - \hbar\mathbf{k}_s. \tag{16}$$

The branch indices (σ) and wavevectors can take all values consistent with these restrictions. Thus the wavevectors need not be small, as in first-order scattering, but can take on values throughout the BZ. The phonon wavevectors are typically three orders of magnitude larger than the light wavevectors over the major part of the zone, and Eq. (16) can be replaced by

$$\hbar \mathbf{q}_\sigma + \hbar \mathbf{q}'_{\sigma'} = 0 \tag{17}$$

to a very good approximation. The vibrational frequencies are the same at wavevectors $+q$ and $-q$, and the energy conservation condition can thus be written

$$\hbar\omega_{\sigma q} + \hbar\omega_{\sigma' q} = \hbar\omega_i - \hbar\omega_s. \tag{18}$$

6.2.3. Lattice Polarizability Including Free Carrier Effects

We now turn our attention to some of the properties of the lattice polarizability (3). In the long-wavelength limit, the optical vibrations of a diatomic lattice correspond to the motion of one type of atom, all in phase, relative to the other kind. In ionic crystals such motion is associated with strong electric moments. If $\mathbf{u}$ represents the displacement of the positive ions relative to the negative ions, a reduced displacement vector $\mathbf{w}$ may be expressed as $\mathbf{w} = \mathbf{u}(\mu/v_s)^{1/2}$, where μ/v_s is the reduced mass per Bravais unit cell. The macroscopic equations describing the polar motions are then

$$\ddot{\mathbf{w}} = b_{11}\mathbf{w} + b_{12}\mathbf{E} \tag{19}$$

and

$$\mathbf{P} = b_{21}\mathbf{w} + b_{22}\mathbf{E}, \tag{20}$$

where $\mathbf{E}$ is the electric field and $\mathbf{P}$ the dielectric polarization. The b's are constants characteristic of the solid—for example, b_{11} represents a force constant; b_{12} $(= b_{21})$, an effective ionic charge (a transverse charge); and b_{22}, a polarizability. The linearity of Eqs. (19) and (20) implies that anharmonicity and higher-order terms in the electric moment are neglected.

Considering the periodic solutions,

$$(\mathbf{w}, \mathbf{E}, \mathbf{P}) = (\mathbf{w}_0, \mathbf{E}_0, \mathbf{P}_0)e^{i\omega t} \tag{21}$$

Equations (19) and (20) are reduced to

$$-\omega^2\mathbf{w} = b_{11}\mathbf{w} + b_{12}\mathbf{E} \tag{22a}$$

and

$$\mathbf{P} = b_{12}\mathbf{w} + b_{22}\mathbf{E}. \tag{22b}$$

Elimination of $\mathbf{w}$ from Eqs. (22a) and (22b) yields

$$\mathbf{P} = \left(b_{22} + \frac{b_{12}b_{21}}{-b_{11} - \omega^2} \right)\mathbf{E}. \tag{23}$$

Substitution of Eq. (23) in Eqs. (1) and (3) readily gives the dielectric response in terms of the b coefficients as

$$\epsilon = 1 + 4\pi b_{22} + \frac{4\pi b_{12}b_{21}}{-b_{11} - \omega^2}. \tag{24}$$

The similarity of Eq. (24) to the infrared dispersion formula

$$\epsilon = \epsilon_\infty + \frac{(\epsilon_0 - \epsilon_\infty)\omega_0^2}{\omega_0^2 - \omega^2} \tag{25}$$

is obvious. In Eq. (25), ω_0 is the dispersion frequency, $\varepsilon_\infty = n^2$, where n is the refractive index for light of $\omega \gg \omega_0$ in a nondispersive region, and ε_0 is the static or low-frequency dielectric constant. The b coefficients may now be obtained by comparing Eq. (24) with Eq. (25). We obtain

$$b_{11} = -\omega_0^2, \tag{26a}$$

$$b_{12} = b_{21} = \omega_0[(\epsilon_0 - \epsilon_\infty)/4\pi]^{1/2}, \tag{26b}$$

and

$$b_{22} = (\epsilon_\infty - 1)/4\pi. \tag{26c}$$

Comparing Eqs. (19) and (20) with Eqs. (26a)–(26c), we see that except for multiplicative constants, b_{11} could be identified with a restoring force constant, b_{12} $(= b_{21})$ with the transverse effective charge on an ion, and b_{22} with the ionic polarizability. The reason for calling the ionic charge the transverse effective charge will become clear as we identify the dispersion frequency.

Since, macroscopically, the crystal is electrically neutral, we can apply Gauss's law

$$\nabla \cdot \mathbf{D} = \nabla \cdot (\mathbf{E} + 4\pi \mathbf{P}) = 0. \tag{27}$$

Also, we have

$$\nabla \times \mathbf{E} = 0. \tag{28}$$

In a cubic crystal $\tilde{\epsilon}$ is a not a tensor, i.e. **D**, **E**, and **P** are all parallel. Thus, for the spatial dependence of these quantities, we can write (5)

$$(\mathbf{D}, \mathbf{E}, \mathbf{P}) = (\mathbf{D}_0, \mathbf{E}_0, \mathbf{P}_0)e^{i\mathbf{k}\cdot\mathbf{r}}. \tag{29}$$

Thus, Eq. (27) reduces to $\mathbf{k}\cdot\mathbf{D}_0 = 0$, which requires that

$$\mathbf{D} = 0 \quad \text{or} \quad \mathbf{D}, \mathbf{E} \quad \text{and} \quad \mathbf{P} \perp \mathbf{k}, \tag{30a}$$

while Eq. (28) reduces to $\mathbf{k} \times \mathbf{E}_0 = 0$, which requires that

$$\mathbf{E} = 0 \quad \text{or} \quad \mathbf{D}, \mathbf{E} \quad \text{and} \quad \mathbf{P} \parallel \mathbf{k}. \tag{30b}$$

In a longitudinal optical mode (LO), **P** is parallel to **k** so that Eq. (30a) requires that $\mathbf{D} = 0$, which from Eq. (2a) leads to

$$\mathbf{E} = -4\pi\mathbf{P}, \qquad \epsilon \equiv 0 \quad (\text{LO}). \tag{31a}$$

On the other hand, in a transverse optical mode (TO), **P** is perpendicular to **k**, which is consistent with Eq. (30b) only if **E** vanishes. This is consistent with Eq. (2a) only if

$$\mathbf{E} = \mathrm{O}, \qquad \epsilon \equiv \infty \quad (\text{TO}). \tag{31b}$$

According to Eq. (25)

$$\epsilon \equiv \infty \quad \text{when} \quad \omega^2 = \omega_0^2, \tag{32}$$

and therefore Eq. (31b) identifies ω_0 as the frequency of the long-wavelength ($q \approx 0$) TO, which we denote as ω_{TO}. The frequency of the LO, ω_{LO}, is determined by $\epsilon = 0$ so that from Eq. (25) we have

$$\omega_{\text{LO}}^2 = (\epsilon_0/\epsilon_\infty)\omega_{\text{TO}}^2. \tag{33}$$

The above relation between ω_{LO} and ω_{TO} is called the Lydanne–Sachs–Teller relationship.

Equation (25) can be generalized to include damping effects, with damping constant Γ

$$\epsilon(\omega) = \epsilon_\infty + \frac{(\epsilon_0 - \epsilon_\infty)\omega_{\text{TO}}^2}{\omega_{\text{TO}}^2 - \omega^2 - i\omega\Gamma}. \tag{34}$$

In homopolar materials such as Si or Ge, in the absence of polar interactions, the atomic motions are determined only by the local elastic restoring forces; therefore, the second term of Eq. (22a) vanishes. But since macroscopically $\mathbf{E} \neq 0$, it follows that

$$b_{12} = 0 \quad \rightarrow \quad \epsilon_0 = \epsilon_\infty. \tag{35a}$$

Thus, for homopolar crystals, Eq. (33) yields

$$\omega_{\mathrm{TO}} = \omega_{\mathrm{LO}} = \omega_0. \tag{35b}$$

In order to account for free carrier effects in ionic materials we consider the motion of a "free carrier" in terms of its effective mass, m^*, and scattering time τ:

$$m^* \ddot{x} + (m^*/\tau)\dot{x} = eE_0 e^{i\omega t}. \tag{36}$$

If there are N "free" carriers per unit volume, then the polarizability, P_f, associated with the free electrons can be written as

$$P_f = Nex = \frac{(Ne^2/m^*)}{-\omega^2 + i\omega/\tau} E_0, \tag{37}$$

and hence the contribution of the free carriers to the dielectric constant is

$$\epsilon_f = -\frac{(4\pi N e^2/m^*)}{\omega(\omega - i/\tau)}. \tag{38}$$

The total dielectric constant can now be written as

$$\epsilon(\omega) = \epsilon_\infty + \frac{(\epsilon_0 - \epsilon_\infty)\omega_{\mathrm{TO}}^2}{\omega_{\mathrm{TO}}^2 - \omega^2 - i\omega\Gamma} - \frac{\omega_p^2 \epsilon_\infty}{\omega(\omega - i/\tau)} \tag{39}$$

where

$$\omega_p^2 = 4\pi N e^2/m^* \epsilon_\infty. \tag{40}$$

For a moment if we neglect Γ and τ in Eq. (39) and look for $\epsilon(\omega) = 0$, it can be shown that

$$\omega_\pm^2 = \tfrac{1}{2}\{(\omega_{\mathrm{LO}}^2 + \omega_p^2) \pm [(\omega_{\mathrm{LO}}^2 + \omega_p^2) - 4\omega_p^2\omega_{\mathrm{TO}}^2]^{1/2}\}. \tag{41}$$

These are "coupled LO–plasmon modes" of the system. We shall discuss their influences later, particularly in determining the free carrier concentration N in ionic-type semiconductors (e.g., zincblende-type materials).

6.2.4. Fano Effect

In heavily doped semiconductors, the large carrier concentration of electrons or holes produces an electronic, broad scattering background that overlaps with the one-phonon Raman lineshape. This overlap results in a Fano-type continuous/discrete interference (6). The resultant Raman lineshape, $I(\omega)$, can be expressed as

$$I(\omega) \propto |Q_F + \eta|^2/(1 + \eta^2), \tag{42}$$

where Q_F is the Fano asymmetry factor and η represents a reduced frequency

$$\eta = (\omega - \omega_0)/\Gamma_F \tag{43}$$

and Γ_F is the Fano broadening factor. In Si it has been shown that Q_F and Γ_F are functions of the carrier concentration and laser wavelength (6).

In group IV semiconductors there is no LO phonon–plasmon coupling since these materials are homopolar, i.e., have no electric dipole moment in the unit cell. The lack of this coupling is a consequence of the fact that b_{12} $(= b_{21}) = 0$ for these materials. Hence the LO phonon–plasmon coupling cannot be used to evaluate the carrier concentration in Si or Ge.

However, in heavily doped diamond-type semiconductors, information about carrier concentration may be obtained from the details of the $q \approx 0$ optic phonon lineshape, i.e., the Fano lineshape mentioned above. We shall return to this point later.

6.2.5. Polarization Selection Rules

The observation of light scattering by optical phonons is frequently done in a geometry where the linearly polarized incident light beam is sent along the z-axis with a given polarization and the scattered beam is received along the y-axis and analyzed along a specific direction (1, 4).

When the crystal is not transparent to the incident radiation (as is the case for most semiconductors), the observation is done in the backscattering geometry; that is, the incident light is along z and the scattered light is collected along $-z$ (denoted as $\bar{z}$). For example, if the incident light is along z, polarized along x, and the scattered light is along $\bar{x}$ and analyzed along y, the designation is $z(x, y)\bar{x}$.

6.2.5.1. Atomic Displacement Effects

We now consider some polarization selection rules. We shall first discuss the "allowed" or atomic displacement (AD) polarization selection rules. These apply when one has a crystal with full translational symmetry (other effects will be considered shortly). The Raman tensors for all the different crystal symmetries have been derived and are given in Hayes and London (1). In this section, however, we will consider only the case for diamond- (point group O_h) and zincblende-type (point group T_d) materials.

The Raman tensor for the $q \approx 0$ phonons for these materials is given by (6, 7)

$$\begin{bmatrix} 0 & 0 & 0 \\ 0 & 0 & d \\ 0 & d & 0 \end{bmatrix} : \quad u_x, \tag{44a}$$

$$\begin{bmatrix} 0 & 0 & d \\ 0 & 0 & 0 \\ d & 0 & 0 \end{bmatrix} : \quad u_y, \tag{44b}$$

and

$$\begin{bmatrix} 0 & d & 0 \\ d & 0 & 0 \\ 0 & 0 & 0 \end{bmatrix} : \quad u_z. \tag{44c}$$

Equations (44) correspond to phonons along x, y, and z, respectively. The LO phonon is along q, where q is defined by Eq. (9b), while the two TO phonons are perpendicular to $\mathbf{q}$.

Consider the case of backscattering along the z ($\langle 001 \rangle$) direction. This means $\mathbf{k}_i$ is along $\langle 001 \rangle$ while $\mathbf{k}_s$ is along $\langle 00\bar{1} \rangle$. Thus $\mathbf{q}$ is along $\langle 001 \rangle$, and we have our choice of two polarization configurations listed below:

$$\begin{aligned} x_1 &= \langle 100 \rangle & x_1' &= (1/\sqrt{2})\langle 110 \rangle \\ y_1 &= \langle 010 \rangle & y_1' &= (1/\sqrt{2})\langle 1\bar{1}0 \rangle \\ z_1 &= \langle 001 \rangle & z_1' &= \langle 001 \rangle. \end{aligned} \tag{45}$$

The selection rules for the intensities using Eq. (13) and the Raman tensors given in Eqs. (44) are listed in Table 6.1 (7). Note that the configuration $z_1(x_i, x_1)\bar{z}_1$ $[z_1'(x_1', x_1')\bar{z}_1']$ gives the same result as $z_1(y_1, y_1)\bar{z}_1$ $[z_1(y_1', y_1')\bar{z}_1]$.

Note that for backscattering along $\langle 001 \rangle$ only LO phonon scattering

Table 6.1. Atomic Displacement Raman Selection Rules for Backscattering from the (001) Surface

Polarization Configuration	Phonon Modes		
	$LO(z_1)$	$TO(x_1)$	$TO(y_1)$
$z_1(x_1, x_1)\bar{z}_1$	0	0	0
$z_1(x_1, y_1)\bar{z}_1$	d^2	0	0
$z_1(y_1, y_1)\bar{z}_1$	0	0	0
	$LO(z_1)$	$TO(x'_1)$	$TO(y'_1)$
$z_1(x'_1, x'_1)\bar{z}_1$	d^2	0	0
$z_1(x'_1, y'_1)\bar{z}_1$	0	0	0
$z_1(y'_1, y'_1)\bar{z}_1$	d^2	0	0

is allowed for the AD mechanism. The TO mode is forbidden for all configurations.

Now consider backscattering from a (110) surface. In this case q is along $\langle 110 \rangle$ and again we have a choice of two polarization configurations. Let us define two new sets of axes:

$$\begin{aligned} x_2 &= \langle 001 \rangle & x'_2 &= (1/\sqrt{6})(\langle \bar{1}12 \rangle \\ y_2 &= (1/\sqrt{2})\langle 1\bar{1}0 \rangle & y'_2 &= (1/\sqrt{3})\langle 1\bar{1}1 \rangle \\ z_2 &= (1/\sqrt{2})\langle 110 \rangle & z'_2 &= (1/\sqrt{2})\langle 110 \rangle. \end{aligned} \tag{46}$$

The Raman tensors relative to the x_2, y_2, and z_2 axes are

$$\begin{bmatrix} 0 & d & 0 \\ d & 0 & 0 \\ 0 & 0 & 0 \end{bmatrix} : \ u_{x_2}(\text{TO}), \tag{47a}$$

$$(1/\sqrt{2})\begin{bmatrix} 0 & 0 & -d \\ 0 & 0 & d \\ -d & d & 0 \end{bmatrix} : \ u_{y_2}(\text{TO}), \tag{47b}$$

and

$$(1/\sqrt{2})\begin{bmatrix} 0 & 0 & d \\ 0 & 0 & d \\ d & d & 0 \end{bmatrix} : \ u_{z_2}(\text{LO}). \tag{47c}$$

Table 6.2. Atomic Displacement Raman Selection Rules for Backscattering from the (110) Surface

Polarization Configuration	Phonon Modes		
	LO(z_2)	TO(x_2)	TO(y_2)
$z_2(x_2, x_2)\bar{z}_2$	0	0	0
$z_2(x_2, y_2)\bar{z}_2$	0	0	0
$z_2(y_2, y_2)\bar{z}_2$	0	d^2	0
	LO(z'_2)	TO(x'_2)	TO(y'_2)
$z_2(x'_2, x'_2)\bar{z}_2$	0	$\frac{2}{3}d^2$	$\frac{1}{3}d^2$
$z_2(x'_2, y'_2)\bar{z}_2$	0	$\frac{1}{3}d^2$	0
$z_2(y'_2, y'_2)\bar{z}_2$	0	0	$\frac{4}{3}d^2$

Similarly the Raman tensor relative to the x'_2, y'_2, and z'_2 axes can be obtained. The polarization selection rules are given for both sets of axes in Table 6.2 (7).

Note that for backscattering from (110) the TO phonon is allowed but scattering from the LO mode is forbidden for the AD mechanism. We shall shortly see that the breakdown of this selection rule can yield valuable information. Since the $\langle 110 \rangle$ axis is a low-symmetry direction, the intensities for $z_2(x_2, x_2)\bar{z}_2$ $[z'_2(x'_2, x'_2)\bar{z}'_2]$ are not equal to those for $z_2(y_2, y_2)\bar{z}_2$ $[z'_2(y'_2, y'_2)\bar{z}'_2]$.

Finally, we consider backscattering from the (111) surface. We define the axes:

$$\begin{aligned} x_3 &= (1/\sqrt{2})\langle 1\bar{1}0 \rangle \\ y_3 &= (1/\sqrt{6})\langle 11\bar{2} \rangle \\ z_3 &= (1/\sqrt{3})\langle 111 \rangle. \end{aligned} \tag{48}$$

Relative to these axes the Raman tensors are

$$(1/\sqrt{2})\begin{bmatrix} 0 & 0 & -d \\ 0 & 0 & d \\ -d & d & 0 \end{bmatrix}: \quad u_{x_3}(\text{TO}), \tag{49a}$$

$$\begin{bmatrix} 0 & -2d & d \\ -2d & 0 & d \\ d & d & 0 \end{bmatrix}: \quad u_{y_3}(\text{TO}), \tag{49b}$$

Table 6.3. Atomic Displacement Raman Selection Rules for Backscattering from the (111) Surface

Polarization Configuration	Phonon Modes		
	$LO(z_3)$	$TO(x_3)$	$TO(y_3)$
$z_3(x_3, x_3)\bar{z}_3$	$\frac{1}{3}d^2$	0	$\frac{2}{3}d^2$
$z_3(x_3, y_3)\bar{z}_3$	0	$\frac{2}{3}d^2$	0
$z_3(y_3, y_3)\bar{z}_3$	$\frac{1}{3}d^2$	0	$\frac{2}{3}d^2$

and

$$(1/\sqrt{3})\begin{bmatrix} 0 & d & d \\ d & 0 & d \\ d & d & 0 \end{bmatrix} : \quad u_{z_3}(\mathrm{LO}). \tag{49c}$$

The polarization selection rules are given in Table 6.3. (7). For backscattering from this direction, both LO phonon scattering and TO phonon scattering are allowed; however, by the appropriate choice of polarizer and analyzer setting, LO may be suppressed. Also, since $\langle 111 \rangle$ is a high-symmetry direction, $z_3(x_3, x_3)\bar{z}_3$ gives the same result as $z_3(y_3, y_3)\bar{z}_3$.

6.2.5.2. *Linear-q and Surface Electric Field-Induced Effects*

As just discussed for a crystal with full translational symmetry (meaning that, in principle, the crystal must be infinite), only $q \approx 0$ phonons are observed in first-order RS with Raman tensors given by Eqs. (44).

There are several ways in which the translational symmetry of the crystal can be broken. For example, when the energy of the incident light is near an interband electronic transition, there is strong absorption. The penetration depth, L_p, of the light is then given by

$$L_p \approx 2\pi/\Delta q, \tag{50}$$

where Δq is given by Eq. (12). In this case the light scattering is from a finite region of the material that, in principle, does not have the full translational symmetry of the "infinite" crystal. Thus there will be a change in the polarization selection rules owing to this "linear-q" (LQ) effect. Also, in the range of strong absorption, any surface electric field (SF) effects will become important, such as those originating from the space charge region, since now L_p may be of the order of the width of the space charge region. Electric fields also destroy the translational symmetry of the material.

It can be shown by group theory that the Raman tensor due to LQ or SF effects is the same (8, 9). We write down the tensor for the electric field $\mathscr{E}_x, \mathscr{E}_y, \mathscr{E}_z$. For LQ effects, q_x replaces $\mathscr{E}_x$, etc. For the sake of completeness we include the AD scattering terms denoted by d:

$$\begin{bmatrix} a\mathscr{E}_x & c\mathscr{E}_y & c\mathscr{E}_z \\ c\mathscr{E}_y & b\mathscr{E}_x & d \\ c\mathscr{E}_z & d & b\mathscr{E}_x \end{bmatrix} : \quad u_{x_1}, \tag{51a}$$

$$\begin{bmatrix} b\mathscr{E}_y & c\mathscr{E}_x & d \\ c\mathscr{E}_x & a\mathscr{E}_y & c\mathscr{E}_z \\ d & c\mathscr{E}_z & b\mathscr{E}_y \end{bmatrix} : \quad u_{y_1}, \tag{51b}$$

and

$$\begin{bmatrix} b\mathscr{E}_z & d & c\mathscr{E}_x \\ d & b\mathscr{E}_z & c\mathscr{E}_y \\ c\mathscr{E}_x & c\mathscr{E}_y & a\mathscr{E}_z \end{bmatrix} : \quad u_{z_1}. \tag{51c}$$

The quantities a, b, c, and d are in general complex. One of the most interesting cases is for backscattering from the (110) surface. As we will see below, the LQ or SF effect will activate the normally forbidden LO phonon.

Referred to the x_2, y_2, and z_2 axes [see Eq. (46)] the LQ-SF Raman tensor becomes, with $\mathscr{E}_x(q_x) = \mathscr{E}_y(q_y) = \sqrt{2}\mathscr{E}(q)$ and $\mathscr{E}_z(q_z) = 0$,

$$\begin{bmatrix} 0 & d & \sqrt{2}c\mathscr{E} \\ d & 0 & \sqrt{2}c\mathscr{E} \\ \sqrt{2}c\mathscr{E} & \sqrt{2}c\mathscr{E} & 0 \end{bmatrix} : \quad u_{x_2}(\mathrm{TO}), \tag{52a}$$

$$(1/\sqrt{2})\begin{bmatrix} \sqrt{2}(a-b)\mathscr{E} & 0 & -d \\ 0 & \sqrt{2}(a-b)\mathscr{E} & d \\ -d & d & 0 \end{bmatrix} : \quad u_{y_2}(\mathrm{TO}), \tag{52b}$$

and

$$(1/\sqrt{2})\begin{bmatrix} \sqrt{2}(a+b)\mathscr{E} & 2\sqrt{2}c\mathscr{E} & d \\ 2\sqrt{2}c\mathscr{E} & \sqrt{2}(a+b)\mathscr{E} & d \\ d & d & \sqrt{2}b\mathscr{E} \end{bmatrix} : \quad u_{z_2}(\mathrm{LO}). \tag{52c}$$

Table 6.4. Surface Field, Linear-*q*, and Atomic Displacement Raman Selection Rules for Backscattering from the (110) Surface

Polarization Configuration	$LO(z_2)$	$TO(x_2)$	$TO(y_2)$
$z_2(x_2, x_2)\bar{z}_2$	$b^2\mathscr{E}^2(q^2)$	0	0
$z_2(x_2, y_2)\bar{z}_2$	0	0	d^2
$z_2(y_2, y_2)\bar{z}_2$	$[(a-b)-2c]^2\mathscr{E}^2(q^2)$	d^2	$(a-b)^2\mathscr{E}^2(q^2)$

It can be shown from Eqs. (46) and (52) that for the configuration $z_2(x_2, x_2)\bar{z}_2$ or $z_2(y_2, y_2)\bar{z}_2$ the LO phonon mode becomes allowed with an intensity that is proportional to $\mathscr{E}^2$ (or q^2). The selection rules are listed in Table 6.4. We shall see later how the LQ–SF-induced LO phonon mode can yield important information about SFs.

6.2.6. Dispersion Relations for Lattice Vibrations

In order to gain some understanding of the dispersion relation for semiconductors such as the diamond- and zincblende-type materials, let us consider a simple linear chain model with two atoms per unit cell (5, 10). As shown in Fig. 6.1 the atoms have masses M_1 and M_2, and we will assume that each atom interacts only with its nearest neighbor. For simplicity we assume only one spring constant, K. (We could put in two different spring constants, but this only complicates things and does not add much to the physical insight.) The parameter a is the repeat distance of the unit cell. The equations of motion can be written as

$$M_1\ddot{u}_s = K(v_s + v_{s-1} - 2u_s) \tag{53a}$$

and

$$M_2\ddot{v}_s = K(u_{s+1} + u_s - 2v_s), \tag{53b}$$

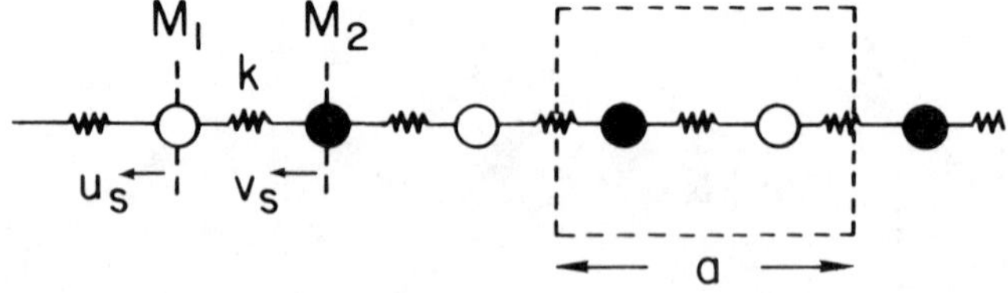

Figure 6.1. One-dimensional diatomic chain with perfect regularity.

where u_s and v_s are displacements of M_1 and M_2, respectively, in the sth primitive cell.

We look for a solution in the form of a traveling wave, with different amplitudes u and v:

$$u_s = ue^{i(\omega t + sqa)} \tag{54a}$$

and

$$v_s = ve^{i(\omega t + sqa)}, \tag{54b}$$

where q is the wavevector and ω is the frequency. On substitution into Eqs. (53) we obtain

$$-\omega^2 M_1 u = Kv(1 + e^{-iqa}) - 2Ku \tag{55a}$$

$$-\omega^2 M_2 v = Ku(1 + e^{iqa}) - 2Kv. \tag{55b}$$

These homogeneous linear equations have a nontrivial solution only if the determinant of the coefficients of u and v vanishes:

$$\begin{vmatrix} 2K - M_1\omega^2 & -K(1 + e^{-iqa}) \\ -K(1 + e^{iqa}) & 2K - M_2\omega^2 \end{vmatrix} \equiv 0, \tag{56}$$

or

$$M_1 M_2 \omega^4 - 2K(M_1 + M_2)\omega^2 + 2K^2(1 - \cos qa) \equiv 0, \tag{57a}$$

which can be rewritten as

$$\omega^4 - \frac{2K}{\mu}\omega^2 + \frac{2K^2}{M_1 M_2}(1 - \cos qa) \equiv 0, \tag{57b}$$

where μ is the reduced mass: $1/\mu = 1/M_1 + 1/M_2$.

Equation (57b) can be solved exactly. The solutions are plotted in Fig. 6.2 as a dispersion curve, i.e., frequency as a function of the wavevector q within the first BZ. The extent of the first BZ is π/a.

Shown in Fig. 6.3(a, b) (7) are the actual dispersion curves for Si and GaAs, respectively, in the $\langle 100 \rangle$ and $\langle 111 \rangle$ directions of the BZ.

In general, a three-dimensional crystal will have three acoustical modes (two transverse and one longitudinal) and $3(M - 1)$ optical modes, where M is the number of atoms per unit cell. Thus, the diamond- and zincblende-type materials have three acoustical and three optical modes.

Note that one of the main differences between the dispersion curve of the

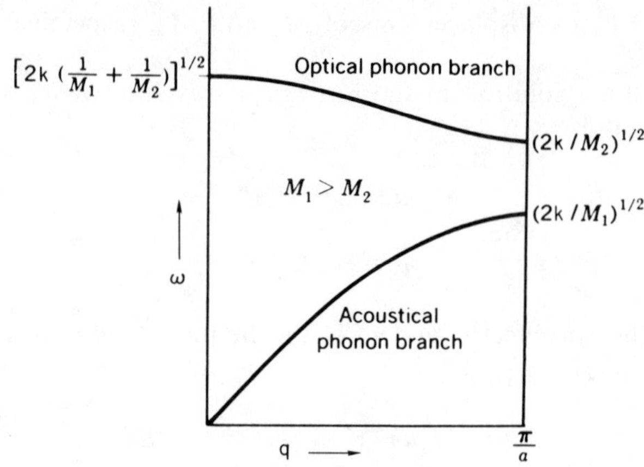

Figure 6.2. Dispersion of optical and acoustical phonon branches from Eq. (57b).

diamond-type material (Si) and the zincblende-type semiconductor (GaAs) is the removal of the degeneracy of the $q \approx 0$ optic mode into the LO and TO modes. This $q \approx 0$, LO–TO splitting is a consequence of the lack of inversion symmetry and the related polar character, and hence the splitting is a measure of the effective charge (or degree of ionicity).

6.3. INSTRUMENTATION

Since RS is a second-order process, its intensity is weak compared to first-order processes such as Rayleigh scattering. Also, the Raman shifts are generally of the order 50–1000 cm^{-1}. This means that the excitation source should be intense with well-defined wavelengths, i.e., a laser. Also the monochromator should have a high rejection ratio (double monochromator), and the detection system should be able to detect low levels (photon-counting systems).

The functional block diagram of the experimental setup for Raman scattering is shown in Fig. 6.4 (7). A complete description of modern Raman instrumentation is given in Chapter 2 by Chase.

There have recently been some interesting new instrumentation developments. These include the use of an optical multichannel analyzer (OMA) or equivalent system and the Raman microprobe (11, 12). An OMA (or equivalent apparatus) is a system for extremely rapid parallel spectral data

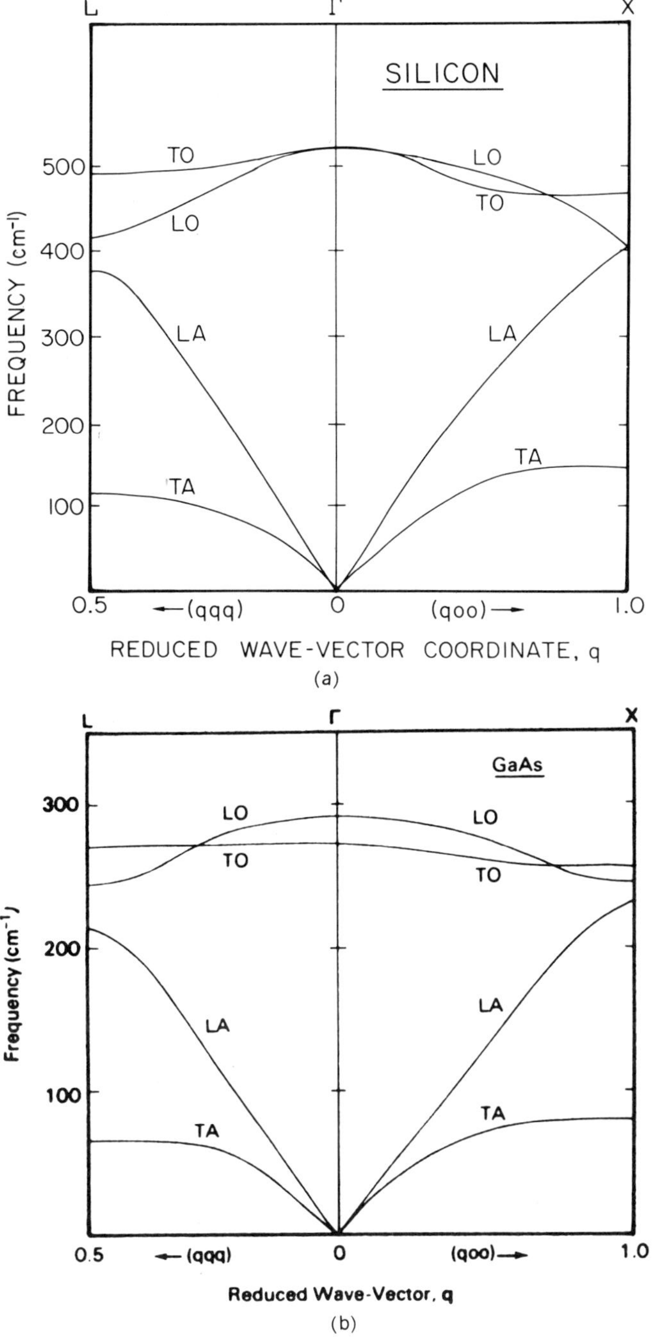

Figure 6.3. (a) Phonon dispersion curves along ⟨100⟩ and ⟨111⟩ for silicon. (b) Phonon dispersion curves along ⟨100⟩ and ⟨111⟩ for GaAs. From Pollak (7).

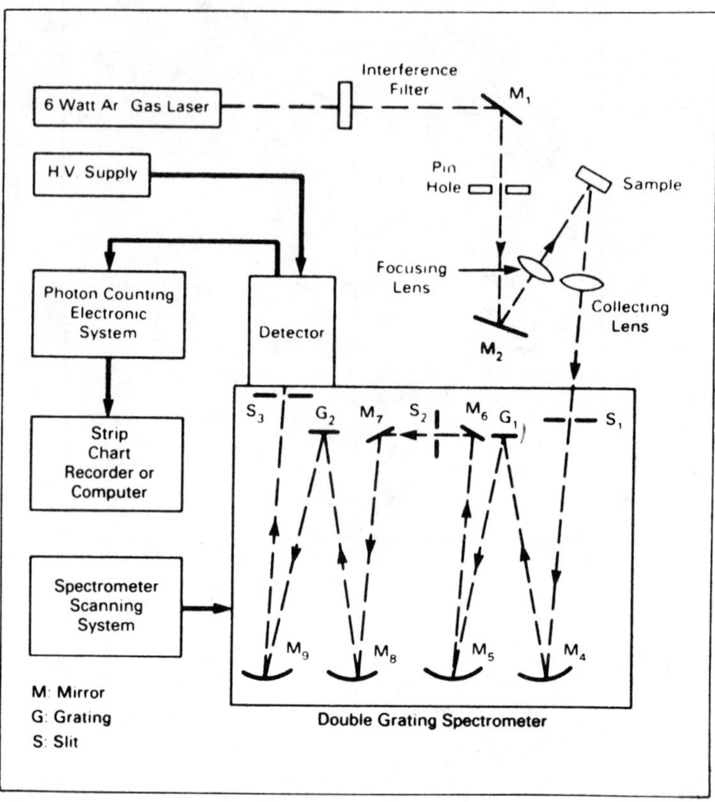

Figure 6.4. Schematic functional block diagram of Raman scattering apparatus. From Pollak (7).

acquisition, analysis, and presentation. The Raman microprobe consists of a conventional microscope that is optically and mechanically coupled to a Raman spectrometer (11, 13). Thus, a focused light spot of spatial dimensions 1–2 μm can be employed for a variety of purposes. Applications for microelectronics such as identification of organic contaminants on Si wafers and detection of elemental Si near the vicinity of an Al wire bounded to Au have been discussed by F. Adar (12). Fauchet and co-workers (14–19) have recently reviewed the use of the Raman microprobe as a quantitative analytical tool to characterize laser-processed semiconductors. Other uses of the Raman microprobe will be discussed in subsequent sections.

One of the drawbacks of the microprobe is the low signal levels due to the low laser power that must be employed. Since the light spot is 1–2 μm in size, the power density can become quite high and possibly damage samples, particularly zincblende-type materials. Thus, the absolute power level must typically be kept to only several milliwatts. In a regular Raman experiment using a cylindrical lens as in Fig. 6.4a, several hundred milliwatts can be employed without harming the material. However, a combination of the OMA (or equivalent) with the microprobe might overcome this problem of low signal-to-noise levels.

6.4. SILICON AND OTHER GROUP IV SEMICONDUCTORS

For silicon samples RS has been used to evaluate a number of important phenomena including microcrystalline particle size, lattice mismatch strains for silicon on various substrates such as Al_2O_3, SiO_2, or stainless steel, properties of ion-damaged and annealed material, and interfacial reactions of Pt, Pd, and Ti on single-crystal and amorphous silicon (20, 21).

In this material, only one peak, at about 522 cm^{-1} (at room temperature), is seen in first-order RS since the optic phonons (LO and TO) are degenerate at $q \approx 0$ [see Fig. 6.3(a)]. This peak has a Lorentzian lineshape with a full width at half maximum (FWHM) of $\Gamma_0 = 3$ cm^{-1}.

6.4.1. Microcrystalline Geometries

The first-order Raman spectrum provides a fast and convenient method to determine whether a semiconductor such as silicon is crystalline or amorphous. For example, as discussed above, the conservation of phonon momentum, q, in crystalline silicon (c-Si) leaves as Raman active only the zone center ($q \approx 0$) optic phonon at 522 cm^{-1} (see Fig. 6.3a), which gives rise to a single line with a Lorentzian lineshape with a FWHM of $\Gamma_0 = 3.0$ cm^{-1}. In amorphous materials such as amorphous silicon (a-Si) the q-vector selection rule does not apply at all owing to the loss of long-range order. The RS of a-Si is shown in Fig. 6.5 (22). All phonons in Fig. 6.3a are therefore allowed, and the RS resembles the phonon density-of- states with prominent humps around 140 and 480 cm^{-1}, the former having its origins in the transverse acoustic (TA) phonon while the latter is related to the TO phonons (see Fig. 6.5) (22). Thus, the observations of these broad features or the sharp line at about 522 cm^{-1} have been used to differentiate between a-Si and c-Si.

Within the past several years a number of investigators (23, 24) have reported spectra intermediate between the two extreme cases described above, involving a shift of the 522 cm^{-1} line toward lower energy accompanied by an asymmetric (low-frequency) broadening. This phenomenon has been

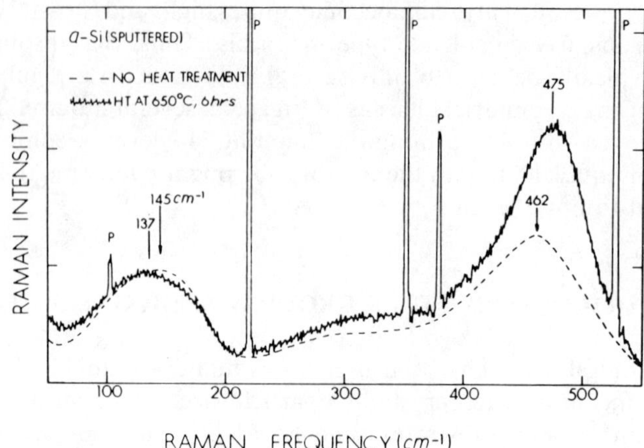

Figure 6.5. Raman scattering intensity (in arbitrary units) vs. Raman frequencies (in cm^{-1}) for a sputtered a-Si sample before and after heat treatment at 650°C for 6 h. The noise and plasma line for the sample before heating have been removed for clarity. From Tsu et al. (22).

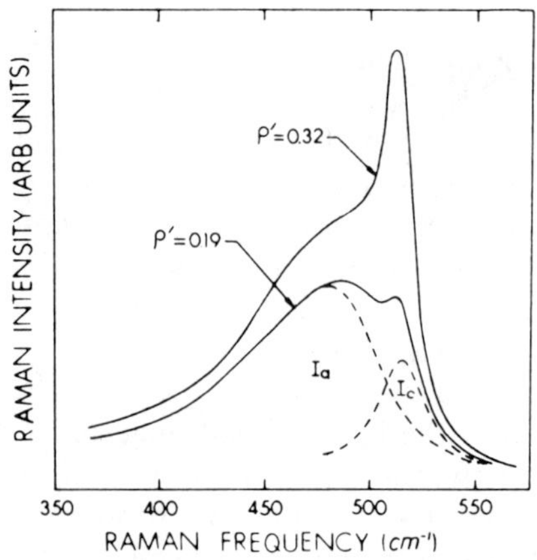

Figure 6.6. Raman intensity versus Raman frequency for a two-phase system, microcrystallites embedded in an amorphous matrix. From Tsu et al. (24).

taken as evidence of a "microcrystalline phase" in Si (μc-Si). This is shown in Fig. 6.6 (24). A similar effect has been reported for μc-Ge and μc-GaP (25).

Richter, Wang, and Ley (23) have developed a detailed model to quantitatively explain this observation in silicon. A similar model was proposed at about the same time by Nemanich et al. (26) to explain microcrystalline effects in boron nitride. A more qualitative model has been presented by Tsu et al. (27). The basic idea behind these approaches is as follows. For a finite material the mode correlation functions of the phonon become finite owing to the lack of translational symmetry. Thus, there is a relaxation of the $q \approx 0$ selection rules. Richter, Wang, and Ley assumed a Gaussian localization factor, i.e.,

$$W(r) \propto \exp(-2r^2/L^2), \tag{58}$$

for the phonon wave function, where L is the diameter of the microcrystalline phase (assumed to be spherical). Upon Fourier transformation this leads to an average over q with a similar weighing factor, $W(q)$, given by

$$W(q) \propto \exp(-q^2L^2/8). \tag{59}$$

Thus the Raman intensity, $I(\omega)$, at frequency ω can be written as

$$I(\omega) \propto \int_{\mathrm{BZ}} \exp(-q^2L^2/8) \frac{d^3q}{[\omega - \omega(q)]^2 + [\Gamma_0/2]^2}, \tag{60}$$

where q is expressed in units of $2\pi/a$ (a is the lattice constant), the parameter Γ_0 is the intrinsic linewidth of c-Si and $\omega(q)$ represents the dispersion of the phonon [see Fig. 6.3(a)].

The above considerations lead to a red shift and asymmetrical broadening

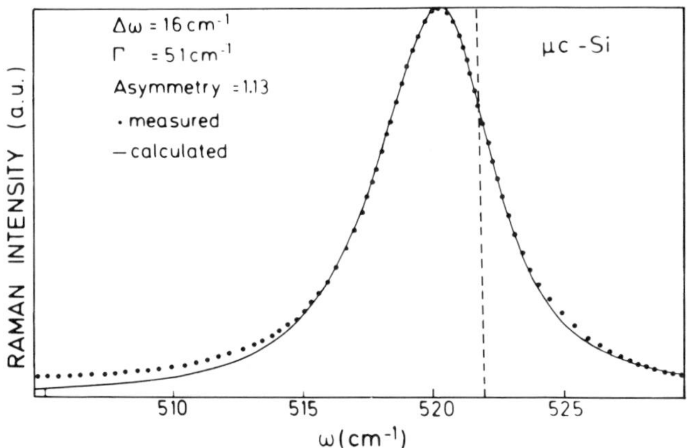

Figure 6.7. Measured and calculated Raman line of μc-Si sample. The dashed line indicates the Raman shift for c-Si. The asymmetry is defined as Γ_a/Γ_b, where Γ_a and Γ_b are the low- and high-frequency half-widths, respectively, of the Raman line. From Richter et al. (23).

with tailing toward low frequencies as the $q \approx 0$ rule is relaxed. Shown in Fig. 6.7 are the results of Richter, Wang, and Ley (23), with $L = 40$ Å for the μc-Si phase. The dashed lines indicate the Raman shift for c-Si (that is, $522\,\text{cm}^{-1}$). The calculation produces a shift of $\Delta\omega = 1.6\,\text{cm}^{-1}$, $\Gamma = 5.1\,\text{cm}^{-1}$, and an asymmetry of 1.13. The asymmetry is defined as Γ_a/Γ_b, where Γ_a and Γ_b are the low-frequency and high-frequency half-widths, respectively, of the Raman line. The agreement between experiment and theory is quite good. Figure 6.8 (23) gives the relationship calculated from Eq. (60) between $\Delta\omega$ and the linewidth Γ calculated for a range of Δq (and hence L) values in silicon (solid line). Also shown are a number of experimental points. The overall agreement between experiment and theory is quite good in both Fig. 6.7 and Fig. 6.8, making it possible to evaluate the μc-Si particle size from the first-order Raman spectra.

Campbell and Fauchet (28) have extended this model to various microcrystalline shapes and have indicated that there are significant differences between spherical, columnar, and thin-slab configuration. This

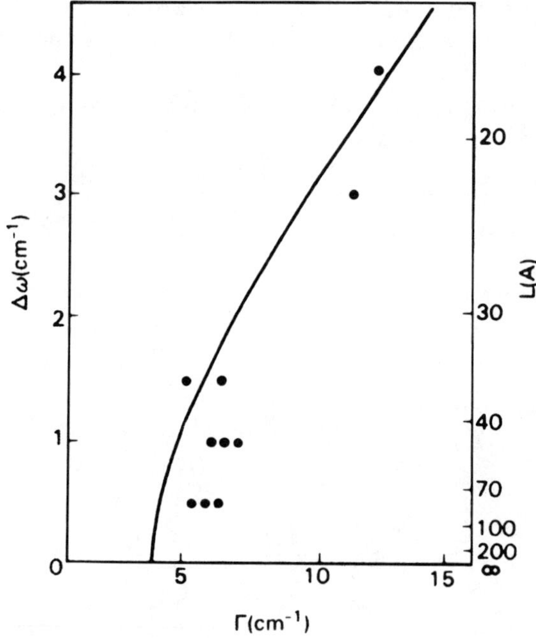

Figure 6.8. Raman shift, $\Delta\omega$, and broadening, Γ, as a function of L for silicon. From Richter et al. (23).

may make it possible to deduce microcrystalline shapes as well as size from the details of the Raman spectrum.

The model developed by Richter, Wang and Ley is being used extensively to analyze microcrystalline effects in Si and Ge. As discussed above, the model has been extended to various shapes (28). Some applications will be discussed in the following section on ion-damage and annealing effects.

This model has recently been applied to evaluate low-pressure chemical vapor deposition (LPCVD) silicon for polysilicon films in integrated circuit manufacture (29). Richter and Ley (30) have studied the transition from amorphous to microcrystalline Si using both photoemission and RS.

Gonzalez-Hernandez et al. (31) have recently reported a comparative study of Raman, transmission electron microscopy (TEM), and conductivity measurements in molecular-beam-deposited (MBD) μc-Si and μc-Ge prepared under various growth conditions. The TEM measurements yield an average grain size L_0 ranging from 200 A, to 1.5 μm. On the other hand, the lineshape of the RS is determined by a mean free path, l, related to the average separation between defects (or impurities), where $l \leqslant 150$ A'. Shown by the dots in Fig. 6.9 (31) are the experimentally determined Γ and Γ_a/Γ_b for a number of samples of μc-Si and μc-Ge. The solid line in Fig. 6.9 is Γ and Γ_a/Γ_b evaluated from an expression similar to Eq. (60) with two differences. The phonon dispersion relation $\omega(q)$ was evaluated from a simple linear chain model:

$$\omega(q)^2 = A + \{A^2 - B[1 - \cos \pi q]\}^{1/2}, \tag{61}$$

where A and B can be evaluated from the dispersion curves of the relevant material. Also Gonzalez-Hernandez et al. (31) explicitly included the instrumental transfer function effects. There is good agreement between experiment and the theoretical curve, thus allowing l to be evaluated. These authors found in a number of cases that the electrical conductivity is determined by l rather than L_0. This experiment is an important demonstration of the significance of RS as a structural characterization method.

Fauchet and co-workers (14–19) have recently performed a number of studies of RS of laser-induced effects, including extensive use of the Raman microprobe. For example, they have reported results on films of silicon on insulator (SOI) grown by LPCVD. These films have columnar, polycrystalline structure and are under tensile stress owing to the difference in thermal expansion coefficients of silicon and the fused quartz of the insulator. The material is processed by picosecond pulses from a frequency-doubled or-tripled Nd/YAG laser focused down to a spot size of 1–2 μm. Regions damaged during processing are first examined using Nomarski microscopy,

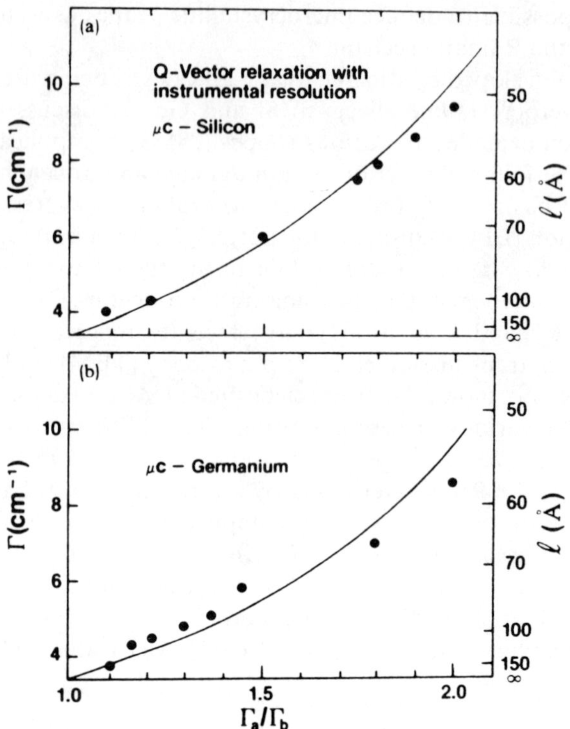

Figure 6.9. Relationship between width Γ and asymmetry of the Raman line in (a) μc-Si and (b) μc-Ge, corrected by instrumental resolution. The points correspond to samples prepared by MBD. From Gonzalez-Hernandez, et al. (31).

and then the Raman microprobe (1 μm spatial resolution) is used to analyze the specific nature of the material.

Shown in Fig. 6.10 (17) is the Stokes Raman spectra of c-Si (fine line) and the SOI (thick line). The former has a narrow, symmetric line centered at $522\,\text{cm}^{-1}$, the linewidth Γ (FWHM) being about $3\,\text{cm}^{-1}$. For the SOI spectrum there is a shift in the peak position, $\Delta\omega$, and an asymmetric broadening that can be characterized by Γ and the asymmetry factor Γ_a/Γ_b. The frequency shift $\Delta\omega$ is caused by both strain (we shall return to this point later) and microcrystalline effects. The latter effect can be evaluated from the microcrystalline particle size, which can be deduced from Γ and Γ_a/Γ_b using Eqs. (60) and (61). Thus the SOI region shown in Fig. 6.10 has an average crystallite size of 200 Å and is under a tensile stress of 5×10^9 dyn/cm^2.

Fauchet et al. (18, 19) have made topographical studies using the Raman

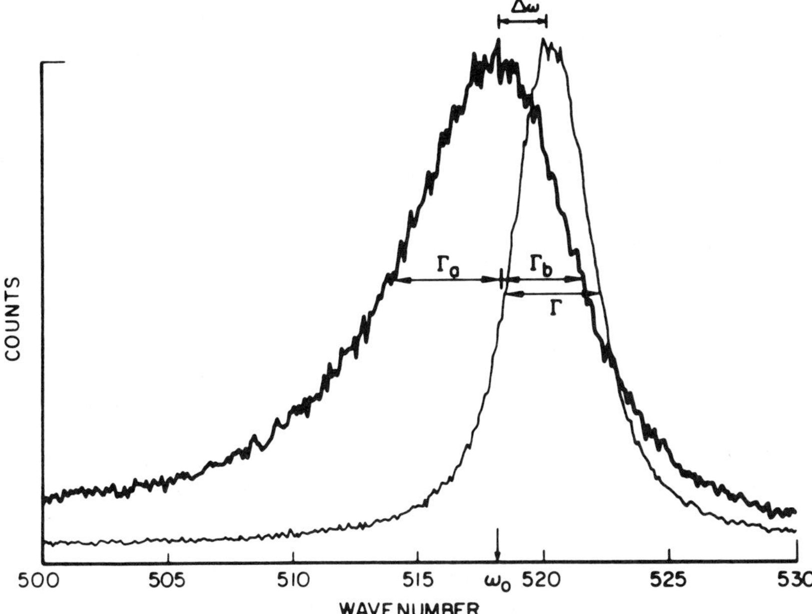

Figure 6.10. Optic phonon Stokes line of single-crystal silicon (fine line) and of polycrystalline silicon thin film (thick line). With the standard operating parameters of the instrument, $\omega_0 = 520\,\text{cm}^{-1}$, $\Gamma \simeq 3.8\,\text{cm}^{-1}$, and $\Gamma_a/\Gamma_b \simeq 1$ for crystalline silicon, and $\omega_0 \simeq 518.5\,\text{cm}^{-1}$, $\Gamma \simeq 9\,\text{cm}^{-1}$, and $\Gamma_a/\Gamma_b \simeq 1.4$ for the polycrystalline silicon thin films. In these films, the exact parameters of the Raman line are not perfectly identical everywhere (for example, ω_0 varies from 518.5 to $519\,\text{cm}^{-1}$ over large distances). These variations reflect stress and homogeneity changes in the film itself. From Campbell et al. (17).

microprobe of composition, stress, and structural variation across laser-damaged silicon.

6.4.2. Structural and Crystallization Effects in a-Si, a-Ge, and a-C

RS measurements provide a means of studying the lattice dynamics of disordered solids thus giving insight into the structure, bonding, and nature of disorder (32–45). Work on a-Si, a-Ge, and their alloys has recently been reviewed by Lannin (32). Various aspects of the Raman spectra have been used to gain information about these microstructural effects. These features include the frequencies of the TO- and TA-like bands, linewidth of the TO-like band, ratio of intensities of the TO- and TA-like bands, depolarization ratio, etc.

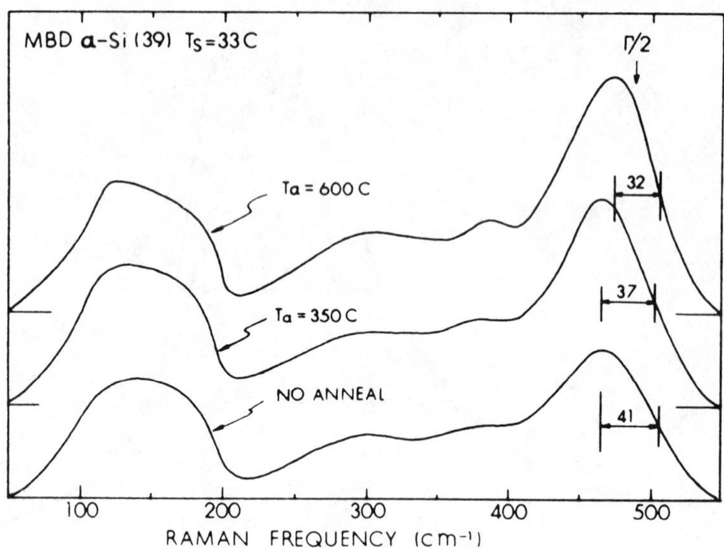

Figure 6.11. Raman spectra for a-Si prepared by MBD. The HWHM, $\Gamma/2$ is shown. Annealing was done in a furnace with flowing nitrogen. From Tsu et al. (33).

Tsu and co-workers have recently conducted detailed experimental (33) and theoretical work (34) on the width of the TO-like band. They report that in a-Si and a-Ge the width of this feature increases linearly with the root mean square (rms) bond-angle distortion $\Delta\theta_b$ of the random network with $7.7° \leqslant \Delta\theta_b \leqslant 10.5°$. Typical Raman spectra for a-Si annealed at various temperatures, T_a, are shown in Fig. 6.11 (33). These MBD samples were deposited in an ultrahigh vacuum (UHV) system at a substrate temperature $T_s = 30\,°\mathrm{C}$ and then annealed at T_a. The half-widths at half maxima (HWHM) are marked by $\Gamma/2 = 32$, 37, and $41\,\mathrm{cm}^{-1}$. The linewidth Γ is taken as indicated because the low-frequency side of the peak involves contributions from the BZ-edge LO phonons and perhaps even some longitudinal acoustic (LA) phonons. Note that Γ decreases from $82\,\mathrm{cm}^{-1}$ for as-deposited a-Si to $64\,\mathrm{cm}^{-1}$ after annealing at $T_a = 600°\mathrm{C}$ for 1 h. Similar narrowing in Γ upon annealing was also observed in a-Ge films prepared by sputtering at $T_s = 30°\mathrm{C}$.

On the basis of model calculations, Beeman et al. (34) have correlated the above-measured Γ (cm^{-1}) and $\Delta\theta_b$ (degrees). They find the relationship for a-Si to be

$$\Gamma = 15 + 6\,\Delta\theta_b. \tag{62}$$

Using arguments based on the deformation potentials of C-Si, Tsu et al. (33) have proposed for a-Si that

$$\Gamma^2 = (32)^2 + (6.75\,\Delta\theta_b)^2, \tag{63}$$

with a similar relationship for a-Ge. However, as pointed out in Beeman et al. (34), over the experimentally accessible range $64 \leqslant \Gamma \leqslant 82\,\mathrm{cm}^{-1}$ Eqs. (62) and (63) are essentially indistinguishable. Thus from Eq. (63) for a-Si the experimentally accessible values of $\Delta\theta_b$ lie in the range $8.2 \leqslant \Delta\theta_b \leqslant 11.2$. For a-Ge Tsu et al. (33) $6.1 \leqslant \Delta\theta_b \leqslant 8.8$. However, Beeman et al. (34) deduce that it is impossible to construct fully bonded amorphous networks with $\Delta\omega \leqslant 6.6$ from a-Si.

Lannin and his group (32, 35–37) have also performed a number of experiments in which the TO-band linewidth was used as a measure of local structural order. However, in contrast to Tsu et al. (33) and Beeman et al. (34), Lannin et al. have evaluated the linewidth from the depolarization spectra i.e., crossed polarizer and analyzer (VH configuration). For example, variations in short-range order of a-Ge films have been determined from a combination of depolarization RS, optical absorption, and radial distribution function (RDF) studies (35). Shown in Fig. 6.12 (35) are the inverse of the full width, $(\Delta_{\mathrm{VH}})^{-1}$, of the TO-like Raman peak and the inverse of twice the high-frequency half-width, $(\Delta'_{\mathrm{VH}})^{-1}$, as compared to the optical (or Tauc) gap, E_0, of a number of a-Ge films. The diamonds indicate rf-sputtered films deposited at low pressure, the circles, evaporated and annealed evaporated films, and the triangle, an ion-beam-sputtered film. As Fig. 6.12 demonstrates, both the aforementioned Raman width parameters are approximately linearly related to the optical gap, thus indicating the dependence of electronic order on the bond width. By comparison with RDF studies, Lannin et al. have deduced a minimum value of $\Delta\theta_b$, obtained for evaporated, anneal-stable a-Ge with $E_0 = 1.0\,\mathrm{eV}$. Also, Lannin et al. (35) report a maximum value of $\Delta\theta_b$ for the most disordered films in Fig. 6.12, that is, $E_0 = 0.6\,\mathrm{eV}$.

Maley et al. (36) have measured the Raman spectra of laser-quenched a-Si (LQ a-Si). From a comparison with chemical-vapor-deposited (CVD) a-Si these authors deduce that the LQ a-Si has $\Delta\theta_b$ equal to, but no greater than, that achieved in CVD a-Si. These results suggest a maximum allowed order that may be achieved in a-Si.

In spite of the differences in the number value of $\Delta\theta_b$ cited by Tsu et al. and Lannin and co-workers, it is clear that the linewidth of the TO-band in a-Si, a-Ge, and related materials can be used as a convenient measure of the degree of local ordering.

Shimada et al. (38) studied the structural properties of the two-phase silicon system (a-Si and μc-Si) by means of RS from the TA-like band. They

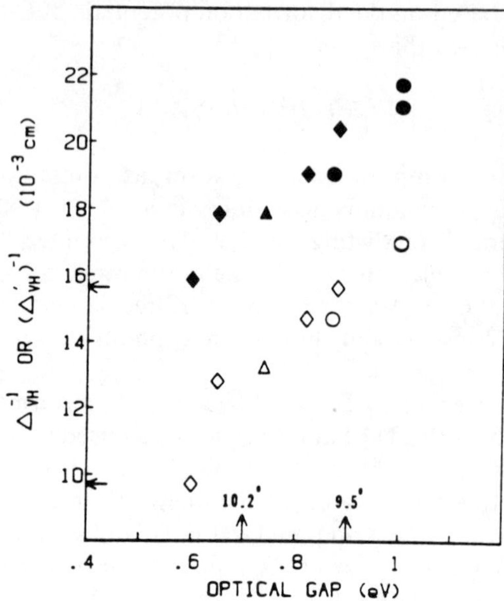

Figure 6.12. Variation of inverse Raman width parameters, Δ_{VH}^{-1} (open) and Δ_{VH}^{-1} (solid) versus the optical gap of a-Ge films. rf-sputtered (diamonds), evaporated (circles), and ion-beam-sputtered (triangles) films. Vertical arrows correspond to optical gap and $\Delta\theta$ values. Horizontal arrows correspond to Raman width parameters of the as-deposited film. From Lannin et al. (35).

mention that this feature is a sensitive probe of the randomness of the Si—Si bonding structure in μc-Si.

The details of the Raman spectra of the μc-Si phase (relatively sharp features in the 500–522 cm^{-1} range) have been employed to study the critical volume fraction of μc-Si phase in Si:F:H alloys (24) and the nucleation and growth rates of a-Si alloys (39). In the former experiment, a number of samples of highly P-doped a-Si:F:H alloys were prepared having a range of volume fraction, ρ, of μc-Si embedded in an a-Si matrix. The integrated RS intensity of the crystalline part, I_c, relative to the intensity of the amorphous component, I_a, is used to deduce ρ. Figure 6.6 shows the typical Raman spectra of two such samples. The curve denoted by $\rho' = 0.19$ has been resolved into two parts, I_a (shaded) and I_c. A computer program was used to fit a typical amorphous spectrum having a maximum located at 480 cm^{-1}. After I_a is subtracted from the total measured scattering, the remaining area under the curve is attributed to I_c, the contribution of the fraction of crystallinity. These authors have found that the measured critical volume fraction at the onset of rapidly increased conduction in this two-phase system of microcrystallites embedded in an amorphous matrix is $\rho' = 0.18$. This value coincides with the theoretical percolation limit and serves to explain the conduction process in these two-phase materials.

Segregated regions of amorphous silicon in as-deposited CVD Si-rich SiO_2 films have been detected by RS (as well as optical transmission) (41). Annealing the films at 1150°C completely crystallizes the a-Si. Annealing at lower temperatures produces films with both amorphous and crystalline regions.

RS has also been utilized to study amorphous and microcrystalline carbon a-C and μc-C (42–46) and related materials such as graphite and coal (41). Several studies show that as-deposited a-C films have only graphite-like bonding, whereas annealing produces μc-C that grows in size with annealing conditions. On the other hand, a-C:H films prepared by glow discharge shows evidence for the μc-C phase with no annealing (42). Dillon et al. (45) have studied carbon films prepared by ion-beam as well as rf-discharge deposition, and annealed at temperatures up to 950°C. The Raman spectra of these films, in the range 1000–1800 cm^{-1}, were analyzed via a best fit to computer-generated line-shapes, used to simulate the diamond (D) and graphite (G) lines. This study shows that crystallite growth is promoted by the higher annealing temperatures. From RS measurements on a larger variety of well-characterized hydrogenated and unhydrogenated ion-beam-deposited a-C films (together with pressure/optical absorption studies), it was shown that neither the μc-G nor the random two-dimensional network model is compatible with the structure of a-C (46). Instead, this study concludes that the data are qualitatively consistent with a network model consisting of both threefold (G) and fourfold (D) coordinated carbon atoms. The role of hydrogenation is to increase the local order in addition to effecting passivation of local bonds.

6.4.3. Ion Damage and Laser Annealing

Considerable work on RS of ion-damaged and annealed Si has been performed by a number of workers (48–60). Some results are shown in Fig. 6.13 for implantation of As at fluences of 10^{14}, 10^{16}, and 10^{17} cm^{-2} and annealing with a frequency-doubled Nd/YAD laser at various power densities, denoted by J. In general, annealing recrystallizes the material as can be seen from the shift of the Raman peak. For 10^{14} cm^{-2}, the peak at 513 cm^{-1} at $J = 0$ is probably a microcrystalline phase. At increased J, the peak shifts to higher frequencies until it reaches 522 cm^{-1} for $J = 0.75$. For higher fluences there are additional complications due to a large number of free carriers causing a Fano shift (Section 6.2.4). For dosage exceeding 4×10^{16} cm^{-2}, Raman peaks due to metallic As have been observed, indicating the presence of clusters of As (20).

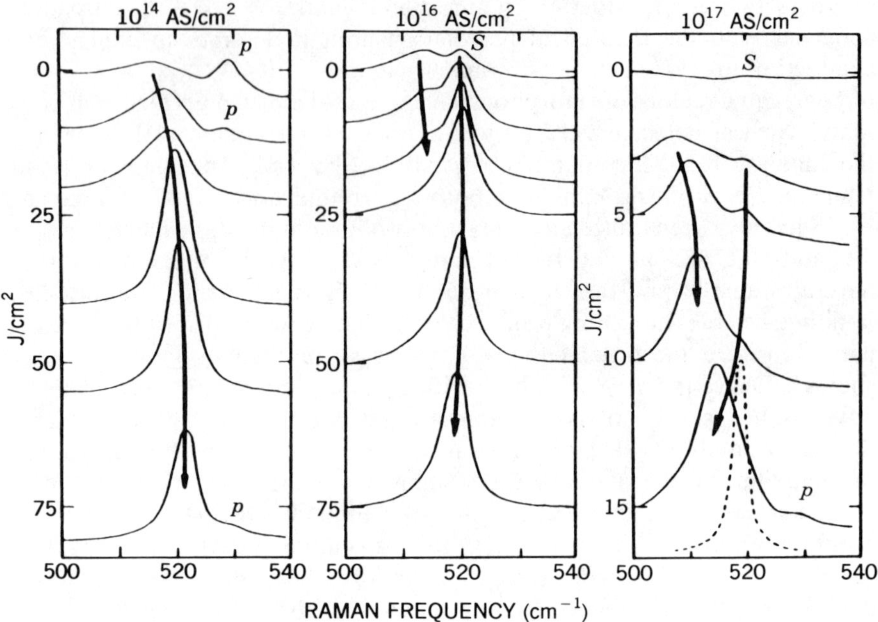

Figure 6.13. Raman scattering for several ion-implanted pulsed-laser-annealed (100) Si samples. A 4800-Å Ar-ion laser is used. From Tsu (49).

Nissim et al. (50, 51) have studied the phase transformation induced by a picosecond laser pulse in implanted amorphized silicon. A Raman microprobe with 1-μm spatial resolution was utilized to investigate the annealed spots. It was found that for high incident energies of the annealing pulse single-crystal silicon is observed in the central spot and in the first recrystallized ring of the annealed area. Nakashima and co-workers (52) have also used a Raman microprobe ($= 1\ \mu$m) to characterize ion-implanted and laser-annealed polycrystalline Si.

Hopkins et al. (53) have recently applied the polarization-selective Raman effect to the detailed nondestructive analysis of the local crystal orientation of polycrystalline silicon samples grown over SiO_2 and laser annealed. The experiment was performed in the microprobe configuration with a spot size of about 1 μm. Intensity measurements taken as a function of input polarization angle were fit to an expression derived from the RS selection rules to calculate the angles by which the crystal structure is twisted within the original substrate plane, as well as the degree of tipping of the crystal plane away from the plane of the silicon substrate. Similar measurements have also been reported by Nakashima et al. (54).

RS has been used to characterize boron-implanted silicon at various depths (55). In the RS, the boron distribution was deduced from the depth dependence of the normalized ratio of the boron local mode ($620\,cm^{-1}$) between various depths and the surface. The results were compared with the concentration profile obtained by secondary ion mass spectroscopy (SIMS).

The annealing behavior of this polycrystalline silicon film damaged by ^{28}Si ions (in two stages) and subsequently annealed at 525°C was studied by RS, TEM, and the optical appearance method (56). Two distinct phases were found to coexist in the as-implanted films: one phase consists of a mosaic of crystallites surrounded by a-Si, and the other consists of only a-Si. It was reported that the RS of the as-implanted films consisted of a broad peak at about $470\,cm^{-1}$ with a FWHM of about $100\,cm^{-1}$, which is characteristic of a-Si. After the material is annealed at 525°C for 48 h, the RS exhibits two features: a feature similar to the one above, and a second at $517\,cm^{-1}$ with a FWHM of $10\,cm^{-1}$ (there was no mention of any low-frequency asymmetry aspects of this peak). The authors deduced that it was characteristic of crystalline material and did not consider that it might be related to microcrystalline effects. After 650°C annealing for 15 min, only the latter peak ($517\,cm^{-1}$, with $10\,cm^{-1}$ FWHM) is observed.

Kirillov et al. (57) have reported the use of RS to study rapid thermal annealing of Si after implantation with 60 keV $^{75}As^{-}$ ions. Spectra were measured at different stages of annealing. Polarization selection rules were used to separate contributions from amorphous and crystalline contributions. Also nondestructive depth profiling was performed by using various lines of an Ar^{+} laser. It was found that the as-implanted amorphous phase was just transformed into the higher-order amorphous phase before transformation into the crystalline phase. This experiment showed that for the amorphous phase after annealing at 600°C for 10 s the TO-type peak at about $470\,cm^{-1}$ narrows from about 105 (FWHM) to $80\,cm^{-1}$, shifts in frequency from 463 to $475\,cm^{-1}$, and grows in intensity relative to the lower-frequency TA-type structure. These spectral changes are all indicative of increased local ordering as discussed in Section 6.4.2. Kirillov et al. (57) also reported that the regrowth of the crystalline phase occurred only from the substrate side and no polycrystalline region could be detected at intermediate stages of annealing. In addition, electrical activation of the implanted As^{-} dopant in the annealed crystalline layer was deduced by the lineshape changes caused by the contribution of free carriers, i.e., the Fano effect (see Section 6.2.4). The concentration of activated free carriers was also determined.

A comparative RS study of ultraheavily-As^{-}-implanted silicon annealed with the 1.06- and 0.53-μm wavelengths of a Q-switched Nd/YAG laser has been made to ascertain the degree of disorder in the supersaturated alloy that shows up in the lineshape asymmetry of the one-phonon optical mode and the scattering amplitude of the two-phonon TA and TO modes. Enhanced diffusion of the As atom in silicon is observed in the Raman spectrum of ultraheavily implanted silicon after pulsed laser annealing with 1.06 μm radiation (58).

An investigation has been made using RS (and electrical characterization) of the effects of laser recrystallization on polycrystalline silicon with three different capping layer structures on separate wafers (59). The capping layer structures include a 60-A, nitride layer, a combination of 640-A, nitride and 200-A, oxide layers, and 500-A, nitride antireflection periodic stripes oriented parallel to the laser scanning. The stress in each case has been characterized using a Raman spectrometer with microprobe.

A Raman study of thermally annealed silicon small particles and evaporated thin films has been performed by Okada et al. (60). For the particles, the changes in the RS upon annealing is slight over the entire temperature range. The spectra are characteristic of amorphous material with a broad peak at about 480 cm^{-1}. On the other hand, for the evaporated films there is a dramatic change for temperatures above 700°C. Below this temperature the RS is characteristic of amorphous material while above this region the RS shows a relatively sharp feature at about 520 cm^{-1}. No detailed analysis of this peak in terms of frequency shift, broadening, or low-frequency asymmetry was reported in order to determine microcrystalline effects. The authors deduce that the silicon particles are not noticeably changed by the heat treatment because of their inner crystalline structure, whereas the silicon films are amorphous-like and the crystallites grow easily. The activation energy of the crystallizaton was determined to be that of vitreous flow, i.e., 0.35 eV.

6.4.4. Reactive Ion Etching

RS has been used to characterize the surface modification introduced into (100)Si by reactive ion etching (RIE) with a CF_4/H_2 plasma (61). Both RS due to the destruction of crystalline long-range order by lattice damage and the stretching modes for protons bonded to Si atoms have been observed without any special sample preparation. These results are found to be consistent with Rutherford backscattering and nuclear reaction profiling studies of similarly prepared samples.

Figure 6.14 (61) shows the Raman spectra for frequency shifts between 50 and 750 cm^{-1} obtained from (100)Si wafers processed in different ways. This spectral range permits study of the Si—Si bonds. Figure 6.14(a) is the Raman spectrum of (100)Si under 300 mW of 5154-A, light s-polarized with the **E** field parallel to a ⟨011⟩ axis and the scattered light polarized perpendicular to the incident light. In this scattering geometry, both first- and second-order

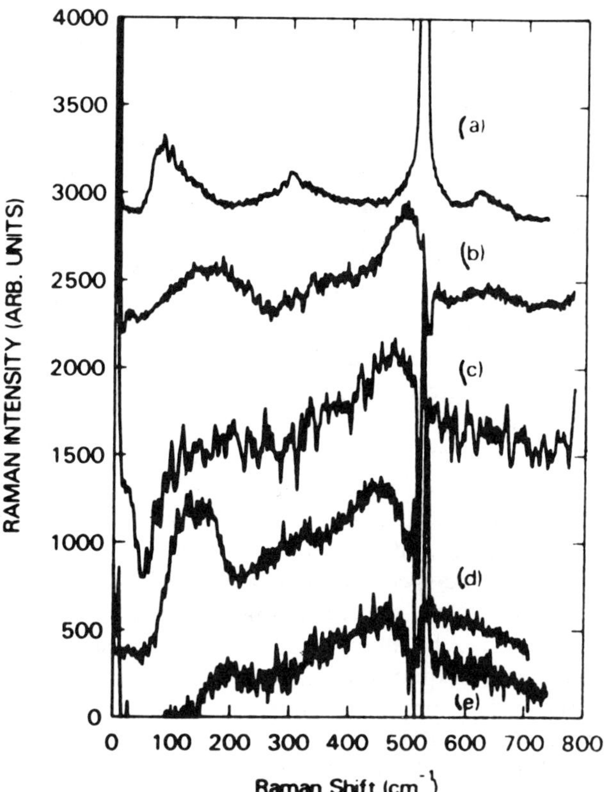

Figure 6.14. Raman spectrum of amorphous or disordered Si on (100)Si. (a) Spectrum of (100)Si for $E_{\text{inc.}}$ parallel to the ⟨011⟩ axis and $E_{\text{scat.}}$ perpendicular to $E_{\text{inc.}}$ (b) Spectrum of 60 Å of hydrogenated amorphous Si on (100)Si. (c) Spectrum of (100)Si overetched for 10 min by RIE. (d) Spectrum of (100)Si after exposure to 10^{18} Ar^+/cm^2 at 500 eV. (e) Spectrum of (100)Si after exposure to 10^{18} protons/cm^2 at 100 eV. Single crystal substrate scattering subtracted from (b) to (e). Derivative-like structure at 522 cm^{-1} is due to the subtraction process. From Tsang et al. (61).

scattering are forbidden by the polarization selection rules for RS. By working in this geometry, the sensitivity to any a-Si layer at the surface is enhanced since the Raman spectrum of a-Si is unpolarized. The sharp line at 522 cm^{-1} is due to $q \approx 0$ crystalline TO phonon scattering in c-Si, while the broad bands near 300 and 630 cm^{-1} are due to second-order scattering. All of these c-Si-derived strucures would be two orders of magnitude stronger in any scattering geometry where they are nominally allowed. In Fig. 6.14(b) is shown the Raman spectrum obtained from a 60-Å layer of hydrogenated a-Si on (100)Si. The first- and second-order Si substrate scattering has been subtracted out. The broad band scattering between 150 and 600 cm^{-1} is due to the vibrational modes of the amorphous Si. The intensity of the 480 cm^{-1} band is four to six times stronger than that of the forbidden second-order scattering around 300 cm^{-1}. Figure 6.14(c) was obtained by taking the Raman spectrum of an RIE-processed (100)Si wafer that was overetched for 10 min and subtracting from it the Raman spectrum of a substrate that had not been RIE processed. The resulting spectrum shows a well-defined, though broad peak at 480 cm^{-1} that is not observed in the substrate scattering and is identical to the disorder-induced structure observed in the Raman spectrum of a-Si. The intensity of the RIE-induced amorphous Raman scattering in Fig. 6.14(c) is comparable to that of the nominally symmetry-forbidden second-order Raman scattering in Fig. 6.14(a) and weaker than the scattering shown in Fig. 6.14(b).

Figure 6.14(d) is the Raman spectrum obtained from a (100)Si wafer after exposure to 10^{18} Ar^+/cm^2 with an energy of 500 V. The spectrum was obtained using the same scattering geometry as shown in Fig. 6.14(a). The symmetry-forbidden first- and second-order Raman spectrum of (100)Si has been subtracted from the spectrum in Fig. 6.14(d). The broad structures in Fig. 6.14(d) correspond to the Si phonon density-of-states features in Fig. 6.14(b, c). However, the intensity of the disorder-induced spectrum due to Ar^+ bombardment is quantitatively stronger than the similar spectrum arising from the RIE processing, dwarfing the two phonon bands near 300 cm^{-1}. In Fig. 6.14(e) is shown the Raman spectrum obtained from a (100)Si surface bombarded by 10^{18} protons/cm^2 at 1000 V. Again the spectrum is obtained by subtracting the spectrum of an unprocessed sample from that of the processed sample. The spectral shape and intensity of the damaged-induced spectrum in this case are similar to that which we observed from the RIE overetched samples but much less intense than the scattering seen in Fig. 6.14(d).

6.4.5. Strain Effects

The Raman spectrum can be used to measure stress in crystalline samples since it is sensitive to the strain in the lattice. Under stress, there are changes in the spring constants (and hence frequencies) that characterize the phonon dispersion of solids (8, 62, 63). The phenomenological coefficients p, q, and r describe the change in spring constant due to an applied strain. The parameters p, q, and r of Cardona et al. (62, 63) are designated K_1, K_2, and K_3, respectively, in Anastassakis (8).

Let us consider a two-dimensional strain in the (100) plane. This situation occurs, for example, when expitaxial silicon films are grown on (1012) sapphire. Such a strain splits the normally triply degenerate $q=0$ phonon of a material such as Si into a doublet and a single (8, 62, 63). In backscattering from (100) only the singlet mode is observed. It can be shown that the frequency shift, $\Delta\omega$, of this mode is given by

$$\frac{\Delta\omega}{\omega_0}=\left[\frac{q}{\omega_0^2}-\left(\frac{C_{12}}{C_{11}}\right)\frac{p}{\omega_0^2}\right]T, \tag{64}$$

where T is the stress, ω_0 is the unperturbed Raman frequency (522 cm^{-1} for Si), and C_{11} and C_{12} are elastic compliance constants. The direction of the shift gives the sign of the stress (tensile or compressive), while the magnitude of the shift yields the magnitude of the strain.

Raman spectroscopy has been used by a number of investigators to study stress effects in Si/SiO_2, Si/Al_2O_3, Si/CaF_2, or other surfaces (64–69). The method is direct and nondestructive and can be applied at various temperatures. For example, Englert, Abstreiter, and Pontcharra determined the built-in stress in silicon on sapphire devices. They found that epitaxially grown silicon films having an orientation (100) on (1012) sapphire substrate develop a built-in compressive stress of 7 kbar at room temperature. Their results are shown in Fig. 6.15 (64). The origin of this stress is the differential thermal expansion coefficient between silicon and sapphire. Since the epitaxial growth is usually at temperatures above 750°C, on cooling the larger thermal expansion of sapphire results in a two-dimensional compressional stress in silicon.

Recently, the Raman microprobe technique has been used to map the local stresses in patterned islands of laser-crystallized silicon on bulk glass

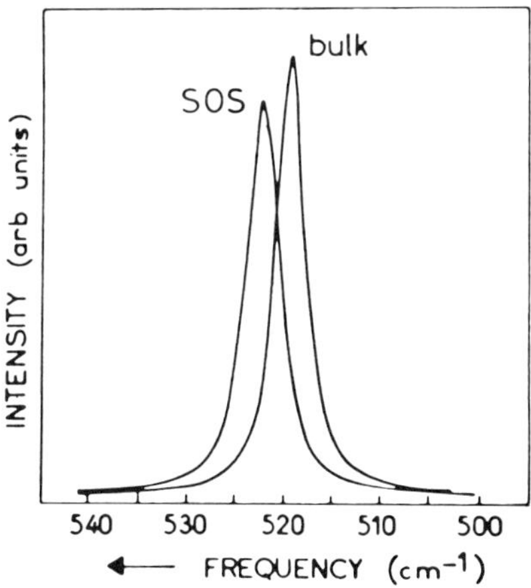

Figure 6.15. Raman spectra of bulk silicon and silicon on sapphire (SOS). The shift is due to built-in stress in the silicon film. From Englert et al. (64).

(70) and on a laser-crystallized, seeded lateral epitaxial silicon film on an oxidized silicon wafer (71). Lyon et al. (70) found that the stress increased from 2 kbar in the seed region to 5 kbar in the SOI region at distances greater than 20 μm from the seed-silicon. boundary. Depth variations of the stress may be obtained by using different wavelengths of the exciting laser lines. This type of Raman measurement adds a new dimension to the study of grain-boundary formation.

A unique application of RS for the study of microadhesion has been recently introduced by Gonzalez-Hernandez, Martin, and Tsu (72). The technique involves the correlation of the measured Raman shift of the zone-center optical phonon frequency with the differential thermal expansion coefficient between the silicon film and the substrate material. It was observed, for example, that silicon films, whether epitaxially grown or in microcrystalline form, may or may not show a shift corresponding to the values compatible with thermal expansion dependent upon the degree of microadhesion. Furthermore, when amorphous silicon is crystallized, the stress-free condition is always reached at the crystallization temperature regardless of the annealing temperature.

As noted previously, Campbell, Fauchet, and Adar (17) have observed a frequency shift $\Delta\omega$ for laser-processed silicon (see Fig. 6.10). A portion of this shift is due to microcrystalline effects, the remaining shift being caused by the differential strain between the silicon and insulating substrate. Campbell et al. (17) have used an analysis similar to Eq. (64) to evaluate this strain.

6.4.6. The Metal/Si Interface

Several groups have recently used RS to probe reactions at the metal/Si interface (73–84). The experimental characterization of such an interfacial layer is sometimes difficult since it is often buried under a much thicker layer of unreacted metal. Thus surface-sensitive methods such as UPS (ultraviolet photoemission spectroscopy), XPS, and Auger spectroscopy are at a disadvantage. However, RS can be employed since the penetration depth of the light for a typical laser such as Ar^+ is several hundred angstroms, even in a metal. The ability to probe more than 100 Å below the surface makes RS a powerful tool facilitating characterization of such silicide interfaces.

Tsang et al. (77) have used multichannel Raman spectroscopy to quantitatively characterize the formation of PtSi at the interface with (100)Si. Raman spectra from PtSi layers as thin as 10 Å and from >40 Å of PtSi under 140 Å of Pt have been observed. The Raman spectra can be used to identify the PtSi layer, estimate its thickness, and describe its crystallographic order.

Figure 6.16 (77) shows the Raman spectrum of PtSi for different Pt

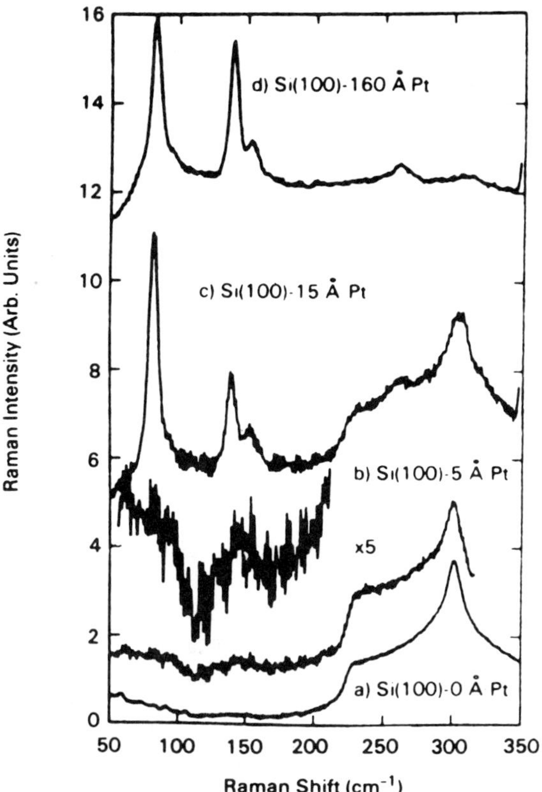

Figure 6.16. The Raman spectra of Pt on Si after reaction for 30 min. (a) (100)Si with no Pt. (b) (100)Si covered with 5 Å of Pt and reacted at 500°C. Center section × 5. (c) Si covered with 15 Å of Pt and reacted at 500°C. (d) (100)Si covered by 160 Å of Pt and reacted at 500°C. Here, a–c are normalized against the intensity of the second-order Raman scattering of Si. From Tsang et al. (77).

thicknesses. Figure 6.16(a) is the Raman spectrum of Si. The structure between 220 and 350 cm^{-1} is due to two-phonon scattering. The weak structure at 150 cm^{-1} has been attributed to defect-induced scattering, while the sharp lines between 50 and 120 cm^{-1} are due to atmospheric scattering. The PtSi spectra shown in Fig. 6.16(b–d) were obtained on samples annealed for 30 min at 500°C. TEM of these samples shows only the presence of PtSi with no other phases (for example, Pt_2Si or pure Pt) detected. Figure 6.16(b) is the Raman spectrum from the reaction of 5 Å of Pt on (100)Si. There is a shoulder superimposed on the edge of the elastic scattering background at about

$85\,cm^{-1}$ and a well-defined line at $140\,cm^{-1}$. These structures are the same as those seen in Fig. 6.16(c), the Raman spectrum of PtSi from 15 Å of Pt on (100)Si after 500°C annealing. These are sharp lines at 82 and $140\,cm^{-1}$, with weaker structures at 150 and $260\,cm^{-1}$. Figure 6.16(d) is the Raman spectrum from (100)Si reacted with 160 Å of Pt optically, essentially bulk PtSi. These phonon energies are identical to those obtained by Nemanich et al. (74, 75) for PtSi, although the relative intensities of various lines differ. The strength of the $82\,cm^{-1}$ mode could mask a weak signal at $88\,cm^{-1}$ from a small amount of Pt_2Si if present. The similarities of the Raman spectra in curves (b)–(d) demonstrate that RS can be used to identify PtSi as thin as 10 Å.

The second-order RS from the Si substrate provides a means to normalize the intensity of the RS from the silicide for an accurate sample-to-sample comparison. By this means, the line near $140\,cm^{-1}$ in Fig. 6.16(c) is five to ten times more intense than in Fig. 6.11(b). Thus, in the absence of optical attenuation due to the PtSi, the normalized RS of the film reacted from 15 Å of Pt should be three times stronger than that of the 5-Å film. The optical attenuation due to the silicide layer will increase this ratio, making it close to that observed experimentally.

Figure 6.17 (77) shows the evolution of the Raman spectrum of PtSi on (100)Si by reaction at the (100)Pt/Si interface. Figure 6.17(a) is obtained from the deposition of 15 Å of Pt at room temperature. Surface electron spectroscopy shows that silicide-like chemical bonds are formed within the first 10–20 Å of Pt deposition. The Raman spectrum shows broadened features between 80 and $150\,cm^{-1}$ indicative of these silicide-like bonds. Figure 6.17(b–e) shows the growth of PtSi by thermally induced contact reaction of a thicker, unreacted Pt metal overlayer. Figure 6.17(b) is the spectrum obtained from the deposition of 160 Å of Pt on Si without annealing. The scattering from the Si substrate and the PtSi interface is greatly reduced by the attenuation of the unreacted Pt. There is little evidence in Fig. 6.17(b) for the initially formed silicide layer seen in Fig. 6.17(a). This stems from the poor signal-to-noise ratio and the discontinuous morphology of the 15-Å Pt film. The Raman spectra obtained from 160 Å of Pt on (100)Si after 30- and 480-min anneals at 230°C are shown respectively in Fig. 6.17(c, d). The emergence of structure at 82 and $140\,cm^{-1}$ in these spectra is observed.

Annealing for 30 min at 300°C produces a silicide Raman spectrum [Fig. 6.17(e)] at least 20 times stronger than the spectrum in Fig. 6.17(d). While TEM observes the PtSi phase only on the 300°C annealed sample, it barely resolves the reacted silicide layer in the sample used for Fig. 6.17(d); it is estimated that the silicide thickness is between 20 and 60 Å, which is consistent with the relative intensities of Fig. 6.17(d, e). The linewidths of the fully reacted

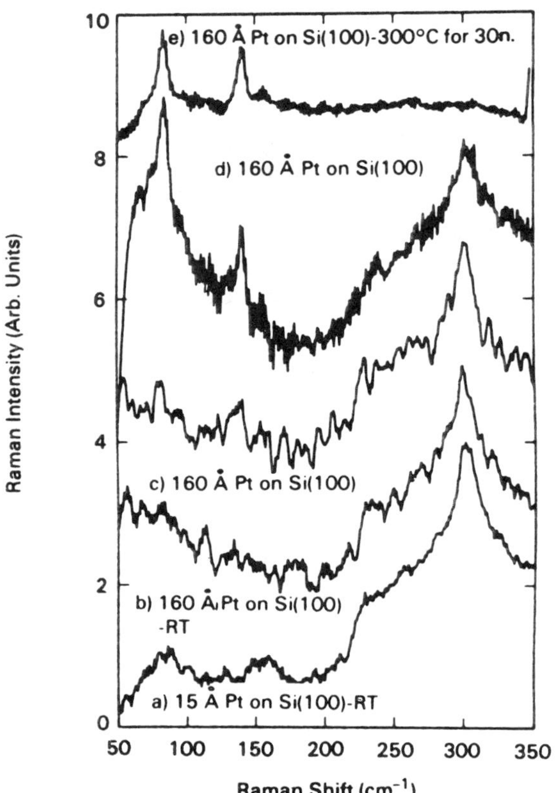

Figure 6.17. The Raman spectrum of (100)Si covered by Pt as a function of annealing temperature. (a) Si covered by 15 Å Pt at room temperature. (b) Si covered by 160 Å of Pt at room temperature. (c) Si covered by 160 Å Pt and annealed at 230°C for 30 min. (d) Si covered by 160 Å Pt and annealed at 230°C for 480 min. (e) Si covered by 160 Å Pt and annealed at 230°C for 30 min. From Tsang et al. (77).

PtSi are less than 6 cm^{-1}. In contrast, the initial room-temperature reaction shows broad Raman lines with a width greater than 20 cm^{-1}. The former widths are characteristic of well-ordered, crystallized materials, whereas the latter are characteristic of materials in which there is short-range, local order but no long-range order. The appearance of sharp lines in Fig. 6.17(d) suggests the silicide growing from the Pt-Si interface in the 160-Å Pt samples has long-range order, even when it is only 20–60 Å thick.

A well-defined Pt_2Si phase is not observed on these samples in TEM and surface spectroscopy or these Raman measurements under the deposition and growth/reaction conditions of the UHV preparation.

Nemanich et al. (82) have recently reported a Raman study of the variations of thin-film titanium on silicon. The experiments involve Ti thin films deposited on (100)p-Si wafers. Deposition was performed by Ar^+ sputtering or e-beam evaporation in vacuum systems with base pressures in the 10^{-7}–10^{-8} Torr range. These substrates were then processed in a rapid thermal anneal system or in an UHV chamber. Raman spectra were obtained after various processing stages *in situ* in the UHV system. The results indicate the simultaneous formation of crystalline Ti_2O_3 and a Ti silicide tentatively identified as TiSi. Higher-temperature annealing to greater than 750°C leads to the formation of TiS_2 and the disappearance of the Ti_2O_3 signal.

The Raman spectra after the various rapid annealing and processing steps are summarized in Fig. 6.18 (82). The as-deposited sample shows sharp peaks

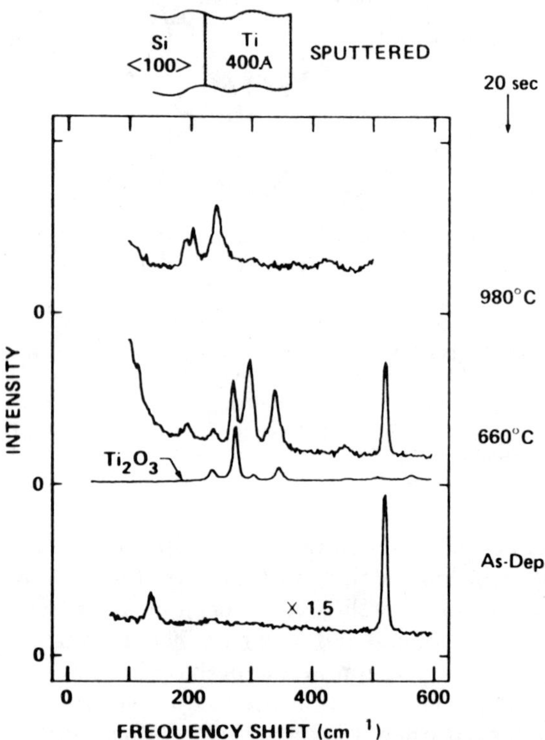

Figure 6.18. Raman spectrum of 400 Å of Ti on (100)Si as deposited and after 660°C and 980°C anneals. The 980°C anneal followed a 650°C anneal and Ti strip. The Raman spectrum of Ti_2O_3 is shown for comparison. From Nemanich et al. (82).

at 138 and 520 cm^{-1}, which are attributed to the Ti film and the underlying Si substrate, respectively. A weak, broad background is attributed to surface oxidation of the titanium. After annealing to above 550°C, the spectrum exhibits a noticeable change (Fig. 6.18). The line due to Ti disappears, and several peaks are observed in the 250–30 cm^{-1} range. A first analysis might attribute the lines to the vibrations of a titanium silicide, but because the Raman spectrum of the titanium silicides has not been previously reported, other possibilities must also be considered. A likely possibility is oxide formation, and shown also in Fig. 6.18 is the Raman spectrum of crystalline Ti_2O_3. It is clear that the lines at 270, 290, and 320 cm^{-1} in the annealed films exhibit the intensity dependence expected for Ti_2O_3. While the strongest line in the spectrum of the reacted TiSi located at about 305 cm^{-1} does correspond to a weak line in the Ti_2O_3 spectrum, the intensity is inconsistent with an assignment to Ti_2O_3. The *in situ* UHV experiments clearly indicated that this line is due to Ti silicide formation.

The only variation in the spectra upon annealing the above samples to temperatures from 55° to 750°C was in the ratio of the new spectral lines to that of Si at 520 cm^{-1}. This change is attributed to an increased thickness of the reacted film. While the thickness of the reacted layer grew, the structure and relative ratio of the two components would seem to have been unaffected. Furthermore, the samples were etched to remove the remaining unreacted Ti, and the etch process did not affect the relative intensities of the silicide and Ti_2O_3 lines.

Upon annealing to temperatures greater than 800°C, a distinct spectral change was observed; 800°C lies in the temperature region where TiS_2 is expected to form. The authors attribute the two sharp peaks at 280 cm^{-1} to the presence of this silicide. More intriguing is the absence of sharp lines due to Ti_2O_3. Furthermore, there was no evidence of TiO_2, which is the fully oxidized state of Ti.

Codella et al. (83) have reported a Raman microprobe analysis of tungsten silicide. Using a MOLE (the commercially available Raman microprobe, Molecular Optics Laser Examiner) on an annealed sample of tungsten deposited over crystalline silicon, they have observed for the first time the RS of WSi_2. A similar examination of thin tungsten lines, 8 μm wide by 200 Å thick, selectively deposited on a c-Si surface using laser-induced CVD techniques, produced an identical spectrum when superimposed with that of the silicon substrate.

A comprehensive study of rapid thermally annealed tungsten silicide formation using RS (microprobe mode), resistivity measurements, Auger and Rutherford backscattering, and scanning electron microscopy has been reported by Kumar et al. (84). The observed changes in the RS spectra have been correlated with the change in resistivity and surface morphology.

6.4.7. Laser Writing of Si Microstructures

Raman microprobe techniques have been used as a real-time probe during local direct laser writing and also as an *in situ* probe after writing (85). The Stokes–Raman emission observed *during* pyrolytic deposition of micrometer-dimension structures of silicon on germanium and vitreous carbon substrates is found to be weaker and more asymmetric and to peak at a smaller Raman shift than the corresponding spectrum of the same structure similarly probed *in situ after* deposition. Results of detailed postdeposition Raman analysis of these silicon microstructures are presented and compared to the Raman spectra of oven-heated silicon. Potential applications of these techniques are discussed in this article (85).

6.4.8. Temperature Probe

RS can also be used as a sensitive probe of local temperatures in Si (or Ge) (86, 87). For example, the frequency of the $q \approx 0$ phonon line ($522\,\mathrm{cm}^{-1}$ in Si at 300 K) recorded during continuous wave (CW) laser-beam heating of Si has been used to characterize the lattice temperature inside the laser spot. Raptis et al. (87) have reported the use of RS in Si to determine temperature inhomogeneities generated by a critically focused CW laser beam. Thus RS can be used not only to evaluate temperature but also to gain information about temperature distribution.

6.5. BINARY ZINCBLENDE SEMICONDUCTORS

A considerable amount of work on characterization of materials has been done in zincblende-type semiconductors (20, 21, 88); binary materials as well as ternary and quaternary alloy systems have been studied. Most of this work has focused on GaAs and related materials such as GaAlAs.

In a zincblende-type material the $q \approx 0$ LO and TO phonons are not degenerate as in the case for binary materials, and hence the first-order Raman spectrum, in general, shows two peaks corresponding to TO and LO phonon scattering. The lower-frequency peak corresponds to the TO phonon, while the higher-frequency structure is scattering from the LO phonon. This identification can be verified by the selection rules discussed above. For example, in backscattering, neglecting LQ and SF effects, only LO is observed for (100), only TO is seen from (100), while both LO and TO are allowed from (111) (see Tables 6.1 to 6.3).

For ternary alloys the situation is somewhat more complex. Most ternary alloys of zincblende-type semiconductors display a "two-mode" behavior.

For example, in GaAlAs two sets of TO and LO phonons are observed, corresponding to the "GaAs-like" modes and "AlAs-like" modes (20, 88). We shall discuss this point in more detail later.

6.5.1. Carrier Concentration and Space/Charge Layer Effects

For polar semiconductors the net electric dipole moment due to the lack of inversion symmetry couples the LO phonons with the longitudinal plasma oscillations of the free carriers (6, 20, 21, 88–91). This coupling is demonstrated in Eq. (39). The frequencies of these coupled plasmon–LO phonon modes are given by ω_+ and ω_- of Eq. (41); the corresponding peaks are generally denoted as L_+ and L_- (or ω_+ and ω_-), respectively.

Shown in Fig. 6.19 (89) is the Raman spectra from the (100) surface of GaAs with three different n-type carrier concentrations using the 5145 Å (2.41 eV) line of an Ar^+ laser. The (100) surface is used since LO phonon

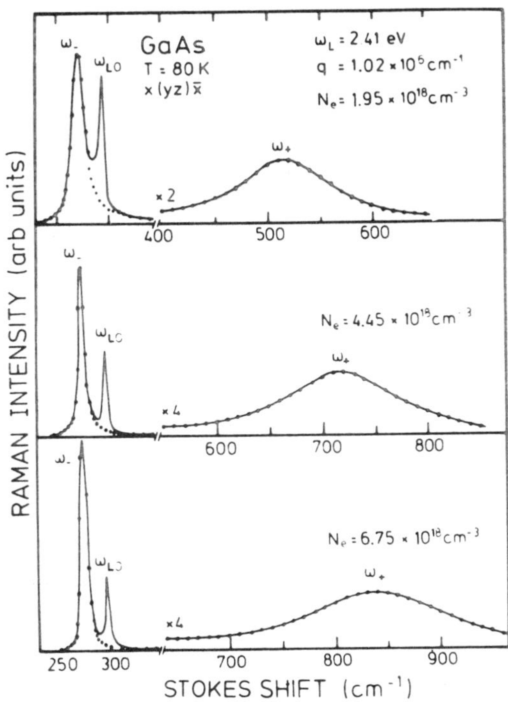

Figure 6.19. Raman spectra of three different n-GaAs samples obtained in backscattering geometry from (100) surfaces. From Abstreiter et al. (89).

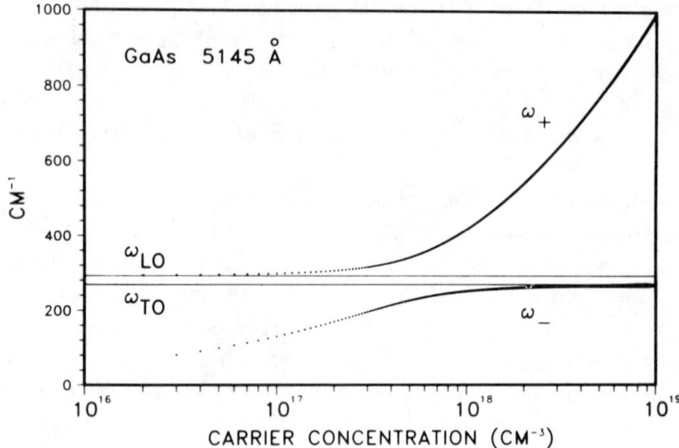

Figure 6.20. Calculated plasmon–phonon mode frequencies ω_+ and ω_- as a function of free-carrier concentration for Raman scattering from GaAs at 5145 Å. The solid lines correspond to the bulk LO and TO modes. The calculations have been corrected for finite q-vector and nonparabolic effective mass effects. From Schwartz et al. (90).

scattering is allowed from this face and it is this phonon (not TO) that couples to the plasmon. The features labeled ω_- and ω_+ are due to the coupled plasmon–LO phonon modes. Note also that the unscreened LO phonon mode at 292 cm^{-1} is also observed (ω_{LO} in Fig. 6.19). We shall return to this important point later. Plotted in Fig. 6.20 (90) are the values of ω_+ and ω_- as a function of the carrier concentration from Eq. (41) for the case of GaAs. From the frequency positions of ω_+ and ω_- modes it is possible to determine the carrier concentration v in a contactless manner. For $N > 1 \times 10^{18}\ \mathrm{cm}^{-3}$ the ω_+ line is more useful since ω_- saturates at the position of the TO phonon mode. Similar curves have been obtained for n-InP (see 90) and n-InAs (see 91).

Note that data for Figs. 6.19 and 6.20 were collected at the 5145 Å excitation laser line. Some care must be taken when one is using other excitation frequencies. Equation (41) is rigorously true only for $q \approx 0$. A more general expression for the plasma frequency is (6, 86)

$$\omega_p(N, q)^2 = (4\pi N e^2/\mathrm{m}^*\varepsilon_\infty) + \tfrac{3}{5}(qv_F)^2, \tag{65}$$

where v_F is the conduction Fermi velocity and q is given by Eq. (10). Thus, the plasma frequency has a q-dependence (and hence a wavelength dependence). Corrections for various commonly used laser frequencies are given in Abstreiter et al. (88).

For carrier concentration $N > 1 \times 10^{18}\,cm^{-3}$ the Fermi level lies high enough into the conduction band so that the nonparabolicity of the conduction-band effective mass (which enters into the plasma frequency ω_p) [see Eq. (40)] must be taken into account (6, 90). This has been done in Schwartz et al. (90).

As noted before, the Raman spectra of Fig. 6.19 display not only ω_+ and ω_- but also the unscreened LO phonon mode. This mode shows up because of the depletion region in *n*-GaAs (and other *n*-type semiconductors), which is related to Fermi-level pinning. For a wavelength of 5145 Å the penetration depth of the light is approximately 1000 Å and hence both the depletion region and the bulk are sampled (89). The unscreened LO phonon mode originates in this depletion region since there are no free carriers there. The ω_+ and ω_- modes arise from the bulk where the free carriers are present. We shall return to this point later.

It should be pointed out that great care must be exercised in evaluating the carrier concentration from RS since too high a laser-power density will create photo-generated carriers, thus leading to spurious results.

Schwartz and co-workers (90) have recently used the positions of ω_+ and ω_- to examine variations in free carrier concentration of S and Sn, and Ge-doped (100)InP and Si-doped (100)GaAs induced by vacuum annealing.

The first-order RS from heavily doped and highly compensated GaAs:Si with a constant impurity concentration of $1.1 \times 10^{19}\,cm^{-3}$ has been reported by Kamijoh et al (92). The damped $q \approx 0$ ω_- and ω_+ branches due to the compensation-limited low mobilities have been observed. These authors also have observed localized vibrational modes from Si-related defects, including Si-on-As sites and Si-on-Ga sites.

Nakamura and Katoda (93) have examined the effects of optically excited carriers on the Raman spectra from InP for various power densities of the incident laser beam. Concentrations of the optically excited excess carriers, ΔN, evaluated from the frequency of the coupled LO phonon–plasmon mode were in the range 0.5×10^{17} to $1.7 \times 10^{17}\,cm^{-3}$ at the laser power density of $5 \times 10^2\,W/cm^2$ for equilibrium carrier concentrations of 0.9×10^{17} to $5.0 \times 10^{17}\,cm^{-3}$. The surface recombination velocities of 1.0×10^3 to $8.0 \times 10^3\,cm/2$ were estimated from ΔN assuming that the bulk lifetime was greater than 10^{-7} s.

For *p*-GaAs and other zincblende semiconductors the situation is much more complex because of the degeneracy of the valence band and the short lifetime of the carriers. Hence, the coupled modes cannot conveniently be used as in the *n*-type materials (6).

Shen et al. (94, 95) have employed RS as a nondestructive contactless method for determining not only the free-carrier concentration ν but also the width of the space-charge layer in (100)*n*-GaAs with $4 \times 10^{17}\,cm^{-3} <$

$N < 1 \times 10^{19}\,\text{cm}^{-3}$, using as an excitation several different wavelengths of an Ar^+ laser. By comparing the intensity at different wavelengths of the uncoupled LO phonon mode (ω_{LO} in Fig. 6.19) originated in the space-charge layer with the signal from a piece of undoped (100) material, it is possible to experimentally evaluate the width of the depletion layer L_s. These authors find that there is very good agreement between the experimental values and those obtained from a generalized theory for both degenerate materials. Thus, these experimental results demonstrate that for (100) III–V semiconductors, RS can be used as a contactless method to determine the width of the space-charge region for carrier concentrations up to $1 \times 10^{19}\,\text{cm}^{-3}$.

Stolz and Abstreiter (96) have used the properties of the [110] surface to study SF effects. As discussed previously, RS from the LO phonon in the backscattering configuration is normally forbidden from the [110] surface. However, SFs or LQ terms can induce LO scattering from the (110) face (see Table 6.4). Once LO phonon scattering is activated it is also possible to see other manifestations of this mode such as coupled plasmon–LO modes. Stolz and Abstreiter were able to utilize these coupled modes to study band bending on UHV-cleaved (110) GaAs surfaces.

RS from heavily doped *n*-GaAs with and without a semitransparent metal film has been investigated by Tsu et al. (97). They have reported a shift of the LO phonon frequency from $292\,\text{cm}^{-1}$ to as high as $302\,\text{cm}^{-1}$ depending on the applied bias voltage. They speculated that the presence of a metal contact may increase the surface recombination time of the laser-excited carriers resulting in a shift to higher frequency due to plasmon–phonon coupling. Although a systematic study of the intensity dependence was not pursued in their work, it seems that such an investigation can further clarify the mechanism as well as the possible application to the field of surface recombination and steady-state carrier injection. Abstreiter (98) also has reported on Raman spectra obtained from highly doped *n*-GaAs with a Ni-Schottky barrier on top of the (100) surface. A strong decrease in the amplitude of the coupled LO-phonon plasmon mode L_- is observed when the backward bias voltage is increased while the unscreened LO phonon mode gains. These effects are due to the increased depletion width caused by the applied bias voltage.

RS at resonance at the E_1 gap (Λ_3–Λ_1 transitions in the BZ) in InAs substrates has been used to obtain information about the space/charge region. A narrow peak just below the unscreened LO peak for both *n*- and *p*-InAs at 77K for zero and positive gate voltages was observed by Ching et al. (99). It has been attributed to scattering by LO phonons coupled to collective inter-subband excitations, which is the counterpart of the coupled plasmon–phonon mode in a two-dimensional electron–plasmon system.

Pinczuk et al. (100) have studied RS from the surface depletion layer of etched (111)B surface of InP before and after evaporation of a semitransparent Ag film. They have observed an increase in the band-bending upon evaporation of the metal film, which then forms a Schottky barrier.

The effects of an amorphous phosphorus (P) overlayer on the intrinsic properties of the surface of *n*-GaAs have been investigated by means of RS and photoluminescence spectroscopy (101). It is reported that the surface barrier is lowered from 0.7 to 0.18 eV and that there is a reduction of one order of magnitude in surface-recombination velocity. These results indicate that the Fermi level is no longer pinned at midgap but is moved closer to the conduction band. A model is proposed in which the P atoms at the interface reduce the number of As^- missing lattice defects that pin the Fermi level at midgap in *n*-GaAs (93).

6.5.2. Microcrystalline Effects

In relation to μc-Si little work has been done on μc-III–V semiconductors. Figure 6.21(a, b) shows the evolution of the TO and LO phonons of GaAs

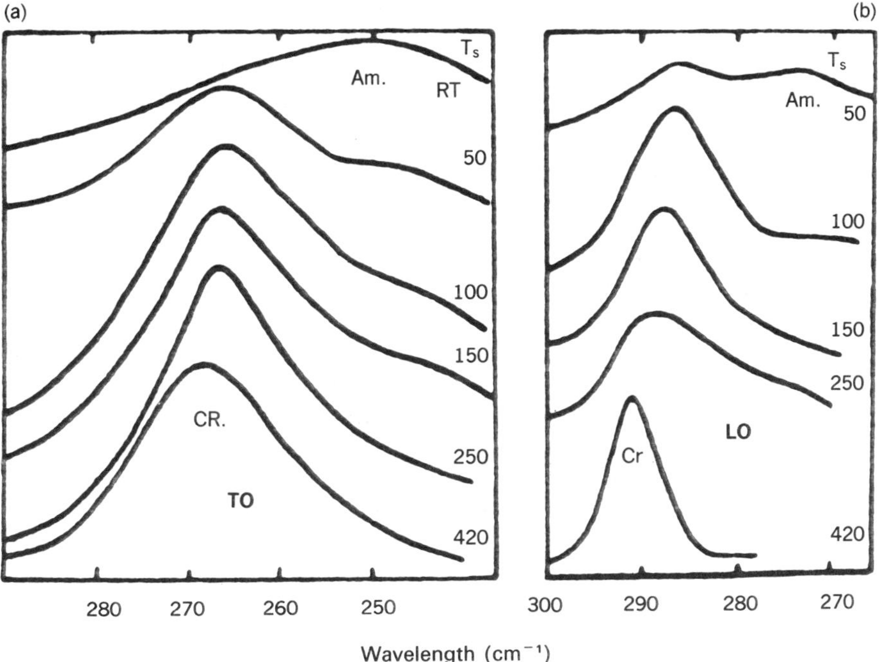

Figure 6.21. (a) Evolution of the TO phonon reflectance spectra of GaAs with T_s. (b) Evolution of the LO phonon reflectance spectra of GaAs with T_s. From Paparodites et al. (102).

(as evaluated from the reststrahlung region of the IR reflection spectra) as a function of substrate temperature T_s. Note that as a function of T_s the LO phonon peak shifts to the blue (and also narrows) while the TO phonon peak position remains almost constant. This is consistent with the model of Eq. (60) since there is a considerable dispersion in the phonon frequency of the LO phonon, whereas the TO phonon frequency is quite flat over the BZ [see Fig. 6.3(b)]. Paparodites et al. (102) did not attempt a fit in order to evaluate L.

6.5.3. Symmetry-Forbidden Transverse Optical Mode Scattering

As discussed before, TO phonon scattering is normally forbidden in back-scattering from the (100) surface (see Table 6.1). It can be shown from Eq. (50) that SF or LQ effects cannot activate the TO phonon from this surface. However, this symmetry-forbidden-TO mode has been observed by many workers from the (100) face of GaAs (see 20, 88, 103–108) or InP (see 108–111). Several investigators have discussed the origins of this mode (20, 88, 111). A very detailed analysis is given by Biellmann et al. (112). The amplitude of the symmetry-forbidden-TO mode in relation to the symmetry-allowed-LO phonon can be used as a measure of crystal quality (20, 103–112).

Raman characterization of twinning in heteroepitaxial semiconductor layers of GaAs/(Ca, Se)F_2 has recently been reported by Landa et al. (108). A detailed analysis of the presence of the TO phonon mode recorded from (100)GaAs grown by molecular beam epitaxy (MBE) on the lattice-matched insulator (Ca, Sr)F_2 gives evidence of internal misorientational effects (twins). Calculations have been performed in order to obtain quantitative evaluations of the misoriented volume amount.

6.5.4. Ion Implantation and Annealing

Ion implantation and annealing (laser, thermal, flash lamp) of semiconductors are processes of considerable technological importance. Raman spectroscopy is an extremely useful tool for obtaining information about the state of implanted or annealed material since, as discussed above, it is sensitive to interactions on the order of a few lattice constants. A number of Raman studies on ion-implanted and/or annealed III–V semiconductors, including GaAs (see 49, 104–107, 113–119), InP (see 110, 111, 113), and GaP (see 113) have been reported.

It has recently been shown that for ion implantation the "spatial correlation" model (SCM) discussed above can be used to analyze the details of the Raman lineshape, i.e., broadening and asymmetry (113). This analysis then yields the average size of the undamaged region.

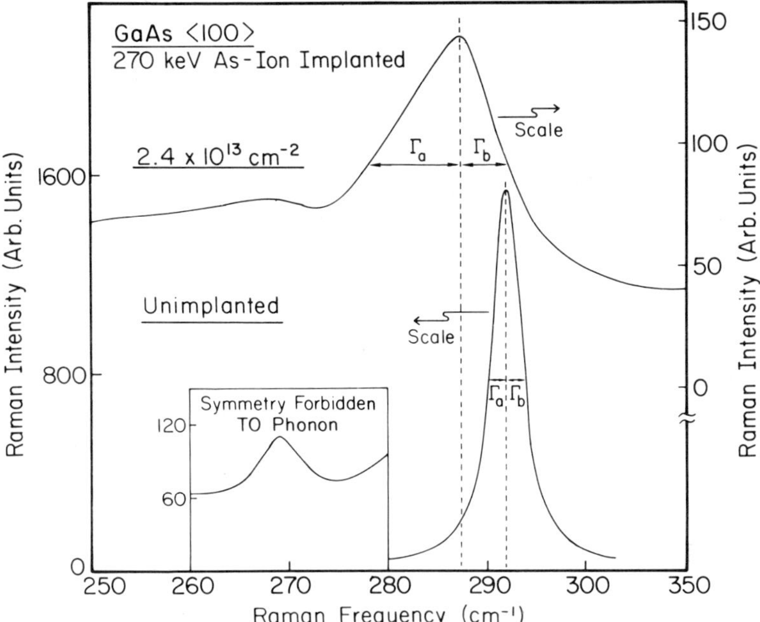

Figure 6.22. Expanded version of spectra of unimplanted and As-ion implanted ($2.4 \times 10^{13}\,cm^{-2}$) (100)GaAs in the range of 250 to $350\,cm^{-1}$. From Tiong et al. (113).

In Fig. 6.22 (113) is shown, from 250 to $350\,cm^{-1}$, an expanded version of the LO and symmetry-forbidden TO phonons for (100)GaAs for both the unimplanted and $2.4 \times 10^{13}\,cm^{-2}\,As^{-}$-implanted samples. The FWHM, $\Gamma = (\Gamma_a + \Gamma_b)$, is indicated on both spectra. The LO phonon line for the implanted sample has (a) shifted to lower frequencies and (b) broadened asymmetrically. However, the TO features show only a small red shift.

In an ideal crystal, the region over which the spatial correlation function of the phonon extends is infinite. This leads to the usual plane wave phonon eigenstates and the $q \approx 0$ momentum selection rule of first-order RS. However, as the crystal is damaged by ion bombardment, the mode-correlation functions become finite owing to the induced defects. Thus, there is a relaxation of the $q \approx 0$ selection rule and, associated with this relaxation, a finite correlation length. As discussed previously, the assumption of a Gaussian attenuation factor, $\exp(-2r^2/L^2)$, where L is the diameter of the correlation region, leads, upon Fourier transformation, to an average over q with a similar weighting factor, $\exp(-q^2L^2/8)$.

It is assumed that there is also a spherical region associated with the finite size of the correlation regions in the damaged material. Then the Raman intensity at a frequency $\omega, I(\omega)$, can be expressed by Eq. (60) with the dispersion given by Eq. (61). For GaAs the parameters $A = 4.26 \times 10^4\ \text{cm}^{-2}$ and $B = 7.11 \times 10^8\ \text{cm}^{-4}$ [where A and B are from Eq. (61)] reproduce quite well with actual dispersion of the LO phonon in GaAs along (100). For simplicity we assume a spherical BZ.

Figure 6.23 (113) shows the shift $\Delta\omega_{\text{LO}}$ of the maximum of $I_{\text{LO}}(\omega)$ away from 292 cm^{-1} and of the LO line as a function of L as evaluated from Eq. (60) (solid line). [Tiong et al. (113) used a slightly different equation for the dispersion relation in comparison to Eq. (61). Thus, there are slight differences in L between Fig. 3 in their paper and Fig. 6.23 herein.] Also shown are the experimental values of $\Delta\omega_{\text{LO}}$ and Γ for various fluences. From the above equations it is also possible to evaluate the asymmetry Γ_a/Γ_b in relation to $\Delta\omega_{\text{LO}}$ for various L [Fig. 4 of Tiong et al. (113)]. The agreement in both cases is quite good. Thus from Fig. 6.23 [and Fig. 4 of Tiong et al. (113)] it is possible to relate the shift, broadening, and asymmetry of the LO phonon for a given fluence to an average undamaged particle size. For example, a fluence of $2.4 \times 10^{13}\ \text{cm}^{-2}$ corresponds to an $L = 45$ Å. Such a relation

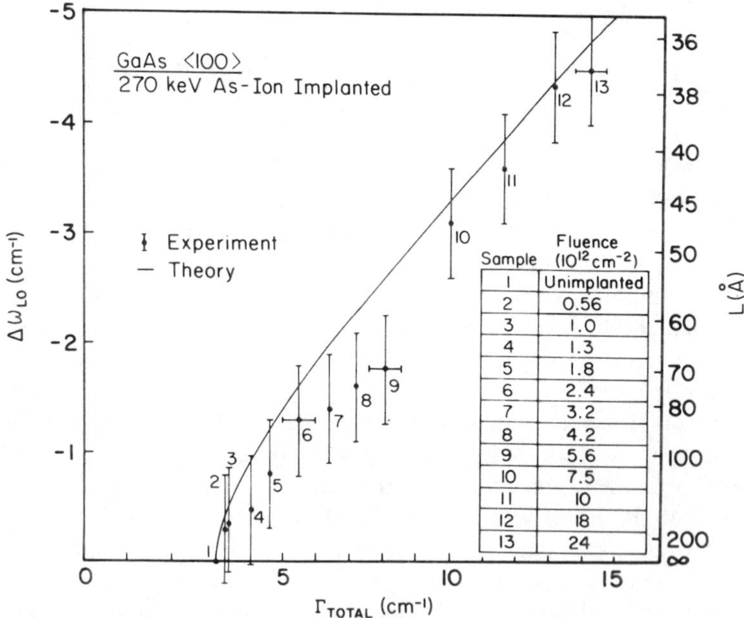

Figure 6.23. The LO phonon shift, $\Delta\omega_{\text{LO}}$, and broadening, Γ, as a function of L as determined from Eqs. (60) and (61) for GaAs (solid line). Also shown are experimental points for various fluences. Modified from Tiong et al. (113); see comments in the text.

between fluence and average undamaged region size is consistent with the results of spectroscopic ellipsometry (120).

The SCM also explains the fact that the TO phonon exhibits only a small red shift. For this mode, in GaAs as in other III–V materials, there is little dispersion: $\Delta\omega_{TO}$ decreases by only about 10–15 cm^{-1} from zone center to the edge. This dispersion, when substituted in Eq. (60), yields only a slight shift to lower frequencies with decreasing L down to about 45 Å. The evolution of zone-edge phonons with increasing fluence is a density-of-states effect that can also be understood on the basis of the SCM. From Eqs. (60) and (61), the intensity corresponding to the LO zone-edge phonon at 247 cm^{-1} is zero for large L but increases monotonically with decreasing L.

The success of this approach, which is quite general, makes it possible to obtain from the Raman spectra of the LO phonon an average size for the undamaged regions for a given fluence. Although the SCM, based on a spherically symmetric Gaussian cutoff, yields good agreement for our experimental conditions, different weighting functions may be appropriate in other cases. Further work is needed in this area.

Nakamura and Katoda (93) have investigated the size of the crystalline particles in Si-implanted GaAs. The minimum size of the crystalline particle has been estimated to be ~45–50 Å based on the SCM discussed above.

There has also been a considerable amount of work done on the RS of annealed III–V semiconductors (20, 104–107, 114, 115), particularly GaAs. Shown in Fig. 6.24 (20) are the Raman features of a piece of undamaged semi-insulating (SI) (100)GaAs (bottom spectrum) and the SI GaAs implanted with Te^+ at 20 keV to a fluence of 5×10^{15} cm^2 (top spectrum). Also shown are the spectra of the implanted material that has been laser annealed with the 5300 Å line of a frequency-doubled Nd/YAG laser with 10-ns pulses at various power densities (expressed as MW/cm^2). In Fig. 6.25 (104) are shown the Raman spectra from different annular regions of a 450-μm spot of ion-damaged GaAs annealed at 1.5 J/cm^2 with 27-ns pulses using the 1.06-μm line of a Nd/YAG laser. The measurements were taken in a microprobe manner, with the Raman laser beam (5145 Å) focused to a spot the size of 1 μm. Spectrum A of Fig. 6.25 is from a (100)GaAs single crystal for reference. Spectrum B was taken at the position of the first annular ring, as shown in the figure. Spectra C, D, and E were taken at the positions shown in the figure. The features of spectra of the annealed material in Figs. 6.24 and 6.25 show considerable similarities. For example, comparison of spectra at 10 MW/cm^2 of Fig. 6.24 and spectra C of Fig. 6.25 to single-crystal material shows that the LO phonon mode is shifted significantly to the red from the undamaged 292 cm^{-1} position and is asymmetrically broadened. The TO mode, although broad, is still fairly symmetric and has shifted very little. Thus, the LO phonon feature of laser-annealed material could be analyzed

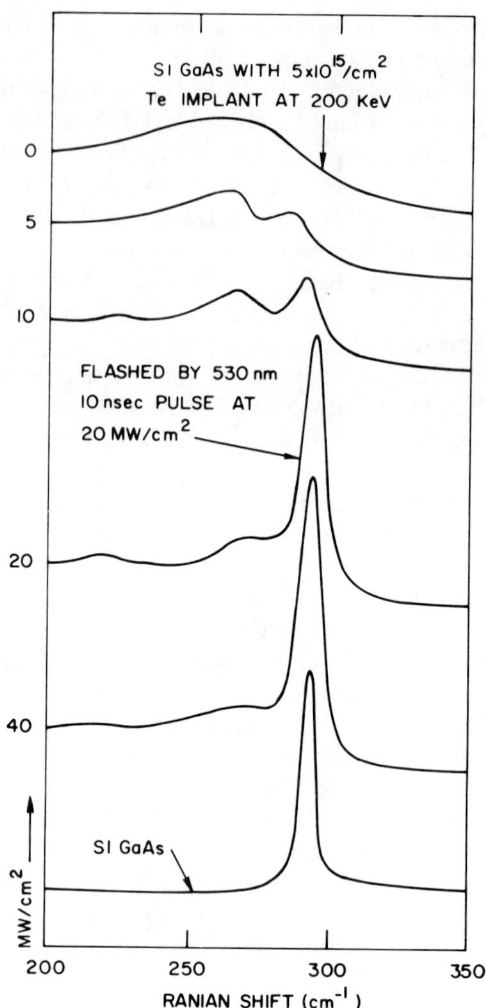

Figure 6.24. Raman spectra of laser-annealed Te-implanted semi-insulating (100)GaAs. From Tsu (20).

on the basis of the SCM in order to extract a measure of the average size of the undamaged region.

Sapriel et al. (104, 105) have also investigated the annealing of the high-dose-implanted (100)GaAs with halogen lamps. Shown in Fig. 6.26 (104) is the Raman spectra of amorphized GaAs irradiated with two 150-W halogen lamps. The dashed line is the RS of an amorphous GaAs sample used for

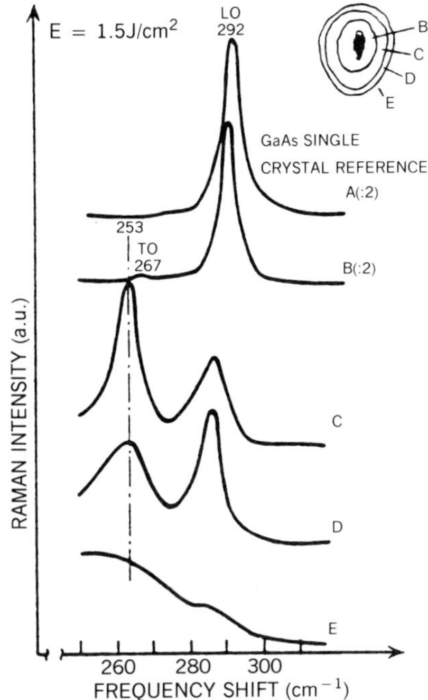

Figure 6.25. Raman spectra of the different regions of a picosecond laser-annealed, ion-implanted amorphized GaAs spot for $E = 1.5\,J/cm^2$ and $\lambda = 1.06\,\mu m$. From Sapriel et al. (104).

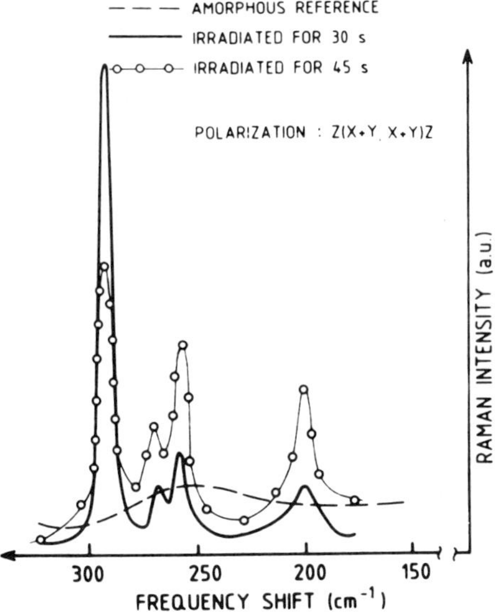

Figure 6.26. RS spectra of samples "over-irradiated." The As peaks are the two low-frequency peaks (199 and $257\,cm^{-1}$). From Sapriel et al. (104).

reference. The solid line is the spectrum irradiated for 30 s. Note that although there is recovery of the LO mode there is still present the symmetry-forbidden TO mode, indicating incomplete recrystallization. Also two peaks due to crystalline As are observed that may be caused by evaporation and recrystallization of As at the sample surface or thermal oxidation with As retained in the interfacial region. Irradiation for 45 s (dot-dashed lines) does not improve the crystal quality and in fact increases the As lines.

Ashokan et al. (106) have reported the results of an RS study of the structural disorder in phosphorus-ion-implanted and pulsed-laser-annealed (PLA) (100)GaAs. The distortion of the LO-photon lineshape in the ion-implanted semiconductor is due to the breakdown of the wavevector selection rules in the disordered material as described by the SCM above. The observed softening of this Raman mode has been attributed to the presence of microcrystallites in the GaAs as discussed in Section 5.4.1 for μc-Si. The oscillator strength of the two-phonon combination spectra has also been used to characterize the disorder. The localized vibrational mode of phosphorus impurities in ion-implanted and PLA GaAs is reported and its intensity variation with fluence has been studied. For ion-implanted PLA material the softening of the LO phonon mode is explained as an inherent impurity effect. The presence of a partially annealed deep layer has been investigated by observing a disorder-activated mode in the RS and by an enhancement TO phonon. The occurrence of defects on the surface after PLA has been investigated by nondestructive profiling of the LO-phonon softening using various laser lines as excitation.

A series of RS experiments were carried out on GaAs implanted with Si^+ and with SiF_3^+, both before and after annealing, for samples subjected to fluences spanning a wide range (119). The implantation-induced amorphization of the damage layer was clearly observed via the evolution, with increasing fluence, of the broad three-band continuum of amorphous GaAs that extends from near zero up to 300 cm^{-1}. As the amorphous bands grow to dominate the spectrum with increasing fluence, the crystal LO phonon line weakens, asymmetrically broadens, and shifts downward in frequency, similar to the results of Fig. 6.24. Annealing recovers the simple-line spectrum of crystalline GaAs, but with a changed LO intensity ratio that indicates a departure from epitaxial regrowth. Three lines observed near 400 cm^{-1} in heavily implanted samples were identified with silicon vibrational local modes. The effect of annealing on these local-mode lines is not an intensity increase but rather a line narrowing that reveals an annealing-induced sharpening of the distribution of local settings sampled by the substitutional silicons. In particular, a line (at 381 cm^{-1}) assigned to the Si-at-a-Ga-site donor impurity is clearly seen before annealing, even though annealing is needed to transform the highly resistive implanted material into semiconducting *n*-type GaAs. Holtz et al. (119) propose that the primary role of annealing in the "electrical activation" of implanted semiconductors is not to shift the impurity atoms into substitutional donor or acceptor sites (they already occupy such sites) but rather to recover the high carrier mobility of the crystalline form. "Healing" and "activation" thus correspond to the same process, the elimination of amorphicity.

Characterization of implantation and thermal annealing of Zn-implanted InP has been performed by Bedel et al. (110). First- and second-order RS by Zn-implanted material has been investigated in order to determine the critical fluences needed for complete disturbance of the lattice and recovery temperature during annealing.

Berg, Yu, and Weber (118) have used RS to study GaAs that has been damaged by irradiation with either high-energy electrons or neutrons. They report the observation of new and relatively sharp peaks that are attributed to vibrational modes of intrinsic point defects, most likely an As vacancy.

6.5.5. Process-Induced Damage

RS is a particularly useful tool for investigating process-induced damage for several reasons. The influence on the Raman spectrum of damage-induced perturbations such as strain and disorder (which produce different effects) is well understood. Nondestructive depth profiling can be performed by using various laser lines. In zincblende-type materials the LO phonon is a singlet; hence, certain complications that occur in electronic transition spectroscopies are avoided, such as strain-induced inter- and intraband splittings.

An investigation of the effects of polish-induced strain on the lineshape of the LO phonon RS of InP and GaAs has recently been reported (121, 122). Nondestructive depth profiling was accomplished by using the various wavelengths of an Ar^+ laser. Measurements were made at room temperature in the backscattering geometry from the (100) and (111) surfaces of InP and GaAs single crystals polished with several grit sizes for different periods of time. The observed lineshape changes were quantitatively accounted for by a model based on the convolution of the penetration depth of the light and the skin depth of the polish-induced surface strain. This study yields the magnitude and sign of the surface strain as well as the strain skin depth for a variety of polishing conditions.

Shown in Fig. 6.27 (122) is the RS in the region of the LO phonon for undamaged (100)InP using the 5145 Å line (spectrum A) and (100) material polished with 0.05-μm grit for 5 min using the 5145, 5017, 4880, 4765, and 4579 Å excitation lines (labeled B, C, D, E, and F, respectively). The undamaged RS is a fairly sharp, symmetrical peak centered at 345.6 cm^{-1}.

Spectrum B has a relatively sharp feature at about 348 cm^{-1} and a broad shoulder at about 355 cm^{-1}. As the penetration depth of the light decreases, the amplitude of the high-frequency shoulder grows relative to the low-frequency feature. Note that the amplitudes of the RS of the damaged material were considerably less than those of the undamaged sample.

Also plotted in Fig. 6.27 are the results for (100)GaAs. Spectrum A is the undamaged material taken with 5145 Å, while spectra B–F are for the case of 0.3-μm grit polish for 60 min using 5145, 5017, 4880, 4765, and 4579 Å, respectively. In this case the high-frequency shoulder in spectrum B has become the dominant feature at the shortest wavelength. As will be demonstrated below, the two structures in the spectra of damaged InP and GaAs do not arise from two separate features but are the consequence of the convolution of the skin depth of the damage-induced strain and the optical-penetration depth for the various wavelengths.

The aforementioned results can be accounted for on the basis of the following considerations. Polish induces strain at the surface of the material,

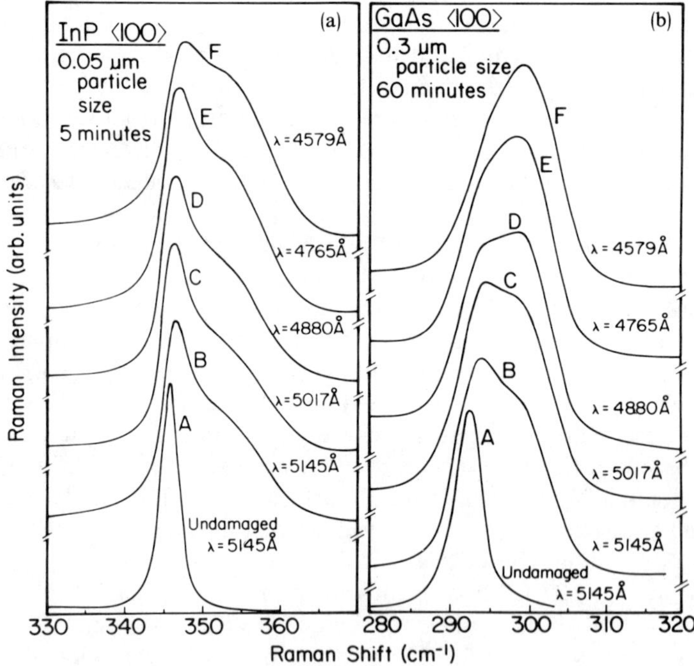

Figure 6.27. Raman scattering from (100)InP and (100)GaAs for undamaged and damaged material. The latter spectra are shown for several excitation lines. From Shen et al. (122).

thus producing a shift in the LO frequency from its unstrained value. This damage-induced strain (and hence RS shift) diminishes as a function of depth into the sample. In addition, the light intensity decays from the sample surface. The observed RS signal is then the convolution of the depth dependence of the LO frequency and the optical skin depth of the light. Assuming an exponential depth dependence for the Raman frequency, $\omega(x)$ (and hence strain), one can write

$$\omega(x) = \omega_0 + \Delta\omega_0 \exp(-x/d_s), \tag{66}$$

where x is the distance from the surface, ω_0 is the unperturbed frequency, $\Delta\omega_0$ is the shift due to surface strain, and d_s is the skin depth of the strain. Thus the Raman intensity $I(\omega)$ is given by

$$I(\omega) \propto \int_0^\infty dx L[\omega - \omega(x)] \exp(-2x/d_0), \tag{67}$$

where d_0 is the optical skin depth and $L(\omega)$ is a lineshape factor which is a convolution of the Lorentzian lineshape of the RS and the instrumental transfer function. Shen et al. (122) have neglected the small differences in the optical skin depth for the incident and scattered radiation. For simplicity they have assumed a triangular instrument transfer function, $T(\xi)$, with FWHM of Γ_I. Thus $L(\omega)$ can be written as

$$L(\omega) = \int_{-\infty}^\infty \frac{(\Gamma/2\pi)}{(\omega - \xi)^2 + (\Gamma/2)} T(\xi)\, d\xi, \tag{68}$$

where Γ is the linewidth of the RS. Equation (68) can be solved to yield the analytical form

$$L(\omega) = (\Gamma/2\pi\Gamma_I)[R(\omega + \Gamma_I) + R(\omega - \Gamma_I) - 2R(\omega)], \tag{69}$$

where

$$R(\omega) = (2\omega/\Gamma)artg(2\omega/\Gamma) - (\tfrac{1}{2})\ln[(2\omega/\Gamma)^2 + 1]. \tag{70}$$

It is also possible to take into account any inhomogeneous strain in the plane perpendicular to the polishing direction (for a given value of depth). This can be done by increasing the broadening parameter, Γ, in Eq. (68) from its intrinsic value (2.4 cm^{-1} for GaAs and 1.2 cm^{-1} for InP) and assuming a depth dependence similar to that of the frequency shift [see Eq. (66)],

$$\Gamma = \Gamma_0 + \Delta\Gamma_0 \exp(-x/d_s), \tag{71}$$

where Γ_0 is the intrinsic value and $\Delta\Gamma_0$ takes into account the inhomogeneous broadening.

In order to correlate the RS peak shift $\Delta\omega$ with surface strain the authors (122) assumed that the polishing produces a two-dimensional strain in the plane perpendicular to the polishing direction. For the (100) case the quantity $\Delta\omega$ is related to the strain T by

$$\left(\frac{\Delta\omega}{\omega_0}\right)_{100} = \left(\frac{q}{\omega_0^2} - \lambda_{100}\frac{p}{\omega_0^2}\right)T = K_{100}T, \tag{72a}$$

where

$$\lambda_{100} = (C_{12}/C_{11}). \tag{72b}$$

For the (111) situation it can be shown that

$$\left(\frac{\Delta\omega}{\omega_0}\right)_{111} = \left[\frac{p+2q}{6\omega_0^2}(2-\lambda_{111}) - \frac{2r}{3\omega_0^2}(1+\lambda_{111})\right]T = K_{111}T \tag{73a}$$

and

$$\lambda_{111} = 2(C_{11} + 2C_{12} - 2C_{44})/(C_{11} + 2C_{12} + 2C_{44}), \tag{73b}$$

where ω_0 is the unperturbed Raman frequency; C_{11}, C_{12}, and C_{44} are elastic stiffness constants, and p, q, and r are coefficients that describe the strain-induced shift of the optic phonon (62, 63). For GaAs, $K_{100} = 1.5$ and $K_{111} = -0.7$. For InP, p, q, and r have not been determined; however, since these parameters do not vary much for the III–V materials, the authors (122) have taken the GaAs values for InP and hence $K_{100} = -1.3$ and $K_{111} = -0.80$.

In Fig. 6.28 (122) are plotted experiment and theory for (100)GaAs at 5145 and 4579 Å damaged with a 0.3-μm grit for 15 min. It is found that at 5145 Å $\Delta\omega = 10.9\ \text{cm}^{-1}$ and $d_s = 260$ Å while at 4579 Å $\Delta\omega = 10.1\ \text{cm}^{-1}$ and $d_s = 230$ Å, again demonstrating consistency. Values of $\Delta\omega$ and d_s (average over the wavelengths) for GaAs for different polishing conditions are listed in Table 6.5.

An investigation of the effects of polishing the (111) surface of GaAs and InP has been performed (122). The results for the (111) situation are summarized in Table 6.6. Considerable difference between (100) and (111) faces has been found. For the former, as discussed above, the surface strain is ~2–3% and has a relatively short penetration depth into the material.

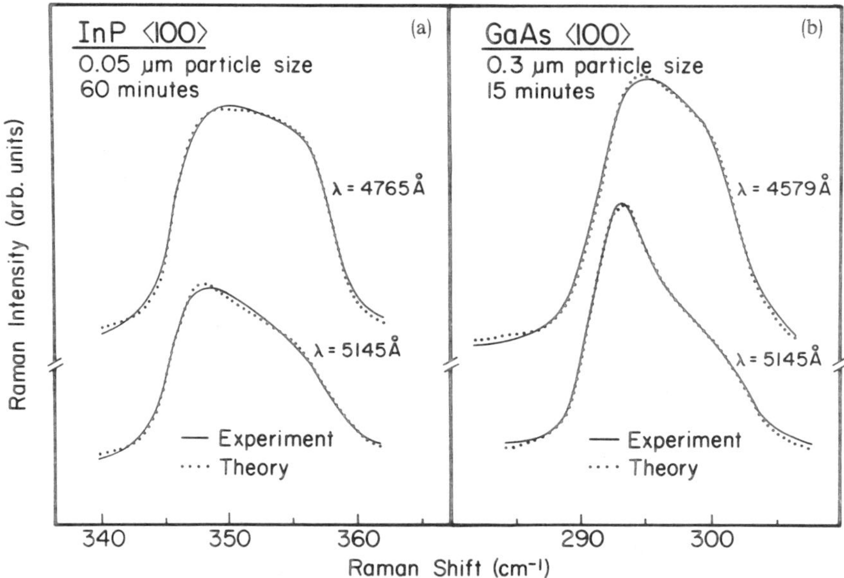

Figure 6.28. Comparison of experiment and theory for damaged (100)InP and (100)GaAs for 5145 Å excitation lines. From Shen et al. (122).

Table 6.5. Values of $\Delta\omega_0$, $\Delta\Gamma_0$, and d_s for Various Polishing Conditions on the (100) Surface of InP and GaAs

	Polish Conditions				
Material	Particle Size (μm)	Time (min)	$\Delta\omega_0$ (cm^{-1})	$\Delta\Gamma_0$ (cm^{-1})	d_s (Å)
Inp	0.05	5	$+12.6 \pm 0.5$	$\leqslant 3.0$	190 ± 25
	0.05	60	$+12.1 \pm 0.5$	$\leqslant 3.0$	300 ± 40
	0.3	30	$+12.5 \pm 0.5$	$\leqslant 3.0$	440 ± 50
GaAs	0.05	60	$+9.1 \pm 0.5$	$\leqslant 3.0$	200 ± 25
	0.3	15	$+10.5 \pm 0.5$	$\leqslant 3.0$	250 ± 30
	0.3	60	$+11.4 \pm 0.5$	$\leqslant 3.0$	360 ± 50
	1.0	2	$+10.4 \pm 0.5$	$\leqslant 3.0$	530 ± 75

However, for the (111) surface it is found that the depth of damage is on the order of the particle size, as found in a number of other works, but the strain at the surface itself is rather small, on the order of 1% [as opposed to 2–3% for (100)]. These differences are of considerable significance from both fundamental and applied points of view and should be explored further.

Table 6.6. Values of $\Delta\omega_0$, $\Delta\Gamma_0$, and d_s for Various Polishing Conditions on the (111) Surface of InP and GaAs

	Polish Conditions				
Material	Particle Size (μm)	Time (min)	$\Delta\omega_0$ (cm^{-1})	$\Delta\Gamma_0$ (cm^{-1})	d_s (Å)
InP	0.05	10	2.8 ± 1.0	4.8 ± 1.0	350 ± 100
	0.05	60	3.3 ± 1.0	7.0 ± 1.0	1100 ± 200
	0.3	10	3.6 ± 1.0	6.0 ± 1.0	3400 ± 500
	0.3	60	3.6 ± 1.0	7.0 ± 1.0	4500 ± 500
GaAs	0.05	10	1.3 ± 0.5	6.5 ± 1.0	240 ± 100
	0.05	60	1.5 ± 0.5	7.3 ± 1.0	440 ± 100
	0.3	10	1.0 ± 0.5	5.5 ± 1.0	1200 ± 200
	0.3	60	1.4 ± 0.5	6.3 ± 1.0	1800 ± 200

Evans and Ushioda (123) have used RS to study the dependence of the $q \approx 0$ TO- and LO-phonon lineshapes on surface preparation in the compounds GaAs, InAs, GaSb, and InP. Polished surfaces have broad phonon lines compared to etched, annealed, or cleaved surfaces. These authors show that phonon-line broadening is a consequence of microscopic disorder generated during the polishing process. Temperature-dependence studies of phonon linewidths in polished and annealed GaAs surfaces demonstrate that the additional broadening observed in the polished surfaces is caused by a temperature-independent process. These authors have concluded that the additional line broadening in the polished surfaces is caused mainly by inhomogeneous shifting of the phonon frequencies resulting from inhomogeneous strain near the sample surface. A secondary mechanism for phonon-line broadening is increased phonon anharmonic decay in polished surfaces caused by local lowering of lattice symmetry in damaged surface layers. These investigators have observed that phonon-lineshapes in cleaved, etched, and 0.05-μm polished GaAs surfaces narrow as the optical skin depth of the exciting laser is increased; phonon lines in similarly prepared InAs surfaces broaden as the skin depth is increased. From this observation it is concluded that the magnitude of the strain in surface layers is peaked at about 500 Å below the surface. The strain decreases beyond this depth and also relaxes to a lower value toward the surface.

Carles et al. (124, 125) have also studied the effects of polish-induced damage on III–V semiconductors InAs, InSb, and GaSb. These authors find that for (110)InAs the effect of the rough polishing is to destroy the $q \approx 0$

wavevector conservation rule and hence a variety of first-order disorder-activated modes is observed.

The effects of various dry-processing procedures have been reported by Chung et al. (126) and Bouadma et al. (127). The former group has studied the influence of plasma exposure on GaAs. The RS shows that the dependence of carrier removal on plasma species and substrate temperatures is consistent with the thermal migration of the plasma species. The free-carrier removal owing to a plasma exposure is apparently dependent on the impurities present in the material. After a hydrogen-plasma exposure, a decrease in carrier concentration and mobility occurs for the S-doped samples, whereas an increase in mobility and a decrease in carrier concentration occurs for the Si-doped samples. The production of surface roughness resulting from various surface treatments was also observed by means of Raman spectroscopy and ellipsometry. The observed narrowing of the LO phonon line and the plasmon–phonon (ω_+) line may be correlated with the degree of annealing. The degradation at the interface between Si–N films and GaAs substrates was observed to depend on the means of film deposition such as plasma-enhanced deposition and CVD. Bouadma et al. have described a new method of Ar^+ etching of InP using a liquid-N_2-cooled sample holder. Smooth and low-damage etched surfaces have been obtained using this technique. Auger electron spectroscopy analysis indicated that the morphology degradation was significantly reduced.

RS has been used to study RIE induced in GaAs [see Kirillov et al. (128)]. The phonon spectra of undoped material allowed an evaluation of damage to the crystal lattice, while the coupled plasmon–phonon spectra of *n*-type material provided a sensitive probe of electrical characteristics. Studies were made of layers exposed to plasmas of Ar, SF_6, and $SiCl_4$. Conditions for low-damage Ar-plasma cleaning and for dielectric cap removal by SF_6 were established. Etching in the $SiCl_4$ plasma generally produced strong damage, although low damage was observed in a few cases.

RS and photoluminescence measurements were performed on semi-insulating (100)InP:Fe samples heat treated with phosphosilicate glass (PSG) protective layers (109). Photoluminescence spectra of PSG-encapsulated InP samples indicated that native phosphorus vacancies were reduced in these samples following heat treatment. The presence of a TO phonon line in Raman spectra, which is related to a modification of structural bonding, was observed in InP samples heat treated with an SiO_2 layer, while the absence of this TO phonon line was observed in InP heat treated with the PSG layer. Activation of Si implanted into the InP:Fe samples increased with increasing phosphorus concentration in the PSG layers that were used as protective layers for the post-ion-implantation annealing.

6.6. ALLOY SEMICONDUCTORS

6.6.1. Alloy Composition

Most alloy semiconductors exhibit what is called a "two-mode" behavior. For example, in $Ga_{1-x}Al_xAs$ $(0 < x < 1)$ both "GaAs-like" and "AlAs-like" phonon modes (LO and TO) are observed with the selection rules derived above (see Tables 6.1 to 6.4). This is illustrated in Fig. 6.29 (20), where the frequencies of the "AlAs-like" modes, designated LO_1 and TO_1, and the "GaAs-like" modes, designated LO_2 and TO_2, are plotted as a function of composition, x. Thus, the frequency position of these modes can be used to nondestructively evaluate the composition x. Two-mode behavior in other crystal systems is discussed by several authors (3, 20, 29, 130). The question of "two-mode," "mixed-mode," and "one-mode" behavior has recently been discussed by Landa et al. (130).

6.6.2. Alloy Potential Fluctuations

6.6.2.1. Optic Phonons

In an alloy semiconductor such as $Ga_{1-x}Al_xAs$ there exist potential fluctuations owing to the microscopic nature of the compositional disorder,

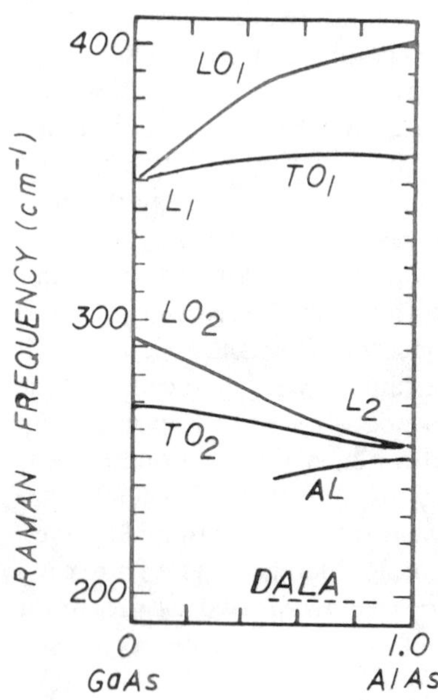

Figure 6.29. Frequencies of the "AlAs-like" (LO_1, TO_1) and "GaAs-like" (LO_2, TO_2) modes of $Ga_{1-x}Al_xAs$ as a function of molar fraction x. From Tsu (20).

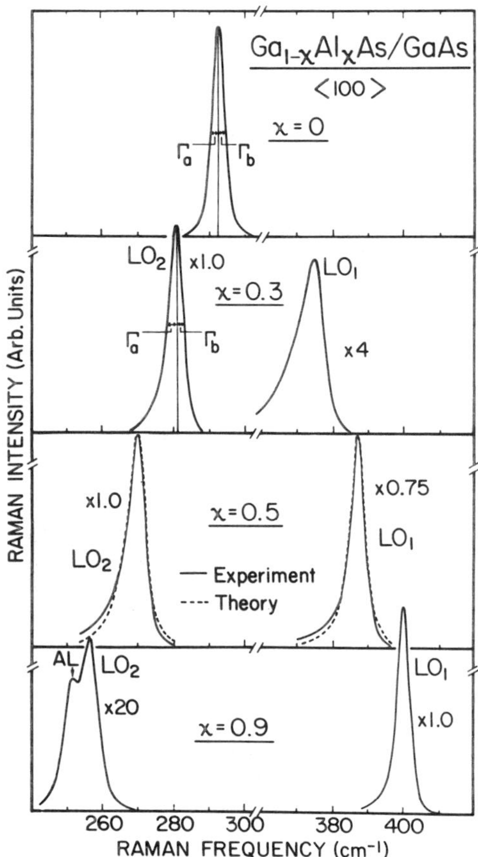

Figure 6.30. Raman spectra from the (100) surface of $Ga_{1-x}Al_xAs/GaAs$ at 300 K for $x = 0.0$, 0.3, 0.5, and 0.9. A small background owing to the disorder-activated TO phonons has been subtracted from the low-energy side of the alloy spectra. The dashed lines show the fit obtained from the theoretical model [Eqs. (60) and (61) using correlation lengths 62 and 80 Å for LO_2 and LO_1, respectively]. From Parayanthal and Pollak (131).

and hence translational invariance is broken. This effect also will manifest itself as a q-vector relaxation process. Indeed, it is found that in ternary alloy semiconductors the LO phonon mode is asymmetric ($\Gamma_a > \Gamma_b$) and broader than that of the binary endpoint materials.

Shown in Fig. 6.30 (131) are the LO phonon spectra from the (100) surface of GaAs ($x = 0$) and $Ga_{1-x}Al_xAs$ for $x = 0.3$, 0.5, and 0.9. This material exhibits a two-mode behavior, i.e., "AlAs-like" (LO_1) and "GaAs-like" (LO_2). For the alloys, both LO_1 and LO_2 are asymmetric ($\Gamma_a > \Gamma_b$) with a total linewidth Γ greater than that of GaAs. For the $x = 0.9$ material the lineshape of LO_2 is complicated by the presence of the AL feature, which has been ascribed to a localized phonon mode resulting from the motion of As atoms about a Ga atom on an Al site. The behavior of the alloy spectra, i.e., broadening and asymmetry; also can be quantitatively accounted for by $q \approx 0$ vector relaxation induced by the microscopic nature of the alloy potential fluctuations. In this case, the variable L can be thought of as a correlation length over which the material is compositionally ordered. Thus, the $q \approx 0$ vector relaxation model of Eq. (60) can be used to fit the details of the lineshape. The appropriate dispersion relationships of Eq. (61) must be used for the "GaAs-like" and "AlAs-like" modes. In Fig. 6.30 the dashed line for the $x = 0.5$ sample has been calculated from Eqs. (60) and (61) using correlation lengths of 62 and 80 Å for LO_2 and LO_1, respectively.

Plotted as the solid line in Fig. 6.31 (131) is the broadening Γ and

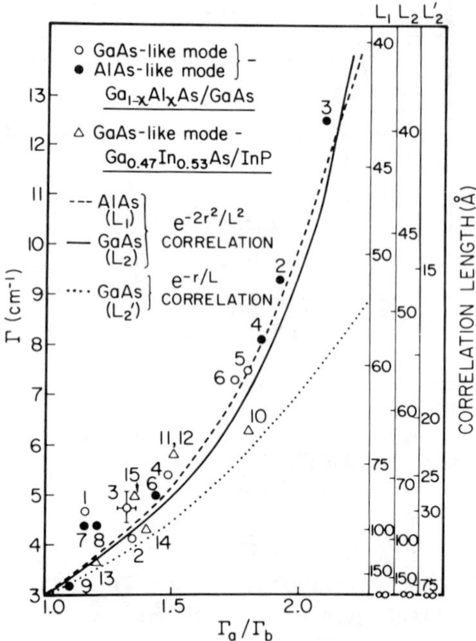

Figure 6.31. The relationship between broadening and asymmetry Γ_a/Γ_b as a function of correlation length L. The dashed line and solid line are calculated from AlAs dispersion (L_1) and GaAs dispersion (L_2), respectively, with use of $\exp(-2r^2/L^2)$ as the correlation function. The dotted line shows the relationship for GaAs dispersion (L_2') using $\exp(-r/L)$ spatial correlation. The experimental points for the various samples are also shown. From Parayanthal and Pollak (131).

asymmetry Γ_a/Γ_b as a function of the correlation length L as determined from Eqs. (60) and (61). Also plotted in this figure are the experimental points for the "GaAs-like" and "AlAs-like" modes of a number of samples of $Ga_{1-x}Al_xAs/GaAs$ (the data for sample 5 is taken from another work) as well as the "GaAs-like" mode for $Ga_{0.47}In_{0.53}As/InP$ (for this material the "InAs-like" mode is not observed). The characteristics of the samples are given in Table I of Parayanhal and Pollak (131).

From Fig. 6.31 we note that in a given crystal of GaAlAs, the "GaAs-like" mode may have a different spatial correlation length from that of the "AlAs-like" mode. In general, as the crystal becomes Al rich, L_1 increases while L_2 decreases.

The GaInAs samples present an interesting case since they all have the same nominal composition but were prepared with different growth parameters. Figure 6.31 shows that there are somewhat different correlation lengths for these samples, depending not only on the growth technique but apparently on the difference in the growth conditions for the same method. For example, samples 13 and 14, both prepared by MBE, have different correlation lengths.

Another interesting preliminary result for the GaInAs system is that samples 13 and 14 have a longer correlation length than the comparable $Ga_{1-x}Al_xAs$ system with $x = 0.5$. This may be related to the recent EXAFS (extended X-ray absorption fine structure) results of Mikkelsen and Boyce (132) concerning the conservation of bond lengths in alloy semiconductors. Thus, for example, in GaInAs the In—As bond length remains approximately that of the end-point material, InAs, the same being true of the Ga—As bond length. The average bond length does, however, follow Vegards' Law. The effects of these observations on the properties of alloy systems has been discussed by several authors, including Fedders and Muller (133) and Muller (134) who discuss the associated elastic energy. Since it takes elastic energy to interchange cations it may be that a system such as GaInAs, where there is about a 7% difference in bond lengths between In—As and Ga—As, can be prepared with reduced alloy potential fluctuations in relation to GaAlAs, where the Ga—As and Al—As bond lengths are almost equal. In the latter system there is little strain energy associated with interchanging a Ga and Al atom and hence it may become more random.

The compositional dependence of the Raman frequencies and linewidths of the optic phonons of $Cd_{1-x}Zn_xTe$ films grown by MBE has been reported by Olego et al. (135). The first-order RS for these ternary samples show two sets of LO and TO modes that arise from "CdTe-like" and "MnTe-like" vibrations. The lineshapes of the LO Raman bands develop asymmetrical broadenings that are observed to increase with LO–TO splitting. These authors find that although the SCM fits the observed LO phonon lineshape for most of the composition range, it does not account for the "ZnTe-like" mode for $x = 0.5$. They have been able to fit the lineshape of all the observed LO peaks using the Fano lineshape of the discrete-continuum interaction that yields Eqs. (42) and (43) for the Raman intensity $I(\omega)$. It should be noted that the Fano model fit of Eqs. (42) and (43) always involves two parameters (Q_F, Γ_F), whereas the SCM of Eq. (60) involves only the parameter L, the spatial correlation length. We shall discuss the relation between the SCM and Fano fit models in more detail below.

Recently, 300K RS and 4K photoluminescence (PL) measurements of a number of samples of AlInAs/InP grown by MBE with substrate temperatures in the range $495°\text{C} < T_s < 550°\text{C}$ have been reported (136, 137). Both PL and RS can be used to study the microscopic alloy potential fluctuations in alloy semiconductors. There is a strong correlation between the details of the RS lineshape and the PL linewidth. At $T_s = 515°\text{C}$ there is a significant decrease in both the RS and PL linewidths, indicating that under this growth condition there is a decrease in the alloy potential fluctuations, i.e., a tendency towards ordering. The RS of this alloy system exhibits a two-mode behavior corresponding to "AlAs-like" and "InAs-like" lattice vibrations. The first-order LO phonon Raman lines exhibit a low-frequency asymmetrically broadened lineshape. This lineshape has been analyzed in terms of the SCM of Eq. (60). The lineshapes of the "AlAs-like" mode for all samples and the "InAs-like" mode for $T_s = 515°\text{C}$ can be fit by the SCM using only one adjustable parameter, i.e., L in Eq. (60). However, for the "InAs-like" mode for $T_s = 515°\text{C}$, both L and the intrinsic linewidth (Γ_0) must be varied to achieve a fit (two-parameter fit). This indicates that for these growth conditions a new scattering mechanism has become available for this mode. Ksendzov et al. (137) discuss the relation between the SCM and Fano-fit model. It is demonstrated that with regard to the quality of the fit, the "two-parameter" SCM procedure (both L and Γ_0 as variables) is equivalent to the Fano-fit procedure, which also involves two parameters (Q_F and Γ_F). These authors point out that the need to vary Γ_0 indicates that new phonon-decay channels have opened up and these channels may also be the process which causes the discrete-continuum interaction of the Fano model. Therefore, this may be the reason why the two-parameter SCM, in which Γ_0 is different (larger) than the end-point material, is equivalent to the Fano model fit (two variables). However, Ksendzov and co-workers (137) also note that in a situation where the lineshape can be fit by the one-parameter SCM, comparison with the Fano model is inappropriate since the latter involves two variables and hence will always fit the spectrum in such a case.

Observations concerning the behavior of the lineshape of the LO phonon in alloy semiconductors have been made by a number of other authors.

Jusserand and Sapriel (138) have reported a Raman investigation of anharmonic and disorder-induced effects in $Ga_{1-x}Al_xAs$ epitaxial layers. RS and PL spectra have been presented for the metastable alloy $GaAs_{1-x}Sb_x$ grown by organometallic vapor-phase epitaxy throughout the miscibility gap extending from $0.2 < x < 0.75$ (139). The optic phonon peak half-widths are found to be broadened by nearly a factor of 2 over the half-widths found in the binary compounds GaAs and GaSb. Phonon lineshapes become more asymmetric in the miscibility gap as the selection rules break down; in addition, a second peak appears for samples grown near the center of the miscibility gap. Lineshapes were analyzed on the basis of the SCM, and the phonon-coherence length was found to be reduced from several hundred angstroms in GaAs to approximately 60 Å in samples grown in the miscibility gap. The compositional dependence of the room-temperature band-gap energy was found to closely follow earlier predictions.

Recently Kamijoh et al. (140) have observed an asymmetric low frequency of the LO phonon lineshape in the $Ga_{1-x}Al_xAs$ ($x = 0.32$) alloys doped with Zn. The spatial correlation lengths are reduced by this doping. The authors speculate that the structural disorder could be caused by electron–phonon interactions leading to spatially inhomogeneous force constants.

6.6.2.2. *Disorder-Activated Zone-Edge Phonons*

As discussed in Section 6.2.1 in a crystal with full translational symmetry only the $q \approx 0$ phonons are observed in first-order RS. Thus, the phonons at the edge of the BZ, as shown in Fig. 6.3, are normally forbidden. However, once this translational symmetry is broken by effects such as alloy potential fluctuations, disorder induced by ion-implantation, or process-induced damage, these modes become Raman allowed. Such zone-edged modes have been designated as disorder-activated modes (141). For example, disorder-activated zone-edge TA modes are designated DATA, while DALA denotes disorder-activated zone-edge LA modes. These disorder-activated modes have been investigated in simple crystals of $Ga_{1-x}Al_xAs$ (see 138, 142–144) and $Ga_{1-x}In_xAs$ (see 143).

These disorder-activated modes also may be used to characterize semiconductor alloys. For example, the Raman spectra of single-crystal $Ga_{0.7}Al_{0.3}As$ grown by LPE encapsulated with Si_3N_4 has been measured (145). The encapsulation is observed to alter certain low-frequency spectral features below $250\,cm^{-1}$ associated with the disorder-activated modes. Shown in Fig. 6.32 (145) are the Raman spectra from (a) as-grown $Ga_{0.7}Al_{0.3}As$, and $Ga_{0.7}Al_{0.3}As$ encapsulated with Si_3N_4 films (b) 2000 Å and (c) 4000 Å thick. Note that the DALA and DATA mode intensities are enhanced by the encapsulations. Kamijoh et al. (145) have attributed this

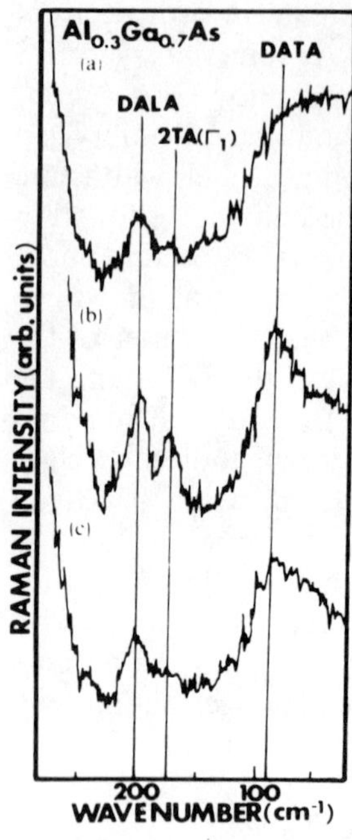

Figure 6.32. Raman spectra from (a) as-grown $Al_{0.3}Ga_{0.7}As$ and (b) $Al_{0.3}Ga_{0.7}As$ encapsulated with Si_3N_4 films 2000 Å and (c) 4000 Å thick. Disorder-activated longitudinal acoustical (DALA) and disorder-activated transverse acoustical (DATA) mode intensities are enhanced by the encapsulations. From Kamijoh et al. (145).

enhancement to breakdown of the symmetry selection rules by the strain field induced by the dielectric overlayer.

6.6.3. Ion Implantation and Annealing

Kakimoto and Katoda (146) have reported the use of RS to study the disorder and annealing behavior of Si-implanted $Ga_{1-x}Al_xAs$. From an analysis of the RS lineshapes and modes these authors have deduced that an increase in Al content suppresses the introduction of damage. However, Al in the $Ga_{1-x}Al_xAs$ also tends to reduce the concentration of Si on lattice sites after annealing.

6.6.4. Carrier Concentration

As discussed in Section 6.2.4, the coupled plasmon–LO phonon modes in binary zincblende-type semiconductors can be used to evaluate the free-carrier concentration for *n*-type material. The situation is more complicated for an alloy system such as GaAlAs because of the two-mode phonon behavior, i.e., two sets of LO and TO phonons, and because the positions of these modes vary with x as well as N. Recently, several groups have approached this problem (147–149). For example, Becker et al. (148) have presented the relation for the frequencies of plasmon-phonon modes in the two-component $Ga_{1-x}Al_xAs$ system. Raman spectra from more than 80 samples of $Ga_{1-x}Al_xAs$ over a wide range of x values and varying electron concentrations have been fitted to predicted curves using this relationship. Data were taken at both room and nitrogen temperature on samples grown by both metal-organic CVD (MOCVD) and MBE.

6.6.5. $Hg_{1-x}Cd_xTe$

Despite the significance of $Hg_{1-x}Cd_xTe$ from both fundamental and applied perspectives and the utility of RS in studying semiconductors, only a few Raman investigations of this material in bulk (or thin-film) form have been reported (150, 151). Mooradian and Harman (150) have used RS to measure the composition dependence of the TO and LO phonon frequencies: $Hg_{1-x}Cd_xTe$ has a two-mode behavior like $Ga_{1-x}Al_xAs$. Even though the spectra for the constituent binaries, HgTe and CdTe, were presented in Mooradian and Harman (150), no spectrum for $Hg_{1-x}Cd_xTe$ was included. Furthermore, the polarization dependence of the various features was not investigated.

The compositional dependence of the TO and LO frequencies has been investigated using far-IR reflectivity (FIR) (152, 153).

A complete polarization-dependent RS study of $Hg_{1-x}Cd_xTe$ ($x = 0.2$) has been reported (151). The experiment was performed at 77K in the backscattering geometry from both (100), (110), and (111) surfaces with various polarization configurations and using several excitation wavelengths near resonance with the E_1 optical gap. The bulk "HgTe-like" and "CdTe-like" TO and LO phonon ($q \approx 0$) features were observed as well as a strong symmetry-forbidden TO mode ("HgTe-like") from the (100) surface. In addition, features were recorded that appear to be related to deviations from the ideal, stoichiometric crystal. For example, a feature at $108\,\text{cm}^{-1}$ was reported, denoted as "defect mode," which has also been seen in HgTe but not CdTe. It has been suggested that it may be due to an antisite defect, i.e., Hg on Te sites. In addition, another structure ("clustering mode") has been tentatively assigned to vacancies or clustering effects that have also been observed in FIR measurements (153).

Amirtharaj et al. (151) also report the observation of coupled LO phonon–intersubband excitation (S_-) due to the inversion layer (which occurs in p-$Hg_{1-x}Cd_xTe$). This S_- feature was found to be extremely sensitive to surface treatment and may prove to be a valuable tool for study of surface conditions and preparation. It has been found that this feature is more surface sensitive than electrolyte electroreflectance at the E_1 optical feature since the penetration depth of the light is about 300 Å in this wavelength region while the depth of the inversion layer is <50 Å.

6.6.6. Heterojunctions

RS may be used to characterize the interface of a heterojunction formed by a GaAlAs epilayer on GaAs substrate (154). In particular, information can be obtained about (a) composition x of the $Ga_{1-x}Al_xAs$ epilayer, (b) crystalline quality of the epilayer, (c) band bending at the interface, (d) carrier depletion from the GaAs substrate, and (e) carrier concentration of the GaAs substrate material (grown by liquid phase epitaxy).

Swaminathan et al. (103) have used RS on angle-lapped samples to investigate the interfaces between different layers on a multilayer structure. Shown in Fig. 6.33 (103) are the RS from n-GaAs($N \approx 10^{18}\,cm^{-3}$)–$n$-$Ga_{0.65}Al_{0.35}As$($N \approx 3 \times 10^{17}\,cm^{-3}$) layers and the interface between them. The two layers were part of a ten-layer structure of alternating GaAs and $Ga_{0.65}Al_{0.35}As$ layers of thickness 3 and 3.5 μm, respectively, grown by MOCVD. To access the different layers the sample was angle lapped to $<1°$. Only two such layers are shown in Fig. 6.33 and qualitatively similar spectra

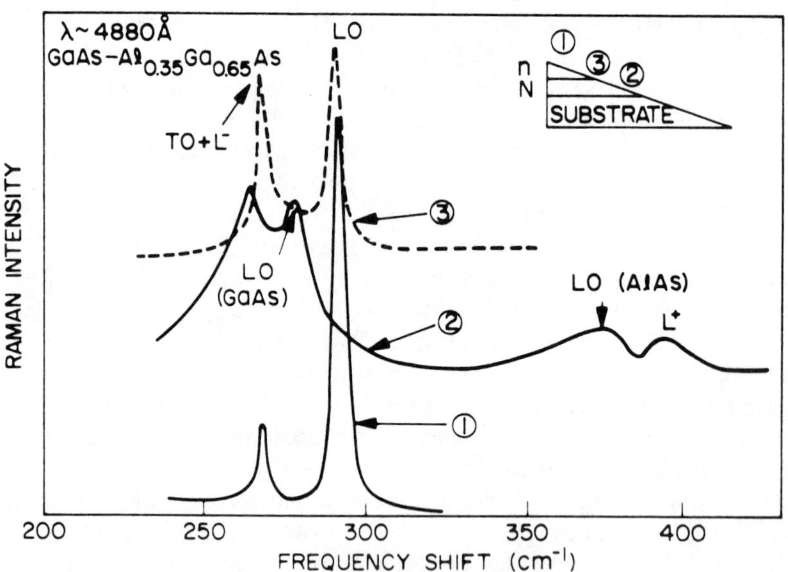

Figure 6.33. Raman spectra from a n-GaAs–n-$Ga_{0.65}Al_{0.35}As$ structure grown by MOCVD. Traces 1, 2, and 3 are obtained respectively from GaAs, $Ga_{0.65}Al_{0.35}As$, and the interface between them as shown in the inset. From Swaminathan et al. (103).

were obtained from other layers as well. The interesting feature in Fig. 6.33 is trace 3 obtained from the interfacial region between GaAs and $Al_{0.35}Ga_{0.65}As$. Surprisingly these authors did not observe the "AlAs-like" LO phonon in this region as observed in the spectrum from the $Ga_{0.65}Al_{0.35}As$ layer. This perhaps indicates that the compositional change across the interface is rather sharp. Notice also the enhancement of the line at about $268\,cm^{-1}$ in the spectrum from the interface region. This enhancement is attributed to both an increase in the coupled mode (ω_- mode of Fig. 6.20) and an increase in the forbidden TO mode. The former suggests the absence of carrier depletion near the interface, and the latter suggests the breakdown of the selection rules perhaps owing to the interfacial disorder.

6.6.7. Two-Phonon Raman Scattering

While most of the attention in RS has been devoted to the description and understanding of the properties of the first-order Raman features of elemental, binary, and alloy semiconductors, much less effort has been spent on studying two-phonon spectra, where the density of states favors near-zone edge phonons (ZEP), e.g., the X and L points in Fig. 6.3(a, b). As mentioned previously, Ashokan et al. (106) have used the properties of the two-phonon spectra to characterize phosphorus-ion-implanted and laser-annealed GaAs.

Recently two works (155, 156) have utilized the properties of the two-phonon Raman spectra for alloy semiconductor characterization. Teicher et al. (155) have investigated the two-phonon spectra of $Ga_{1-x}Al_xAs$ and $GaP_{1-x}As_x$. The other work (156) has studied the crystalline quality of GaAs layers in GaAs–$Ga_{1-x}Al_xAs$ superlattices by comparing the DATA modes and the second-order Raman lines.

Teicher et al. (155) report that the two-phonon spectra show coupled ZEP, which are "GaAs-like" and also combinations of GaAs and AlAs. Figure 6.34 (155) shows the two-phonon spectra of GaAs, AlAs, and several samples of $Ga_{1-x}Al_xAs$. As long as the Al content is small ($x < 0.3$) the two-phonon spectrum is very much like that of GaAs, the peaks tending to blur with increasing x. The intensity of the two-phonon spectrum, normalized to that of LO, remains roughly constant. When $x = 0.365$ the two-phonon RS is composed of four bands: the A band ($420\,cm^{-1} < \omega < 480\,cm^{-1}$), the B band ($480\,cm^{-1}$), the C band ($460\,cm^{-1} < \omega < 670\,cm^{-1}$), and the D band ($670\,cm^{-1} < \omega < 800\,cm^{-1}$). The B band is "GaAs-like" but without the structure seen in the pure material. The D band is "AlAs-like," but the sharp transverse optic overtone (2TO) (X, K) peak has disappeared. In addition to these two bands, which are found in any two-mode, behavior-mixed crystals, two additional bands are seen. The one labeled as C between the "GaAs-like"

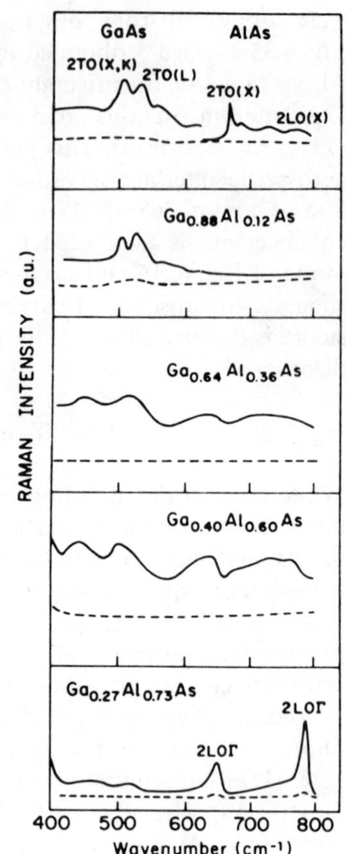

Figure 6.34. Second-order Raman spectra of GaAs, AlAs, $Ga_{0.88}Al_{0.12}As$, $Ga_{0.64}Al_{0.36}As$, $Ga_{0.40}Al_{0.60}As$, and $Ga_{0.27}Al_{0.73}As$. From Teicher et al. (155).

and the "AlAs-like" two-phonon bands is clearly due to combinations of GaAs and AlAs optical ZEP. Prior to this report such a combination band has not been seen in any nonresonant two-mode behavior-mixed crystal. The A band, which does not appear in GaAs, is due to combinations of acoustical and optical phonons of GaAs and AlAs. The intensity of the B band, normalized to the LO one, does not vary with composition, x. For $x = 0.60$ the spectrum is similar to that of $Ga_{0.73}Al_{0.27}As$, with the four bands clearly defined. For $GaP_{1-x}As_x$ the A and C bands discussed previously are not seen, i.e., only ZEP associated with the GaP and GaAs modes.

These results have been qualitatively explained by assuming a random distribution of Ga and Al atoms in the $Ga_{1-x}Al_xAs$ crystals and a distortion of the lattice structure, as well as a departure from a random distribution of atoms (i.e., clustering) in the $GaP_{1-x}As_x$ system.

Sapriel et al. (156) have investigated the relative intensities of the DATA mode and the second-order transverse acoustic overtone (2TA) mode in GaAs from the zone edge of the unfolded BZ. Since the amplitude of the DATA mode is clearly less than that of the 2TA structure, they conclude that the disorder in the GaAs layers is quite small. The good crystalline quality of the superlattices has also been verified by the high degree of polarization of the Raman features.

6.7. SURFACE FILMS INCLUDING OXIDES

There has been considerable work on the use of backscattering Raman spectroscopy as an optical probe for monitoring the presence and growth of oxide films (thermal, anodic, plasma) (157–163) or deposits of elemental materials on the surface of zincblende-type semiconductors (binary, ternary, quaternary) (164, 165).

Figure 6.35 (158) shows Raman spectra from GaAs wafers with 1000 Å oxide films grown via electrochemical and plasma [radio frequency (RF) and direct current (DC)] oxidation processes. Crystalline As in its A7 structure displays sharp A_{1g} and E_g modes around 257 and 200 cm^{-1}, respectively. Amorphous As (a-As) is characterized by a broad band between 180 and 260 cm^{-1} and weaker structure between 100 and 200 cm^{-1}. Comparison of the oxidized wafer spectra with that of the bare substrate in panel (a) indicates that elemental As deposits are not detected in the electrochemically grown film, whereas both RF and DC plasma films show an additional mode around 205 cm^{-1}. The DC plasma film also shows evidence for some additional featureless scattering contribution between 200 and 260 cm^{-1}. Definitive identification of this structure is complicated owing to the GaAs substrate background in the form of the LO and TO modes and 2TA scattering that overlaps the signals of interest.

When the oxidized wafers were thermally aged in vacuum at 400°C for 1 h, the generation of both crystalline and amorphous allotropes of As becomes apparent (Fig. 6.36) (158). The As generated during this thermal aging results from an interfacial reaction between As_2O_3 and substrate according to

$$As_2O_3 + 2GaAs \rightarrow Ga_2O_3 + 4As$$

The variation in the width and position of the A_{1g} and E_g crystalline As modes reflects a distribution of crystallite sizes in which these modes are

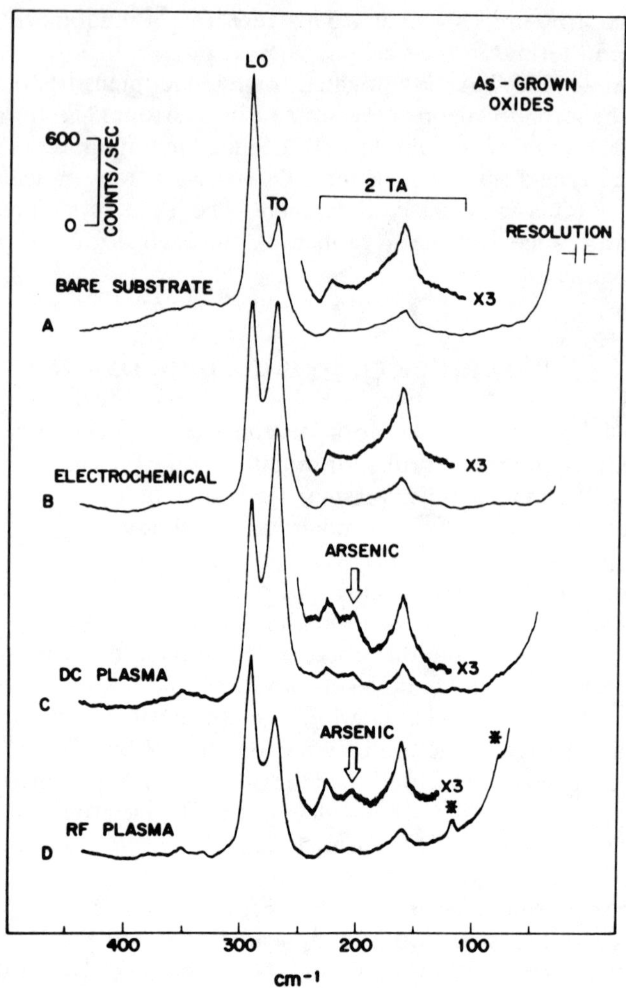

Figure 6.35. Raman spectra from anodic and plasma grown films. Asterisks indicate plasma lines. From Schwartz (158).

broadened and shifted for particles comparable in dimension to the coherence length associated with the RS process.

Interfacial chemical reactions occurring on the surface of thermally annealed single-crystal GaAs have been studied by both RS and Auger spectroscopy (160). McDevitt and Solomon (160) have used this information to deduce that oxygen plays a more significant role in the formation of crystalline As in the oxide film than in the compound As_2O_3. The Auger data demonstrates that As is distributed throughout the film.

The effects of adding Al or In to GaAs on precipitates and their distribution in anodic oxides were studied using RS (161). Crystalline As precipitated from $Ga_{1-x}Al_xAs$ ($0 < x < 0.54$) and $Ga_{1-x}In_xAs$ ($0 < x < 0.15$), whereas no

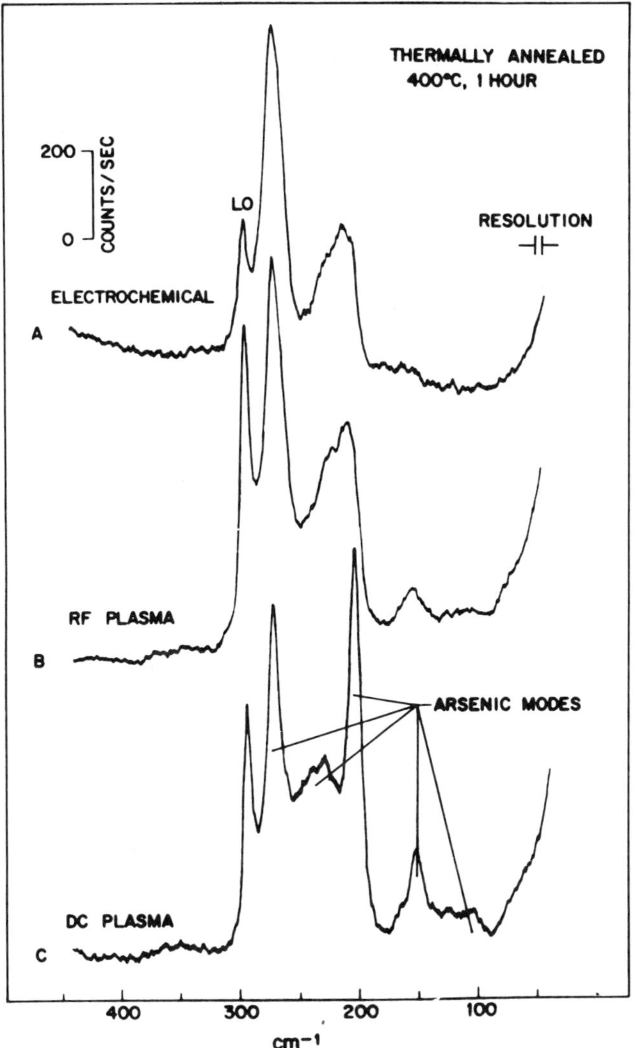

Figure 6.36. Raman spectra from thermally treated anodic and plasma films. From Schwartz (158).

species precipitated from $Ga_{1-x}In_xAs$ ($0.53 < x < 1.0$) after annealing in a nitrogen atmosphere at a temperature higher than 430°C. On the other hand, amorphous species precipitated from $Ga_{1-x}In_xAs$ ($0 < x < 1$) and no species precipitated from $Ga_{1-x}Al_xAs$ ($0.22 < x < 0.54$) after annealing in a hydrogen atmosphere at temperatures higher than 400°C. The crystalline As is produced by a solid-state interfacial reaction at the oxide/semiconductor interface, while the amorphous species is produced by the reduction of the oxides.

Adar et al. (162) have measured RS from native oxide films on $Hg_{0.7}Cd_{0.3}Te$. They have observed features that can be correlated with the Raman peaks of $HgTeO_3$, $CdTeO_3$, and TeO_2. Sakashita et al. (159) have recently reported results on anodic oxide films on HgTe. They find a complex oxide of $HgTe_2O_5$ on an amorphous oxide of TeO_2 that might be formed in the anodic oxidation.

Amirtharaj and Pollak (164) reported a RS study of the properties and removal of excess Te on CdTe surfaces. This investigation shows that a 0.1% Br/methanol etch or chemo-mechanical polish leaves a thin residual layer of polycrystalline Te of thickness 10–40 A‘ (under a tensile stress of about 8 kbar). It was found that this Te film can be removed by a rinse in a 1 N KOH/methanol solution.

Shin et al. (165) have observed RS from Te precipitates under compressive stress in CdTe crystals.

REFERENCES

1. See, for example, W. Hayes and R. Loudon, *Scattering of Light by Crystals.* Wiley, New York, 1978.
2. See, for example, M. Cardona, ed., *Light Scattering in Solids*, Vol. 1. Springer-Verlag, New York, 1975; M. Cardona and G. Guntherodt, eds., *Light Scattering in Solids*, Vols. 2–4. Springer-Verlag, New York, 1982–1984.
3. See, for example, S. S. Mitra and N. E. Massa, in *Handbook on Semiconductors* (T. S. Moss, ed.), Vol. 1, p. 82. North-Holland Publ., New York, 1982.
4. See, for example, M. Balkanski, in *Handbook on Semiconductors* (T. S. Moss, ed.), Vol. 2, p. 497. North-Holland Publ., New York, 1980.
5. See, for example, N. W. Ashcroft and N. D. Mermin, *Solid State Physics.* Saunders, Philadelphia, Pennsylvania, 1976.
6. G. Abstreiter, M. Cardona, and A. Pinczuk, in *Light Scattering in Solids* (M. Cardona and G. Guntherodt, eds.), Vol. 4, p. 5. Springer-Verlag, New York, 1984.
7. F. H. Pollak, *Test Meas. World* May, p. 120 (1985).
8. See, for example, E. Anastassakis, in *Dynamic Properties of Solids* (G. K. Horton and A. A. Maradudin, eds.), p. 157. North-Holland Publ., New York, 1980.
9. G. W. Rubloff, E. Anastassakis, and F. H. Pollak, *Solid State Commun.* **13**, 1755 (1973).

10. See, for example, C. Kittel, in *Introduction to Solid State Physics*, 5th ed., p. 105. Wiley, New York, 1976.
11. Spex Industries, Metuchen, New Jersey 08840.
12. F. Adar, c/o Instruments, SA, Inc., Metuchen, New Jersey 08840.
13. Yvon-Jobin is marketed by Instruments, SA, Inc., Metuchen, New Jersey 08840.
14. P. M. Fauchet, *Scanning Electron Microsc.* **II**, 425 (1986).
15. P. M. Fauchet, *IEEE Circuits Devices Mag.* January, p. 37 (1986).
16. P. M. Fauchet, *Mater. Res. Soc. Symp. Proc.* **51**, 149 (1986).
17. I. H. Campbell, P. M. Fauchet, and F. Adar, *Mater. Res. Soc. Symp. Proc.* **53**, 311 (1986).
18. P. M. Fauchet, I. H. Campbell, and F. Adar, *Proc. 17th Annu. Symp. Opt. Mater. High Power Lasers, Boulder, Colorado, 1985.*
19. P. M. Fauchet, I. H. Campbell, and F. Adar, *Appl. Phys. Lett.* **47**, 479 (1985).
20. See, for example, R, Tsu, *Proc. SPIE—Int. Soc. Opt. Eng.* **276**, 78 (1981), and references therein.
21. See, for example, F. H. Pollak and R. Tsu, *Proc. SPIE—Int. Soc. Opt. Eng.* **452**, 26 (1984), and references therein.
22. R. Tsu, J. Gonzalez-Hernandez, J. Dohler, and S. R. Ovshinsky, *Solid State Commun.* **46**, 79 (1983).
23. See, for example, H. Richter, Z. P. Wang, and L. Ley, *Solid State Commun.* **39**, 625 (1981), and references therein.
24. R. Tsu, J. Gonzalez-Hernandez, S. S. Chao, S. C. Lee, and K. Tanaka, *Appl. Phys. Lett.* **40**, 534 (1982).
25. S. Hayashi and H. Kanamori, *Physica (Amsterdam)* **177B/118B**, 520 (1983).
26. R. J. Nemanich, S. A. Solin, and R. M. Martin, *Phys. Rev. B: Condens. Matter* [3] **23**, 6348 (1981).
27. R. Tsu, S. S. Chao, M. Izu, S. R., Ovshinsky, G. J. Jan, and F. H. Pollak, *J. Phys. (Orsay, Fr.)* **42**, 269 (1981).
28. I. H. Campbell and P. M. Fauchet, *Solid State Commun.* **58**, 739 (1986).
29. G. Harbeke, L. Krausbauer, E. F. Steigmeir, A. E. Widmer, H. F. Koppert, and G. N. Neugebauer, *J. Electrochem. Soc.* **131**, 675 (1984).
30. H. Richter and L. Ley, *J. Phys. (Orsay, Fr.)* **42**, C4–C261 (1981).
31. J. Gonzalez-Hernandez, G. H. Azarbayejani, R. Tsu, and F. H. Pollak, *Appl. Phys. Lett.* **47**, 1350 (1985).
32. J. Lannin, in *Semiconductors and Semimetals* (J. Pankove, ed.), Vol. 21B, p. 159, and references therein. Academic Press, Orlando, Florida, 1984.
33. R. Tsu, J. Gonzalez-Hernandez, and F. H. Pollak, *Solid State Commun.* **54**, 447 (1985); *J. Non-Cryst. Solids* **66**, 109 (1984).
34. D. Beeman, R. Tsu, and M. F. Thorpe, *Phys. Rev. B: Condens. Matter* [3] **32**, 874 (1985).
35. J. S. Lannin, N. Meley, and S. T. Kshirsagar, *Solid State Commun.* **53**, 939 (1985).
36. N. Maley, J. S. Lanin, and A.G. Culles, *Phys. Rev. Lett.* **53**, 1571 (1984).

37. N. Maley, L. J. Pillone, S. T. Kshirsager, and J. J. Lanin, *Physica (Amsterdam)* **117B/118B**, 880 (1983).

38. T. Shimada, Y. Katagama, K. Nokagawa, H. Matsubara, M. Migiroka, and E. Maruyoma, *J. Non-Cryst. Solids* **59/60**, 783 (1983).

39. J. Gonzalez-Hernandez and R. Tsu, *Appl. Phys. Lett.* **42**, 90 (1983).

40. R. Tsu, J. Gonzalez-Hernandez, and F. H. Pollak, *J. Non-Cryst. Solids* **66**, 109 (1984).

41. A. Hartstein, J. C. Tsang, D. J. DiMaria, and D. W. Dong, *Appl. Phys. Lett.* **36**, 836 (1980).

42. N. Wada, P. J. Gaczi, and S. A. Solin, *J. Non-Cryst. Solids* **15/16**, 543 (1980); S. A. Solin, N. Wada, and J. Wong, *Conf. Ser.—Inst. Phys.* **43**, 721 (1979).

43. R. D. Dillon, A. Khan, F. G. Ullman, J. R. Hardy, V. Kotbanant, J. A. Wollam, and B. Banks, *Bull. Am. Phys. Soc.* [2] **28**, 563 (1983).

44. F. J. Kampas and F. H. Pollak, *Bull. Am. Phys. Soc.* [2] **28**, 565 (1983).

45. R. D. Dillon, J. Woollam, and V. Katkanant, *Phys. Rev. B: Condens. Matter* [3] **29**, 3482 (1984).

46. S. K. Hark, M. A. Machonkin, F. Jansen, M. L. Slade, and B. A. Weinstein, *AIP Conf. Proc.* **120**, 465 (1984).

47. R. Tsu, J. Gonzalez-Hernandez, I. Hernandez-Calderon, and C. A. Luengo, *Solid State Commun.* **24**, 809 (1977); R. Tsu, J. Gonzalez-Hernandez, and I. Hernandez-Calderon, *ibid.* **27**, 507 (1978).

48. Y. F. Morhange, G. Kanelis, and M. Balkanski, *Solid State Commun.* **31**, 805 (1979).

49. R. Tsu, in *Defects in Semiconductors* (J. Narayan and T. Y. Tan, eds.), p. 445. North-Holland Publ., New York, 1981.

50. Y. I. Nissim, J. Sapriel, and J. L. Oudar, *Appl. Phys. Lett.* **42**, 504 (1983).

51. J. Sapriel, Y. I. Nissim, and J. L. Oudar, *J. Phys. (Orsay, Fr.)* **44**, C5–C187 (1983).

52. S. Nakashima, Y. Inoue, M. Miyauchi, A. Mitsuishi, T. Nishimura, T. Fukumato, and Y. Akasska, *Appl. Phys. Lett.* **41**, 524 (1982); *J. Appl. Phys.* **54**, 2611 (1983).

53. J. B. Hopkins, L. A. Farrow, and G. J. Fisanick, *Appl. Phys. Lett.* **44**, 535 (1984).

54. S. Nakashima, Y. Inoue, and A. Mitsuishi, *J. Appl. Phys.* **56**, 2989 (1984).

55. P. T. T. Wong and M. Simard-Normandin, *J. Electrochem. Soc.* **132**, 980 (1985).

56. P. Kwizers and R. Reif, *Thin Solid Films* **100**, 1227 (1983).

57. D. Kirillov, R. A. Powell, and D. T. Hodul, *J. Appl. Phys.* **58**, 2174 (1985).

58. A. K. Shukla and K. P. Jain, *Phys. Rev. B: Condens. Matter* [3] **34**, 8950 (1986).

59. H. E. Lu, J. J. Boyd, H. E. Jackson, and J. L. Janning, *J. Appl Phys.* **60**, 4273 (1986); H. E. Jackson, J. J. Boyd, S. Dasgupta, H. E. Lu, and T. O. Mantei, *Proc. SPIE—Int. Soc. Opt. Eng.* **794**, 170 (1987).

60. T. Okada, T. Iwaki, H. Kasahara, and K. Youmamoto, *Solid State Commun.* **52**, 363 (1984).

61. J. C. Tsang, G. S. Oehrlein, I. Haller, and J. S. Custer, *Appl. Phys. Lett.* **46**, 589 (1985).

62. E. Anastassakis, A. Pinczuk, E. Burstein, F. H. Pollak, and M. Cardona, *Solid State Commun.* **8**, 133 (1970).
63. F. Cerdeira, C. J. Buchenauer, F. H. Pollak, and M. Cardona, *Phys. Rev. B: Solid State* [3] **5**, 580 (1971).
64. T. Englert, G. Abstreiter, and J. Pontcharra, *Solid State Electron.* **23**, 31 (1980).
65. Y. Ohmura, T. Inoue, and T. Yoshii, *Solid State Commun.* **37**, 538 (1981).
66. B. Y. Tsaur, J. C. C. Fan, and M. W. Geis, *Appl. Phys. Lett.* **40**, 322 (1982); S. R. J. Brueck, B. Y. Tsaur, J. C. C. Fan, D. V. Murphy, T. F. Deutch, and D. J. Silversmith, *ibid.* p. 895.
67. Y. Kobayashi, M. Nakamura, and T. Suzuki, *Appl. Phys. Lett.* **40**, 1040 (1982).
68. F. Moser and R. Beserman, *J. Appl. Phys.* **54**, 1033 (1983).
69. M. B. Stern, T. R. Harrison, V. D. Archer, P. F. Liao, and J. C. Bean, *Solid State Commun.* **51**, 221 (1984).
70. S. A. Lyon, R. J. Nemanich, N. M. Jobinson, and D. D. Biegelsen, *Appl. Phys. Lett.* **40**, 316 (1982).
71. P. Zorabedian and F. Adar, *Appl. Phys. Lett.* **43**, 177 (1983).
72. J. Gonzalez-Hernandez, J. G. Martin, and R. Tsu, *Proc. SPIE—Int. Soc. Opt. Eng.* **452**, 44 (1984).
73. C. C. Tsai, R. J. Nemanich, and T. W. Sigmon, *J. Phys. Soc. Jpn.* **49**, Suppl. A, 1265 (1980).
74. R. J. Nemanich, C. C. Tsai, M. J. Thompson, and T. W. Sigmon, *J. Vac. Sci. Technol.* **19**, 685 (1991).
75. R. J. Nemanich, T. W. Sigmon, N. M. Johnson, M. D. Moyer, and S. S. Lau, in *Laser and Electron-Beam Solid Interactions and Materials Processing* (J. F. Gibbons, L. D. Hess, and T. W. Sigmon, eds.), p. 541. North-Holland Publ., New York, 1981.
76. R. J. Nemanich, C. C. Tsai, and T. W. Sigmon, *Phys. Rev. B: Condens. Matter* [3] **23**, 6828 (1981).
77. J. C. Tsang, Y. Yokota, R. Matz. Y. Yokota, and G. W. Rubloff, *J. Vac. Sci. Technol., A* [2] **2**, 556 (1984).
78. R. J. Nemanich, B. L. Stafford, J. R. Abelson, and W. T. Sigmon, *Proc. Int. Conf. Phys. Semicond. 17th, San Francisco, 1984*, p. 155 (1985).
79. R. J. Nemanich and C. M. Doland, *J. Vac. Sci. Technol., B* [2] **3**, 1142 (1985).
80. R. J. Nemanich, R. T. Fulks, B. L. Stafford, and H. A. Vander Plas, *J. Vac. Sci. Technol., A* [2] **3**, 938 (1985).
81. R. J. Nemanich, M. J. Thompson, W. B. Jackson, C. C. Tsai, and B. L. Stafford, *J. Vac. Sci. Technol., B* [2] **1**, 519 (1983).
82. R. J. Nemanich, R. T. Fulks, B. L. Stafford, and H. A. Vander Plas, *Appl. Phys. Lett.* **46**, 670 (1985).
83. P. J. Codella, F. Adar, and Y. S. Liu, *Appl. Phys. Lett.* **46**, 1076 (1985).
84. S. Kumar, S. Dasgupta, H. E. Jackson, and J. T. Boyd, *Appl. Phys. Lett.* **50**, 323 (1987).

85. F. Magnotta and I. P. Herman, *Appl. Phys. Lett.* **43**, 195 (1986).
86. D. Kirillov and J. L. Merz, *Mater. Res. Soc. Symp. Proc.* **17**, 95 (1983).
87. J. Raptis, E. Liarokapis, and E. Anastassakis, *Appl. Phys. Lett.* **44**, 125 (1984).
88. G. Abstreiter, E. Bauser, A. Fischer, and K. Ploog, *Appl. Phys.* **16**, 345 (1978).
89. G. Abstreiter, R. Trommer, M. Cardona, and A. Pinczuk, *Solid State Commun.* **30**, 703 (1979).
90. G. P. Schwartz, G. J. Gualtieri, L. H. Dubois, W. A. Bonner, and A. A. Ballman, *J. Electrochem. Soc.* **131**, 1716 (1984).
91. C. K. N. Patel and R. E. Slusher, *Phys. Rev. Lett.* **21**, 1563 (1968); *Phys. Rev.* **167**, 413 (1968).
92. T. Kamijoh, A. Hashimoto, H. Takano, and M. Sakuta, *J. Appl. Phys.* **59**, 2382 (1986).
93. T. Nakamura and T. Katoda, *J. Appl. Phys.* **55**, 3064 (1984).
94. H. Shen, F. H. Pollak, and R. N. Sacks, *Appl. Phys. Lett.* **47**, 891 (1985).
95. H. Shen, F. H. Pollak, and R. N. Sacks, *Proc. SPIE—Int. Soc. Opt. Eng.* **524**, 145 (1985).
96. H. J. Stolz and G. Abstreiter, *J. Vac. Sci. Technol.* **19**, 380 (1981); *J. Phys. Soc. Jpn.* **49**, Suppl. A, 1101 (1980).
97. R. Tsu, H. Kawamura, and L. Esaki, *Solid State Commun.* **15**, 321 (1974).
98. See, for example, G. Abstreiter, M. Cardona, and A. Pinczuk, in *Light Scattering in Solids* (M. Cardona and G. Guntherodt, eds.), Vol. 4, p. 5. Springer-Verlag, New York, 1984.
99. L. Y. Ching, E. Burstein, S. Buchner, and H. H. Weider, *J. Phys. Soc. Jpn.* **49**, Suppl. A, 951 (1980).
100. A. Pinczuk, A. A. Ballman, R. E. Nahory, M. A. Pollock, and J. M. Worlock, *J. Vac. Sci. Technol.* **16**, 1168 (1979).
101. D. J. Olego, R. Schachter, and J. A. Baumann, *Appl. Phys. Lett.* **45**, 1127 (1984).
102. G. Paparodites, A Rideau, G. Mannon, and P. Goucherd, *Physica (Amsterdam)* **117B/118B**, 922 (1983).
103. V. Swaminathan, A. Jayaraman, J. L. Zilko, and R. A. Stall, *Mater. Lett.* **3**, 325 (1985).
104. J. Sapriel, Y. I. Nissim, B. Joukoff, J. L. Oudar, S. Abraham, and R. Beserman, *J. Phys. (Orsay, Fr.)* **45**, C5–C75 (1984).
105. Y. I. Nissim, B. Joukoff, J. Sapriel, and N. Duhamel, *J. Phys. (Orsay, Fr.)* **44**, C5–C247 (1983).
106. R. Ashokan, K. P. Jain, H. S. Mavi, and M. Balkanski, *J. Appl. Phys.* **60**, 1985 (1986).
107. H. D. Yao, A. Compaan, and E. B. Hale, *Solid State Commun.* **56**, 677 (1985).
108. G. Landa, R. Carles, J. B. Renucci, C. Fontaine, E. Bedel, and A. Munoz-Yague, *J. Appl. Phys.* **60**, 1025 (1986); *Proc. Int. Conf. Phys. Semicond., 18th, 1986* p. 1361 (1987).

109. T. Kamijoh, H. Takano, and M. Sakerta, *J. Appl. Phys.* **55**, 3756 (1984).

110. E. Bedel, G. Landa, R. Carles, J. B. Renucci, J. M. Roquais, and P. N. Favennec, *J. Appl. Phys.* **60**, 1980 (1986).

111. D. Kirillov and J. L. Merz, in *Energy Beam-Solid Interactions and Transient Thermal Processing* (J. C. C. Fan and N. M. Johnson, eds.), p. 707. Elsevier North-Holland, New York, 1984.

112. J. Biellmann, B. Prévot, and C. Schwab, *J. Phys. C* **16**, 1135 (1983).

113. See, for example, K. K. Tiong, P. M. Amirtharaj, F. H. Pollak, and D. E. Aspnes, *Appl. Phys. Lett.* **44**, 122 (1984), and references therein.

114. M. Erman, P. Cambon, B. Prévot, and C. Schwab, *J. Phys.* (*Orsay. Fr.*) **44**, C10–C261 (1983).

115. P. Chambon, M. Erman, J. B. Theeten, B. Prévot, and C. Schwab, *Appl. Phys. Lett.* **45**, 390 (1984).

116. T. Nakamura and T. Katoda, *J. Appl. Phys.* **57**, 1084 (1985).

117. T. Nakamura and T. Katoda, *Jpn. J. Appl. Phys.* **23**, L552 (1984).

118. R. S. Berg, P. Y. Yu, and E. R. Weber, *Appl. Phys. Lett.* **47**, 515 (1985).

119. M. Holtz, R. Zallen, A. E. Gerssberger, and R. A. Sadler, *J. Appl. Phys.* **59**, 1946 (1986).

120. D. E. Aspnes, S. M. Kelso, C. G. Olson, and D. W. Lynch, *Phys. Rev. Lett.* **48**, 1863 (1982).

121. H. Shen and F. H. Pollak, *Appl. Phys. Lett.* **45**, 692 (1984).

122. H. Shen, Z. Hang, and F. H. Pollak, *Proc. SPIE—Int. Soc. Opt. Eng.* **524**, 118 (1985); *J. Appl. Phys.* **46**, 3233 (1988).

123. D. J. Evans and S. Ushioda, *Phys. Rev. B: Solid State* [3] **9**, 1638 (1974).

124. R. Carles, J. B. Renucci, A. Zwick, and M. A. Renucci, *J. Phys.* (*Orsay, Fr.*) **43**, C9–C363 (1982).

125. R. Carles, N. Saint-Cricq, A. Zwick, M. A. Renucci, and J. B. Renucci, *J. Phys. Soc. Jpn.* **49**, Suppl. A, 665 (1980).

126. Y. Chung, D. W. Langer, R. Becker, and D. Look, *IEEE Trans. Electron Devices* **ED32**, 40 (1985); Y. Chung, C. Y. Chen, D. W. Langer, and Y. S. Park, *J. Vac. Sci. Technol., B* [2] **1**, 799 (1983).

127. N. Bouadma, P. Devoldere, B. Jusserand, and P. Ossart, *Appl. Phys. Lett.* **48**, 1285 (1986).

128. D. Kirillov, C. B. Cooper, III, and R. A. Powell, *J. Vac. Sci. Technol., B* [2] **4**, 1316 (1986).

129. B. Jusserand, D. Paquet, and K. Kunc, *Proc. Int. Conf. Phys. Semicond., 17th, San Francisco, 1984* p. 1165 (1985).

130. G. Landa, R. Carles, and J. B. Renucci, *Proc. Int. Conf. Phys. Semicond., 18th, Stockholm, 1986* p. 1361 (1987).

131. P. Parayanthal and F. H. Pollak, *Phys. Rev. Lett.* **52**, 1822 (1984).

132. J. C. Mikkelsen, Jr. and J. B. Boyce, *Phys. Rev. Lett.* **49**, 1412 (1982); *Phys. Rev. B: Condens. Matter* [3] **28**, 7130 (1983).

133. P. A. Fedders and M. W. Muller, *J. Phys. Chem. Solids* **45**, 685 (1984).
134. M. W. Muller, *Phys. Rev. B: Condens. Matter* [3] **30**, 6169 (1984).
135. D. J. Olego, P. M. Raccah, and J. P. Faurie, *Phys. Rev. B: Condens. Matter* [3] **33**, 3819 (1986).
136. D. F. Welch, G. W. Wicks, L. F. Eastman, P. Parayanthal, and F. H. Pollak, *Appl. Phys. Lett.* **46**, 169 (1985).
137. A. Ksendzov, P. Parayanthal, F. H. Pollak, D. Welch, G. W. Wicks, and L. F. Eastman, in *Proceedings of the Seventh International Conference on Ternary and Multiternary Compounds* (S. K. Deb and A. Zunger, eds.), p. 371. Mater. Res. Soc., Pittsburgh, Pennsylvania, 1987; *Phys. Rev.* **B37**, 4044 (1988).
138. B. Jusserand and J. Sapriel, *Phys. Rev. B: Condens. Matter* [3] **24**, 7194 (1981).
139. R. M. Cohen, M. J. Chernog, R. E. Benner, and G. B. Stingfellow, *J. Appl. Phys.* **57**, 4817 (1985).
140. T. Kamijoh, A. Hashimoto, N. Watanabe, and M. Sakuta, in *Proceedings of the International Conference on Defects of Semiconductors* (H. J. von Bardeleben, ed.), p. 1201. Trans Tech Publ., Aedermanns, Switzerland, 1986.
141. H. Kawamura, R. Tsu, and L. Esaki, *Phys. Rev. Lett.* **29**, 1397 (1972).
142. N. Saint-Cricq, R. Carles, J. B. Renucci, A. Zwick, and M. A. Renucci, *Solid State Commun.* **39**, 1137 (1981).
143. R. Carles, N. Saint-Cricq, H. Zwick, M. A. Renucci and J. B. Renucci, *J. Phys. (Orsay, Fr.)* **42**, C6–C105 (1981); *Nuovo Cimento Soc. Ital. Fis., D* **2D**, 1712 (1983).
144. W. Xiao-jun and Z. Xin-yi, *Solid State Commun.* **59**, 869 (1986).
145. T. Kamijoh, A. Hashimoto, H. Takano, and M. Sakuta, *Appl. Phys. Lett.* **44**, 1084 (1984).
146. K. Kakimoto and T. Katoda, *J. Appl. Phys.* **59**, 1477 (1986).
147. T. Yuasa, S. Naritsuka, M. Mannoh, K. Shinozaki, K. Yamanaka, Y. Nomura, M. Mihara, and M. Ishii, *Appl. Phys. Lett.* **46**, 176 (1985).
148. R. J. Becker, P. F. Luehrmann, and D. W. Langer, *Appl. Phys. Lett.* **47**, 513 (1985).
149. D. Kirillov, Y. Chai, C. Webb, and G. Davis, *J. Appl. Phys.* **59**, 231 (1986).
150. A Mooradian and T. C. Harman, in *Proceedings of the Conference on Physics of Semimetals and Narrow-Gap Semiconductors* (D. L. Carter and R. J. Bate, eds.), p. 297. Pergamon, New York, 1971.
151. P. M. Amirtharaj, K. K. Tiong, and F. H. Pollak, *J. Vac. Sci. Technol., A* [2] **1**, 1944 (1983); K. K. Tiong, P. M. Amirtharaj, P. Parayanthal, and F. H. Pollak *Solid State Commun.* **50**, 891 (1984); P. M. Amirtharaj, K. K. Tiong, P. Parayanthal, F. H. Pollak, and J. K. Furdyna, *J. Vac. Sci. Technol., A* [2] **3**, 226 (1985).
152. J. Baars and F. Sorger, *Solid State Commun.* **10**, 875 (1972).
153. S. P. Kozyrev, L. K. Vodopyanov, and R. Triboulet, *Solid State Commun.* **45**, 383 (1983).
154. P. Parayanthal, F. H. Pollak, and J. M. Woodall, *Appl. Phys. Lett.* **41**, 961 (1982).

155. M. Teicher, R. Beserman, M. V. Klein, and H. Morkoc, *Phys. Rev. B: Condens. Matter* [3] **29**, 4652 (1984).
156. J. Sapriel, J. C. Michel, J. C. Toledano, R. Vocher, J. Kervarec, and A. Regreny, *Phys. Rev. B: Condens. Matter* [3] **28**, 2007 (1983).
157. R. L. Farrow, R. K. Chang, S. Mroczkoski, and F. H. Pollak, *Appl. Phys. Lett.* **31**, 768 (1977).
158. See, for example, G. P. Schwartz, *Proc. SPIE—Int. Soc. Opt. Eng.* **276**, 72 (1981).
159. G. P. Schwartz, B. V. Dutt, M. Malyj, J. E. Griffiths, and G. P. Gualtieri, *J. Vac. Sci. Technol., B* [2] **1**, 254 (1983).
160. N. T. McDevitt and J. Solomon, *J. Electrochem. Soc.* **133**, 1913 (1986).
161. K. Egucki and T. Katoda, *Jpn. J. Appl. Phys.* **24**, 1043 (1985).
162. F. Adar, R. E. Kvaas, and D. R. Rhiger, in *Advances in Materials Characterization*. Plenum, New York 1983.
163. M. Sakashita, T. Ohtsuba, and N. Sato, *J. Electrochem. Soc.* **132**, 1864 (1985).
164. P. M. Amirtharaj and F. H. Pollak, *Appl. Phys. Lett.* **45**, 789 (1984).
165. S. H. Shin, J. Bajaj, L. A. Moudy, and D. T. Cheung, *Appl. Phys. Lett.* **43**, 68 (1983).

CHAPTER

7

POLYMER APPLICATIONS

BERNARD J. BULKIN

BP Research Centre
Sunbury-on-Thames
Middlesex, England

7.1. INTRODUCTION

Raman spectroscopy plays an important role in the characterization of synthetic organic polymers. It has done so for many decades, even before the advent of laser-excited Raman spectra. In the laser decades, this role has grown, as the great diversity of sampling techniques has allowed Raman spectroscopy to tackle a very varied set of problems in polymer science.

The original basis for this role was due to the sensitivity of the Raman spectra to various functional groups in polymers, much as vibrational spectroscopy has been applied to organic chemistry. However, this has been expanded far beyond the determination of the presence or absence of functional groups to identify a specific type of polymer. Extensive work has been published on end group determination, structure (especially crystallinity), conformation, and chain orientation. Considerable effort has been devoted to connecting Raman spectra, particularly low-frequency vibrations of chains, with mechanical properties of polymers.

Raman studies of polymers have been carried out over a wide range of temperatures and under reaction conditions. The ease of handling samples in many different physical forms, e.g., powders, injection molded pieces, pipe, tubing, film, sheets, and fibers, makes Raman spectroscopy an excellent technique for polymer characterization, The biggest limitation to its application has been fluorescence, often coming from fillers, traces of catalyst, monomers, other impurities, or in some cases from the polymer itself. Even this limitation has been overcome, in many examples, by excitation in the

Analytical Raman Spectroscopy, Edited by Jeanette G. Grasselli and Bernard J. Bulkin. Chemical Analysis Series, Vol. 114.
ISBN 0-471-51955-3

near infrared (NIR), usually with FT (Fourier transform)-Raman spectroscopy. As the use of NIR excitation of Raman spectra becomes more widespread, we may envisage even greater use of Raman spectroscopy in polymer science and engineering. Still, not every intractable sample will become manageable just because NIR excitation is used. Particularly for deeply colored and filled samples, thermal emission curves often become the dominant feature obscuring Raman spectra.

7.2. RAMAN SPECTRUM OF POLY(ETHYLENE TEREPHTHALATE)

Because of the very large number of studies that have been carried out, an entire book would be necessary to comprehensively review the application of Raman spectroscopy to polymers. This chapter introduces the subject by looking in detail at one important, reasonably complex polymer, poly(ethylene terephthalate) (known as PET or 2GT), having the basic chemical structure shown in Fig. 7.1. The studies on this polymer illustrate most of the key aspects of the use of Raman spectroscopy in polymer science. A brief section at the end of this chapter is a guide to literature covering other applications of Raman spectra to polymer problems.

Figure 7.2 shows two Raman spectra of PET in the 200–1800 cm^{-1} region. Both are spectra of solid samples. The upper spectrum is of a sample that has been quenched from the melt; the lower spectrum is of the same sample after isothermal annealing at 160°C for 1 h. Thermal analysis, X-ray diffraction, and other techniques show that the upper spectrum is of a highly amorphous sample, whereas the lower one shows considerable crystallinity. There are many differences between these spectra. Most obvious is the growth in intensity of bands near 1092, 1000, 850, and 273 cm^{-1}, and the decrease in width of the band near 1725 cm^{-1} when the sample is annealed.

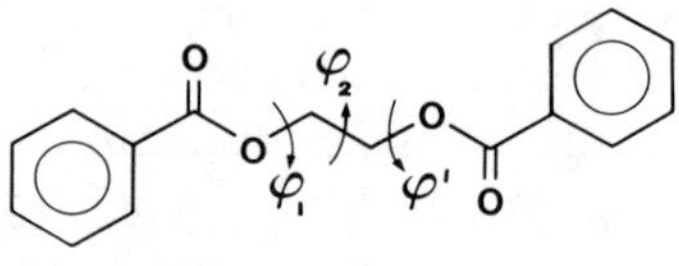

PET

Figure 7.1. Basic chemical structure of PET. The dihedral angle ϕ_2 is the one generally discussed as being gauche in the amorphous phase and trans in the crystalline. However, the outer dihedral angles ϕ_1 and ϕ'_1 can also adopt trans and gauche conformations. If the conformation about ϕ_2 is designated as G or T, and the outer angles as g, t, and g', t', then the "all-trans" crystalline phase is tTt', while the amorphous phase contains this and all other possible conformers in varying proportions according to thermal history.

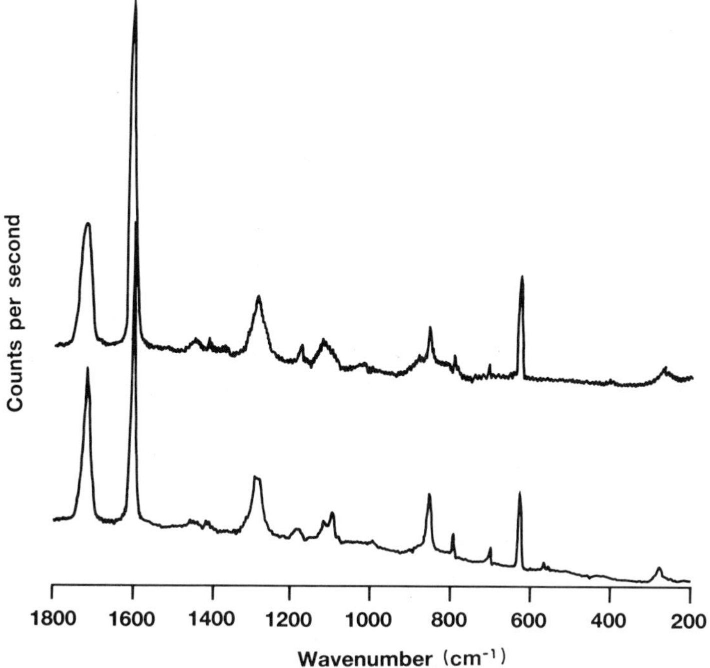

Figure 7.2. Raman spectra of PET. The upper spectrum is of a sample quenched from the melt, while the lower spectrum is of the same sample following thermal annealing.

If we can understand these spectral differences at the molecular level, they can provide us with considerable information about the fundamental nature of crystallization in PET. Moreover, the time dependence of the Raman spectrum during annealing can reveal the kinetics of crystallization, an area of considerable importance in polymer science. Because Raman spectroscopy can deal with many physical forms of a polymer, one can examine this crystallization process in powders, films, fibers, etc.

Two explanations were initially offered for these changes. The first relates the spectral differences to the conversion of gauche conformers in the ethylene glycol units to trans conformers during crystallization. This was originally proposed by Ward (1) and was taken up by many others. Subsequently Liang and Krimm (2) argued that the key change affecting the spectra was a change in symmetry around the monomer units as a result of rotation of the carbonyl groups out of the plane of the phenyl ring.

Qualitatively, Ward's view would lead us to look for the disappearance of bands associated with gauche conformers as annealing takes place and the growth of bands for trans conformers. Liang and Krimm's explanation suggests that we should focus on the comparison of IR and Raman spectra in order to identify bands that are forbidden, as a result of a center of symmetry around the ring center, from appearing in both IR and Raman spectra. These bands would be present in the amorphous sample but would decline in intensity as crystallinity increased.

Such qualitative approaches to understanding changes in Raman spectra of polymers are rarely successful, usually leading to ambiguous results. Often bands are too weak to ascertain the presence or absence of key groups conclusively. Selection rules are for isolated species, in solution or gas phase, and may be greatly altered by intermolecular or intrachain effects. Such selection rules only indicate whether a band may appear and give no indication of what its intensity, if any, may be. And even the most crystalline forms of polymers like PET are rarely more than 60% crystalline (although, to be sure, there are other polymers that are very highly crystalline).

7.3. VIBRATIONAL ANALYSIS OF POLYMERS AS A FOUNDATION TO UNDERSTANDING

What is needed, then, is a detailed normal coordinate vibrational analysis to shed light on the origin of various modes and how we expect these to change. A frequent criticism of such analyses is that there are insufficient constraints on the analysis to distinguish between alternatives. Nonetheless, experience shows that vibrational analysis is essential to illuminating the application of Raman spectroscopy to polymer science. The approach to carrying out such analyses is described in detail, with many examples, in the books by Zbinden (3) and Painter, Coleman, and Koenig (4). Zbinden's book is particularly valuable as a guide to treating the low-frequency vibrational modes of polymers in the context of the vibrational analysis of solids, whereas Painter et al. set the analysis in the context of small-molecule vibrational analyses.

In a vibrational analysis, Boerio, Bahl, and McGraw (5) studied PET as well as deuterated derivatives of PET—one in which the four ring protons are deuterated; the other with the four methylene protons deuterated—and a series of low-molecular-weight aromatic esters. (Only by such means can the origins of vibrations be established with reasonable certainty.) To relate this to the crystallization process, it was necessary to examine the normal modes of PET over a range of conformations.

This series of detailed calculations was probably beyond the available computational capacity of any laboratory in the late 1950s when the original proposals were made, and such computation was still a major but tractable undertaking in 1976 when Boerio et al. published their work. The advent of very powerful workstations for molecular modeling (some of which incorporate high-speed programs for vibrational analysis) makes this a much more feasible approach in the 1990s and should allow many possible structural variations to be examined for their effect on the vibrational spectrum.

The conclusion from the vibrational analyses of PET, as well as its building blocks and derivatives, is that the crystallization process in PET is complex, involving both trans–gauche isomerism in the ethylene glycol segments and changes in the symmetry and resonance characteristics of the substituted benzenoid structure. Indeed, we shall see in a subsequent section that the vibrational spectrum is sensitive to conformational change, as well as to the development of the longer range order that is crystallinity, and that these are distinguishable. Moreover, when there is orientation of the chains, as for example in PET fibers, this is also manifested in the vibrational spectrum. Changes in orientation can be distinguished from conformational changes and crystallinity changes.

Figure 7.3 defines key internal coordinates of PET, and Table 7.1

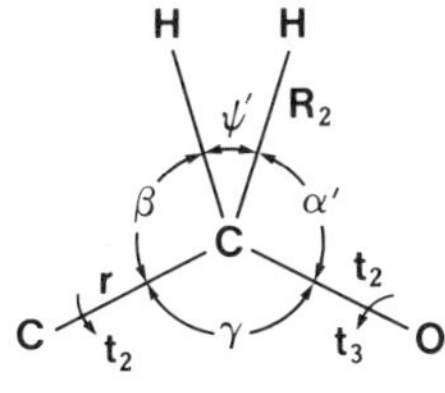

Figure 7.3. Internal coordinates used to describe the vibrational modes of PET. [From Boerio et al. (5).]

Table 7.1. Calculated and Observed Frequencies and Potential Energy Distribution (PED) for PET[a,b]

Symmetry Species	ν_{obs} (cm^{-1}) IR	R	ν_{calc} (cm^{-1})	PED
A_g		3085	3074	$k(99)$
		3085	3072	$k(99)$
		2912	2889	$R_2(99)$
		1730	1723	$S(90)$
		1615	1606	$T(74)$, $\phi(25)$
			1576	$T(82)$, $\phi(13)$
		1462	1452	$\psi'(58)$, $\alpha(44)$
		1418	1395	$\beta(62)$, $\alpha(28)$
		1310	1316	$\phi(71)$
		1295	1273	$T'(37)$, $t_1(32)$, $\phi(18)$, $\theta(18)$
		1192	1178	$\phi(72)$, $T(15)$
		1119	1118	$t_1(38)$, $r(35)$, $\gamma'(15)$
		1096	1100	$T(31)$, $r(28)$, $t_1(10)$
		1000	1002	$t_2(80)$, $r(21)$
		857	844	$T(27)$, $t_1(15)$, $\theta(12)$, $T'(11)$, $r(10)$
		701	693	$\Omega(36)$, $\theta(22)$, $T'(14)$
		626	628	$\Omega(60)$, $\phi(15)$
			559	$\gamma(37)$, $\phi(22)$
			373	$\gamma(23)$, $\phi(21)$, $\Omega(10)$
		278	282	$\theta(33)$, $\phi(28)$, $\delta(21)$
			151	$\delta(44)$, $r(13)$, $\Phi(11)$
B_g		2968	2967	$R_2(9)$
			1285	$\alpha(62)$, $\beta(48)$
			1174	$\beta(43)$, $\alpha(39)$
			972	$\mu_2(136)$
			859	$\mu_2(114)$
		800	799	$Z(48)$, $\mu_3(48)$, $M(40)$
			673	$Z(86)$, $\mu_3(39)$
			245	$M(77)$
			119	$\tau_3(89)$
			63	$\tau_1(60)$, $\tau_2(37)$
			37	$\tau_2(60)$, $\tau_1(34)$
$A_u(\perp)$	2962		2957	$R_2(100)$
			1271	$\alpha(94)$
			991	$\mu_2(126)$
	875		871	$\mu_2(81)$, $M(24)$, $Z(15)$
	845		837	$\beta(98)$
	727		723	$\mu_3(60)$, $\mu_2(43)$
			470	$Z(89)$, $M(79)$

Table 7.1 (*Continued*)

Symmetry Species	ν_{obs} (cm^{-1}) IR	R	ν_{calc} (cm^{-1})	PED
	430		411	$Z(135)$
			141	$\tau_4(64)$, $\tau_2(18)$
			80	$Z(29)$, $M(27)$, $\mu_2(17)$
			54	$\tau_2(49)$, $\tau_1(32)$
$B_u(\parallel, \perp)$	3081		3074	$k(99)$
	3067		3072	$k(99)$
	2889		2882	$R_2(100)$
	1727		1721	$S(93)$
	1504		1504	$\phi(54)$, $T(39)$
	1475		1465	$\Psi'(62)$, $\alpha(35)$
	1410		1407	$T(46)$, $\phi(39)$
	1337		1334	$\beta(55)$, $\alpha(41)$
			1293	$T(138)$
	1263		1254	$t_1(50)$, $T'(38)$, $\theta(23)$
	1126		1126	$\Omega(26)$, $t_1(26)$, $\phi(19)$, $T(12)$
	1109		1102	$\phi(56)$, $T(33)$
	1018		1019	$\Omega(37)$, $T(28)$, $\phi(27)$
	973		971	$t_2(89)$
			868	$t_1(29)$, $\delta(23)$, $\theta(14)$
	502		506	$T'(37)$, $\theta(19)$
	438		459	$\delta(31)$, $\gamma(27)$, $\theta(16)$
	382		384	$\gamma(29)$, $\gamma'(18)$, $\theta(17)$
	145		135	$\phi(46)$, $\gamma(30)$

Source: From Boerio et al. (5).

[a]Because of contributions from off-diagonal force constants, the potential energy distribution as listed may be greater or less than 100%.

[b]Observed frequencies in parentheses may be assigned to more than one calculated mode.

summarizes the observed IR and Raman spectra and the calculated potential energy distribution (PED) in terms of these internal coordinates. This type of result forms the basis for interpretation of spectral changes. But a most important cautionary note is needed here. Few of the observed modes are pure—that is, they have significant contributions from several internal coordinates.

The purest modes occur at the high- and low-frequency ends of the spectra, and include the C—H and carbonyl stretching modes at the higher frequency end and the torsional modes at the low-frequency end. Some of the well-known out-of-plane ring modes in the 800–1000 cm^{-1} region are also factored from other coordinates. For all other cases we can expect a fairly complex behavior, one that is sensitive to many sorts of changes in the structure or molecular environment.

7.4. SENSITIVITY OF THE RAMAN SPECTRUM TO CONFORMATIONAL CHANGE

We return now to a brief discussion of two of the major changes in the Raman spectrum, illustrative of the sensitivity of Raman spectroscopy to polymer conformational change. The carbonyl stretching vibration at $1725\,\mathrm{cm}^{-1}$ is characterized by a broad band (full width at half maximum, FWHM, is $\sim 26\,\mathrm{cm}^{-1}$) in the amorphous solid, and narrows (FWHM is as small as $\sim 12\,\mathrm{cm}^{-1}$) in the annealed samples. The peak position is constant to within 1–$2\,\mathrm{cm}^{-1}$. Melveger (6) pointed out that this half-width in PET correlated with sample density in a linear relationship with only moderate scatter and over a wide range of PET samples. This change in density is often associated with crystallinity but is in fact really due to conformational changes in the polymer that are part of the annealing process.

Kim et al. (7) pointed out that the carbonyl stretching band is somewhat asymmetrical in the amorphous samples. This asymmetry can be seen in the spectrum of PET published by Melveger (6), although it was not remarked upon. A convenient way of quantifying the degree of asymmetry is the asymmetry ratio. $\mathrm{SI} = B/A$, where SI is the symmetry index, calculated as shown in Fig. 7.4. Amorphous PET shows values of SI of 1.3. After annealing, the SI for PET falls to 1.0. Kim et al. pointed out that for poly(propylene

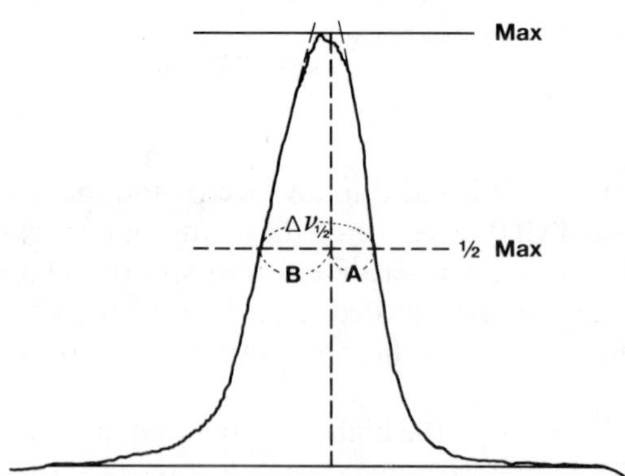

Figure 7.4. Asymmetry in the carbonyl band of PET. The symmetry index is defined as the ratio B/A.

terephthalate) SI for the amorphous material is also near 1.3, but for the crystalline material it rises to 1.8.

The width and shape of the carbonyl stretching band in the Raman spectrum of PET reflects the conformational distribution in the sample. As crystallization occurs through annealing, this distribution narrows, and it converges to a particular, all-trans conformation. The band narrows and becomes more symmetrical as other conformations disappear from the sample.

The other major change in the Raman spectrum observed and discussed by Melveger (6) is the growth of the new band at $1092\,cm^{-1}$. One of the problems with comparing Raman spectra of different samples in order to understand and use the spectral differences to quantify differences in physical properties is the many factors that can give rise to absolute differences between spectra. Half-bandwidths—for example, of the carbonyl band—are the best such parameters, as they are an absolute property of the sample and should be independent of most instrumental variation so long as the resolution and sampling interval are properly chosen. Intensities of individual bands, however, depend on laser power, scattering efficiency of the particular small volume of sample being studied, etc. To circumvent this problem, one often uses relative intensities. In the case of the $1092\ cm^{-1}$ band of PET, Melveger examined the ratio of this band to a band at $632\,cm^{-1}$, chosen because the latter is a sharp, well-separated band in the spectrum that does not appear to change with annealing. For the same range of samples on which he had studied the carbonyl band, the $1092/632\ cm^{-1}$ intensity ratio did not appear to correlate nearly as well with density as did the carbonyl band. In fact, Bulkin et al. (8) later showed that the origin of this difference is in the sensitivity of the $632\,cm^{-1}$ band to orientation, rather than in any property of the $1092\,cm^{-1}$ band. This illustrates well both the difficulty in using intensity ratios and the sensitivity of the spectrum to different effects. When the intensity ratio used is $1092/1117\,cm^{-1}$ instead, Bulkin et al. showed that the spectral changes are identical to those of the carbonyl half-width, for a range of samples having different crystallinity and orientation. Thus both the $1092\,cm^{-1}$ band and the carbonyl half-width correlate with conformational reorganization in the polymer.

7.5. LOW-FREQUENCY ($<200\,cm^{-1}$) SPECTRA

Thus far we have only discussed spectra above $200\,cm^{-1}$. One of the advantages of Raman spectroscopy as a characterization technique when compared with IR spectroscopy, for example, is the ability to study low-frequency vibrations as well as higher frequency modes using the same

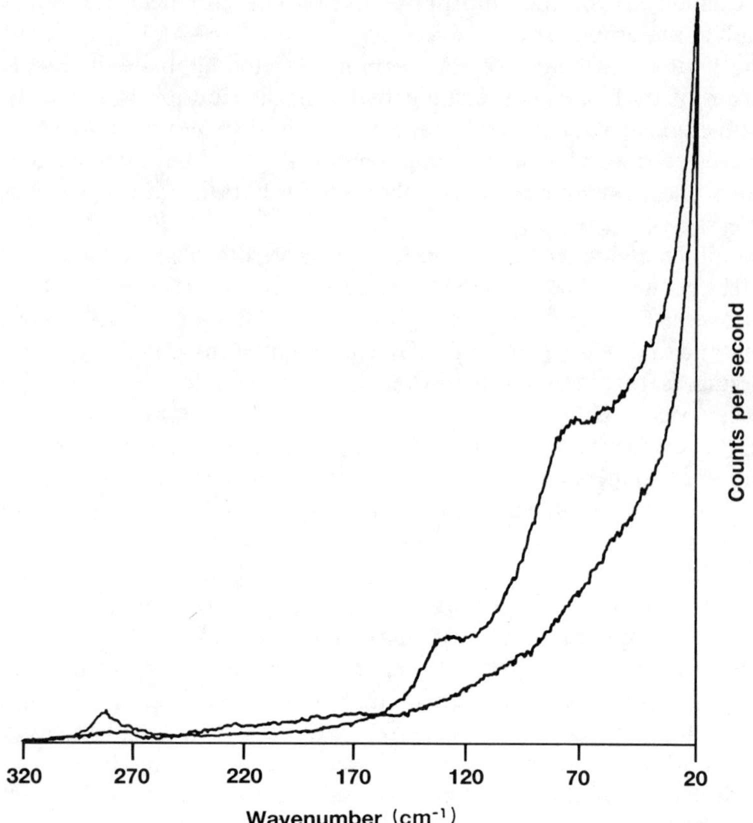

Figure 7.5. Low-frequency Raman spectrum of PET, same samples as in Fig. 7.2. The upper trace, showing relatively well-defined peaks, is of the annealed sample. Spectra have been arbitrarily normalized to the 20 cm^{-1} intensity.

instrument. Figure 7.5 shows the Raman spectra of PET (9), for the same quenched amorphous and annealed samples, between 20 and 300 cm^{-1}. The change in the 273 cm^{-1} band has already been mentioned in Section 7.2, but now we see that many other changes occur in this region of the spectrum. The spectrum of the annealed sample clearly shows the development of features at 129 and 73 cm^{-1}. In addition, there are probably several other bands between 120 and 75 cm^{-1} and below 70 cm^{-1} that cannot be easily distinguished from background scattering. The spectra have been normalized to make the scattering level at 20 cm^{-1} the same.

In addition to these changes, the amorphous sample has a broad band

($\sim 70\,cm^{-1}$ half-width) centered at $215\,cm^{-1}$. This broad band probably derives its intensity from the vibrational modes associated with the bands present in the crystalline material but absent in the amorphous. The vibrational analysis supports this view, even though at the time of that analysis there had been no observations in this region. The calculations predict the band at $282\,cm^{-1}$ as an A_g mode that can be approximately described as a symmetrical, in-plane deformation of all the angles of the ester linkage. The calculation also predicts B_g modes at 119 and $63\,cm^{-1}$ that are torsional modes involving both the ester and glycol units. In the amorphous material, these modes are at once free to mix, because of the reduced symmetry, but are more localized to individual monomer units. These monomer units have, as we have already seen in the analysis of the carbonyl band, a distribution of conformations and environments. The result is the broad scattering in the $200\,cm^{-1}$ region.

From the spectra of these two samples of PET, we see that there is sensitivity to both conformation and crystallinity, that there are methods whereby we can analyze and in fact quantify this sensitivity, and that vibrational analysis gives us insight into the origin of the spectral changes.

7.6. FEATURES IN THE SPECTRA OF FIBERS

PET is, of course, an important fiber material, and we turn now to the information content of the Raman spectra of polymer fibers. Melveger (6), Jarvis and co-workers (10), and Purvis and Bower (11) carried out a number

Table 7.2. Orientation Parameters from Data for the $1732\,cm^{-1}$ Line from Raman Spectra of PET

Sample Draw Ratio	$\langle P_2 \rangle$	$\langle P_4 \rangle$
1.25	0.02	−0.09
1.87	0.05	−0.00
2.19	0.09	−0.19
2.50	0.13	−0.09
2.75	0.17	0.01
3.16	0.22	−0.08
3.82	0.27	0.05
3.25	0.31	0.02
4.40	0.43	0.04
4.85	0.43	0.14
5.85	0.43	−0.13

of investigations of such fibers and tapes using Raman polarization data to understand the origin of various modes in the spectrum. Moreover, it has been demonstrated that Raman data can give quantitative measure of orientation. By example, Table 7.2 shows order parameters $\langle P_2 \rangle$ and $\langle P_4 \rangle$ (relating to $\langle \cos^2 \theta \rangle$ and $\langle \cos^4 \theta \rangle$ orientation terms) for the 632 cm^{-1} band

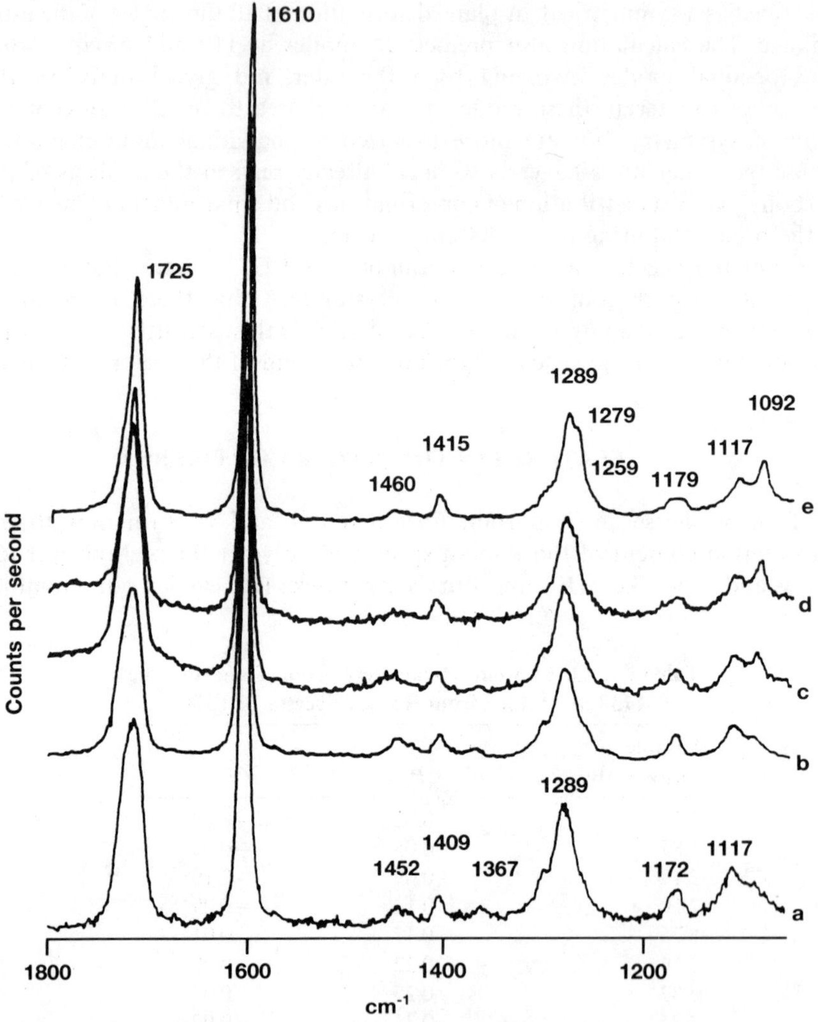

Figure 7.6. Raman spectra of five PET fibers (1100–1800 cm^{-1} region). (a) Original PET yarn prepared with take-up speed of 3.3 km/min. (b) Sample annealed with ends free at 70°C for 60 min. (c) Sample annealed with ends fixed at 86°C for 2.5 min. (d) Sample annealed with ends fixed at 122°C for 5 min. (e) Sample annealed with ends fixed at 152°C for 5 min.

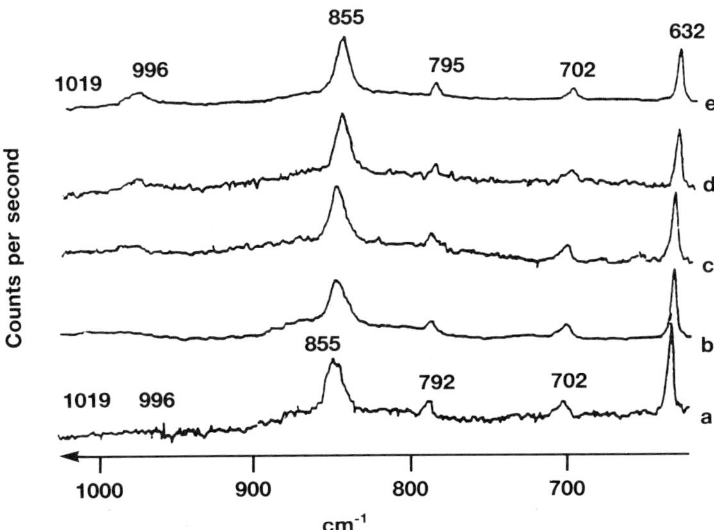

Figure 7.7. Raman spectra of five PET fibers (600–1100 cm^{-1} region). See Fig. 7.6 for sample descriptions.

as a function of draw ratio for a PET tape. The sensitivity of this band to orientation that we referred to earlier is clearly shown by these data. A more detailed discussion of the analysis of data from Raman spectra of oriented polymers is found in this volume in Chapter 8 by Rabolt.

In the processing of fibers, properties such as conformational change, chain orientation, and crystallinity do not generally vary independently. Bulkin et al. showed the Raman spectra of five PET fibers (Figs. 7.6–7.8) treated under various annealing conditions to produce a variety of properties. Here, we see that for fibers such as these virtually every band in the spectrum is sensitive to the changes taking place. These changes are clearly more complex than those in bulk samples, and the Raman spectrum is mirroring that complexity.

7.7. CONSIDERATIONS IN SELECTING SPECTROSCOPIC PARAMETERS FOR COMPARING SPECTRA

Given that a vibrational analysis of the polymer already exists, the first task in analyzing these changes is to establish quantities that are good spectroscopic parameters characterizing each change. Such parameters should be independent of the quantity of scatterers illuminated, any multiple

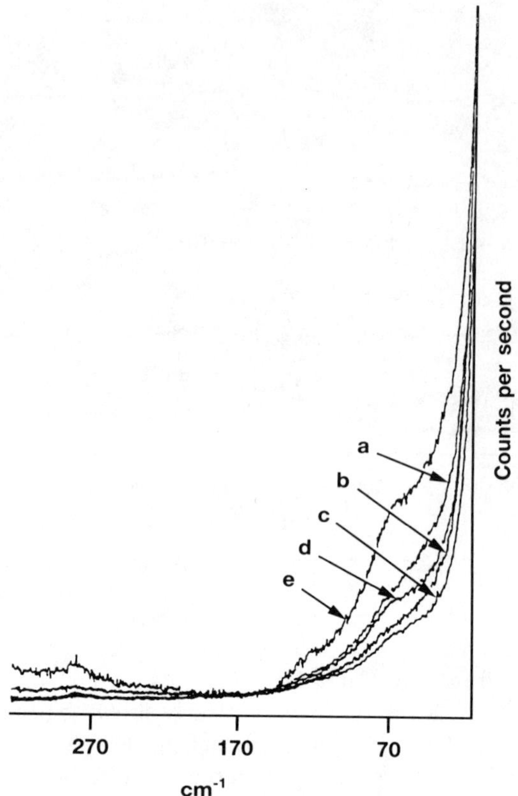

Figure 7.8. Raman spectra of five PET fibers (20–300 cm^{-1} region). Spectra have been normalized to 20 cm^{-1} intensity. See Fig. 7.6 for sample descriptions.

scattering due to refractive index discontinuities, and other such sampling variations. For any such quantitative spectroscopic analysis, selecting these parameters is a very important process, and the PET case illustrates most of the important considerations.

We have already discussed the utility of the carbonyl half-width and the 1092/1117 cm^{-1} intensity ratio, as well as the approach of normalizing low-frequency spectra to scattering levels at 20 cm^{-1}. The authors also noted that the weak band at 1452 cm^{-1} shifts its peak intensity to 1460 cm^{-1} in the most highly annealed samples. To characterize this change, the ratio of intensities at these two wavenumbers was used. Many similar coupled pairs seem to be present in the spectra, and other intensity ratios at 1415/1409, 1279/1259, 996/1019, 795/792, and 282/273 can be followed.

When bands are very weak and/or highly overlapped, it may be difficult either to locate frequency maxima or to measure intensities very well. Such cases may be regarded as a variant of the carbonyl stretching band changes and characterized by a half-width.

The bands at 632 and 855 cm^{-1} change in relative intensity during the annealing process, but there is no obvious neighbor band that is coupled to the band in question. In these cases the relative intensity of the band to that at 1289 cm^{-1} was used. This decision is made because the 1289 cm^{-1} band appears to be the most constant. The 1610 cm^{-1} band is also nearly constant in intensity and half-width, but its intensity is so large that its use as the denominator in an intensity ratio produces serious round-off errors.

Note that while the intensity of a band in the Raman spectrum should be proportional to concentration, these parameters, selected somewhat arbitrarily, need not be directly proportional to concentration. However, as we will see, when these parameters are examined in comparison to each other and to other data on these samples (e.g., thermal analysis, density, birefringence, and X-ray scattering), the results can be quite enlightening as to the origin of the observed changes. Thus we do not necessarily expect or find linear behavior when various spectroscopic changes are plotted against each other or against concentration, but we do look for trends or changes in the same direction, albeit with a sometimes complex proportionality relationship.

7.8. ORIENTATION, CONFORMATION, AND CRYSTALLINITY

Having defined these spectroscopic parameters, we examine what information can be derived from them. The PET fiber used in this study was partially oriented by using a take-up speed of 3300 m/min when the fiber was made. This yields what is known as a partially oriented yarn (POY), oriented but still amorphous, with considerable conformational disorder. When partially oriented PET samples are annealed, several things can happen. The relative amounts of amorphous and crystalline phases change. This is both preceded and accompanied by conformational change within the chains. The degree of orientation of the amorphous and crystalline phases also changes. The nature of these changes depends on temperature and time of annealing, and whether the yarn is allowed to shrink or is held fixed.

To analyze the changes, we first look at internal correlations between different spectroscopic changes. This allows one to examine whether these spectroscopic changes seem to be measuring the same molecular phenomena. To assign the spectroscopic changes to particular events, other data are used.

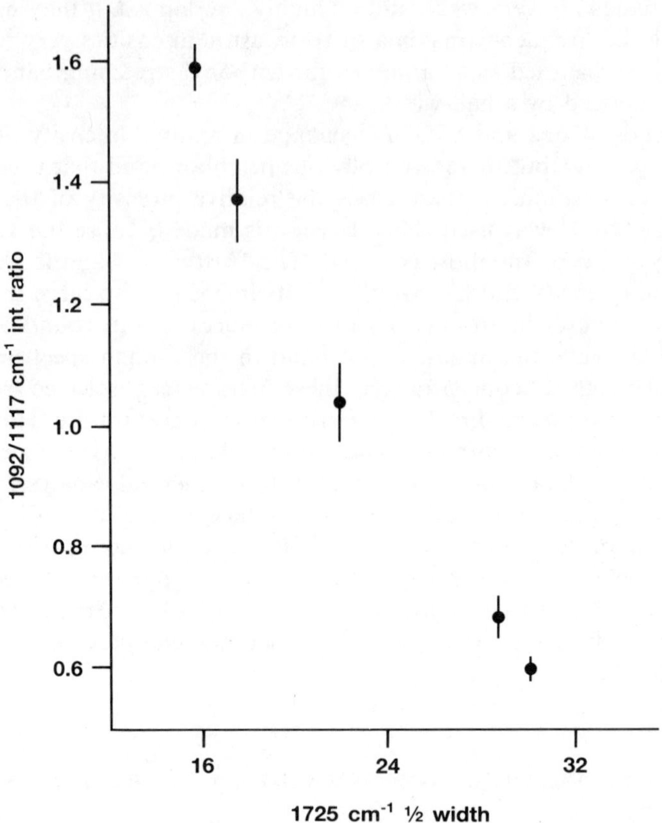

Figure 7.9. Correlation of the intensity ratio I_{1092}/I_{1117} cm^{-1} with the half-bandwidth of the carbonyl stretching mode. Error bars shown in this and subsequent figures are calculated by assuming that the measured intensities have standard deviations proportional to the square root of the total number of counts. Propagation of errors is then used to calculate the standard deviation of the computed result.

These are called external correlations. Most of the external data are straightforwardly interpreted and used, in contrast to the Raman data.

Figure 7.9 shows a plot of the intensity ratio 1092/1117 cm^{-1} versus the FWHM of the carbonyl stretching mode. There is a very linear relationship here (correlation coefficient 0.99 for a least squares straight line). It must be assumed that these two spectroscopic changes measure the same molecular phenomenon. This sort of correlation is also found when the intensity ratios 1452/1460 and 1415/1409 are plotted versus carbonyl band FWHM. These four changes then form a first group.

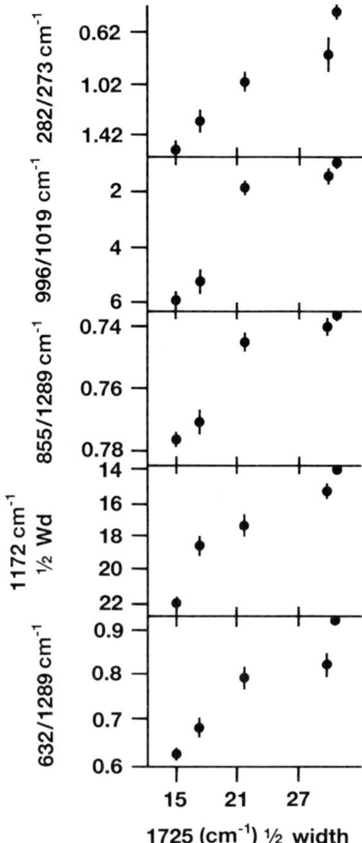

Figure 7.10. Intensity ratios and bandwidths for five sets of bands correlated with the half-bandwidth of the carbonyl stretching mode.

In Fig. 7.10 we see what happens when the 632/1289 cm^{-1} intensity ratio is plotted versus FWHM of the carbonyl band. This strange dependence on annealing is also found for other cases as shown in the figure. These curves may be best characterized by noting the discontinuity between the rightmost two samples, which are quite amorphous, and the three more crystalline ones that fall to the left in each part of the figure.

The third and final group is the 73 cm^{-1} intensity and the intensity ratios 1279/1259 cm^{-1} and 795/792 cm^{-1} shown in Fig. 7.11. The essential point here is that the two amorphous samples give the same value in each case.

These groupings would be of only passing interest if not for a striking fact: when they are viewed together with the vibrational analysis of PET, one sees that they also represent groupings of molecular vibrational motions.

The most clear-cut of these is group 2, which represents almost completely

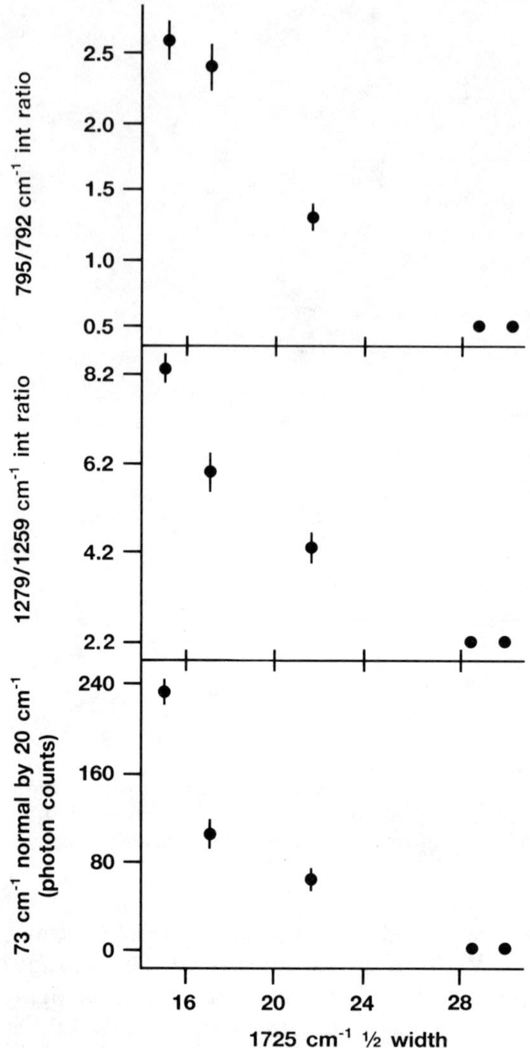

Figure 7.11. Intensity ratios for three sets of bands correlated with the half-bandwidth of the carbonyl stretching vibration.

modes that are localized to the phenyl rings. These modes are not sensitive to annealing in the bulk sample of PET, and the sensitivity only develops in oriented materials. The rigidity of the phenyl rings thus allows one to follow orientation independent of conformational change.

The first group represents modes that, by contrast, are most sensitive to conformational change. They involve the stretching and out-of-plane deformation modes of the glycol and ester linkage. Any conformational change in this linkage must result in a change in the potential energy distribution of these modes as the relevant bond angles and electron distributions change. The carbonyl stretching mode falls into this group as the change in conformation alters its force constant through interaction with the pi system of the ring.

Group 3 is small but consists of two in-plane motions of the ester linkage, and the low-frequency mode, which is probably a torsional mode. This is an important group because it seems to be insensitive to changes that are due to orientation only.

If we turn to the external correlations, the origins of the groupings and their significance is clear. We have already pointed out that the modes in group 1 correlate with density. These are the modes that are sensitive to conformational change.

It is suggested that group 2 modes follow orientation. This is confirmed by the observed correlation of these modes with birefringence, illustrated in Fig. 7.12.

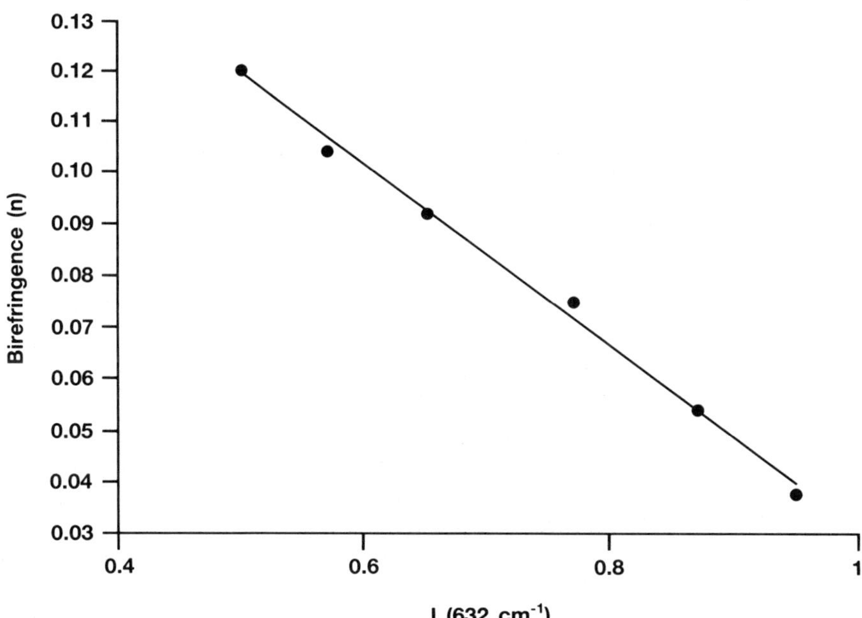

Figure 7.12. Correlation between the birefringence measurements of the five fibers and the 632/1289 cm^{-1} intensity ratio.

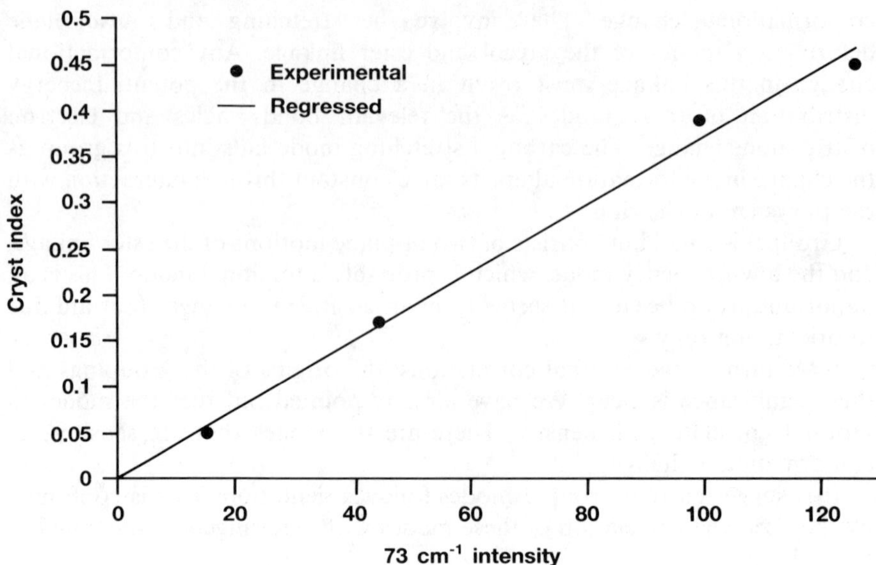

Figure 7.13. Correlation of the WAXS (wide-angle X-ray scattering) crystallinity index with the 73 cm^{-1} intensity.

The modes in group 3 are unchanged for two of the samples but grow when the samples are annealed at higher temperatures. Figure 7.13 shows that these changes correlate well with an index of crystallinity developed from wide-angle X-ray scattering data.

7.9. AMORPHOUS ORIENTATION AND LOW-FREQUENCY SPECTRA

We return now to consider further aspects of the information that can be derived from the low-frequency spectra. Figure 7.14 compares the Raman spectra of amorphous bulk and fiber samples. Superficially they are similar; in fact, on closer examination, it is clear that the spectra are different. There is a small but real intensity difference between the two spectra, which can be seen by plotting a difference spectrum as in Fig. 7.15. The difference spectrum shows an apparent band, with a maximum at 34 cm^{-1}, tailing off slowly to 170 cm^{-1}. The rapid rise is perhaps exaggerated by the normalization of the data at 20 cm^{-1}, but in any case the difference spectrum must be asymmetrical to the high-frequency side.

DeBlase et al. (9) associated the broad scattering in the difference spectrum

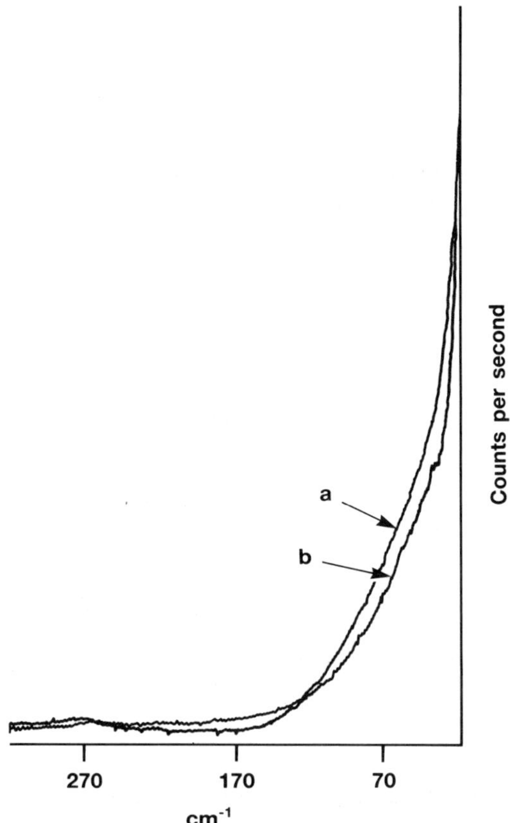

Figure 7.14. Low-frequency Raman spectra of PET. (a) Original sample of fiber yarn that is partially oriented (spectrum a of Fig. 7.8). (b) Amorphous sample quenched from melt.

with the oriented amorphous phase. The authors suggested that the band was of intermolecular origin,. involving loosely coupled chain segments. Such scattering is reminiscent of broad, low-frequency scattering seen in Raman spectra of liquid crystalline phases.

Raman spectra can be taken with the aid of a microscope, and this has been advantageously applied to PET fibers by Adar and Noether (12). It has been known for some time that as PET fibers are formed a thin skin forms on the outside of each filament. Using the same ideas described above to isolate conformational change from orientation, the authors were able to use Raman microscopy to characterize the filament skin as distinct from the bulk.

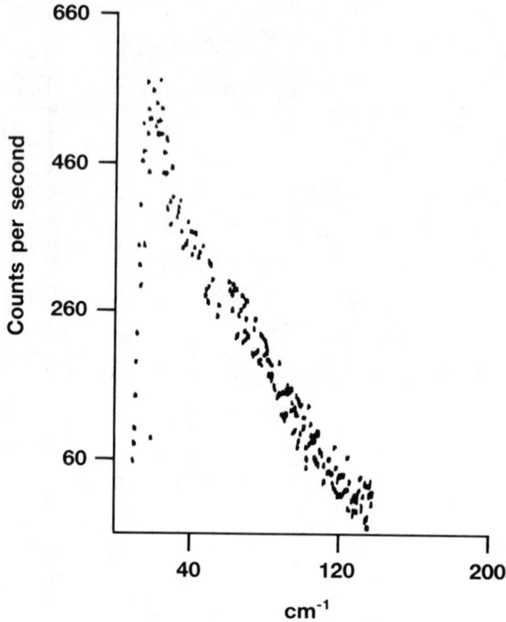

Figure 7.15. Difference spectrum calculated from Fig. 7.14. Sharp drop on low-frequency side is exaggerated by normalization to 20 cm^{-1} intensity.

7.10. DERIVATION OF POLYMER SPECIFIC HEATS

The broad scattering of amorphous polymers in the low-frequency region has been studied by various workers, notably Viras and King (13). These investigators developed a methodology for deriving at least relative specific heat information from the low-frequency Raman scattering. This approach will now be described, and illustrated by its application to PET.

The vibrational component of the specific heat of both amorphous and crystalline solids is calculable from the vibrational spectra through statistical thermodynamic considerations. Because these calculations involve a summation over exponential terms in the vibrational frequencies, the low-frequency modes are the major component of the resulting specific heat (kT is $\sim 200\,cm^{-1}$ at room temperature). The vibrational density of states can be considered as arising from the distribution of phonons within the solid. This phonon distribution is obtained, for both amorphous and crystalline materials, from the low-frequency Raman spectrum. Because of the disorder in even the more crystalline of these materials, all the low-frequency modes can be assumed to be Raman active. Furthermore, since the phonons associated with any material can be considered to be a system of noninteracting bosons, by invoking the Bose–Einstein equation the vibrational component of the specific heat can be calculated. The expression of the relation between the specific heat and the density of vibrational states, derived from either the quasi-harmonic theory or the application of Bose–Einstein statistics, is the same. This expression is

$$C_v^{\text{vib}} = k \int_{\text{min}}^{\text{max}} g(\nu)(h\nu c/kT)^2 [e^{-h\nu c/kT}/e^{-h\nu c/kT} - 1]^2 \, d\nu,$$

where ν = wavenumber and $g(\nu)$ = density of vibrational states.

In order to calculate the specific heat from the low-frequency Raman spectrum, the exact relationship between the density of vibrational states and the Raman scattering intensity is needed. According to the the theory of Shuker and Gammon (14), a reduced Stokes low-frequency intensity I_r should be used to calculate specific heat from the Raman scattering. This reduced intensity takes into account the effect of thermal fluctuations on intensity. The observed Stokes Raman intensity I_{obs} is related to the vibrational density of states by

$$I_{\text{obs}}^{(\nu)} = \sum_b O_b\left(\frac{1}{\nu}\right)(1 - e^{-h\nu c/kT})g_b^{(\nu)},$$

where O_b equals the optical coupling coefficient, summed over all the bands (b). This gives the reduced intensity as

$$I_r = I_{\text{obs}}(\nu)(1 - e^{-h\nu c/kT})^{-1}.$$

In this expression, the contribution to the first-order scattering from thermodynamic fluctuations is accounted for by the Bose factor:

$$1 - e^{-h\nu c/kT}.$$

The expression for I_{obs} given above holds if the scattering is due to vibrations of short correlation range compared to the optical wavelength. This is a reasonable assumption for polymer systems with imperfect crystalline and amorphous domains. This theory is also based on the scattering being first order, which is true for polymers and inorganic glasses to which the approach has been applied.

If we rearrange the expression to get the desired quantity, g_b in terms of

I_{obs}, we obtain

$$g_b(\nu) = \left[\sum_b O_b\left(\frac{1}{\nu}\right)\right]^{-1} I_{obs}(\nu)(1 - e^{-h\nu c/kT})^{-1}$$

or, in terms of $I_r(\nu)$,

$$\sum_b g_b(\nu) = [I_r(\nu)](\nu)\left[\sum_b O_b\right]^{-1}.$$

This expression is similar to the theory of Debye stating that the density of vibrational states is proportional to the frequency squared.

From our expression for g_b, it is now possible to write the vibrational component of the specific heat as

$$C_v^{vib} = k \int_{min}^{max} (N/n) I_r(\nu)\nu O_b^{-1}(h\nu c/kT)^2 [e^{-h\nu c/kT}/e^{-h\nu c/kT} - 1]^2 \, d\nu,$$

where N/n represents the number of vibrational units per gram; and the relationship between the Raman intensity and the density of states is used.

While the expressions for the density of states and C_v^{vib} hold for both amorphous and crystalline samples, the nature of the optical constant O_b is less clear. The value of O_b may vary, both between amorphous and crystalline samples, and even between vibrational bands. In spite of these complications, Viras and King (13) have calculated the density of vibrational states for several polymer systems by assuming that the O_b optical coupling constant is proportional to $1/\nu^2$. This gives the density of vibrational states as

$$g_b(\nu) = I_r/\nu^2.$$

It may be argued, however, that this assumption violates Debye theory and should not be considered as a general approach. An alternative is to consider the constant O_b as an unknown and calculate the relative values of specific heats between different samples. If some additional information is available concerning the calorimetric value of the specific heat, the O_b values may be estimated in order to fit the spectroscopic results to the calorimetric values.

It has already been pointed out that the phonon vibrational unit can be associated with one or more repeat units depending on the nature of the vibration. In a study of polycarbonate. Viras and King (13) considered the low-frequency molecular vibrations of the polymer to be associated with one repeat unit. In that study the origin of the Raman scattering was related to

Table 7.3. C_v^{vib} Data

Sample Fibers[a]	C_v^{vib} (Viras and King Method) (n = No. of rpts vibrational unit) $n=1$	$n=2$	$n=3$
2GT	3.21	1.61	1.07
$2GT_{70}$	2.94	1.50	0.98
$2GT_{86}$	2.95	1.47	0.98
$2GT_{122}$	2.91	1.45	0.97
$2GT_{152}$	2.88	1.44	0.96
3.0 km/min	2.96	1.48	0.99
3.5 km/min	3.09	1.54	1.03
4.0 km/min	3.16	1.58	1.05
4.5 km/min	3.39	1.69	1.13
5.0 km/min	3.32	1.66	1.11
5.5 km/min	2.92	1.46	0.97
Bulk			
Amorphous	2.86	1.43	0.95
Crystalline	2.80	1.40	0.93

[a]2GT = PET.

the pendant aromatic groups of the main polymer backbone. However, one polymer repeat unit is not a strict requirement for every polymer system, and low-frequency vibrational units with greater delocalization surely exist.

Bulkin et al. (15) applied this approach to studying the specific heat of polyester fibers. The results for a series of eleven fibers and two bulk samples are shown in Table 7.3. The first five fiber samples are those discussed previously, as are the bulk amorphous and crystalline samples. The remaining six samples are included to show the effect of varying the take-up speed from 3 to 5.5 km/min.

The three columns in Table 7.3 show the effect of changing the number of monomer units assumed to be responsible for the delocalization. This is done as follows: From the expressions already given,

$$(N/n)I_r(v)/v^2 = g(v),$$

the expression for C_v can be written as

$$C_v^{vib} = k\int_{min}^{max} (N/n)I_r/v^2(hvc/kT)^2[e^{-hvc/kT}/e^{-hvc/kT} - 1]\,dv.$$

Table 7.4. C_p Data

	kJ/kg-K
Undrawn yarn	1.138
Cold-drawn yarn	1.154
Commercially drawn yarn	1.1419
Annealed from melt	1.100
Annealed at temperature below T_m	1.114

Source: From C. W. Smith and M. Dole (15a).

For the samples of PET used in this work, the intrinsic viscosity $[\eta]$ is 0.64, corresponding to a molecular weight M_n of 21,059 as calculated from the equation for PET:

$$[\eta] = KM_n^a,$$

where $K = 3 \times 10^{-4}$ and $a = 0.77$, and the degree of polymerization is 109. the number of chains per gram will accordingly be 3×10^{19}, and the number of monomeric units per gram N will be 3.27×10^{21}. This is the N used in the expression above.

The C_v values were calculated with $n = 1$, 2, and 3 as shown in Table 7.3. The specific heat decreased with increase in crystallinity of the samples studied, and increased with large changes in orientation associated with the transition between bulk amorphous material and oriented amorphous fibers.

Table 7.4 shows the only data in the literature on PET specific heat. Clearly, the values of C_v^{vib} must be lower than the total C_p. To achieve this, the calculations show that n must be at least 3. This allows us to set a lower limit on the extent of delocalization of the low-frequency modes.

7.11. A GUIDE TO OTHER RAMAN STUDIES OF POLYMERS

We have illustrated many of the principles of Raman spectroscopy as applied to polymers in the discussion of PET. In this section some leading references will be cited to indicate areas and systems that have been intensively investigated. Early work on laser-excited Raman spectra was reviewed by Koenig (16). The review by Gerrard and Maddams (17) is a good introduction to the literature from the time of Koenig's paper through the early 1980s. It contains 250 references, and these are highlights pointing to the literature on key applications. Biennial reviews in *Analytical Chemistry* on Raman spectroscopy (even-numbered years) (18) and *Polymers* (odd years) (19) are excellent summarries of the current literature.

Koenig (16) summarized assignments of Raman spectra and vibrational analyses done up to 1970. Subsequently, there was extensive work on polyethylene [see, e.g., Snyder (20)] and polypropylene. Other well-studied systems for which analysis is less well established include polystyrene, poly(vinyl chloride) (PVC), and polychloroprene. A detailed set of references can be found in the review of Gerrard and Maddams (17). In Painter et al. (4) there are a number of worked examples that are invaluable to anyone studying these polymers in an industrial context, as well as to anyone thinking of attempting a normal coordinate vibrational analysis of a polymer. Table 7.5 lists some polymers for which analyses are given in this book.

Table 7.5. Selected List of Polymers for Which a Normal Coordinate Analysis Has Been Applied

Polyolefins:
Polypropylene
Poly(ethyl ethylene)
Poly(alkyl ethylenes)
Haloethylene polymers:
Poly(vinyl chloride)
Poly(vinylidene chloride)
Poly(vinylidene fluoride)
Polytetrafluoroethylene
Polydienes and polyalkenylenes:
Syndiotactic 1,2-polybutadiene
trans-1,4-Polybutadiene
cis-1,4-Polybutadiene
trans-1,4-Polychloroprene and *trans*-1,4-poly(2,3-dichlorobutadiene)
trans-1,4-Polyisoprene and *trans*-1,4-poly(2,3-dimethylbutadiene)
Poly(1-*trans*-pentenylene) and poly(1-*trans*-heptenylene)
Polymers containing aromatic rings:
Polystyrene
Poly(ethylene terephthalate)
Polyamides, polypeptides, and proteins:
Polyglycine I

Source. From Painter et al. (4).

Crystalline and amorphous phases have also been thoroughly investigated for polyethylene, with special emphasis on the low-frequency longitudinal acoustic modes (LAMs). These bands were originally recognized from studies of the low-frequency vibrational spectra of *n*-paraffins, in which a band appears that is inversely proportional to chain length. A model was soon developed relating the frequency of the LAM to Young's modulus of the chain. Many papers have been devoted to testing and extending this model. An excellent summary of both the model and the application literature is found in the review of Gerrard and Maddams (17). Some of the other polymers studied are polypropylene, poly(methyl methacrylate), and polystyrene.

The long controversy in the literature about the nature of chain folding in polyethylene has been extensively linked with Raman spectroscopic investigations. This work is reviewed in the aforementioned paper of Gerrard and Maddams as well.

A useful aid employed in studies of polyethylene, as well as other polymers, has been the fiber-optic probe. If one has a fiber-optic bundle in which one fiber, usually at the center, carries the exciting radiation and surrounding fibers collect the scattered radiation, this can be a very efficient technique for obtaining Raman spectra of polymers under difficult conditions. In this way, for example, the Raman spectrum of molten polyethylene has been investigated (17).

An aspect not touched on in the discussion of PET is the Raman spectra of polymers that absorb light in the region of excitation, i.e., the resonance Raman spectra of polymers. Of the many that fall into this class, the most extensive investigations have been of polyacetylene. Indeed, during the 1980s, work on polyacetylenes dominated the Raman literature on polymers. This declined toward the end of the decade as interest in these polymers as conductors declined. The *Analytical Chemistry* reviews give numerous references in this area. A useful review was given by Leising et al. (21). Related to the polyacetylenes is the formation of resonantly scattering groups in degradation products of PVC. Extremely low quantities of such groups can be detected by resonance Raman spectroscopy.

In Section 7.1, the role of FT-Raman spectroscopy using NIR excitation was mentioned as being of importance in cases where fluorescence is a problem. Many examples of this have been published. Williams and Mason (22) have given a number of examples of industrially important samples where Raman spectroscopy excited in the NIR is effective while visible-excited Raman spectra are only obtainable for purified or ideal samples. These authors also described the use of fiber-optic probes in conjunction with an FT-Raman instrument. Agbenyega et al. (23) examined the FT-Raman spectra of nylons in some detail. Such spectra were previously almost unknown before this technique was used.

But FT-Raman spectroscopy is limited in its ability to access the low-frequency region. This is a consequence of the need to effectively filter the exciting frequency from the interferometer. As a result of this limitation, FT-Raman spectroscopy has not been able to reveal information on LAM modes. This situation will change when NIR-excited spectra are obtained by conventional dispersive instruments with high-quality array detectors.

7.12. CONCLUSION

Raman spectroscopy is now being applied and has been used for a broad range of polymer science problems. Its best features are ease of sample handling, particularly for processed materials, ability to perform equally well for any sample phase, and molecular specificity. The biggest limitation, fluorescence, has largely been overcome. What remains is the considerable skill required to operate a spectrometer at high performance and to interpret the results. There is some evidence that FT-Raman instruments are easier to operate at peak performance than the dispersive instruments commercially available. Raman spectroscopy of polymers is almost always carried out by scientists with advanced education in both the spectroscopic and polymer area. Interpretation is generally indirect and often involves substantiating calculations. Nonetheless, those laboratories in academia, industry, and government that have dedicated themselves to obtaining and interpreting such spectra have reaped ample rewards.

REFERENCES

1. I. M. Ward, *Chem. Ind.* (*London*) p. 905 (1956); p. 1102 (1957).
2. C. Y. Liang and S. Krimm, *J. Mol. Spectrosc.* **3**, 554 (1959).
3. R. Zbinden, *Infrared Spectroscopy of High Polymers.* Academic Press, New York, 1964.
4. P. C. Painter, M. M. Coleman, and J. L. Koenig, *The Theory of Vibrational Spectroscopy and Its Application to Polymeric Materials.* Wiley, New York, 1982.
5. F. J. Boerio, S. K. Bahl, and G. E. McGraw, *J. Polym. Sci., Polym. Phys. Ed.* **14**, 1029 (1976), and references therein to earlier work by these authors.
6. A. J. Melveger, *J. Polym. Sci., Part A-2* **10**, 317 (1972).
7. J. S. Kim, M. Lewin, and B. J. Bulkin, *J. Polym. Sci., Polym. Lett. Ed.* **24**, 1783 (1986).
8. B. J. Bulkin, M. Lewin, and F. J. DeBlase, *Macromolecules* **18**, 2587 (1985).
9. F. J. DeBlase, M. L. McKelvy, M. Lewin, and B. J. Bulkin, *J. Polym. Sci., Polym. Lett. Ed.* **23**, 109 (1985).

10. See, for example, D. A. Jarvis, I. J. Hutchinson, D. I. Bower, and I. M. Ward, *Polymer* **21**, 41 (1980).
11. J. Purvis and D. I. Bower, *J. Polym. Sci., Polym. Phys. Ed.* **14**, 1461 (1976).
12. F. Adar and H. Noether, *Microbeam Anal.* **20**, 41 (1985).
13. F. Viras and T. A. King, *Polymer* **25**, 899 (1984).
14. R. Shuker and R. W. Gammon, *Phys. Rev. Lett.* **25**, 222 (1970).
15. B. J. Bulkin, F. DeBlase, and M. Lewin, *Proc. SPIE—Int. Soc. Opt. Eng.* **665**, 234 (1986).
15a. C. W. Smith and M. Dole, *J. Polym. Sci.* **20**, 37 (1956).
16. J. L. Koenig, *Appl. Spectrosc. Rev.* **4**, 233 (1971).
17. D. L. Gerrard and W. F. Maddams, *Appl. Spectrosc. Rev.* **22**, 251 (1986).
18. A recent example is D. L. Gerrard and H. J. Bowley, *Anal. Chem.* **60**, 368R (1988).
19. A recent example is C. G. Smith, R. A. Nyquist, S. J. Martin, N H. Mahle, P. B. Smith, and A. J. Pasztor, Jr., *Anal. Chem.* **62**, 214R (1989).
20. R. G. Snyder, *J. Chem. Phys.* **76**, 3921 (1982).
21. G. Leising, H. Kahlert, and O. Leithner, *Springer Ser. Solid-State Sci.* **63**, 56 (1985).
22. K. P. J. Williams and S. M. Mason, *Spectrochim. Acta* **46A**, 187 (1990).
23. J. K. Agbenyega, G. Ellis, P. J. Hendra, W. F. Maddams, C. Passingham, H. A. Willis, and J. Chalmers, *Spectrochim. Acts* **46A**, 17 (1990).

CHAPTER

8

ANISOTROPIC SCATTERING PROPERTIES OF UNIAXIALLY ORIENTED POLYMERS: RAMAN STUDIES

JOHN F. RABOLT

IBM Research Division, Almaden Research Center
San Jose, California

8.1. INTRODUCTION

The analysis of anisotropic Raman scattering from oriented molecules in organic and inorganic single crystals has played a major role (1) in unraveling the complex band assignments of symmetry species in these highly ordered systems. Raman spectroscopy has played a successful role because both the polarization of the laser used as the excitation source and the polarization of the scattered light that is detected can be varied. Hence the ability exists to select subsets of the polarizability components simply by choosing the appropriate polarization experiment.

Such polarization analyses on polymers before this decade (2, 3) have been rare primarily owing to the lack of highly oriented visibly transparent specimens. Although it is well known (4) that a single-crystal morphology exists for polymers grown isothermally from solution, the size of such structures are in the micrometer regime, certainly inadequate for routine polarized Raman-scattering (RS) analysis. On the other hand, polymers isothermally crystallized from the melt are at best semicrystalline and often lack sufficient orientation to make polarization measurements practical. Furthermore, many polymers, when melt crystallized, form organized domain structures, e.g., spherulites, whose size (5) is comparable to the wavelength of visible light giving rise to the "milky" appearance often observed in these semicrystalline materials. Multiple scattering in such samples gives rise to internal polarization scrambling, thus rendering polarized Raman studies impossible or, at best, inaccurate.

Analytical Raman Spectroscopy, Edited by Jeanette G. Grasselli and Bernard J. Bulkin. Chemical Analysis Series, Vol. 114.
ISBN 0-471-51955-3

Recent improvements in polymer-processing technology have, however, made available uniaxially oriented monofilaments and yarns that are highly oriented and a significant improvement over stretched films used for previous studies (2, 3). Hence, over the past decade, Raman studies of transparent uniaxially oriented filaments have appeared (5–10) with increasing regularity.

Along with the increased amount of experimental Raman studies, there has been a deepening need to develop the theoretical framework required to make polarization analyses possible. Snyder (11) was the first to present expressions relating Raman scattering to symmetry species for the case where the unique symmetry axis (z) of the polymer was parallel to the direction of orientation in a uniaxially oriented material. His results were used to analyze isotactic polypropylene (12), polyethylene (5), polytetrafluoroethylene (6), and an alternating copolymer of ethylene and tetrafluoroethylene (7). When the unique symmetry axis (z) is perpendicular to the chain backbone and thus perpendicular to the orientation direction, these expressions can no longer be used. A new set (13) has recently been derived for this case. These new expressions have been used to analyze polarized RS from a uniaxially drawn filament of poly(vinylidene fluoride) (PVF_2) (10), which X-ray diffraction indicated existed in the planar zigzag form. In cases like PVF_2, which is known to exist in several different conformational structures (14) (all having different symmetries) depending on the thermal and mechanical processing history, it is important to first establish the direction of the unique symmetry axis relative to the orientation direction before beginning a polarization analysis. This sometimes requires the use of other analytical techniques to establish both the direction and extent of orientation. Later in this chapter a specific polarization analysis of oriented planar zigzag PVF_2 will be given as an example.

8.2. EXPERIMENTAL SCATTERING GEOMETRY

In general, the expressions for the RS activities in molecular coordinates (xyz) can be related to a fixed laboratory coordinate system (XYZ) as shown in Fig. 8.1. In the case where the unique molecular symmetry axis (z) is perpendicular to the orientation direction, the sample orientations are (A) x parallel to Y; (B) x parallel to X; and (C) x parallel to Z. In the case where z is parallel to the orientation direction, three analogous sample orientations result, with (A) z parallel to Y; (B) z parallel to X; and (C) z parallel to Z. It should be pointed out at this stage that these two cases are not equivalent since in uniaxial orientation only one molecular axis is oriented; the other two are randomly oriented in the plane perpendicular to it. Hence, orientation averaging in this plane must be done in order to get the true contribution

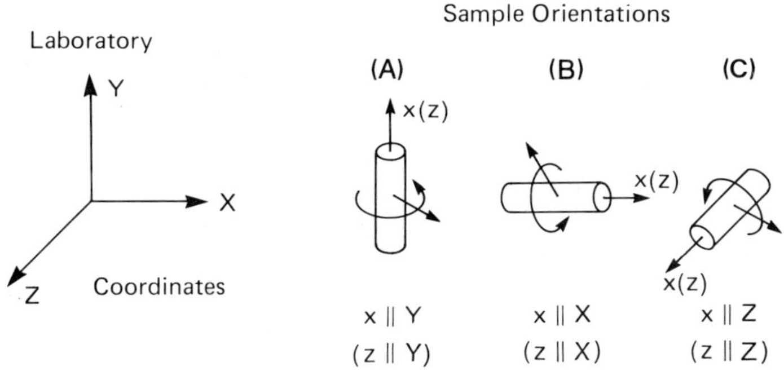

Laboratory Coordinates

Sample Orientations

(A) $x \parallel Y$ ($z \parallel Y$)

(B) $x \parallel X$ ($z \parallel X$)

(C) $x \parallel Z$ ($z \parallel Z$)

Figure 8.1. The laboratory-fixed coordinates are shown on the left (X, Y, Z). The sample coordinates (x, y, z) are attached to the uniaxially oriented sample, with x (z) parallel to the uniaxial orientation direction. The three orthogonal orientations of the sample in the laboratory coordinate system are shown and defined as (A) $x \parallel Y$; (B) $x \parallel X$; and (C) $x \parallel Z$ when the unique molecular symmetry axis (z), is perpendicular to the orientation direction. From Schlotter and Rabolt (13). The orientations become (A) $z \parallel Y$; (B) $z \parallel X$; and (C) $z \parallel Z$ when the unique molecular symmetry axis (z) is parallel to the orientation direction. From Snyder (11).

to the observed polarizability. Thus, one averages about z when it is in the direction of orientation, while (as shown in Fig. 8.1) when z is perpendicular to this orientation direction the averaging is done about x. This subtle difference has caused numerous errors in analysis to appear in the literature. By convention, as shown in Fig. 8.2, the laser is chosen to travel in the X

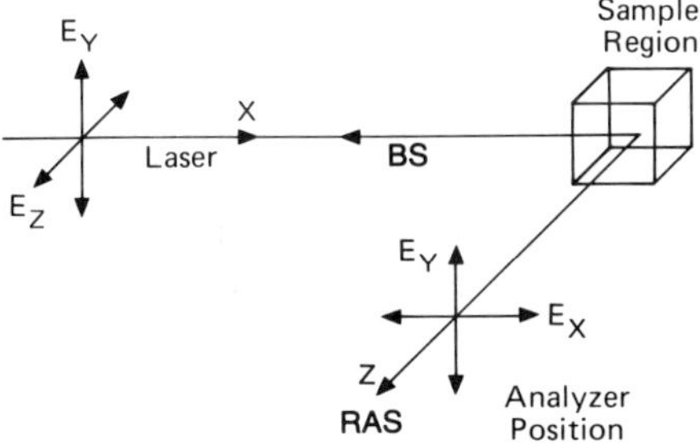

Figure 8.2. The experimental geometry in the laboratory-fixed coordinate system is shown. The laser travels in the X direction with polarizations E_Y or E_Z. In right-angle scattering (RAS) the scattered Raman signal is collected along Z and analyzed for polarizations E_Y and E_X. In backscattering (BS), the signal if collected in the $-X$ direction and analyzed for polarizations E_Y and E_Z.

direction with an electric field polarization along either Z or Y. In the backscattering (BS) geometry, the scattered radiation can be analyzed by a polarizer placed parallel to either Z or Y. On the other hand, in the right-angle scattering (RAS) geometry, the polarization analyzer can be placed parallel to either the X or Y direction. Varying both the input laser polarization and the analyzer position gives four possible combinations, e.g., in RAS they are YY, YX, ZY, and ZX. The designation for each Raman spectrum that incorporates the above polarization information was first introduced by Damen et al. (15). The designation uses the form $A(BC)D$, where A and D are the propagation directions of the incident and scattered radiation while B and C indicate the incident polarization direction and the analyzer position, respectively. If we use the RAS polarization designations and the RAS geometry shown in Fig. 8.2, the designations to describe the resulting Raman spectra are $X(YY)Z$, $X(YX)Z$, $X(ZY)Z$, and $X(ZX)Z$.

As shown in Fig. 8.1, these four designations describe Raman experiments that can be run for three distinct positions of the sample, giving a total of 12 measurements. When added to the 12 possible using BS, a total of 24 spectra can be obtained most of which will not give independent information except in cases of extremely high symmetry. Polymer chains typically have quite low symmetry, resulting in little independent information from the 24 possible Raman measurements. However, some additional qualitative information can be obtained in uniaxially oriented polymers. For example, a comparison of two of these 24 spectra that are predicted to be "identical" will indicate the extent of orientation of the polymer backbone relative to the macroscopic filament axis. This information is invaluable when assigning the symmetry of weaker bands in the presence of intensity leakage due to other polarizations resulting from imperfect uniaxial orientation. This will be illustrated experimentally in the discussion of PVF_2 later in this chapter.

8.3. GEOMETRIC CONSIDERATIONS

Before a detailed analysis of the normal mode symmetries is undertaken, the polarizability elements in the molecular coordinate system must be related to those in the laboratory coordinate system (11, 13). This can be done through the following transformation:

$$\mathbf{F} = \mathbf{\Phi} \mathbf{g}, \tag{1}$$

where the tensor $\mathbf{\Phi}$ is formed from the direction cosines between the two coordinate systems, $\mathbf{F}$ and $\mathbf{g}$. The polarizabilities are then related by the expression:

$$\boldsymbol{\alpha}[F] = \boldsymbol{\Phi}\boldsymbol{\alpha}[g]\boldsymbol{\Phi}^T. \tag{2}$$

Since the relationship between the two systems is not fixed in a uniaxially oriented material but varies about a shared axis, orientation averaging must also be done in the plane perpendicular to the unique axis. The final result for an individual component in the space-fixed frame is

$$\langle \alpha_{FF'}^2 \rangle = \left\langle \left(\sum_{gg'} \Phi_{Fg} \Phi_{F'g'} \alpha_{gg'} \right)^2 \right\rangle. \tag{3}$$

For more explicit details of the form of the transformation matrices, the interested reader is referred to original publications of Snyder (11) or Schlotter and Rabolt (13), where the two individual cases discussed previously are treated in detail.

8.4. SYMMETRY CONSIDERATIONS

When one is analyzing polarized Raman measurements of uniaxially oriented polymers, it is especially convenient to have tabular results that can be used simply by matching the laboratory coordinate systems with those listed in the tables. These tabulations will follow for all point groups having a unique symmetry axis, therefore excluding $C_1, C_i, T, T_d, T_h, O, O_h, G$, and G_h. The remaining point groups can be divided up into five classes, G_1 through G_5, with the generalized irreducible expression A indicating the totally symmetric representation and γ_i, the ith nontotally symmetric representation. These classes are listed in Table 8.1 (13). This classification of point groups is identical whether the unique molecular symmetry axis is parallel or perpendicular to the orientation direction. Depending on its structure, a polymer can be classified by symmetry into one of the point groups that fall under these classes. Once the class is established, say, G_1 or G_3, then a series of Raman polarization measurements can be made using either BS or RAS geometry and the observed band intensities compared with the predicted symmetry species expected (Tables 8.2–8.11).

The results for the case where the unique symmetry axis is perpendicular to the orientation direction are presented in Tables 8.2–8.6, while those where the unique symmetry axis is parallel to the orientation direction are contained in Tables 8.7–8.11. Results for both RAS and BS geometries are given. In the left-hand columns the irreducible representation and the incident polarization of the laser are listed. Entries in the top rows are the sample orientations (A, B, or C) and the direction of polarization of the analyzer.

Table 8.1. Classification of Point Groups According to Their Raman Activities

Class	Point Group								
G_1	C_s	C_2	C_{2h}	G_2	C_3	D_3	C_{3v}	D_{3d}	S_6
Irreducible Symmetry Representations	Symmetry Species								
A	A'	A	A_g	A	A	A_1	A_1	A_{1g}	A_g
γ	A''	B	B_g	γ	E	E	E	E_g	E_g

Class G_3	C_4	C_5	C_6	D_5	D_6	C_{5v}	C_{6v}	C_{3h}	C_{4h}	C_{5h}	C_{6h}
A	A	A	A	A_1	A_1	A_1	A_1	A'	A_g	A'	A_g
γ_1	B	E_2	E_2	E_2	E_2	E_2	E_2	E'	B_g	E'_2	E_{2g}
γ_2	E	E_1	E_1	E_1	E_1	E_1	E_1	E''	E_g	E''_1	E_{1g}

Class G_3 (Cont.)	D_{3h}	D_{5h}	D_{6h}	D_{4d}	D_{5d}	D_{6d}	S_4	S_8	$C_{\infty v}$	$D_{\infty h}$
A	A'_1	A'_1	A_{1g}	A_1	A_{1g}	A_1	A	A	A_1	Σ_g
γ_1	E'	E'_2	E_{2g}	E_2	E_{2g}	E_2	B	E_2	E_2	Δ_g
γ_2	E''	E''_1	E_{1g}	E_3	E_{1g}	E_5	E	E_3	E_1	Π_g

Class G_4	D_4	C_{4v}	D_{4h}	D_{2d}	Class G_5	D_2	C_{2v}	D_{2h}
A	A_1	A_1	A_{1g}	A_1	A	A	A_1	A_g
γ_1	B_1	B_1	B_{1g}	B_1	γ_1	B_1	A_2	B_{1g}
γ_2	B_2	B_2	B_{2g}	B_2	γ_2	B_2	B_1	B_{2g}
γ_3	E	E	E_g	E	γ_3	B_3	B_2	B_{3g}

Source: From Schlotter and Rabolt (13).

Table 8.2. Expressions for Cases Where the Unique Symmetry Axis Is Perpendicular to Orientation Direction[a]

Note: $G_1\{C_s, C_2, C_{2h}\}$ Class G_1 contains the point groups C_s, C_2, C_{2h}.

A $\alpha_{xx}, \alpha_{yy}, \alpha_{zz}, \alpha_{xy}$ Irreducible representations.

γ α_{yz}, α_{zx}

Class/Scattering Geometry		Sample Orientation Direction (Fig. 8.1)					
		A		B		C	
G_1	RAS	Direction of Polarization of Analyzer[b]					
Irreducible Representation	Incident Polarization of Laser	X	Y	X	Y	X	Y
A	Y	$\frac{1}{2}l^2$	$A1$	$\frac{1}{2}l^2$	$C1$	$B1$	$C1$
	Z	$B1$	$\frac{1}{2}l^2$	$\frac{1}{2}l^2$	$B1$	$\frac{1}{2}l^2$	$\frac{1}{2}l^2$
γ	Y	$\frac{1}{2}m^2$	0	$\frac{1}{2}m^2$	$\frac{1}{2}n^2$	$\frac{1}{2}n^2$	$\frac{1}{2}n^2$
	Z	$\frac{1}{2}n^2$	$\frac{1}{2}m^2$	$\frac{1}{2}m^2$	$\frac{1}{2}n^2$	$\frac{1}{2}m^2$	$\frac{1}{2}m^2$

		A		B		C	
G_1	BS	Y	Z	Y	Z	Y	Z
A	Y	$A1$	$\frac{1}{2}l^2$	$C1$	$B1$	$C1$	$\frac{1}{2}l^2$
	Z	$\frac{1}{2}l^2$	C_1	$B1$	$C1$	$\frac{1}{2}l^2$	$A1$
γ	Y	0	$\frac{1}{2}m^2$	$\frac{1}{2}n^2$	$\frac{1}{2}n^2$	$\frac{1}{2}n^2$	$\frac{1}{2}m^2$
	Z	$\frac{1}{2}m^2$	$\frac{1}{2}n^2$	$\frac{1}{2}n^2$	$\frac{1}{2}n^2$	$\frac{1}{2}m^2$	0

[a] $a^2 = r^2 = (\alpha_{xx} + \alpha_{yy})^2$ $e^2 = s^2 = (\alpha_{xx} - \alpha_{yy})^2$

$b^2 = (\alpha_{xx} - \alpha_{yy})^2 + 4\alpha_{xy}^2$ $f^2 = m^2 = \alpha_{xz}^2$

$c^2 = \alpha_{yz}^2 + \alpha_{zx}^2$ $g^2 = n^2 = \alpha_{yz}^2$

$d^2 = p^2 = \alpha_{zz}^2$ $h^2 = l^2 = \alpha_{xy}^2$

[b] $A1 = \frac{1}{4}(r + s)^2$

$B1 = \frac{1}{32}[r^2 + s^2 + 4p^2 - 2rs - 4rp - 4sp]$

$C1 = \frac{1}{32}[3r^2 + 3s^2 + 12p^2 + 4rp - 6rs - 4sp]$

Table 8.3. Unique Symmetry Axis Perpendicular to Orientation Direction[a]

$G_2\{C_3, D_3, C_{3v}, D_{3d}, S_6\}$

$A \quad \alpha_{xx} + \alpha_{yy}$

$\gamma \quad (\alpha_{xx} - \alpha_{yy}, \alpha_{xy}), (\alpha_{yz}, \alpha_{xz})$

		A		B		C	
G_2	RAS	X	Y	X	Y	X	Y
A	Y	0	$\frac{1}{4}r^2$	0	$E2$	$A2$	$E2$
	Z	$A2$	0	0	$A2$	0	0
γ	Y	$B2$	$D2$	$B2$	$F2$	$C2$	$F2$
	Z	$C2$	$B2$	$B2$	$C2$	$B2$	$B2$

		A		B		C	
G_2	BS	Y	Z	Y	Z	Y	Z
A	Y	$\frac{1}{4}r^2$	0	$E2$	$A2$	$E2$	0
	Z	0	$E2$	$A2$	$E2$	0	$\frac{1}{4}r^2$
γ	Y	$D2$	$B2$	$F2$	$C2$	$F2$	$B2$
	Z	$B2$	$F2$	$C2$	$F2$	$B2$	$D2$

[a] $A2 = \frac{1}{32}(r^2 - 4rp + 4p^2)$ $\quad D2 = \frac{1}{4}(s^2 + 2rs)$

$B2 = \frac{1}{2}(l^2 + m^2)$ $\quad E2 = \frac{1}{32}(3r^2 + 4rp + 12p^2)$

$C2 = \frac{1}{32}(s^2 - 2rs + 4sp + 16n^2)$ $\quad F2 = \frac{1}{32}(3s^2 - 6rs - 4sp + 16n^2)$

Table 8.4. Unique Symmetry Axis Perpendicular to Orientation Direction[a]

$G_3\{C_4, C_5, C_6, D_5, D_6, C_{5v}, C_{6v}, C_{3h}, C_{4h}, C_{5h}, C_{6h}, D_{3h}, D_{5h}, D_{6h}, D_{4d}, D_{5d}, D_{6d}, S_4, S_8, C_{\infty v}, D_{\infty h}\}$

A $\alpha_{xx}+\alpha_{yy}, \alpha_{zz}$

γ_1 $\alpha_{xx}-\alpha_{yy}, \alpha_{xy}$

γ_2 $(\alpha_{yz}, \alpha_{xz})$

		A		B		C	
G_3	RAS	X	Y	X	Y	X	Y
A	Y	0	$\frac{1}{4}r^2$	0	D3	A3	D3
	Z	A3	0	0	A3	0	0
γ_1	Y	$\frac{1}{2}l^2$	C3	$\frac{1}{2}l^2$	E3	B3	E3
	Z	B3	$\frac{1}{2}l^2$	$\frac{1}{2}l^2$	B3	$\frac{1}{2}l^2$	$\frac{1}{2}l^2$
γ_2	Y	$\frac{1}{2}m^2$	0	$\frac{1}{2}m^2$	$\frac{1}{2}n^2$	$\frac{1}{2}n^2$	$\frac{1}{2}n^2$
	Z	$\frac{1}{2}n^2$	$\frac{1}{2}m^2$	$\frac{1}{2}m^2$	$\frac{1}{2}n^2$	$\frac{1}{2}m^2$	$\frac{1}{2}m^2$

		A		B		C	
G_3	BS	Y	Z	Y	Z	Y	Z
A	Y	$\frac{1}{4}r^2$	0	D3	A3	D3	0
	Z	0	D3	A3	D3	0	$\frac{1}{4}r^2$
γ_1	Y	C3	$\frac{1}{2}l^2$	E3	B3	E3	$\frac{1}{2}l^2$
	Z	$\frac{1}{2}l^2$	E3	B3	E3	$\frac{1}{2}l^2$	C3
γ_2	Y	0	$\frac{1}{2}m^2$	$\frac{1}{2}n^2$	$\frac{1}{2}n^2$	$\frac{1}{2}n^2$	$\frac{1}{2}m^2$
	Z	$\frac{1}{2}m^2$	$\frac{1}{2}n^2$	$\frac{1}{2}n^2$	$\frac{1}{2}n^2$	$\frac{1}{2}m^2$	0

[a] $A3=\frac{1}{32}[r^2-4rp+4p^2]$ $D3=\frac{1}{32}[3r^2+4rp+12p^2]$

$B3=\frac{1}{32}[s^2-2rs+4sp]$ $E3=\frac{1}{32}[3s^2-6rs-4sp]$

$C3=\frac{1}{4}[s^2+2rs]$

Table 8.5. Unique Symmetry Axis Perpendicular to Orientation Direction[a]

$G_4\{D_4, C_{4v}, D_{4h}, D_{2d}\}$

A $\alpha_{xx}+\alpha_{yy}, \alpha_{zz}$

γ_1 $\alpha_{xx}-\alpha_{yy}$

γ_2 α_{xy}

γ_3 $(\alpha_{yz}, \alpha_{xz})$

		A		B		C	
G_4	RAS	X	Y	X	Y	X	Y
A	Y	0	$\frac{1}{4}r^2$	0	$D4$	$A4$	$D4$
	Z	$A4$	0	0	$A4$	0	0
γ_1	Y	0	$C4$	0	$E4$	$B4$	$E4$
	Z	$B4$	0	0	$B4$	0	0
γ_2	Y	$\frac{1}{2}l^2$	0	$\frac{1}{2}l^2$	0	0	0
	Z	0	$\frac{1}{2}l^2$	$\frac{1}{2}l^2$	0	$\frac{1}{2}l^2$	$\frac{1}{2}l^2$
γ_3	Y	$\frac{1}{2}m^2$	0	$\frac{1}{2}m^2$	$\frac{1}{2}n^2$	$\frac{1}{2}n^2$	$\frac{1}{2}n^2$
	Z	$\frac{1}{2}n^2$	$\frac{1}{2}m^2$	$\frac{1}{2}m^2$	$\frac{1}{2}n^2$	$\frac{1}{2}m^2$	$\frac{1}{2}m^2$

		A		B		C	
G_4	BS	Y	Z	Y	Z	Y	Z
A	Y	$\frac{1}{4}r^2$	0	$D4$	$A4$	$D4$	0
	Z	0	$D4$	$A4$	$D4$	0	$\frac{1}{4}r^2$
γ_1	Y	$C4$	0	$E4$	$B4$	$E4$	0
	Z	0	$E4$	$B4$	$E4$	0	$C4$
γ_2	Y	0	$\frac{1}{2}l^2$	0	0	0	$\frac{1}{2}l^2$
	Z	$\frac{1}{2}l^2$	0	0	0	$\frac{1}{2}l^2$	0
γ_3	Y	0	$\frac{1}{2}m^2$	$\frac{1}{2}n^2$	$\frac{1}{2}n^2$	$\frac{1}{2}n^2$	$\frac{1}{2}m^2$
	Z	$\frac{1}{2}m^2$	$\frac{1}{2}n^2$	$\frac{1}{2}n^2$	$\frac{1}{2}n^2$	$\frac{1}{2}m^2$	0

[a] $A4 = \frac{1}{32}[r^2 - 4rp + 4rp^2]$ $D4 = \frac{1}{32}[3r^2 + 4rp + 12p^2]$

$B4 = \frac{1}{32}[s^2 - 2rs + 4sp]$ $E4 = \frac{1}{32}[3s^2 - 6rs - 4sp]$

$C4 = \frac{1}{4}[s^2 + 2rs]$

Table 8.6. Unique Symmetry Axis Perpendicular to Orientation Direction[a]

$G_5\{D_2, C_{2v}, D_{2h}\}$

A $\alpha_{xx}, \alpha_{yy}, \alpha_{zz}$

γ_1 α_{xy}

γ_2 α_{xz}

γ_3 α_{yz}

		A		B		C	
G_5	RAS	X	Y	X	Y	X	Y
A	Y	0	$B5$	0	$C5$	$A5$	$C5$
	Z	$A5$	0	0	$A5$	0	0
γ_1	Y	$\frac{1}{2}l^2$	0	$\frac{1}{2}l^2$	0	0	0
	Z	0	$\frac{1}{2}l^2$	$\frac{1}{2}l^2$	0	$\frac{1}{2}l^2$	$\frac{1}{2}l^2$
γ_2	Y	$\frac{1}{2}m^2$	0	$\frac{1}{2}m^2$	0	0	0
	Z	0	$\frac{1}{2}m^2$	$\frac{1}{2}m^2$	0	$\frac{1}{2}m^2$	$\frac{1}{2}m^2$
γ_3	Y	0	0	0	$\frac{1}{2}n^2$	$\frac{1}{2}n^2$	$\frac{1}{2}n^2$
	Z	$\frac{1}{2}n^2$	0	0	$\frac{1}{2}n^2$	0	0

		A		B		C	
G_5	BS	Y	Z	Y	Z	Y	Z
A	Y	$B5$	0	$C5$	$A5$	$C5$	0
	Z	0	$C5$	$A5$	$C5$	0	$B5$
γ_1	Y	0	$\frac{1}{2}l^2$	0	0	0	$\frac{1}{2}l^2$
	Z	$\frac{1}{2}l^2$	0	0	0	$\frac{1}{2}l^2$	0
γ_2	Y	0	$\frac{1}{2}m^2$	0	0	0	$\frac{1}{2}m^2$
	Z	$\frac{1}{2}m^2$	0	0	0	$\frac{1}{2}m^2$	0
γ_3	Y	0	0	$\frac{1}{2}n^2$	$\frac{1}{2}n^2$	$\frac{1}{2}n^2$	0
	Z	0	$\frac{1}{2}n^2$	$\frac{1}{2}n^2$	$\frac{1}{2}n^2$	0	0

[a] $A5 = \frac{1}{32}[(r-s)^2 - 4(r-s)p + 4p^2]$

$B5 = \frac{1}{4}(r+s)^2$

$C5 = \frac{1}{32}[3(r-s)^2 + 4(r-s)p + 12p^2]$

Table 8.7. Expressions For Cases Where the Unique Symmetry Axis Is Parallel to Orientation Direction[a]

$G_1\{C_s, C_2, C_{2h}\}$

A $\alpha_{xx}, \alpha_{yy}, \alpha_{zz}, \alpha_{xy}$

γ α_{yz}, α_{zx}

		A		B		C	
G_1	RAS	X	Y	X	Y	X	Y
A	Y	0	d^2	0	L1	$\frac{1}{8}b^2$	L1
	Z	$\frac{1}{8}b^2$	0	0	$\frac{1}{8}b^2$	0	0
γ	Y	$\frac{1}{2}c^2$	0	$\frac{1}{2}c^2$	0	0	0
	Z	0	$\frac{1}{2}c^2$	$\frac{1}{2}c^2$	0	$\frac{1}{2}c^2$	$\frac{1}{2}c^2$

		A		B		C	
G_1	BS	Y	Z	Y	Z	Y	Z
A	Y	d^2	0	L1	$\frac{1}{8}b^2$	L1	0
	Z	0	L1	$\frac{1}{8}b^2$	L1	0	d^2
γ	Y	0	$\frac{1}{2}c^2$	0	0	0	$\frac{1}{2}c^2$
	Z	$\frac{1}{2}c^2$	0	0	0	$\frac{1}{2}c^2$	0

[a] $L1 = \frac{1}{8}(2a^2 + b^2)$

Table 8.8. Unique Symmetry Axis Parallel to Orientation Direction

$G_2\{C_3, D_3, C_{3v}, D_{3d}, S_6\}$

A $\alpha_{xx} + \alpha_{yy}, \alpha_{zz}$

γ $(\alpha_{xx} - \alpha_{yy}, \alpha_{xy}), (\alpha_{yz}, \alpha_{xz})$

		A		B		C	
G_2	RAS	X	Y	X	Y	X	Y
A	Y	0	d^2	0	$\frac{1}{4}a^2$	0	$\frac{1}{4}a^2$
	Z	0	0	0	0	0	0
γ	Y	$\frac{1}{2}c^2$	0	$\frac{1}{2}c^2$	$\frac{1}{8}b^2$	$\frac{1}{8}b^2$	$\frac{1}{8}b^2$
	Z	$\frac{1}{8}b^2$	$\frac{1}{2}c^2$	$\frac{1}{2}c^2$	$\frac{1}{8}b^2$	$\frac{1}{2}c^2$	$\frac{1}{2}c^2$

		A		B		C	
G_2	BS	Y	Z	Y	Z	Y	Z
A	Y	d^2	0	$\frac{1}{4}a^2$	0	$\frac{1}{4}a^2$	0
	Z	0	$\frac{1}{4}a^2$	0	$\frac{1}{4}a^2$	0	d^2
γ	Y	0	$\frac{1}{2}c^2$	$\frac{1}{8}b^2$	$\frac{1}{8}b^2$	$\frac{1}{8}b^2$	$\frac{1}{2}c^2$
	Z	$\frac{1}{2}c^2$	$\frac{1}{8}b^2$	$\frac{1}{8}b^2$	$\frac{1}{8}b^2$	$\frac{1}{2}c^2$	0

Table 8.9. Unique Symmetry Axis Parallel to Orientation Direction

$G_3\{C_4, C_5, C_6, D_5, D_6, C_{5v}, C_{6v}, C_{3h}, C_{4h}, C_{5h}, C_{6h}, D_{3h}, D_{5h}, D_{6h}, D_{4d}, D_{5d}, D_{6d}, S_4, S_8, C_{\infty v}, D_{\infty h}\}$

A $\alpha_{xx} + \alpha_{yy}, \alpha_{zz}$

γ_1 $\alpha_{xx} - \alpha_{yy}, \alpha_{xy}$

γ_2 $(\alpha_{yz}, \alpha_{xz})$

		A		*B*		*C*	
G_3	RAS	*X*	*Y*	*X*	*Y*	*X*	*Y*
A	*Y*	0	d^2	0	$\frac{1}{4}a^2$	0	$\frac{1}{4}a$
	Z	0	0	0	0	0	0
γ_1	*Y*	0	0	0	$\frac{1}{8}b^2$	$\frac{1}{8}b^2$	$\frac{1}{8}b^2$
	Z	$\frac{1}{8}b^2$	0	0	$\frac{1}{8}b^2$	0	0
γ_2	*Y*	$\frac{1}{2}c^2$	0	$\frac{1}{2}c^2$	0	0	0
	Z	0	$\frac{1}{2}c^2$	$\frac{1}{2}c^2$	0	$\frac{1}{2}c^2$	$\frac{1}{2}c^2$

		A		*B*		*C*	
G_3	BS	*Y*	*Z*	*Y*	*Z*	*Y*	*Z*
A	*Y*	d^2	0	$\frac{1}{4}a^2$	0	$\frac{1}{4}a^2$	0
	Z	0	$\frac{1}{4}a^2$	0	$\frac{1}{4}a^2$	0	d^2
γ_1	*Y*	0	0	$\frac{1}{8}b^2$	$\frac{1}{8}b^2$	$\frac{1}{8}b^2$	0
	Z	0	$\frac{1}{8}b^2$	$\frac{1}{8}b^2$	$\frac{1}{8}b^2$	0	0
γ_2	*Y*	0	$\frac{1}{2}c^2$	0	0	0	$\frac{1}{2}c^2$
	Z	$\frac{1}{2}c^2$	0	0	0	$\frac{1}{2}c^2$	0

Table 8.10. Unique Symmetry Axis Parallel to Orientation Direction

$G_4\{D_4, C_{4v}, D_{4h}, D_{2d}\}$

A $\alpha_{xx} + \alpha_{yy}, \alpha_{zz}$

γ_1 $\alpha_{xx} - \alpha_{yy}$

γ_2 α_{xy}

γ_3 $(\alpha_{yz}, \alpha_{xz})$

		A		B		C	
G_4	RAS	X	Y	X	Y	X	Y
A	Y	0	d^2	0	$\frac{1}{4}a^2$	0	$\frac{1}{4}a^2$
	Z	0	0	0	0	0	0
γ_1	Y	0	0	0	$\frac{1}{8}e^2$	$\frac{1}{8}e^2$	$\frac{1}{8}e^2$
	Z	$\frac{1}{8}e^2$	0	0	$\frac{1}{8}e^2$	0	0
γ_2	Y	0	0	0	$\frac{1}{2}h^2$	$\frac{1}{2}h^2$	$\frac{1}{2}h^2$
	Z	$\frac{1}{2}h^2$	0	0	$\frac{1}{2}h^2$	0	0
γ_3	Y	$\frac{1}{2}c^2$	0	$\frac{1}{2}c^2$	0	0	0
	Z	0	$\frac{1}{2}c^2$	$\frac{1}{2}c^2$	0	$\frac{1}{2}c^2$	$\frac{1}{2}c^2$

		A		B		C	
G_4	BS	Y	Z	Y	Z	Y	Z
A	Y	d^2	0	$\frac{1}{4}a^2$	0	$\frac{1}{4}a^2$	0
	Z	0	$\frac{1}{4}a^2$	0	$\frac{1}{4}a^2$	0	d^2
γ_1	Y	0	0	$\frac{1}{8}e^2$	$\frac{1}{8}e^2$	$\frac{1}{8}e^2$	0
	Z	0	$\frac{1}{8}e^2$	$\frac{1}{8}e^2$	$\frac{1}{8}e^2$	0	0
γ_2	Y	0	0	$\frac{1}{2}h^2$	$\frac{1}{2}h^2$	$\frac{1}{2}h^2$	0
	Z	0	$\frac{1}{2}h^2$	$\frac{1}{2}h^2$	$\frac{1}{2}h^2$	0	0
γ_3	Y	0	$\frac{1}{2}c^2$	0	0	0	$\frac{1}{2}c^2$
	Z	$\frac{1}{2}c^2$	0	0	0	$\frac{1}{2}c^2$	0

Table 8.11. Unique Symmetry Axis Parallel to Orientation Direction

$G_5\{D_2, C_{2v}, D_{2h}\}$

A	$\alpha_{xx}, \alpha_{yy}, \alpha_{zz}$
γ_1	α_{xy}
γ_2	α_{xz}
γ_3	α_{yz}

		A		B		C	
G_5	RAS	X	Y	X	Y	X	Y
A	Y	0	d^2	0	$L5$	$\frac{1}{8}e^2$	$L5$
	Z	$\frac{1}{8}e^2$	0	0	$\frac{1}{8}e^2$	0	0
γ_1	Y	0	0	0	$\frac{1}{2}h^2$	$\frac{1}{2}h^2$	$\frac{1}{2}h^2$
	Z	$\frac{1}{2}h^2$	0	0	$\frac{1}{2}h^2$	0	0
γ_2	Y	$\frac{1}{2}f^2$	0	$\frac{1}{2}f^2$	0	0	0
	Z	0	$\frac{1}{2}f^2$	$\frac{1}{2}f^2$	0	$\frac{1}{2}f^2$	$\frac{1}{2}f^2$
γ_3	Y	$\frac{1}{2}g^2$	0	$\frac{1}{2}g^2$	0	0	0
	Z	0	$\frac{1}{2}g^2$	$\frac{1}{2}g^2$	0	$\frac{1}{2}g^2$	$\frac{1}{2}g^2$

		A		B		C	
G_5	BS	Y	Z	Y	Z	Y	Z
A	Y	d^2	0	$L5$	$\frac{1}{8}e^2$	$L5$	0
	Z	0	$L5$	$\frac{1}{8}e^2$	$L5$	0	d^2
γ_1	Y	0	0	$\frac{1}{2}h^2$	$\frac{1}{2}h^2$	$\frac{1}{2}h^2$	0
	Z	0	$\frac{1}{2}h^2$	$\frac{1}{2}h^2$	$\frac{1}{2}h^2$	0	0
γ_2	Y	0	$\frac{1}{2}f^2$	0	0	0	$\frac{1}{2}f^2$
	Z	$\frac{1}{2}f^2$	0	0	0	$\frac{1}{2}f^2$	0
γ_3	Y	0	$\frac{1}{2}g^2$	0	0	0	$\frac{1}{2}g^2$
	Z	$\frac{1}{2}g^2$	0	0	0	$\frac{1}{2}g^2$	0

[a] $L5 = \frac{1}{8}(2a^2 + e^2)$

8.5. EXAMPLE: PLANAR ZIGZAG POLY(VINYLIDENE FLUORIDE)

Perhaps the best way to illustrate the use of the foregoing tables is to take an explicit example and demonstrate how the symmetry species of the bands observed in a series of polarized Raman measurements are assigned. For planar zigzag PVF_2, the factor group of the line group is isomorphic to the C_{2v} point group (10). The normal modes are distributed among the symmetry species as follows: $5A_1 + 2A_2 + 3B_1 + 4B_2$, with all but the A_2 (IR inactive) species being both IR and Raman active. The unique symmetry axis (z) of the planar backbone is perpendicular to the polymer chain axis and hence perpendicular to the orientation direction, indicating that Tables 8.2–8.6 should be used for analysis. If we consult Table 8.1, it is apparent that the point group C_{2v} belongs to class G_5 and that A, γ_1, γ_2, and γ_3 correspond to the symmetry species A_1, A_2, B_1, and B_2. Class G_5 is contained in Table 8.6, and thus this table will be used for the symmetry analysis. Initially the sample was aligned with its macroscopic filament axis parallel to the Y laboratory axis corresponding to orientation A of Fig. 8.1. Reviewing the entries under orientation A, we note that, with the input laser polarization along Y and the analyzer also positioned parallel to Y, the only contribution to the observed Raman spectrum is from the A symmetry species since the entries for each γ_i are zero for the YY experiment. Hence, the spectrum designated $X(YY)Z$ is due to the A_1 symmetry species ($X(YY)Z \rightarrow A_1$); likewise

$$X(YX)Z \rightarrow A_2 + B_1,$$

$$X(ZY)Z \rightarrow A_2 + B_1,$$

and

$$X(ZX)Z \rightarrow A_1 + B_2.$$

The resulting spectra are shown in Fig. 8.3. The observed variation in intensity allows some straightforward symmetry assignments to be made. When the $X(YY)Z$ spectrum, which selects out only the A_1 symmetry species, is taken in conjunction with the $X(ZX)Z$ spectrum, which has contributions from both A_1 and B_2 bands, it is possible to assign the latter since they will have been absent in the $X(YY)Z$ spectrum. Thus the bands at 876 and 442 cm^{-1} belong to the B_2 symmetry species, whereas those at 1432, 1277, and 840 cm^{-1} exhibit A_1 symmetry. It is also tempting to assign the bands at 610 and 800 cm^{-1} to the A_1 symmetry species, but results from another set of experiments using orientation B indicate otherwise. Listed in Table 8.12 are the explicit polarizability contributions to the A_1 and the B_2 symmetry species for two experiments carried out in the alternative orientation B. These

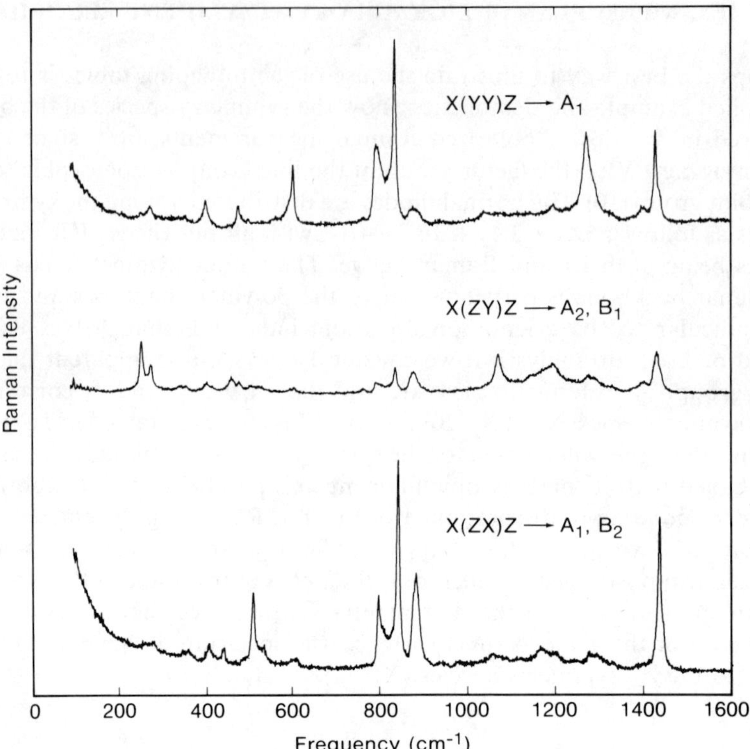

Figure 8.3. Polarized Raman measurements of PVF_2 obtained using orientation A with the laser incident along X and the scattered radiation collected along Z. The spectra have been normalized to the low-frequency lines (50–100 cm^{-1}) of N_2 appearing in the spectra since the incident laser beam is focused at the sample. From Lauchlan and Rabolt (10).

Table 8.12. Contributions to the Raman Scattering Activities for Planar Zigzag Structure of PVF_2 (Orientation B)

Polarization Experiment	Symmetry Species	
	A_1	B_2
$X(YY)Z$	$\frac{1}{32}(12\alpha_{yy}^2 + 24\alpha_{yy}\alpha_{zz} + 12\alpha_{zz}^2)$	$\frac{1}{2}\alpha_{yz}^2$
$X(ZY)Z$	$\frac{1}{32}(4\alpha_{yy}^2 - 8\alpha_{yy}\alpha_{zz} + 4\alpha_{zz}^2)$	$\frac{1}{2}\alpha_{yz}^2$

also come from Table 8.6 but have been explicitly listed in terms of their polarizability contributions, α. Although the polarizability contribution (and thus the intensity) of the B_2 bands remains constant in both measurements, that of the A_1 bands will increase in the $X(YY)Z$ spectrum. As observed in Fig. 8.4, the A_1 bands increase slightly in intensity in going from the $X(ZY)Z$ to the $X(YY)Z$ spectrum, while the B_2 bands remain unchanged. Therefore the medium intensity band observed at $510\,\text{cm}^{-1}$ can also be assigned to the A_1 symmetry species. It is also apparent from Fig. 8.4 that the $610\,\text{cm}^{-1}$ band is absent, whereas that at $800\,\text{cm}^{-1}$ exhibits A_1 behavior. Based on a more detailed analysis (10), both these bands were attributed to a small amount of oriented PVF_2 having a $TGTG'$ conformational structure introduced during the filament-drawing process.

Assignment of the B_1 modes is somewhat straightforward by comparison of the results with those of IR studies since these modes are both IR and Raman active. Similar assistance with the A_2 assignments is not possible since these bands are IR inactive. In comparing the $X(YY)Z$ spectrum with the $X(ZY)Z$ spectrum shown in Fig. 8.3, it is obvious that there is residual intensity of the intense A_1 modes in the latter owing to a lack of perfect orientation or a physical misalignment of the filament in the laser beam. The

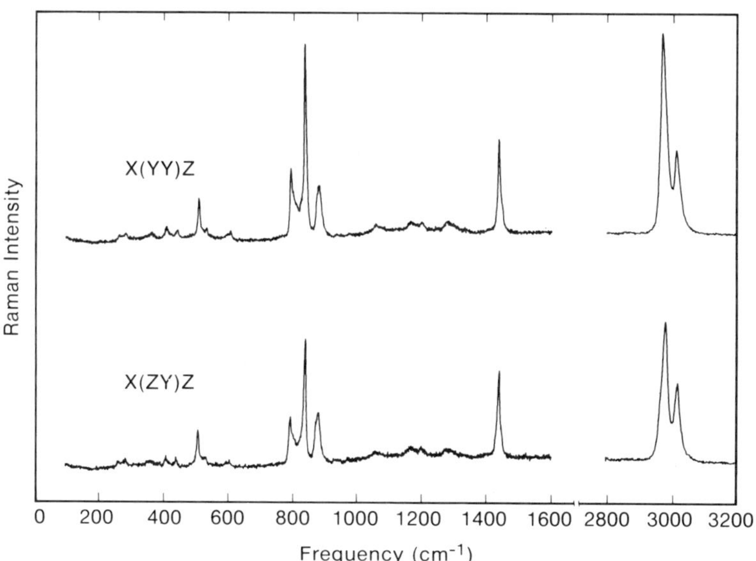

Figure 8.4. Polarized Raman measurements of planar zigzag PVF_2 in orientation B. Spectra have been normalized so that a direct comparison of intensities can be made. From Lauchlan and Rabolt (10).

last possibility can be discounted by a simple comparison of two Raman spectra obtained from different polarization experiments but which should give identical results since they have the same polarizability contributions to each symmetry species. As shown in Table 8.6 under orientation *A*, both the $X(YX)Z$ and the $X(ZY)Z$ spectra should have identical contributions from bands with A_2 symmetry and identical contributions from the B_1 symmetry species. The extent of deviation from equivalent intensities is indicative of sample misalignment in the beam, which, according to Fig. 8.5, is minimal. In addition to the appearance of weak bands from other polarizations owing to a lack of perfect uniaxial orientation, several new bands at 1400, 1200, 1076, and 263 cm^{-1} can be attributed to either the A_2 or the B_1 symmetry species. Detailed assignments of the bands shown in Figs. 8.3–8.5 have been made after consideration of their corresponding IR activity, and these are listed in Table 8.13 together with results from normal coordinate calculations that have appeared in the literature (16).

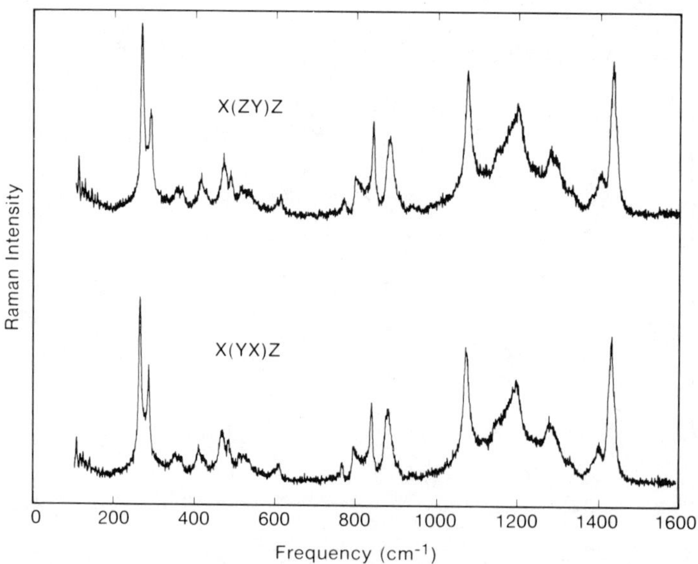

Figure 8.5. Polarized Raman spectra of planar zigzag PVF_2 using orientation *A*. For an oriented filament perfectly aligned perpendicular to the incident laser, the $X(ZY)Z$ and $X(YX)Z$ spectral intensities would be identical. Any deviation is indicative of macroscopic misalignment. Spectra have been normalized to low-frequency N_2 lines. From Lauchlan and Rabolt (10).

Table 8.13. Symmetry Assignment of the Observed Bands from Figs. 8.3–8.5

Symmetry Species	Frequency (cm^{-1}) Raman	Infrared[a]	Calculated[a]	Assignment[a,b]
A_1	2976	2982	2980	$\nu_s(CH_2)$
	1432	1432	1434	$\delta(CH_2)$
	1277[c]	1278	1185	$\delta(CCC) + \nu_s(CF_2) + \nu_s(CC)$
	840[c]	842	886	$\nu_s(CF_2)$
	510	511	516	$\delta(CF_2)$
A_2	1200	—	991	$t(CH_2)$
	263	—	270	$t(CF_2)$
B_1	1400	1403	1398	$w(CH_2) + \nu_a(CC)$
	1076	1074	1063	$\nu_a(CC) + w(CH_2) + w(CF_2)$
	467	470	484	$w(CF_2)$
B_2	3020	3019	3020	$\nu_a(CH_2)$
	1170[c]	1185	1257	$\nu_a(CF_2) + r(CF_2) + r(CH_2)$
	876[c]	882	843	$r(CH_2) + \nu_a(CF_2)$
	442	448	483	$r(CF_2) + \tau(CC) + r(CH_2)$

[a]From Bachmann and Koenig (16).
[b]ν, stretch; δ, bend; t, twist; r, rock; w, wag; τ, torsion.
[c]From Bachmann and Koenig (16), but mode assignment changed.

ACKNOWLEDGMENT

The author would like to thank N. E. Schlotter, Bellcore, for his many contributions to this work.

REFERENCES

1. D. A. Long, *Raman Spectroscopy*. McGraw-Hill, London, 1977.
2. M. Kobayashi, K. Tashiro, and H. Tadokoro, *Macromolecules* **8**, 158 (1975).
3. F. J. Boerio and J. L. Koenig, *J. Polym. Sci., Part A-1* **9**, 1517 (1971).
4. B. Wunderlich, *Macromolecular Physics*, Vols. 1–3. Academic Press, New York, 1973.
5. R. T. Bailey, A. J. Hyde, J. J. Kim, and J. McLeish, *Spectrochim. Acta* **33A**, 1053 (1977).
6. J. F. Rabolt and B. Fanconi, *Macromolecules* **11**, 740 (1978).
7. K. Zabel, N. E. Schlotter, and J. F. Rabolt, *Macromolecules* **16**, 446 (1983).
8. K. Holland-Moritz, *J. Appl. Polym. Sci.: Appl. Polym. Symp.* **34**, 49 (1978).

9. M. Abenoza and A. Armengaud, *Polymer* **22**, 1341 (1981).
10. L. Lauchlan and J. F. Rabolt, *Macromolecules* **19**, 1049 (1986).
11. R. G. Snyder, *J. Mol. Spectrosc.* **37**, 353 (1971).
12. R. T. Bailey, A. J. Hyde, and J. J. Kim, *Spectrochim. Acta* **30A**, 91 (1974).
13. N. E. Schlotter and J. F. Rabolt, *Polymer* **25**, 165 (1984).
14. R. Hasegawa, Y. Takahashi, Y. Chatani, and H. Tadokoro, *Polymer J.* **3**, 600 (1972).
15. T. C. Damen, S. P. S. Porto, and B. Tell, *Phys. Rev.* **142**, 570 (1966).
16. M. A. Bachmann and J. L. Koenig, *J. Chem. Phys.* **74**, 5896 (1981).

CHAPTER

9

ORGANIC AND PETROCHEMICAL APPLICATIONS OF RAMAN SPECTROSCOPY

DONALD L. GERRARD

BP Research Centre
Sunbury-on-Thames
Middlesex, England

9.1. INTRODUCTION

The development of reliable continuous wave (CW) gas lasers, notably the He–Ne, Kr^+, and Ar^+ lasers, which offered an intense and reliable light source, gave a new lease of life to Raman spectroscopy in the early 1970s. Despite this revival, however, the technique is still not firmly established as a generally applicable, broad-based analytical tool in the same way as other diagnostic spectroscopies such as infrared (IR), nuclear magnetic resonance (NMR), and mass spectrometry. There are several reasons for this, some historical and some practical, and it is worthwhile briefly considering these in order to see why the enormous potential of Raman spectroscopy, particularly in the areas of organic chemistry and petrochemicals, has only been recognized by a small number of groups.

Historically, the close relationship between Raman spectroscopy and IR absorption spectroscopy has led to the widely held view that the greatest value of the Raman technique is as a complement to IR studies. In many cases, particularly organic systems, the same qualitative information is available from both Raman and IR spectra. This means that for purposes of characterization and identification of unknown materials there is normally little specific advantage in using the Raman spectrum. This is particularly the case for organic compounds where the commercially available collections of standard spectra are extremely comprehensive, whereas those relating to Raman spectra are very limited in their scope. The reason for this is the long head start that the IR technique had, which still makes it the preferred

Analytical Raman Spectroscopy, Edited by Jeanette G. Grasselli and Bernard J. Bulkin. Chemical Analysis Series, Vol. 114.
ISBN 0-471-51955-3

method for identifying unknown organic compounds. By the time CW gas lasers were generally available there was no realistic possibility of Raman displacing IR for characterization/identification of organic materials.

It is true that the Raman spectrum provides additional structural information, particularly with respect to nonpolar groups in organic molecules, which is not available in the IR spectrum, but in most cases can also be obtained via NMR or mass spectrometry measurements. Hence, as funds are always limited in any organization Raman spectrometers are usually purchased only when a full complement of other spectroscopic equipment is available. In order to make a significant impact in any analytical area, but particularly in the field of organic and petrochemical analyses, which are so well served by other spectroscopic, chemical, and physical techniques, the Raman spectroscopist is forced to look to the particular advantages that the Raman effect confers. If Raman spectroscopy is merely presented as an additional technique for obtaining a small amount of extra analytical information, it will only be the preserve of the highly funded groups. If, however, its unique advantages are used effectively, then it can be applied to a very wide range of analytical problems that are not readily solved by other methods. Although in this chapter we shall be limiting ourselves essentially to the consideration of organic systems, this concept is equally applicable to other areas of analysis. The specific advantages that can be of value to the organic petrochemical analyst are as follows:

1. There is versatility and simplicity in sample handling because—
   a. the Raman effect is the result of a molecular scattering process, rather than an absorption, thus allowing many kinds of materials to be studied with little or no preparation;
   b. both water and glass exhibit only weak Raman scattering, allowing for the use of durable glass sample containers and the analysis of aqueous systems;
   c. excitation wavelengths are normally in the visible region of the spectrum (hence water and glass can be effectively used in the sample system), there is high detection sensitivity, and there is ease in sample alignment.
2. In general, nonpolar groups exhibit strong Raman scattering, which allows for the identification of organics and petrochemicals such as polymers and aromatics. In addition, low-frequency vibrations can be studied without the need for any additional spectroscopic hardware, which simplifies the study of polymers.
3. The intensity of the Raman signal is directly proportional to the concentration of the molecular group producing that signal, which allows for quantification of organic/petrochemical systems.

4. The use of lasers confers a high degree of spatial resolution, allowing small parts of samples to be studied.
5. Under certain conditions the intensity of the Raman signal can be greatly enhanced, allowing the analysis of organic compounds present at low levels.
6. The use of multichannel detection gives Raman spectroscopy a considerable advantage in time-resolved studies, particularly those involving reaction kinetics (important in studying the chemical and physical changes in systems, including catalyzed and uncatalyzed polymerizations).

Each of these advantages will be considered in some detail later in this chapter, with particular reference to the way in which it can be used effectively in the analysis of organic and petrochemical materials. The term "analysis" will be considered in its broadest possible sense, i.e., to cover qualitative and quantitative studies, micro and macro samples, and systems that are static or dynamic in both physical and chemical senses.

The main disadvantage of Raman spectroscopy, and the one which still restricts its application to many industrial problems, particularly in the petroleum and petrochemical industries, is that of sample fluorescence. Often, the lasers commonly used in Raman studies are at wavelengths that stimulate fluorescence in many of the samples of interest in the oil industry. In terms of photon yield, the efficiency of the fluorescence process is normally much greater than that of the Raman effect. Consequently, it may well be the case that the fluorescence is sufficiently intense that it totally obscures the Raman signal. There are two alternative approaches to this problem. One is to adjust the experimental technique such that fluorescence does not occur, or is much reduced, and the other is to extract the Raman signal contained within the fluorescence signal, usually by mathematical techniques.

If the problem of fluorescence can be successfully overcome, and there are many indications that this should be possible in the near future, there is no doubt that the technique of Raman spectroscopy will enter a new phase of development and will experience a second revival that should finally lead to its acceptance in its own right. The various techniques currently being employed to reduce or eliminate fluorescence backgrounds will therefore be discussed and their effectiveness considered.

9.2. ORGANIC AND PETROCHEMICAL SAMPLE-HANDLING TECHNIQUES

9.2.1. Versatility and Simplicity Owing to the Scattering Process

The main advantage conferred by the fact that the Raman effect is the result of a molecular-scattering process is that it provides considerable versatility in terms of sample handling. This, together with its other advantages, makes Raman spectroscopy the most adaptable of the spectroscopic techniques with respect to the variety of samples that can be analyzed. If we are trying to identify a stable, colorless, organic liquid, a range of techniques, including IR, NMR, and mass spectrometry, are available to us, and Raman does not normally have any particular advantage over the others. However, from the point of view of characterization and identification of unknown materials Raman comes into its own when the sample to be identified is an opaque solid and, in particular, when it is awkwardly shaped, such as a fabricated article made from or coated with a polymer. In principle, it is only necessary to direct the laser beam at the sample to be analyzed and then collect the scattered light containing the Raman signal into the monochromator.

Provided that the sample compartment on the spectrometer is sufficiently large to accommodate the sample, there is no specific restriction on size or shape. Similarly, the material to be analyzed can be any color from colorless to black and will provide no sampling difficulties. An extreme example of this is illustrated in Fig. 9.1, which shows the spectrum obtained from a large piece of natural graphite, where bands due to the amorphous and crystalline structures at ~ 1350 and $\sim 1590\,cm^{-1}$, respectively, can be clearly seen. This ability to handle solid samples of a variety of shapes and sizes is particularly valuable in identifying the polymer composition of fabricated articles without

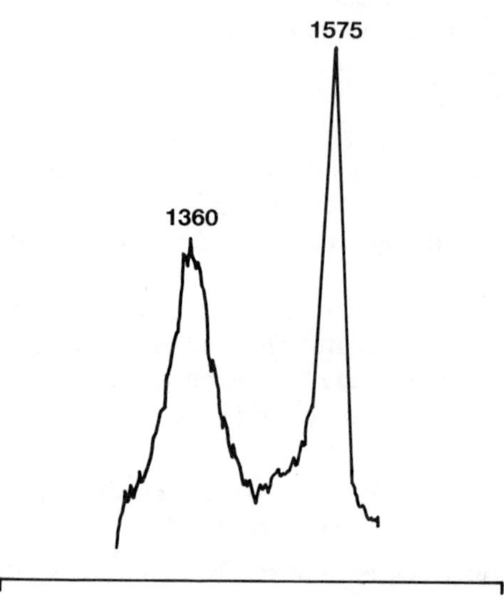

Figure 9.1. Raman spectrum of natural graphite showing amorphous and crystalline bands.

any pretreatment or damage to the sample. Note also that, with the exception of occasional laser damage, which is always avoidable with care, there is no damage caused to the sample as a result of the analysis. Similarly, highly opaque solids, either single compounds, polymer composites (even those filled with carbon black), or naturally opaque materials such as coal, can be examined without difficulty, as supplied, without any need for preliminary preparation, which in some cases can change the feature being studied.

In the petrochemical industry products often need to be analyzed at various stages during the production process. This may be for purposes of process monitoring or for troubleshooting, when the process is not proceeding in the normal way. Petrochemical reaction systems are frequently catalyzed, either homogeneously or heterogeneously, with inorganic or organometallic compounds. These catalysts are often present in the crude product at levels that make IR spectroscopy extremely difficult unless the catalyst is first removed; it is generally a highly opaque material. Raman spectroscopy, however, can be used with comparative ease for the examination of intensely colored, turbid samples without the need for separating the reaction product or reactants from the catalyst.

In the case of heterogeneously catalyzed reactions the catalyst is normally present in the form of small pellets, and it is often necessary to examine these directly, either to identify any product that has formed on the surface during the reaction or to examine changes in the structure of the catalyst itself. Again, it is relatively simple to obtain the Raman spectrum of a catalyst pellet as it stands, and the ability of the Raman technique to obtain useful data on either organic or inorganic systems can be used to advantage in such cases.

Another feature of commercial production methods, particularly in the petrochemical industry, is that crude products are often subjected to filtration and/or decolorization processes, typically involving the use of carbon black or a kaolin. It may be necessary to analyze the product at this stage, and the presence of such materials, which form a turbid slurry with the compound to be analyzed, creates no particular problem for the Raman technique, whereas the sampling problems for IR spectroscopy may be considerable.

Other advantages of the Raman effect, notably the ability to handle samples in sealed glass systems (see Section 9.2.2, below), mean that air-sensitive or water-sensitive catalysts can be examined with equal ease.

9.2.2. Glass Sample Containers and Aqueous Systems

The Raman spectrum of glass is broad, largely featureless, and weak. This, together with the fact that most Raman studies use visible laser excitation, means that, if the sample needs to be contained, then glass is normally the ideal material for this purpose. This has several important implications for the analysis of a wide range of materials, and some of these are discussed separately in Section 9.7. The ability to use glass to contain the sample means, for example, that air-sensitive and water-sensitive materials can be examined in sealed glass tubes under an appropriate atmosphere or under vacuum.

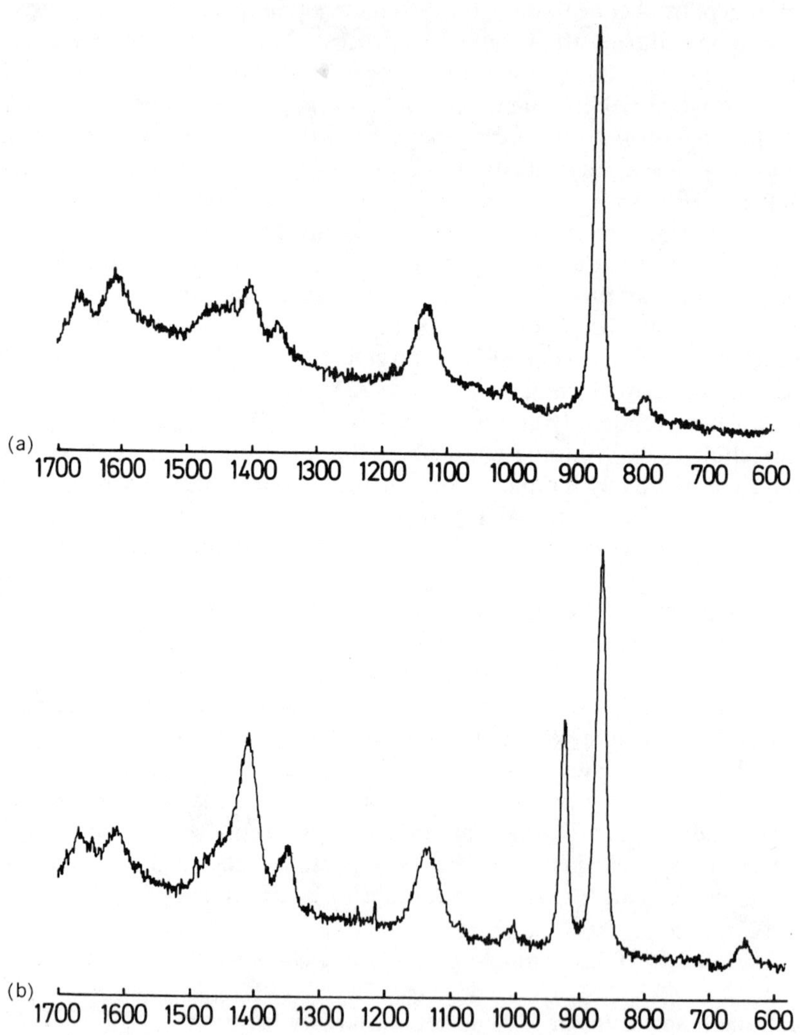

Figure 9.2. Hydrolysis of polyamide, 1% solution in seawater: (a) initial sample; (b) after 7 days at 120°C; (c) after 90 days at 120°C.

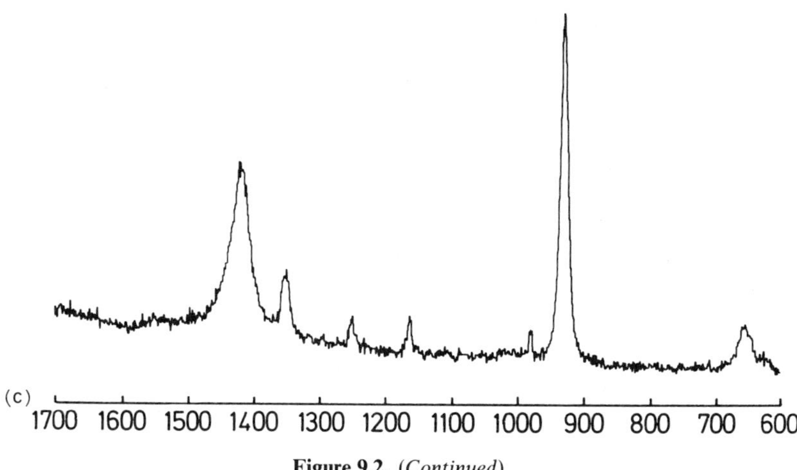

Figure 9.2. (*Continued*)

One area where this has proved particularly useful is in the study of air-sensitive electroactive polymers, where the Raman technique has provided valuable structural information (1–4).

If necessary, a complete reaction system, including reactants, reaction products, and catalyst (if appropriate), can be studied *in situ* in a glass container. The ability of glass to withstand moderately high temperatures and pressures means that the construction of specialized Raman cells for studies at nonambient temperatures and pressures is relatively simple. The implications of this from the point of view of reaction kinetics of organic reactions are considerable and are considered in detail in Section 9.7. In general, the main value of glass to the Raman spectroscopist is its use as a sample container enabling difficult systems to be studied with relative ease. Because of this, Raman can be used to study reacting systems under a very wide range of conditions.

Water, which is also a very weak Raman scatterer, particularly below $3100\,cm^{-1}$ (the area of greatest interest to the organic chemist) is of value because it is an ideal solvent for Raman studies. For example, the study of hydrolysis reactions by Raman spectroscopy is extremely simple; this is illustrated in Fig. 9.2, which shows the study of the hydrolysis of a commercial polyamide. Figure 9.2(a) shows the Raman spectrum of a 1% wt/vol solution of the polyamide in water at room temperature. The spectrum was obtained from a sample sealed in a thick (~ 3 mm) glass ampule. The spectrum shows the presence of a very strong band at $865\,cm^{-1}$, characteristic of the amide group. The solution can then be heated in the sealed tube for as long a period as required for hydrolysis at temperatures up to about 200°C. Above this temperature the vapor pressure of the water usually causes the ampule to fracture and more specialized cells have to be used.

Figure 9.2(b) shows the spectrum of the aqueous polyamide solution after heating for 7 days at 120°C. This shows a much decreased band at 865 cm^{-1} and a new band at 925 cm^{-1}. Further heating, for a total of 90 days, results in the total disappearance of the 865 cm^{-1} band, which shows that the hydrolysis reaction

$$R\cdot CONH_2 \xrightarrow[H_2O]{\text{Heat}} R\cdot COONH_4$$

is now complete. Amides of this type are often used as surfactants in large-scale, commercial operations, often involving moderately high temperatures and pressures. Hence a study of this type, where the amide or compound of interest can be monitored over a long period without having to remove it from the sealed ampule, is beneficial. The same method can be used to study the hydrolysis, either neutral or acid- or base-catalyzed, of organic esters.

Other aqueous systems of interest to the organic chemist that may be conveniently studied by Raman spectroscopy include emulsion polymerization reactions and organic pollutants in water, which require both qualitative and quantitative study. The fact that water is transparent to most of the wavelengths commonly used in Raman spectroscopy also allows one to do *in situ* studies in water. This technique has been widely used by inorganic chemists to study corrosion products of iron, but it also has considerable potential for the study of the action of corrosion inhibitors, surfactants, scale inhibitors, and weld-impression resins.

The use of glass also enables highly corrosive systems involving concentrated acids and molten salts to be studied without difficulty. Also, glass and water can be used to advantage in the study of electrochemical reactions. The laser beam can be passed through the glass walls of the electrochemical cell, through the aqueous electrolyte solution, and then focused on, or close to, the electrode surface. This technique has been successfully used to monitor the reduction of carbon dioxide to formate and hydrocarbon (5, 6).

9.2.3. Excitation Using Visible Lasers

The weak Raman scatter of water and glass discussed in the previous section can only be used to advantage when the laser wavelength being used to excite the Raman signal is not absorbed by the water or glass. Essentially, this requires that the wavelength lies in the visible region of the spectrum. The wavelengths commonly used, which are provided by Ar^+ and Kr^+ lasers, lie in the range from 454.5 to 799.3 nm, and tunability within this range can be obtained by the use of a dye laser and the appropriate dyes. Not only does the use of visible radiation mean that water and glass can be used to advantage but it also means that detectors available to cover this region of the spectrum (photomultiplier tubes and intensified diode arrays) are extremely sensitive. This is a particularly valuable advantage because the Raman effect is intrinsically weak and it is only the high sensitivity of such detectors, compared with those available for IR work, that enables Raman spectroscopy to match the IR technique in terms of detection limits.

Also, because the laser beam is readily observable by the operator, sample alignment can be achieved with a high degree of accuracy even without the use of a microscope attachment (see Section 9.5). In addition, the beam diameter is very narrow (typically $\sim 50\,\mu m$). The value of these properties is that small samples, or small parts of samples that may be of particular interest, can be readily studied even when large, complex cells or reaction systems are studied.

9.2.3.1. Use of Fiber Optics

The most significant feature of the use of visible radiation, and one which is only just beginning to be realized, is the potential offered by fiber optics. Since the laser light can be readily transferred through a fiber-optic probe (and typically focused by using a microscope objective on the end of the fiber-optic probe), the sample can be irradiated remotely from the laser. Similarly, the Raman signal is also in the visible region of the spectrum. It is convenient, therefore, to collect the Raman signal and take it to the spectrometer via one or more fibers, which can be placed around the single fiber carrying the laser beam. A schematic diagram for a fiber-optic probe of this type, based on the work of McCreery (1), is shown in Fig. 9.3. The advantage of such a probe is that it can be used effectively to take the Raman spectrometer to the system to be analyzed. Systems that have not previously been amenable to analysis (some of which are considered below) can now often be studied.

The realistic maximum length of such fiber-optic probes is still a matter of conjecture and will obviously depend to a large extent on the particular material being analyzed. For strong Raman scatterers such as alkanes, reasonably good-quality spectra can be obtained using probes at least as long as 100 m and probably significantly longer. Figure 9.4 shows a spectrum obtained from a polyethylene pellet using a 100-m fiber optic and a spectrometer fitted with a diode array detector. This spectrum was obtained using a total exposure time of 2 s and shows the C—H deformation region, with the typical fine structure due to the crystalline, intermediate, and amorphous phases of the polymer.

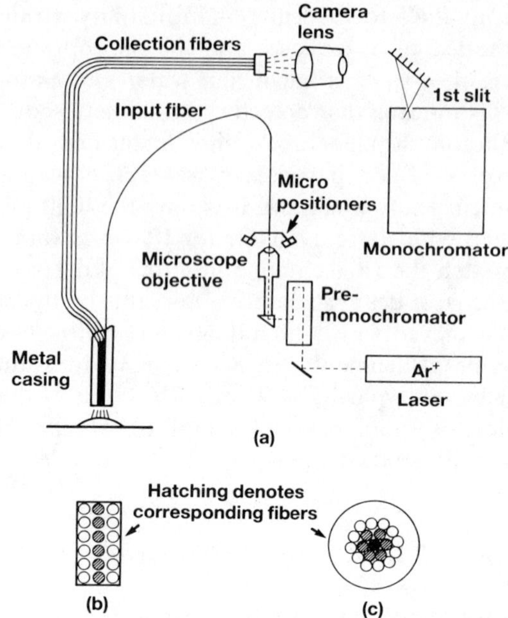

Figure 9.3. Schematic of fiber-optic probe: (a) general layout of experiment; (b) arrangement of collection fibers in front of camera; (c) probe end (● = input fiber). From McCreery et al. (7).

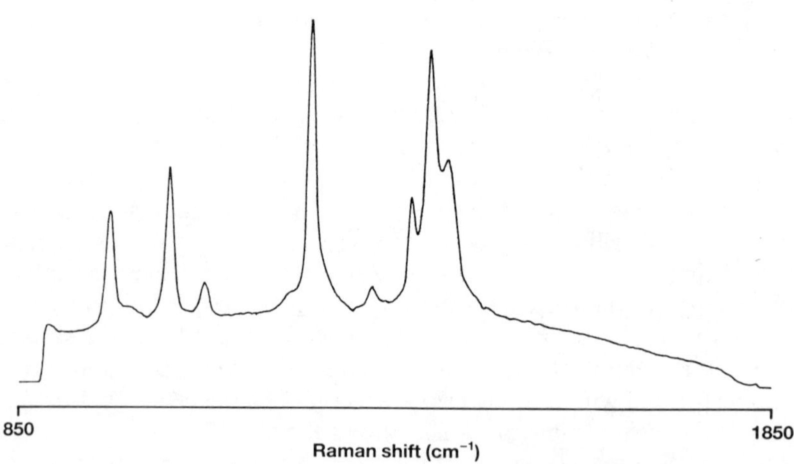

Figure 9.4. Spectrum of polyethylene pellet obtained using 100-m fiber optic in 2 s.

This ability to obtain spectra from samples situated remotely from the spectrometer is of particular value to the organic chemist because it gives the capability of studying large reacting systems *in situ* at high temperatures and pressures without having to use special Raman cells to simulate the reaction conditions on a smaller scale. It also provides the ability to study samples in "hostile" environments, for example in toxic, explosive, or corrosive atmospheres. The end of the probe can be protected by a glass sheath which can be inserted into systems which might otherwise damage the probe. Figure 9.5 shows the spectrum of molten polyethylene obtained by placing a fiber-optic probe protected by such a sheath directly into the molten polymer. In this way it is possible to study changes in the polymer structure upon cooling or heating. In principle, this applies to any system where the Raman spectrum changes as a function of temperature, pressure, or time. The use of fiber-optic probes of this type opens up considerable possibilities for the analytical chemist and means that reacting systems, especially large-scale ones, can be studied without the need for periodic removal of samples for analysis.

Another area where fiber-optic probes should prove valuable is in the study of electrochemical reactions. It is a relatively simple matter to place the probe into the electrolyte, very close to the electrode surface, and hence to monitor reactions occurring on, or close to the electrode. Good-quality spectra have been obtained by Schwab and McCreery (8) of electrochemical systems using a simple probe which uses one fiber to take the laser to the electrode surface and another to collect the backscattered Raman signal.

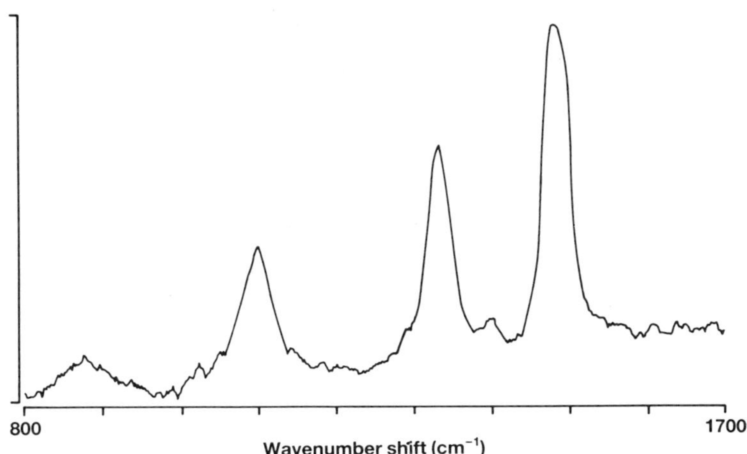

Figure 9.5. Spectrum of molten polyethylene obtained using fiber-optic probe inserted into the polymer melt.

9.3. CHARACTERIZATION OF NONPOLAR GROUPS

Another feature of the Raman effect is that it depends on a change in the induced dipole of the particular functional group exhibiting the molecular vibration. This means that generally the nonpolar species in a molecule produce strong Raman scattering but exhibit only weak IR absorption. For example, homoatomic species such as C—C, C=C, C≡C, O—O, N—N, and S—S, all exhibit strong characteristic Raman scattering (RS) but have a much weaker effect in the IR. Conversely, of course, highly polar species such as C=O and COO^- exhibit very strong IR absorption but are not normally particularly strong in the Raman spectrum. A wide range of functional groups that exhibit strong RS are of interest to organic chemists for the structural evaluation of molecules. In this sense the Raman technique is complementary to the IR—a complete vibrational picture of the molecule only being available by a combination of the two methods.

A large proportion of the petrochemical industry is concerned either with the production of polymerizable monomers or with the reaction of these monomers to form polymers. These polymers may either be single-component systems (homopolymers), two-component systems (copolymers), or multi-component systems. In all cases, however, the most important functional group present is the C=C group. The reason for this is that polymerization takes place through this group according to the general scheme

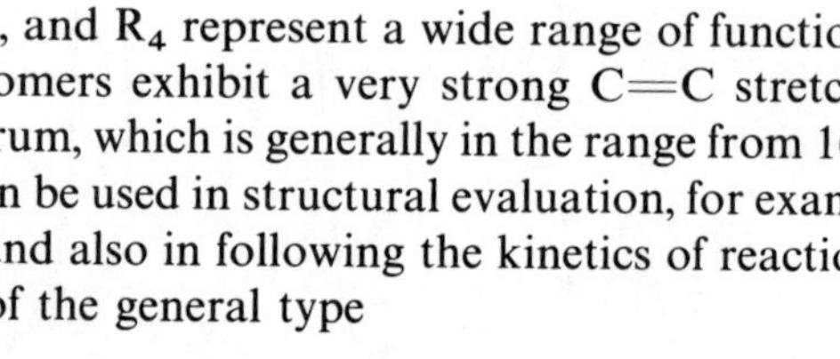

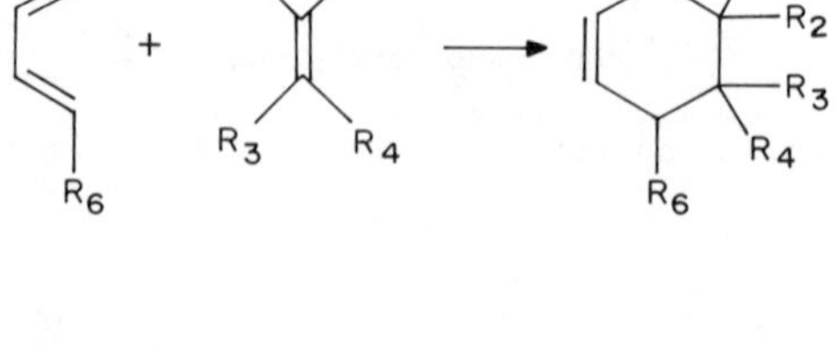

where R_1, R_2, R_3, and R_4 represent a wide range of functional groups. In all cases, such monomers exhibit a very strong C=C stretching vibration in the Raman spectrum, which is generally in the range from 1620 to 1680 cm^{-1}. This vibration can be used in structural evaluation, for example, to determine cis/trans ratios, and also in following the kinetics of reactions such as Diels-Alder reactions of the general type where R_1 through R_6 may be H or other functional groups. More importantly, from the commercial point of view, this C=C stretching vibration can be used to give valuable information about the kinetics of polymerization reactions; this is considered in some detail in Section 9.7.

Strong RS is also exhibited by cyclic systems, both aromatic and non-aromatic, and such ring modes can be successfully used to elucidate the composition of petroleum fractions in the liquid or gas phase. For example, substituted cyclopentanes and cyclopentenes (9–11) can be identified and distinguished from the corresponding cyclohexanes and cyclohexenes (9, 12), all of these groups of compounds being of importance in evaluating petroleum fractions.

The Raman spectra of aromatic compounds can also be particularly informative, especially with respect to the nature and location of substituents, and there is a wealth of information available on substituted benzenes in the literature (9, 13–33). Even more information is available on heterocyclic compounds (9, 34–46), most of these exhibiting very strong Raman spectra. For a comprehensive correlation of data relating to the Raman spectra of the whole range of organic compounds, the reader is referred to the excellent book by Dollish, Fateley, and Bentley (9). This book has sections on the following subjects:

Alkanes
Haloalkanes
Aliphatic alcohols, ethers, and related compounds
Aliphatic amines, nitroalkanes, and their derivatives
Organosulfur compounds
Alkenes
Aliphatic aldehydes and ketones
Carboxylic acids and derivatives
Amides, amino acids, peptides, and proteins
C=N and N=N compounds
Cumulenes and heterocumulenes
C≡C and C≡N compounds
Benzene and its derivatives
Heterocyclic compounds

This impressive list suggests that the Raman spectrum should be just as useful as the IR spectrum for purposes of characterization and identification, but the vastly superior data collections available for the IR technique suggest that Raman will not be used to any great extent for this purpose except where sampling difficulties for IR make Raman the preferred technique.

However, Raman spectroscopy has one major advantage that does have some value in the area of characterization/identification: its ability to observe low-frequency vibrations without any difficulty. If low frequencies ($<200\ cm^{-1}$) are to be studied by IR absorption spectroscopy, expensive additional hardware is necessary. The region below $200\ cm^{-1}$ can, however, be routinely covered by a standard Raman spectrometer. Modern monochromators have sufficiently good stray light rejection that bands as low as 5 or $6\ cm^{-1}$ from suitable solid or liquid samples can be studied. Stray light is much less of a problem in the gas phase, and so Raman bands appearing at less than $2\ cm^{-1}$ can be obtained in gas-phase studies. The use of low-frequency vibrations has proved of most use in studies of polymer structure, notably of the various forms of polyethylene; this, however, is outside the scope of this chapter. In the general area of organic Raman spectroscopy, low-frequency modes are observed in alkanes where CH_2—CH_2 torsional modes and C—C—C deformations (9) can occur at less than $200\ cm^{-1}$. Haloalkenes are another group of compounds exhibiting low-frequency modes that enable them to be distinguished. On the whole, these low-frequency vibrations are of more use to the polymer chemist and polymer morphologist than to the organic or petrochemical chemist.

9.4. QUANTITATIVE STUDIES

It is probably a fair description of Raman spectroscopy to say that it is not intrinsically quantitative. Since it is only a single-beam technique, unlike conventional IR absorption spectroscopy where a reference beam is used and where the absorption can be quantified, the intensity of the detected Raman signal is not absolute. It can vary with several parameters, including sample position relative to the collection optics in either the horizontal or vertical planes. Similarly, the nature of the sample, whether it is a solid, liquid, or gas, and whether it is clear or opaque, smooth or roughened, will affect the observed intensity of the Raman signal (i.e., the number of Raman photons scattered by the sample that can actually be measured). Other parameters that affect the recorded photon count are the spectrometer slit-width, laser power, laser wavelength, laser beam diameter, and photomultiplier tube response. It is at best very difficult, and normally not possible, to obtain the same intensity of recorded Raman signal at any particular wavelength from the same sample on two different occasions. Even when this can be done, it is so time consuming and involves the adjustment of so many parameters that it is not worthwhile.

Hence, although the actual Raman signal from any molecule has an absolute value relating to its scattering cross section, for experimental purposes, particularly for analytical work, this is of no value since the proportion of Raman photons being collected is always an unknown quantity. Hence it is unrealistic to try to assign molecular constants, such as the molar extinction coefficients that are so widely and effectively used in IR and UV/visible spectroscopy to quantify these techniques. This is not to say that Raman spectroscopy cannot be used quantitatively but, rather, that some considerable care must be exercised in order to make the technique reliably quantitative. One major advantage that the Raman effect gives to the analyst with respect to quantitative studies is that the intensity of the Raman signal is directly proportional to the concentration of the species producing it. With absorption spectroscopy this relationship is logarithmic, and so as less radiation passes through the sample the absorption measurement becomes less reliable. If we need to determine the ratio of a strong IR band to a weak one in the same sample, it may well be that, in order to obtain a measurable signal from the weak band, the intensity of the strong band cannot be measured reliably. This does not occur with Raman spectroscopy, where the linearity of the effect means that very strong bands can be measured with comparable accuracies.

9.4.1. Internal Standard Method

Traditionally, the way around the problem of not being able to make absolute intensity measurements is to use an internal standard, and some considerable thought needs to be given to the choice of this standard. The principle of the internal standard is that whatever changes are made to the various parameters of the instrument, laser, and sample system being studied, or whatever changes occur to that system as a result of internal or external physical and chemical changes, the measured intensity of the internal standard will change by the same amount (in percentage terms) as that of the particular Raman band we wish to measure. This means, then, that from the ratio of the standard Raman band to that of the unknown material, the concentration of the unknown may be calculated, provided that the concentration of the standard is known and that a calibration has already been carried out. This is obviously much less convenient than obtaining a molar extinction coefficient from a reference book and calculating concentration directly from observed peak intensity as can be done with absorption spectroscopy. Hence, quantitative studies of this type, whereby the concentration of one or more components in a mixture is required for a static system, are normally better carried out by other methods.

It is sometimes the case, however, that Raman spectroscopy provides the only means available for a particular analysis; in that event we must resort to the internal standard method. Obviously it is necessary that any such standard needs to be inert in the context of the particular system being studied. Also the Raman band of the internal standard needs to be as close to that of the species being measured as possible to compensate for the nonlinearity of the detector response.

One example of the quantitative use of Raman spectroscopy employing an internal standard is the determination of thiourea in aqueous solution in the presence of sulfur dioxide and metal salts (47). This is a system in which the thiourea cannot be determined by conventional methods, chemical or spectroscopic, and Raman spectroscopy provides the only convenient method of quantification. In this case the thiourea solution contains sulfuric acid, sulfur dioxide, various inorganic salts, and oxidation products of the thiourea. Hence the best standard for this particular application is one which is very water soluble with a strong Raman band close to the strongest band of the thiourea (C=S stretching mode at 735 cm^{-1}), but separated from the Raman bands of the other components. The standard must also be stable in sulfuric acid. A common internal standard for aqueous systems is sodium sulfate, because of its very stable nature, high solubility, and very strong Raman band, but in this system it is inappropriate because of the presence of sulfuric acid. The most readily available compound that meets all of the criteria is acetic acid, which has a strong γ(OCO) mode at 880 cm^{-1}. Calibration with known standards shows that the concentration of thiourea is reliably estimated from the ratio of the intensities of the 735 cm^{-1} (thiourea) and 880 cm^{-1} (acetic acid) bands. This example illustrates the great care that must be taken when one is choosing the internal standard and shows that the Raman technique is worthwhile only if it provides the only or most accurate solution.

9.4.2. Reacting Systems

A quantitative study of a reacting system in which the Raman band being monitored changes with time was presented by Gulari et al. (48). They were following the polymerization of styrene

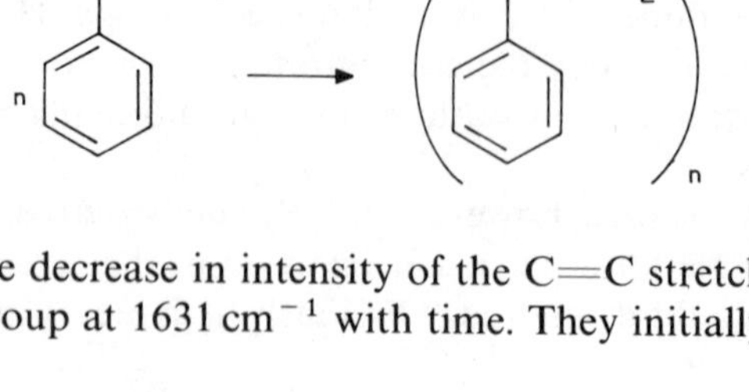

$$n\ \mathrm{HC{=}CH_2}\ (\text{on benzene ring}) \longrightarrow \left(\mathrm{HC{-}CH_2}\ (\text{on benzene ring})\right)_n$$

by monitoring the decrease in intensity of the C=C stretching mode of the exocyclic vinyl group at 1631 cm^{-1} with time. They initially tried to use the

$1603 \, cm^{-1}$ ring mode of styrene as the internal standard but found that its intensity changed as the polymerization proceeded. In order to quantify their measurements they used instead a simple but very effective technique based on the ability of modern gas lasers to give a highly stable output over a long period. Modern CW gas lasers have the capability of operating in a so-called light mode. In this mode the output of the laser in terms of photon flux is kept constant. Modern lasers can give a stable photon output for periods of several hours. Hence, by keeping all of the other parameters relating to the sample, the spectrometer, and the laser constant during the course of the reaction, the change in the observed intensity of the C═C stretching mode can be used to determine the rate of polymerization of the styrene.

It is this capability of being able to follow reaction kinetics by using the stabilized output of a CW gas laser that promises to be one of the most important applications of the Raman technique in the next few years. The fact that reacting systems can be studied quantitatively, together with the other advantages that the Raman effect confers, means that Raman spectroscopy should, at least in the near future, be the preferred technique for studying a wide range of chemical reactions *in situ*. Some of these reactions will be considered in Section 9.7.

9.5. SPATIAL RESOLUTION: RAMAN MICROSCOPY

In Section 9.1 we stressed the value of CW gas lasers from the point of view of supplying a continuous source of intense, coherent, monochromatic light in a reliable manner. On the whole, most Raman spectroscopists have been happy with this situation and have not looked at the second major advantage of such lasers—spatial resolution. Because CW lasers give emissions in the visible region of the spectrum, it is relatively simple to focus the laser beam down to a diameter of about $0.5 \, \mu m$. Thus, it is possible to examine very small samples or very small areas of samples that may be of particular interest. Although it is a relatively simple matter to connect a standard optical microscope to the monochromator of a Raman spectrometer [and this was reported in the literature as long ago as 1979 (49)], there are still relatively few groups who are using the technique of Raman spectroscopy on a regular basis for such microanalytical purposes.

In Raman microscopy the laser beam is passed down through what is essentially a conventional optical microscope and focused on the sample. The sample and the laser beam can be viewed on a screen or via a TV camera on a TV monitor. In this way the laser can be lined up on a part of the sample. The backscattered Raman signal is then collected via the microscope objective and is directed into the monochromator, where it is analyzed in the

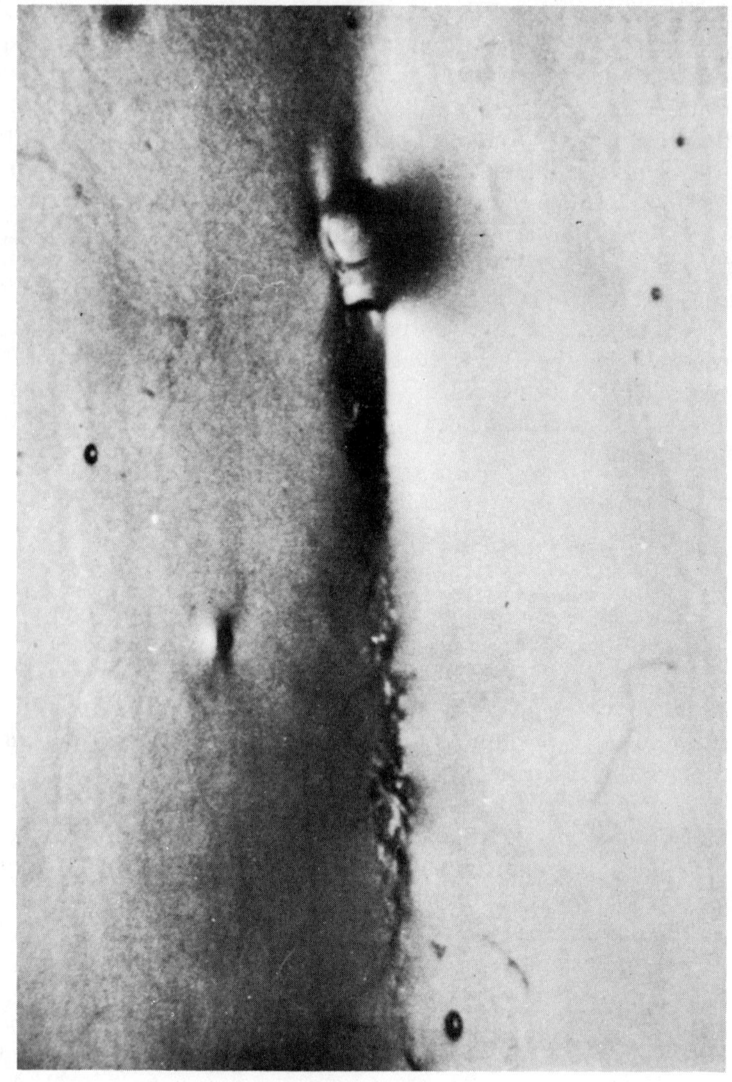

Figure 9.6. Photograph of inclusion in polyethylene film.

9.6. METHODS OF SIGNAL ENHANCEMENT

One of the major limitations of Raman spectroscopy is caused by the very weak nature of the Raman effect—that is, its lack of sensitivity. In many cases the limit of detection is significantly worse than with the IR technique, despite the relatively poor response of IR detectors. The only readily available technique for increasing the Raman signal is to alter the instrumental parameters, such as laser power, slitwidth, spectral accumulation time, and laser wavelength. If the laser power is increased too much this may result in photochemical or thermal damage to the sample. Opening the slits of the monochromator will result in poorer resolution. Shorter laser wavelengths give a greater Raman signal, but with conventional CW gas lasers the shortest wavelength lines are normally of relatively low power and, being of higher energy, are more likely to produce photochemical damage.

On the whole, therefore, the Raman technique cannot be applied to the analysis of organic compounds that are present in a system at very low levels. There are, however, two useful exceptions to this rule. At the moment both of these exceptions, although extremely useful when they can be applied, are relatively limited in their application. Advances in laser technology should, in the near future, expand their application. In particular, the development of tunable UV laser sources should be of great benefit to the organic analyst. This is considered in some detail below. The two methods of increasing the signal of a species present at very low levels in the presence of one of more other components (e.g., a solvent) are resonance-enhanced Raman spectroscopy (RRS) and surface-enhanced Raman spectroscopy (SERS). In each case the Raman signal specific to one component is greatly enhanced, often by several orders of magnitude, and the two methods are considered separately below.

9.6.1. Resonance-Enhanced Raman Spectroscopy and Applications in Organic and Petrochemical Systems

For a detailed account of the theory of RRS the reader is referred to the excellent articles by Behringer (72) and Clark (73). In simple terms, when the laser wavelength being used to obtain the Raman spectrum lies close to or within the absorption profile of an electronic transition of a particular chromophore within the molecule being irradiated, then the Raman signal of that chromophore, and of that chromophore only, may be enhanced by several orders of magnitude. Obviously, where this occurs, the sensitivity of the technique is increased dramatically. The degree of resonance enhancement depends on how close the laser wavelength is to the wavelength of maximum absorption and also on the nature of the electronic transition producing that absorption.

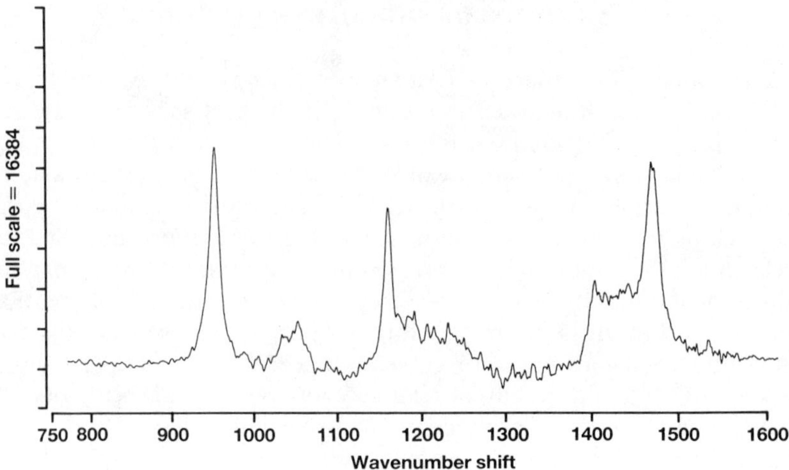

Figure 9.8. Spectrum of β-carotene solution in tetrahydrofuran (10^{-4} M).

One of the best examples of the use of RRS in organic chemistry occurs with conjugated polyenes of the general type

$$\left[CH{=}CH \right]_n$$

The greater the value of n, the longer the wavelength of maximum absorption, and so these compounds are ideal for RRS using visible laser radiation. Figure 9.8 shows the spectrum of a solution of β-carotene (10^{-4} M) in tetrahydrofuran (THF). The bands at 1158 and 1527 cm^{-1} are the resonance-enhanced bands of the C—C and C=C stretching modes, respectively, of the conjugated polyene system. These two species constitute the chromophore that gives rise to the visible absorption of the β-carotene, and so it is these modes that are enhanced in the Raman spectrum, other Raman-active vibrations in the molecule remaining at their normal strength and being unobservable in the presence of the intense resonance-enhanced bands. Other bands in the spectrum are due to the THF.

The ability to detect molecules containing long polyene sequences at extremely low levels has been used to advantage in a wide range of biological studies, notably of bacteriorhodopsin systems (74–78). One of the most valuable areas of study involving RRS of conjugated polyene systems has been the degradation of certain vinyl polymers. Homopolymers or copolymers of vinyl chloride, vinyl bromide, vinyl acetate, and vinyl alcohol (79–87) have all been studied. In each case the polymer can degrade, either thermally, photochemically, or chemically via the mechanism shown here (A).

$$\left[\begin{array}{c}H\quad X\\ |\quad\;\; |\\ -C-C-\\ |\quad\;\; |\\ H\quad H\end{array}\right]_x \xrightarrow{-HX} \underset{a_2}{\left[\begin{array}{c}H\quad X\\ |\quad\;\; |\\ -C-C-\\ |\quad\;\; |\\ H\quad H\end{array}\right]_m} - \begin{array}{c}H\\ |\\ C=C\\ \quad\;\; |\\ \quad\;\; H\end{array} - \underset{a_3}{\left[\begin{array}{c}H\quad X\\ |\quad\;\; |\\ -C-C-\\ |\quad\;\; |\\ H\quad H\end{array}\right]_{x-m-1}} \qquad \text{(A)}$$

a_1

The isolated double bond formed in the polymer backbone acts as the preferred location for subsequent elimination of HX to give a long conjugated polyene sequence in the polymer chain, as shown (B). When these conjugated

$$\underset{a_2}{\left[\begin{array}{c}H\quad X\\ |\quad\;\; |\\ -C-C-\\ |\quad\;\; |\\ H\quad H\end{array}\right]_m} - \begin{array}{c}H\\ |\\ C=C\\ \quad\;\; |\\ \quad\;\; H\end{array} - \underset{a_3}{\left[\begin{array}{c}H\quad X\\ |\quad\;\; |\\ -C-C-\\ |\quad\;\; |\\ H\quad H\end{array}\right]_{x-m-1}} \xrightarrow{-HX}$$

$$\underset{a_4}{\left[\begin{array}{c}H\quad X\\ |\quad\;\; |\\ -C-C-\\ |\quad\;\; |\\ H\quad H\end{array}\right]_m} - \underset{a_5}{\left[\begin{array}{c}H\quad\;\;\\ |\quad\;\;\\ -C=C-\\ \quad\;\; |\\ \quad\;\; H\end{array}\right]_y} - \underset{a_6}{\left[\begin{array}{c}H\quad X\\ |\quad\;\; |\\ -C-C-\\ |\quad\;\; |\\ H\quad H\end{array}\right]_{x-m-y}} \qquad \text{(B)}$$

systems reach a certain length, which seems to be dictated by steric considerations and depends on the nature of X, a cross-linking reaction occurs. The mechanism of this reaction is still not completely characterized, but that shown here (C), put forward by Rabek and his co-workers (88), is

$$-\dot{C}H{+}CH{=}CH{+}_y \rightarrow {+}CH{=}CH{+}_b-\dot{C}H{+}CH{=}CH{+}_{y-b}$$

$$+$$

$$-\dot{C}H{+}CH{=}CH{+}_z \rightarrow {+}CH{=}CH{+}_c-\dot{C}H{+}CH{=}CH{+}_{z-c}$$

$$\downarrow \qquad \text{(C)}$$

$$\begin{array}{c}{+}CH{=}CH{+}_b-CH{+}CH{=}CH{+}_{y-b}\\ |\\ {+}CH{=}CH{+}_c-CH{+}CH{=}CH{+}_{z-c}\end{array}$$

one of the most probable. The overall result of this cross-linking reaction is that long polyenes combine with each other to give a larger number of shorter polyenes. In the foregoing case the maximum reduction in sequence length will occur when $b = y/2$ and $c = z/2$.

As already mentioned, conjugated polyenes of the general type

$$\text{+CH=CH}\text{+}_x$$

absorb strongly in the visible and UV region of the spectrum, the wavelength of maximum absorption becoming longer as the value of x increases. This is very well illustrated by the work of Sondheimer et al. (89), who prepared a range of such compounds up to 1,3,5,7,9,11,13,15,17,19-eicosadecaene ($x = 10$) and showed that at $x = 3$ the wavelength of maximum absorption is at 257 nm while for $x = 10$ the value is 447 nm. As a consequence of this UV/visible absorption, these compounds exhibit very strong RRS signals when a suitable laser wavelength is used, i.e., one which is close to the absorption maximum. Such polyenes give resonance Raman spectra consisting of two bands near 1100 cm^{-1} (C—C, ν_1) and 1500 cm^{-1} (C=C, ν_2). These bands correspond to the +C—C=C+ chromophore, which, as with β-carotene, gives rise to the UV/visible absorption, and the values of ν_1 and ν_2 vary depending on the value of x. As x increases the values of ν_1 and ν_2 (expressed in Δcm^{-1}) decrease. Hence, for a conjugated polyene of unknown sequence length the value of ν_2, as determined from the resonance Raman spectrum, can be used to determine the value of x, and expressions have been deduced suggesting values of x up to approximately 60 can be distinguished in this way (90–92).

In the case of degraded vinyl polymers of the aforementioned type, notably poly(vinyl chloride), it is found that the values obtained for ν_1 and ν_2 vary with the laser frequency being used to obtain the spectrum. This variation can reasonably be attributed to the fact that the degraded polymer contains a range of conjugated sequences of different x values. The reason for this is illustrated in Fig. 9.9 and is considered in detail below.

The total intensity $I_{m,n}$ of light of frequency ν scattered by one freely orientable molecule passing from state m to state n is given by the expression

$$I_{m,n} = \frac{2^7\pi^5}{3^2C^4} I_0\nu^4 \sum_{\rho,\sigma} |\alpha\rho\sigma, mn|^2,$$

where I_0 = intensity of incident radiation; C = velocity of light; and $\alpha\rho\sigma, mn$ = the $\rho\sigma$ component ($\rho\sigma = X, Y, Z$ space-fixed Cartesian coordinate system) of the scattering tensor.

Since $\alpha\rho\sigma, mn$ is inversely proportional to $|\nu_{rm} - \nu_0|$, where ν_{rm} is a vibronic transition frequency and ν_0 is the frequency of the exciting line, it can be seen that, as ν_0 approaches ν_{rm}, the value of $\alpha\rho\sigma, mn$ and hence of $I_{m,n}$ will

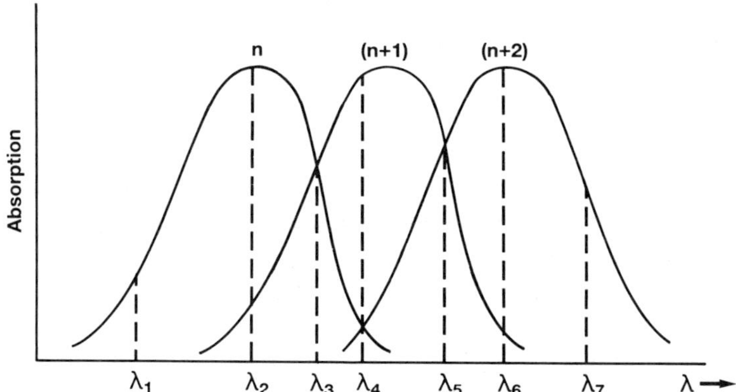

Figure 9.9. Effect of exciting wavelength (λ) on the observed value of ν_2 for a three-component polyene system.

be considerably enhanced. In physical terms what this means is that the polarizability of the system, and hence the Raman intensity, increases very rapidly when $\nu_0 \rightarrow \nu_{rm}$. It does not tend to infinity because, as a more detailed theoretical treatment shows, a damping term must be included in the equation. Figure 9.9 illustrates the effect of the exciting wavelength on a three-component system with overlapping absorption bands. It can be seen that, as the exciting wavelength approaches an absorption maximum, the contribution of the particular species producing that maximum will be more enhanced than those of the other species whose absorption profiles overlap. For example, if we consider the seven exciting wavelengths illustrated in Fig. 9.9 we can see the following:

1. For λ_1 there will be resonance enhancement only of the species with x double bonds. Hence the observed value of ν_2 will be equal to ν_2^1 and the band will be symmetrical.
2. For λ_2 there will be a very strong contribution from the molecule containing n double bonds but also some contribution from $(n + 1)$. Hence the observed value of ν_2 will be slightly lower than ν_2^1 and there will be an asymmetry on the low $\Delta\,\text{cm}^{-1}$ side of the band.
3. For λ_3 there is an approximately equal contribution from n and $(n + 1)$, and so the observed value of ν_2 will be approximately equal to $(\nu_2^1 + \nu_2^{11})/2$ and the band will be more or less symmetrical.
4. For λ_4 there will be a contribution from all three species. The greatest contribution will obviously be from $(n + 1)$. Lesser contributions will be from n and $(n + 2)$, which will tend to cancel each other out from

the point of view of the observed value of v_2 but will have the effect of broadening the band in a more or less symmetrical manner.

5. For λ_5 there is an approximately equal contribution from $(n+1)$ and $(n+2)$. Hence the observed value of v_2 will be approximately equal to $(v_2^{11} + v_2^{111})/2$ and the band will be nearly symmetrical.
6. For λ_6 the major contribution will be from $(n+2)$, but there will be a small contribution from $(n+1)$. The observed value of v_2 will, therefore, be slightly greater than v_2^{111} and there will be an asymmetry on the high $\Delta\,\mathrm{cm}^{-1}$ side of the bond.
7. For λ_7 only the $(n+2)$ species will be resonance enhanced, and so the observed value of v_2 will be equal to v_2^{111} and the band will be symmetrical.

Obviously, the foregoing illustration is very limited in its scope, since it covers only three components with overlapping absorption bands and the intensities of these three absorptions are all equal. The real situation with degraded vinyl polymers is much more complex, involving more components with different absorption intensities; in addition, particularly at the longer-wavelength end of the spectrum, the overlap of absorption bands is considerable. Nonetheless, by changing the exciting wavelength over a wide range by small increments (typically ~5 nm) using both conventional CW gas lasers and a dye laser, it is possible to obtain a great deal of valuable information about such systems. The position of the v_2 band and the asymmetry of its profile yield a great deal of information about the distribution of polyene sequences in the degraded polymer. This information can be used to obtain a reasonably reliable estimate of the level of HX loss, which in itself is very valuable and cannot be obtained for a naturally weathered material by any other method. Even more importantly, however, by monitoring the changes in this distribution the onset of the cross-linking reaction can be observed, since it corresponds to a decrease in the longest polyene sequences and an increase in those of intermediate length. Work of this kind has yielded much useful information, both from the academic standpoint, concerning the nature of degradation of such materials, and for the industrial chemist, where it has helped to establish the best polymer formulation for the particular conditions in which the polymer is to be used.

While the most significant application of RRS in the fields of organic and petroleum chemistry has been the study of polyene systems in vinyl polymers, it is a technique with vast potential in this area. The main problem that has limited its wider application is the limited availability of laser wavelengths below about 410 nm. While a large number of organic compounds exhibit absorptions above or around this value, the use of lasers tunable down to about 220 nm would open up virtually the whole of organic chemistry to the application of RRS. The sensitivity and specificity of RRS would be beneficial, for example, in the analysis of petroleum fractions and in the identification and estimation of low levels of pollutants, particularly in aqueous systems. CW lasers are not yet available to cover this region of the spectrum. Attempts have been made to apply RRS to organic compounds that absorb below the region covered by CW lasers, by complexing the compounds with materials such as tetracyanoethylene (93). This produces a charge-transfer complex with an absorption usually well into the visible region. While this technique has met with some success, the level of resonance enhancement achieved has usually not been particularly great.

The best solution is undoubtedly to use a tunable laser in the UV region, and in the past few years a considerable degree of success has been achieved by using pulsed laser systems, based either on a neodymium YAG or an excimer laser. Although such lasers have complete tunability over the range from 190 nm to 4 μm, their pulse rate is only $\sim$150 Hz at best and the duration of each pulse is typically only a few nanoseconds. Hence in normal operation the detector of the spectrometer would only be receiving Raman photons for about 1–2 μs is every second, and for the remainder of this time would be producing background noise. This means that, in order for such a system to be of any use, the detector has to be open only for the period when the sample is actually being irradiated. If a photomultiplier tube is being utilized as the detector, this necessitates the use of a BOXCAR device, but the preferred method, and one that has been employed very successfully, is to use a gated diode array as the detector. Asher and his co-workers (94–97) have produced some excellent resonance Raman spectra of a wide range of organic compounds, mostly aromatic, that absorb in the UV region.

Although such laser systems are expensive, they open up a huge area of organic and petroleum chemistry to qualitative, quantitative, and time-resolved study with a high degree of sensitivity and specificity. One advantage of short laser pulses is that they can be used to study short-lived species such as unstable radicals and reaction intermediates. There is no doubt that the increasing use of these tunable laser systems and their growing reliability will prove to be one of the major areas of interest in Raman spectroscopy in the near future, but particularly in the study of organic systems.

9.6.2. Surface-Enhanced Raman Spectroscopy and Applications

In 1974 Fleischmann, Hendra, and McQuillan (98) reported very intense Raman bands from pyridine adsorbed onto an anodized silver surface and attributed this strong signal to the presence of a large number of pyridine molecules present at the highly roughened electrode surface. By 1977 other workers had noticed the increase in Raman band intensity of pyridine adsorbed onto silver surfaces and had discovered that an intrinsic surface-enhancement effect plays a fundamental role in producing the Raman spectrum (99, 100). This early work marked the advent of what is known today as the SERS effect. It was initially thought to be highly specific to the silver/pyridine system, but subsequent work has shown it to be a more general phenomenon in terms of the characteristics of both the substrate and the adsorbate. Several hundred papers have now been published concerning both the theoretical and experimental aspects of SERS in an attempt to elucidate the precise origins of the effect.

One of the major considerations for the formation of a SERS surface is surface roughness, and this has been achieved in various ways. The most commonly used approach for generating the SERS surface employs various electrochemical techniques (101–103). This approach usually involves subjecting the electrode to an oxidation/reduction cycle to roughen the surface. Other surfaces capable of exhibiting SERS effects have included mechanically polished surfaces (104), silver- and gold-island films (105), matrix-isolated microclusters (106), silver films sputtered using ultra-high vacuum (107), a photodecomposed silver iodide film (108), and colloids (109, 110). Vo-Dinh and co-workers have recently shown that SERS effects can be achieved from silver-coated spheres ranging from 38 to 1000 nm in diameter (111).

The SERS effect is exhibited most strongly by four metal substrates—silver, gold, copper, and nickel—although weaker enhancement has been achieved using platinum (112), cadmium (113), aluminum (114), mercury (115), and palladium (116). The wavelength of the incident laser beam has also been identified as an important factor in optimizing the SERS enhancement. For silver surfaces green laser light (typically 514.5 nm) is best, whereas for copper and gold lower-energy laser lines (e.g., 647.1 nm) give the best results. However, the overall frequency dependance is a complex function of surface roughness and substrate dielectric properties. In this context the use of tunable UV lasers should extend the number of surfaces that exhibit SERS.

The origin of the phenomenon is still not fully understood despite intensive theoretical and experimental research, and it is beyond the scope of this chapter to consider all the theories in any detail. The theories put forward fall into two categories: the first says that the phenomenon is due to electromagnetic interactions occurring in the presence of the incident laser light (117–126), and the second that interactions, including chemical ones, that occur in the absence of the incident field are responsible (127–134). The most consistently supported theory is the surface plasmon theory (117–120). It predicts that the use of shorter-wavelength lasers should extend the SERS effect to metals such as chromium and the alkali metals. The large number of theories that still have some credibility indicates the high level of confusion that still exists when precise mechanisms are discussed. However, the arguments generated between the theoreticians do not detract from the very interesting experimental data recorded or from the potential analytical value of the technique.

The most studied example of SERS so far has been the pyridine/silver system, but the analytical value for this system is clearly limited. Other examples of SERS are discussed briefly in the following four subsections.

9.6.2.1. Electrochemical Applications

Organic adsorbates that display the SERS effect on silver electrode surfaces include azides (135), formates (136), cyclohexane (137), acetonitrile (137), benzene (138), cyanopyridines (139), pyridine (140), phenol (141), pyrazine (99), aniline (99), benzylamine (99), methylpyridines (99) and other substituted pyridines (99), dithiozone (142), diazines (143), triphenylphosphines (144), ethylenediaminotetraacetate (EDTA) (145), isoquinolines (146), biliverdine (147), pyrromethenone (147), cytochrome *c* (148), oxyhemoglobin (148), myoglobin (148), purine (149), adenine (149), adenosine (149), various nucleic acid components (149), acetylene (150), ethylene (150), and propene (151). The technique can also be used in the detection of saturated hydrocarbons formed during the electrochemical reduction of CO_2 at a silver electrode.

9.6.2.2. Colloids

It has been amply demonstrated that both silver and gold colloids exhibit SERS effects very similar to those observed from electrochemically roughened silver and gold electrode surfaces. Kerker (152) has investigated a range of macromolecules on colloidal silver and Hilderbrandt and Stockburger (110) and Lee and Meisel (153) have studied dye molecules adsorbed on colloidal silver. The use of colloids in SERS studies has been reviewed in depth by Creighton et al. (154), so in the context of this chapter it is sufficient to point out that the general applicability of metal colloids to enhance the sensitivity of the Raman effect is important from an analytical viewpoint.

9.6.2.3. Metal-Island Films

Various metals have been used as island films and have been shown to exhibit SERS effects. The metals concerned are usually deposited under vacuum as thin films ($\sim 10\,\mu m$) onto tin oxide glass. The compounds to be examined are then either deposited as films, also under vacuum, or are coated on the metal substrate using an *in situ* dipping method. As with other SERS techniques, the use of metal-island films has shown that Raman spectra can be obtained from sub-monolayer coverage of a wide range of organic compounds.

For example, Jennings et al. (155) have studied metal phthalocyanines on silver. Such compounds are important because of the catalytic role they play in electrochemical processes. Among other systems studied are tetrathiafulvalene (105), dye molecules (156), organic sulfides (157), and *p*-nitrobenzoates (158).

9.6.2.4. General Analytical Applications for Organic Materials

The analytical scope of SERS is only just starting to be realized. Vo-Dinh et al. (111) have been among the first to publish a general analytical paper based on SERS phenomena. In their article they describe a generally applicable method which uses silver-coated submicrometer-sized spheres bonded to a filter paper substrate. Traces of organic molecules such as benzoic acid ($\sim 3.6\ \mu g$) delivered as a 3-μl spot of solution and dried were detected and identified by reference to their normal Raman spectra.

Other workers have used SERS effects in water-pollutant-monitoring experiments (159), and a method has also been developed for the rapid detection and identification of algae in water (160). The technique has also been applied to the determination of trace amounts of antitumor agents (161), metalloporphyrins (162), and conjugated polyenes (163).

Both RRS and SERS provide methods for making Raman spectroscopy extremely sensitive for the detection and identification of trace levels of organic materials. Both techniques have their advantages and their limitations, but both provide applications that will expand considerably in scope with continuing developments in tunable lasers.

9.7. TIME-RESOLVED STUDIES: EXAMPLES OF *IN SITU* STUDIES OF REACTING SYSTEMS

This aspect of the Raman spectroscopy of organic and petrochemical materials has been left as the final one to be considered because, more than any other application, it can make use of practically every advantage of the Raman effect that has been considered so far in this chapter. Most time-resolved Raman studies reported in the literature use short laser pulses for the study of transient species and often make use of nonlinear effects. While this type of study is of considerable value, it lies outside the scope of the present work and the reader is referred to the many excellent reviews on the subject (164–169). The main value of time-resolved studies in the context of the petrochemical industry relates to work on a rather longer time scale using, in most cases, CW lasers. It is an area only just beginning to be appreciated and applied, but the potential benefits to the industry are considerable. It can essentially be described as the *in situ* study of reacting systems. For this purpose, the use of multichannel detectors, notably diode arrays, and charge-coupled devices has recently opened up many new possibilities (although many reacting systems are amenable to study using conventional scanning spectrometers and photomultiplier tubes as detectors), some of which will be considered below.

Many organic compounds exhibit sufficiently strong Raman scattering that, using a multichannel detector, a good-quality spectrum, covering a range of approximately $1000\,cm^{-1}$, can be obtained in only a few seconds. This makes it possible to obtain a large number of consecutive spectra in a relatively short time and thus enables reactions to be studied in a period of time ranging from a few minutes to several hours. It is worth considering at this stage, before discussing specific systems in detail, how some of the features of the Raman effect mentioned previously prove to be of use in this type of study.

a. The very versatile nature of Raman spectroscopy in terms of sample handling means that specialized cells for working at high or low temperatures and pressures can readily be accommodated in the sample compartment of the spectrometer. This enables reactions to be studied under a wide range of conditions such as those used at a large-scale chemical plant. A range of cells that can be used for this type of study is illustrated in Fig. 9.10. The weak Raman scatter of glass and the fact that CW lasers used in this type of study are normally in the visible region of the spectrum means that the high/low temperature/pressure cells can be fitted with thick glass windows capable of maintaining high pressures without fracture while still enabling the laser light and the Raman signal to pass through them.

b. As mentioned in Section 9.3 above, the strongest Raman signals tend to be generated by nonpolar, particularly homoatomic, species within molecules. The C=C group normally exhibits very strong scattering and is the most important in terms of reactions of commercial interest, especially polymerization reactions. Other functional groups that give strong Raman scattering and are of interest in terms of their reactivity include C≡C, C—N, C—S, S—S, O—O, N—N, C—O, C—C, and cyclic structures (e.g., cyclopentadiene or norbornenes).

c. If very large reaction vessels, which cannot be accommodated in the sample compartment of the spectrometer, are to be used, then fiber-optic probes can be employed that effectively take the spectrometer/laser system to the sample. This technique may well be applied to highly toxic, corrosive, or explosive systems where it is necessary to isolate the system to be studied well away from the spectrometer. This means, for example, that in the event of the system exploding it is only the end of the fiber-optic probe that is damaged and not the much more expensive Raman spectrometer.

Figure 9.10. Cells for use at nonambient temperatures and pressures and for electrochemistry: (a) high-temperature cell; (b) high-pressure/high-temperature cell; (c) polymerization cell; (d) electrochemical cell. Pictures by courtesy of Ventacon Ltd.

(c)

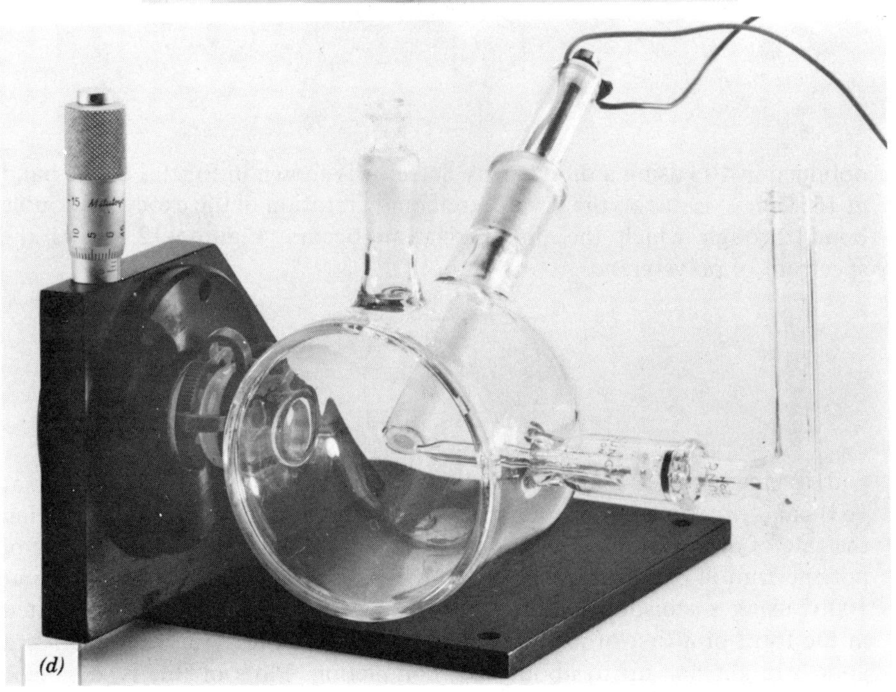

(d)

d. The highly stable output of CW gas lasers, in terms of photon flux, that can now be achieved means that Raman spectroscopy can be used to follow quantitatively a reaction occurring over several hours. This opens up the possibility of studying reaction kinetics in a very accurate and reproducible manner. The linear relationship between concentration and intensity means that most reactions can conveniently be followed to 100% completion.

e. The considerable degree of spatial resolution that the highly focusable CW lasers confer means that the technique can be used to follow reactions occurring, for example, on a small portion of catalyst in a heterogeneously catalyzed reaction.

When the aforementioned advantages are considered together with the ability to obtain good-quality spectra very rapidly using a diode array detector, it can be seen what an immensely powerful technique Raman spectroscopy is for the study of reacting systems. Two polymerization reactions of commercial significance will now be considered in some detail.

9.7.1. Homopolymerization of Styrene

The Raman spectrum of unstabilized styrene monomer

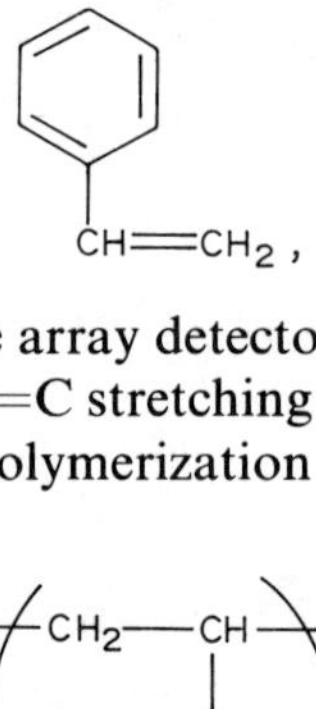

C_6H_5–$CH{=}CH_2$,

obtained in 10 s using a diode array detector is shown in Fig. 9.11. The band at 1631 cm^{-1} is due to the C=C stretching vibration of the exocyclic double bond through which the polymerization occurs. Figure 9.12 shows the spectrum of polystyrene

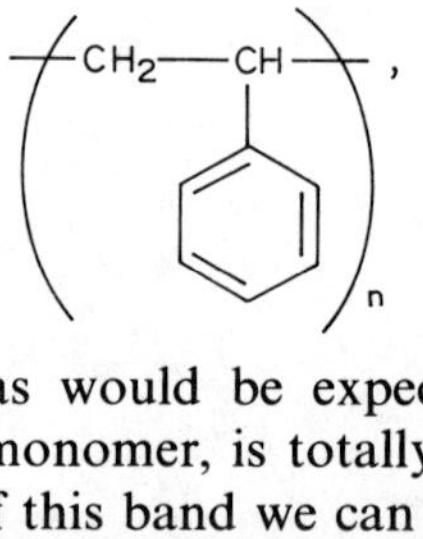

$\left(CH_2-CH(C_6H_5) \right)_n$,

and it can be seen that, as would be expected, the C=C vibration at 1631 cm^{-1}, present in the monomer, is totally absent. Hence, by following the rate of disappearance of this band we can obtain a value for the rate of polymerization of styrene. Data from the bulk polymerization of styrene at 100°C using α-azoisobutyronitrile (AIBN) as the initiator have been plotted in the form of a first-order rate plot as shown in Fig. 9.13. The result is a good straight line up to about 80% completion. Plots of this type can be used to obtain rate constants for homopolymerization reactions, including

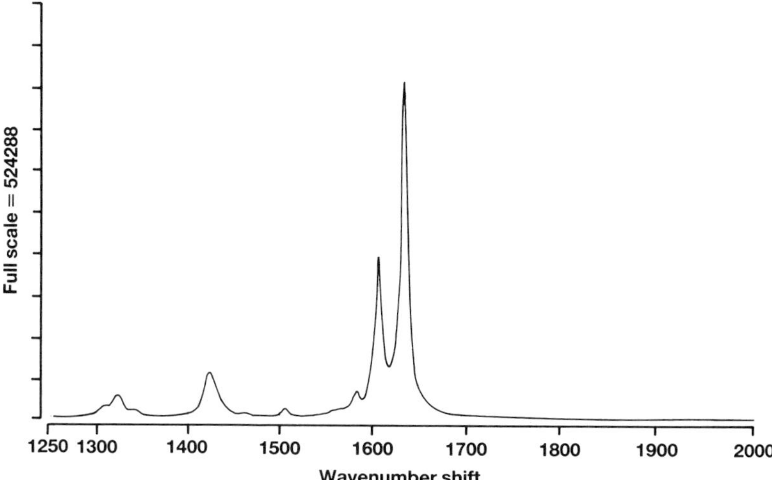

Figure 9.11. Raman spectrum of styrene monomer.

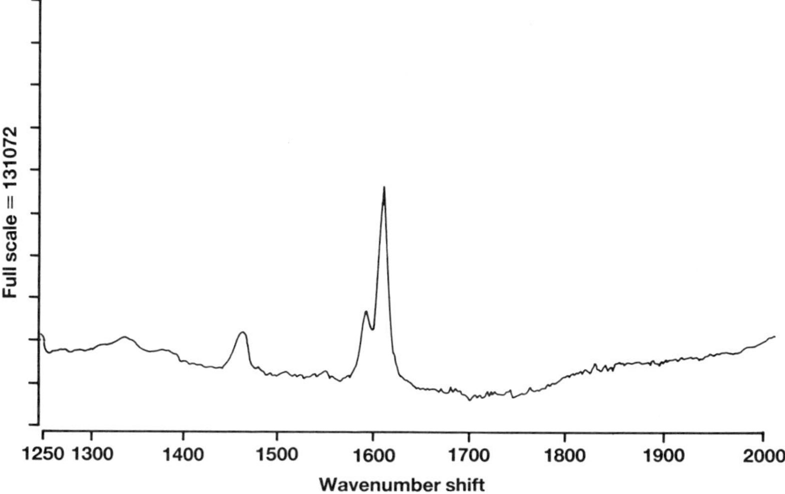

Figure 9.12. Raman spectrum of polystyrene.

those carried out under commercial reaction conditions. Thermodynamic data such as activation energy can also be obtained by carrying out the same reaction at a range of temperatures.

The traditional method for obtaining such information is by dilatometry, and while this has proved extremely successful from both a scientific and commercial point of view, it does have some specific limitations that the

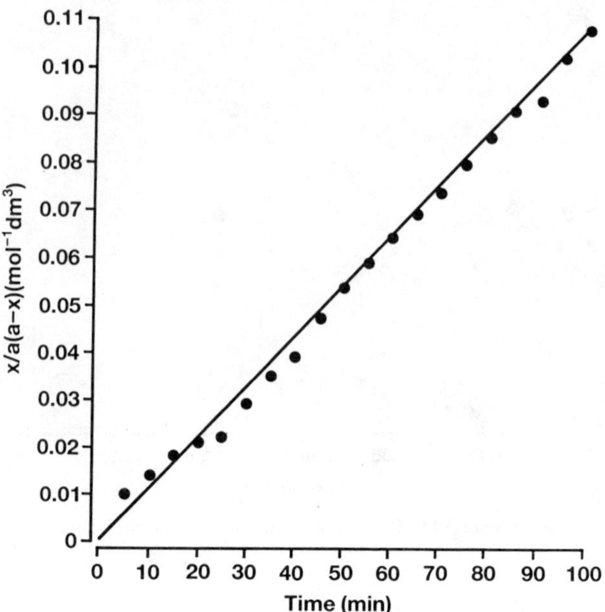

Figure 9.13. First-order rate plot for the polymerization of styrene.

Raman technique does not. For example, dilatometry cannot be used to follow reactions to very high levels of completion, and it is very limited in the range of temperatures and pressures that can be employed. It does not have the capability, therefore, of studying such systems under the conditions used on the commercial scale. However, most of the data available to industrial chemists and chemical engineers relating to polymerization reactions has been obtained by dilatometry. Therefore the use of Raman spectroscopy for this purpose promises to have a considerable impact on commercial polymerization studies.

9.7.2. Homopolymerization of Methyl Methacrylate

As with the polymerization of styrene, the homopolymerization of methyl methacrylate can be followed by monitoring the disappearance of the C=C stretching mode due to the vinyl group and centered at $1642\,cm^{-1}$. In this case, the first-order rate plot for a bulk polymerization carried out at 80°C using AIBN as initiator is shown in Fig. 9.14. Polymerization proceeds at a constant rate until about 40% completion. At this point there is a sudden rapid increase in the reaction, a situation not observed with styrene. This rapid increase is due to an effect known as the Trommsdorff effect and is a

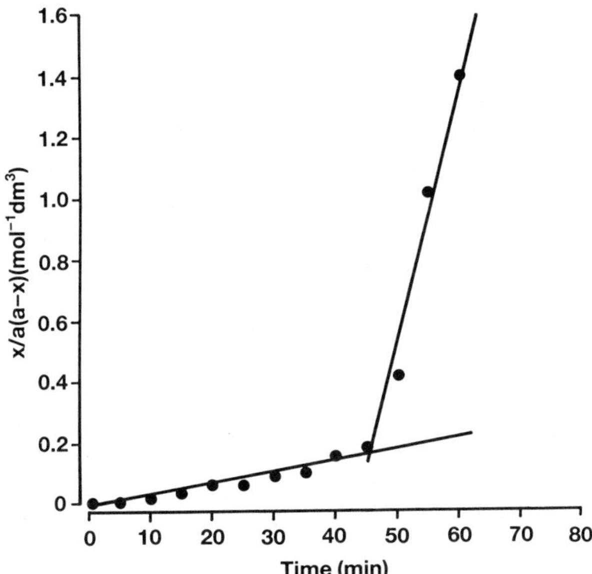

Figure 9.14. First-order rate plot for the polymerization of methyl methacrylate.

result of the polymer precipitating out of solution in the monomer. This prevents growing polymer chains from diffusing away from the solid bulk but still allows monomer units to diffuse in, and so the reaction speeds up dramatically. The ability to observe the onset of this effect is obviously of considerable importance when one is considering reaction conditions for commercial-scale polymerizations.

9.7.3. Copolymerizations and Multicomponent Polymerizations: The Use of Mathematical Deconvolution

While information on homopolymerization reactions of the type just described can be of great use to the industrial chemist, of much more interest, both academically and industrially, is the study of multicomponent polymerization reactions. Increasingly, there is a requirement for highly specialized polymers containing several (as many as eight) components. In order to obtain the correct structure, in terms of both the proportion of the separate components and the degree of randomness of the polymer, a knowledge of the relative rates of polymerization of the various components in the presence of each other and under the reaction conditions being used is essential. This information has only been readily available for two-component systems and has been obtained by dilatometry measurements.

The inability to obtain data under commercial plant conditions has meant that the preparation of such polymers has largely been carried out on an empirical basis.

The use of *in situ* Raman spectroscopy employing the type of cells illustrated in Fig. 9.10 could well allow the whole area of speciality copolymers and multicomponent polymers to be placed on a much sounder, more scientific footing. The way in which such measurements can be made is to follow the rate of disappearance of the C=C stretching modes of the individual monomers separately. Obviously there will be many cases where these bands will exhibit considerable overlap. In such cases it will not be possible—by eye—to distinguish the separate components in the overall C—C envelope. However, there are now available highly sophisticated computer programs for carrying out mathematical deconvolution processes that are capable of separating such overlapping bands. A relatively simple and successful method is that of Fourier self-deconvolution (FSD) (170). This can be carried out rapidly and will successfully resolve several components provided the signal-to-noise ratio (S/N) is reasonably good.

An alternative approach that holds even more promise in the long term is the maximum entropy method (MEM) (171). This requires considerably more computer capacity than FSD but will still function efficiently in conditions of low S/N. Both of these techniques are described fully in the literature, to which the reader is referred; and although they are still in their infancy from the point of view of their application to Raman spectroscopy, they promise to make a major contribution in the area of reaction kinetics of complex systems.

9.7.4. Other Applications

The Raman technique can be applied to virtually any type of chemical reaction in gas, liquid, or solid phase, to catalyzed or uncatalyzed reactions, and to electrochemical reactions. In the case of catalyzed reactions, these can be either homogeneously or heterogeneously catalyzed.

In addition to monitoring chemical reactions, the same basic technique can be applied to monitoring changes in structure. This can be either chemical structure, such as cis/trans isomerism, or tautomerism, or physical structure such as crystallinity changes in polymers under processing conditions. In this last-named application the use of fiber-optic probes holds considerable promise.

9.8. FLUORESCENCE IN ORGANIC SYSTEMS

The previous sections of this chapter have extolled the virtues of Raman spectroscopy for the study of a wide range of applications in the general area of organic chemistry and petrochemistry. Despite all of these applications, however, there is still one major drawback that applies in general to any application of the technique but is most significant with respect to organic systems, particularly complex ones. This drawback is fluorescence. It is often the case that absorption of the laser beam by the sample will give rise to fluorescence. It is also often the case that the intensity of this fluorescence is so great that it totally obscures the Raman signal. This problem has restricted the general application of Raman spectroscopy, particularly with respect to the analysis of oils and oil-based materials. It is appropriate, therefore, when discussing the application of Raman spectroscopy to organic systems, to consider the techniques that have been applied to overcoming fluorescence and discuss their relative merits. The methods employed fall into two main categories: those attempting to reduce or eliminate the fluorescence itself, and those attempting to extract the Raman signal from the fluorescence background. These categories will be considered in the next two subsections.

9.8.1. Fluorescence Reduction

9.8.1.1. Sample Purification

Fluorescence background produced by a sample is often due to the presence of traces of highly fluorescing impurities or additives. Since the quantum efficiency of the fluorescence process is normally considerably greater than that of the RS process, the effect of only a very small quantity of a highly fluorescing material can completely obscure the Raman signal of the bulk. It is sometimes possible, in such cases, to remove or reduce these impurities. One of the best applications of this method occurs with polymers or macromolecules, which often contain traces of impurities of fluorescing aromatic additives as a result of the preparation or added at the processing stage. These can often be removed by dissolving the polymer in a suitable solvent and then re-precipitating with methanol, which usually leaves the additive or impurity still in solution. The precipitated polymer is then washed with methanol, and the Raman spectrum normally shows a significant improvement. This is quite a good technique where identification of the polymer is required but may, of course, result in the removal of additives that need to be identified. It is also necessary that the polymer be soluble in readily available solvent.

9.8.1.2. "Burning Out" with the Laser Beam

The phrase "burning out" is applied to a technique that was widely used in the early days of laser Raman spectroscopy and is one that can still occasionally be used to advantage. It is a method that involves leaving the fluorescing sample in the laser beam for a period of time, normally at least 1 h and sometimes as long as 24 h. During this time the fluorescence may be considerably reduced if the sample is suitable. The laser powers used in this process have to be considerably greater than those normally used to obtain the Raman spectrum and are often in excess of 500 mW. The burning out process is still not well understood but is popularly thought to result from absorption of the very intense laser beam by the traces of fluorescing impurity, followed by the decomposition of this material owing to the heating effect resulting from the absorption. It is a technique that, for organic samples, only works where the impurities are present in very low concentrations, since the heating effect can damage the sample itself. Similarly, if there is any absorption exhibited by the bulk of the sample at the laser wavelength being used for burning out, then decomposition of the bulk can occur. Another major drawback to this method is the fact that it is only applicable to solid samples, since the same small area of sample must be irradiated for the whole period and then the Raman spectrum must be obtained for this area.

9.8.1.3. The Use of UV and Near-IR Lasers

The fluorescence background produced by most organic samples comprises a broad envelope covering most of the visible region of the spectrum, reaching a maximum near the center of the visible region. It is possible, by the use of laser lines that lie well outside this envelope, to obtain the Raman spectrum without entering the fluorescence profile, especially if only a limited region of the Raman spectrum is required. This can be achieved by using lines from (UV) (96) or Kr^+ (near-IR) (172) lasers. The disadvantage of using UV excitation is that the high energy of the photons may cause photodecomposition of the sample. Also, since many organic materials absorb strongly in this region of the spectrum, the sample may be liable to thermal decomposition resulting from absorption of the laser beam. Recent work by Asher and Johnson (96) has shown that good-quality, fluorescence-free spectra can be obtained from coal oils by using a tunable pulsed UV laser.

At the other end of the spectrum the 752.5 and 799.3 nm lines of the Kr^+ laser can be used very successfully to reduce or eliminate fluorescence. The much lower energies of the photons at these wavelengths compared with shorter wavelengths means that relatively high powers can be used without any danger of photodecomposition of the sample. Also, this is an area of the spectrum where few compounds absorb strongly, and so thermal decomposition is not a problem. Satisfactory spectra have been obtained, using these lines, for a wide range of samples exhibiting very severe fluorescence, including organic dyes. Recent developments in laser technology have extended the operating range of lasers well into the near-IR region ($\sim$900 nm) by the use of suitable dyes, and this could well provide the most satisfactory short-term solution to the fluorescence problem.

The major drawback with the use of these lines is that most Raman spectrometers are optimized for the 514.5 nm Ar^+ line. This imposes a limit on the operating range of the instrument determined by the tracking range of the gratings in the monochromator. This particular problem has been elegantly solved by Hirschfeld and Chase (173) who have recorded fluorescence-free Raman spectra using the 1.064 μm line of a CW YAG laser, the Raman signal being detected by using the emission port of an FT-IR (Fourier transform-infrared) spectrometer. Although the Raman signal in this region of the spectrum is very weak, high laser powers can be used, and if the technique can be made routinely applicable it promises to provide the simplest and most effective solution to the fluorescence problem.

9.8.1.4. Fluorescence-Quenching Agents

In some cases, where the sample being analyzed is an organic species in solution, it is possible by the addition of certain compounds to suppress, or at least reduce, the fluorescence. Kasha (174) first showed that the fluorescence of solutions of aromatic hydrocarbons is largely quenched in halogenated solvents as the result of spin–orbit coupling effects that lead to intersystem crossing. Subsequent workers have studied this effect in detail, and it is known that the quenching effect increases with increasing atomic number of the halogen. However, for strongly fluorescent aromatic hydrocarbons these compounds do not reduce the fluorescence sufficiently to make it possible to obtain a Raman spectrum (175).

Other more effective fluorescence-quenching agents are strong electron acceptors and include butane 2,3-dione, nitrobenzene, picric acid, chloranil, tetracyanoethylene, and tetracyanoquinodimethane, and some of these have been used very effectively to prevent fluorescence of normally highly fluorescing aromatic hydrocarbons. The main disadvantages of such reagents, however, are that they can only be used in solution and they have to be employed at such high concentrations in order to complex all of the fluorescing material that they often make a major contribution to the overall Raman spectrum. Another problem is that some of these compounds are so reactive they may actually react with the system being studied. It seems, therefore, that their main use is in RRS where the Raman signal of the species being studied (usually present at low concentrations) is sufficiently intense to obscure the contribution from the quenching agent.

9.8.1.5. Anti-Stokes Raman Spectroscopy

The Raman effect involves the interaction of a beam of photons of a single, fixed frequency with the vibrational energy levels of the particular molecule being studied. These interactions result in a change in energy of the incident photons, which produces a range of frequencies in the scattered beam constituting the Raman spectrum of the molecule. Some of the photons lose energy and are shifted to lower frequency, and these give rise to the Stokes spectrum. A much smaller number of the incident photons gain energy and are shifted to higher frequencies to give a mirror-image anti-Stokes spectrum. This is normally weaker than the Stokes spectrum by a factor of at least 10. Since fluorescence is normally red-shifted from the exciting line, it interferes mostly with the Stokes spectrum and comparatively little with the anti-Stokes. It is possible, therefore, with some samples to obtain a Raman spectrum by using the anti-Stokes signal when the fluorescence obscures the Stokes spectrum. Before the advent of diode array detectors this method was of very limited use because of the very weak nature of the anti-Stokes spectrum. However, the capability of diode arrays to accumulate a large number of spectra in a short time now make it much more applicable (178).

9.8.2. Extraction of the Raman Signal from the Fluorescence Background

9.8.2.1. Background Subtraction

Where the fluorescence background is not too severe and the Raman signal intensity is greater than the fluorescence noise level, it is possible, by simulating the fluorescence background, to remove it from the overall spectrum, leaving the Raman spectrum. The simplest way of doing this is to use a computer to generate a mathematical function that simulates the fluorescence profile. This will often prove sufficient to extract the Raman signal from the background, especially if only a limited portion of the spectrum is required, when the background often approximates to a quite simple function. Normally, for this type of approach to be successful the Raman signal needs to be sufficiently strong that the major peaks are observable above the fluorescence signal.

9.8.2.2. Use of Picosecond Pulsing and Gating

The Raman signal is generated very rapidly, and although it is not clear just how fast the process is, it probably occurs in less than 1 fs. The buildup of the fluorescence signal is a much slower process, varying with the nature of the sample from several picoseconds to as long as 50 ns. It is theoretically possible, therefore, by using picosecond pulsed lasers and electronic gating to obtain a Raman spectrum from the pulse to discriminate against the slower fluorescence signal. Although this approach probably offers the best long-term solution to the problem, it is not widely applicable at the present time because of the lack of suitable fast detectors and gating systems. The technique has, however, been applied successfully to systems where the fluorescence rise time is relatively slow (179–181).

9.8.2.3. Nonlinear Raman Spectroscopy

With the use of two or more high-power lasers, usually pulsed, it is possible to obtain Raman signals designated as nonlinear that are several orders of magnitude stronger than the normal spontaneous Raman signal. The various types of nonlinear Raman signals that can be generated in this way have been discussed extensively in the literature (182, 183). From the point of view of fluorescence rejection the most important feature of this nonlinear process is that it produces a coherent signal, unlike the normal Raman signal, produced over 4π steradians, and the fluorescence signal, also produced over 4π steradians. The very intense coherent nature of the nonlinear Raman signal means that, by collecting over a very small solid angle, the effect of fluorescence can be virtually eliminated. The main disadvantage of most nonlinear techniques, however, is that owing to the extremely high laser powers that have to be used their application is normally limited to gases and clear liquids because of sample decomposition problems.

9.8.3. Conclusions

Of all the alternatives currently available for reducing fluorescence problems, none is entirely suitable. The most generally valuable is the use of near-IR exciting lines, and the one with most ultimate potential is the use of gated detectors with pulsed lasers. While the fluorescence problem still imposes some limitations on the organic and petrochemical applications of Raman spectroscopy, it promises to be overcome in the near future. In the meantime the applications for which the technique can be used are sufficient to make it at least as valuable for this type of study as any other spectroscopic technique, and for many applications its unique advantages provide the only solution currently available.

REFERENCES

1. H. Shirakawa, T. Ito, and F. Ikeda, *Polym. J.* **4**, 460 (1973).
2. L. S. Lichtmann, A. Sarhangi, and D. B. Fitchen, *Chem. Scr.* **17**, 149 (1981).

3. F. B. Schugerl and H. Kuzmany, *J. Chem. Phys.* **74**, 953 (1981).
4. H. Kuzmany, *Chem. Scr.* **17**, 155 (1981).
5. M. R. Mahoney, M. W. Howard, and R. P. Cooney, *Chem. Phys. Lett.* **71**, 59 (1980).
6. R. P. Cooney, M. R. Mahoney, and M. W. Howard, *Chem. Phys. Lett.* **76**, 448 (1980).
7. R. L. McCreery, M. Fleischmann, and P. Hendra, *Anal. Chem.* **55**, 146 (1983).
8. S. D. Schwab and R. L. McCreery, *Anal. Chem.* **56**, 2199 (1984).
9. F. R. Dollish, W. G. Fateley, and F. F. Bentley, *Characteristic Raman Frequencies of Organic Compounds*. Wiley, New York, 1974.
10. L. Piaux, *Ann. Chem. (Paris)* [11] **4**, 147 (1935).
11. L. M. Sverdlov, and E. P. Krainov, *Opt. Spectrosc.* **6**, 214 (1959).
12. G. Chiurdoglu, and A. Guillemonat, *Bull. Soc. Chim. Fr.* [5] **13**, 222 (1946).
13. G. Varsanyi, *Vibrational Spectra of Benzene Derivatives*. Academic Press, New York, 1969.
14. A. M. Bogomolov, *Opt. Spectrosc.* **9**, 162 (1960).
15. A. M. Bogomolov, *Opt. Spectrosc.* **10**, 162 (1961).
16. A. M. Bogomolov, *Opt. Spectrosc.* **12**, 99 (1961).
17. A. M. Bogomolov, *Opt. Spectrosc.* **13**, 90 (1962).
18. A. M. Bogomolov, *Opt. Spectrosc.* **13**, 183 (1962).
19. J. H. S. Green, W. Kynaston, and A. S. Lindsey, *Spectrochim. Acta* **17**, 486 (1961).
20. J. H. S. Green, *Spectrochim. Acta* **17**, 607 (1961).
21. J. H. S. Green, *Spectrochim. Acta* **18**, 39 (1962).
22. J. H. S. Green, *Spectrochim. Acta* **24A**, 1627 (1968).
23. J. H. S. Green, and W. Kynaston, *Spectrochim. Acta* **25A**, 1677 (1969).
24. J. H. S. Green, *Spectrochim. Acta* **26A**, 1503 (1970).
25. J. H. S. Green, D. J. Harrison, W. Kynaston, and D. W. Scott, *Spectrochim. Acta* **26A**, 1515 (1970).
26. J. H. S. Green, *Spectrochim. Acta* **26A**, 1523 (1970).
27. J. H. S. Green, *Spectrochim. Acta* **26A**, 1913 (1970).
28. J. H. S. Green, and D. J. Harrison, *Spectrochim. Acta* **26A**, 1925 (1970).
29. J. H. S. Green, D. J. Harrison, and W. Kynaston, *Spectrochim. Acta* **27A**, 793 (1971).
30. J. H. S. Green, D. J. Harrison, and W. Kynaston, *Spectrochim. Acta* **27A**, 807 (1971).
31. J. H. S. Green, and H. A. Louwers, *Spectrochim. Acta* **27A**, 817 (1971).
32. J. H. S. Green, D. J. Harrison, and W. Kynaston, *Spectrochim. Acta* **27A**, 2199 (1971).
33. J. H. S. Green, D. J. Harrison, and W. Kynaston, *Spectrochim. Acta* **28A**, 33 (1972).

34. K. W. F. Kohlrausch, and A. W. Reitz, *Z. Phys. Chem. Abt. B* **45**, 249 (1940).
35. H. Tschamber, and H. Voetter, *Monatsh. Chem.* **83**, 302 (1952).
36. P. Klaeboe, *Acta Chem. Scand.* **22**, 369 (1968).
37. W. H. Green, and A. B. Harvey, *Spectrochim. Acta* **25A**, 723 (1969).
38. A. E. Parsons, *J. Mol. Spectrosc.* **6**, 201 (1961).
39. S. A. Barker, E. J. Bourne, R. M. Pinkard, and D. H. Whiffen, *J. Chem. Soc.* p. 802 (1959).
40. J. J. Peron, P. Saumagne, and J. M. Lebas, *Spectrochim. Acta* **26A**, 1651 (1970).
41. J. J. Peron, P. Saumagne, and J. M. Lebas, *C. R. Hebd. Seances Acad. Sci., Ser. B* **264**, 797 (1967)
42. D. A. Long and W. O. George, *Spectrochim. Acta* **19**, 1777 (1963).
43. K. C. Medhi, *Opt. Spectrosc.* **19**, 24 (1965).
44. D. A. Long and R. T. Bailey, *Trans. Faraday Soc.* **59**, 599 (1963).
45. D. A. Long and D. Steele, *Spectrochim. Acta* **19**, 1791 (1963).
46. M. Ito, R. Shimada, T. Kuraishi, and W. Mizushima, *J. Chem. Phys.* **25**, 597 (1956).
47. H. J. Bowley, E. A. Crathorne, and D. L. Gerrard, *Analyst* **111**, 539 (1986).
48. E. Gulari, K. McKeigue, and K. Y. S. Ng, *Macromolecules* **17**, 1822 (1984).
49. B. W. Cook and J. D. Loudon, *J. Raman Spectrosc.* **8**, 249 (1979).
50. B. W. Cook and G. D. Ogilvie, *Proc. Annu. Conf.—Microbeam Anal. Soc.* **17**, 294 (1982).
51. F. Adar and H. Noether, *Proc. Annu. Conf.—Microbeam Anal. Soc.* **18**, 269 (1983).
52. M. E. Andersen, *Adv. Chem. Ser.* **203**, 383 (1983).
53. E. Payen, M. C. Dhamelincourt, P. Dhamelincourt, J. Grimblot, and J. P Bonnelle, *Appl. Spectrosc.* **36**, 30 (1982).
54. F. Adar, M. Leclerq, and R. E. Grayzel, *Am. Lab. (Fairfield, Conn.)* **14**, 59 (1982).
55. C. A. Johnson and K. M. Thomas, *Fuel* **63**, 1073 (1984).
56. M. C. Dhamelincourt and P. Dhamelincourt, *Appl. Spectrosc.* **37**, 512 (1983).
57. C. Beny, N. Guilhaumou, and J. C. Touray, *Chem. Geol.* **37**, 113 (1982).
58. R. Farhi, G. Petot-Ervas, C. Petot, G. Dhalenne, and F. J. Le Guern, *Solid State Chem.* **41**, 147 (1982).
59. J. C. Touray, C. Beny-Bassez, J. Dubessy, and N. Guilhaumou, *Scanning Electron Microsc.* **1**, 103 (1985).
60. J. C. Touray and N. Guilhaumou, *Bull. Mineral.* **107**, 181 (1984).
61. J. Dubessy, N. Guilhaumou, J. Mullis, and M. Pagel, *Bull. Mineral.* **107**, 189 (1984).
62. F. J. Purcell and W. B. White, *Proc. Annu. Conf.—Microbeam Anal. Soc.* **18**, 289 (1983).

63. H. Boyer and D. C. Smith, *Proc. Annu. Conf.—Microbeam Anal. Soc.* **19**, 107 (1984).
64. P. Pineau, R. Bernard, M. Audebrand, P. Freor, J. F. Tessier, and J. G. Faugère, *Raman Spectrosc. Proc. Int. Conf., 8th, 1982* p. 813 (1982).
65. H. Rosen, and T. Novakov, *Nature* (*London*) p. 768 (1984).
66. K. Tateishi, K. Furuya, T. Kikushi, H. Ishida, and A. Ishitani, *Bunseki Kagaku* **30**, 774 (1981).
67. T. W. Carr, *Annu. Proc. Reliab. Phys.* (*Symp.*) **18**, 186 (1980).
68. P. Fraundof, R. Patel, P. Swan, R. M. Walker, F. Adar, and J. J. Freeman, *Proc. Annu. Conf.—Microbeam Anal. Soc.* **17**, 188 (1982).
69. T. E. Doyle and J. L. Alvarez, *Proc. Annu. Conf.—Microbeam Anal. Soc.* **18**, 277 (1983).
70. D. J. Gardiner, E. Baird, A. C. Gorvin, W. E. Marshall, and M. P. Dare-Edwards, *Wear* **91**, 111 (1983).
71. D. J. Gardiner, M. Bowden, J. Daymond, A. C. Gorvin, and M. P. Dane-Edwards, *Appl. Spectrosc.* **38**, 282 (1984).
72. J. Behringer, in *Raman Spectroscopy: Theory and Practice* (H. A. Szymanski, ed.), p. 168. Plenum, New York, 1967.
73. R. J. H. Clark and T. J. Dines, *Angew. Chem.* **98**, 131 (1986).
74. D. Oesterhelt and W. Stoeckenius, *Proc. Natl. Acad. Sci. U.S.A.* **70**, 2853 (1973).
75. A. Lewis, J. Spoonhower, R. A. Bogomoline, R. H. Lozier, and W. Stoeckenius, *Proc. Natl. Acad. Sci. U.S.A.* **71**, 4462 (1974).
76. A. Campion, J. Terner, and M. A. El-Sayed, *Nature* (*London*) **265**, 659 (1977).
77. J. Terner, C. L. Hsien, A. R. Burns, and M. A. El-Sayed, *Proc. Natl. Acad. Sci. U.S.A.* **76**, 3046 (1979).
78. M. A. Marcus and A. Lewis, *Biochemistry* **17**, 4722 (1978).
79. S. A. Liebman, C. R. Foltz, J. F. Reuwer, and R. J. Obremski, *Macromolecules* **4**, 134 (1971).
80. D. L. Gerrard and W. F. Maddams, *Macromolecules* **8**, 54 (1975).
81. D. L. Gerrard and W. F. Maddams, *Macromolecules* **10**, 1221 (1977).
82. G. Martinez, C. Mijangos, J. L. Millan, D. L. Gerrard, and W. F. Maddams, *Makromol. Chem.* **180**, 2937 (1979).
83. I. S. Biggin, D. L. Gerrard, and G. E. Williams, *J. Vinyl Technol.* **4**, 150 (1982).
84. G. Martinez, C. Mijangos, J. L. Millan, D. L. Gerrard, and W. F. Maddams, *Makromol. Chem.* **185**, 1277 (1984).
85. H. J. Bowley, D. L. Gerrard, W. F. Maddams, and M. R. Paton, *Makromol. Chem.* **186**, 695 (1985).
86. H. J. Bowley, D. L. Gerrard, and W. F. Maddams, *Makromol. Chem.* **186**, 707 (1985).
87. H. J. Bowley, D. L. Gerrard, and W. F. Maddams, *Makromol. Chem.* **186**, 715 (1985).

88. J. F. Rabek, G. Canback, J. Lucky, and B. Ranby, *J. Polym. Sci.* **14**, 1447 (1976).
89. F. Sondheimer, D. A. Ben-Efraim, and R. Wolvsky, *J. Am. Chem. Soc.* **83**, 1675 (1961).
90. L. S. Lichtmann and D. B. Fitchen, *Synth. Met.* **1**, 139 (1979).
91. H. Kuzmany, *Phys. Status Solidi* B **97**, 521 (1980).
92. D. L. Gerrard and W. F. Maddams, *Macromolecules* **16**, 578 (1983).
93. K. H. Michaelian, K. E. Rieckhoff, and E.-M. Voigt, *Proc. Natl. Acad. Sci. U.S.A.* **72**, 4196 (1975).
94. S. A. Asher, C. R. Johnson, and J. Murtaugh, *Rev. Sci. Instrum.* **54**, 1657 (1983).
95. S. A. Asher and C. R. Johnson, *J. Phys. Chem.* **89**, 1375 (1985).
96. S. A. Asher and C. R. Johnson, *Anal. Chem.* **56**, 2258 (1984).
97. S. A. Asher, *Anal. Chem.* **56**, 720 (1984).
98. M. Fleischmann, P. J. Hendra, and A. J. McQuillan, *Chem. Phys. Lett.* **26**, 163 (1974).
99. D. C. Jeanmaire and R. P. Van Duyne, *J. Electroanal. Chem.* **84**, 1 (1977).
100. J. A. Creighton and G. Albrecht, *J. Am. Chem. Soc.* **99**, 5213 (1977).
101. G. G. Atkinson, D. A. Guzonas, and D. E. Irish, *Chem. Phys. Lett.* **75**, 557 (1980).
102. B. Pettinger and H. Wetzel, *Chem. Phys. Lett.* **78**, 398 (1981).
103. A. Regis and J. Corset, *Chem. Phys. Lett.* **70**, 305 (1980).
104. A. Otto, *Surf. Sci.* **75**, C392 (1978).
105. C. J. Sandroff, D. A. Weltz, J. C. Chung, and D. R. Herschback, *J. Phys. Chem.* **87**, 2127 (1983).
106. W. Krasser, U. Kettler, and P. S. Bechthold, *Chem. Phys. Lett.* **86**, 223 (1982).
107. E. Pockrand and A. Otto, *Solid State Commun.* **38**, 1159 (1981).
108. J. E. Rowe, D. A. Zwemer, C. A. Murray, and C. V. Shank, *Bull. Am. Phys. Soc.* [2] **25**, 424 (1980).
109. R. L. Garrell, K. D. Shaw, and S. Krimm, *J. Phys. Chem.* **75**, 415 (1981).
110. P. Hilderbrandt and M. Stockburger, *J. Phys. Chem.* **88**, 5935 (1984).
111. T. Vo-Dinh, M. Y. K. Hiromoto, G. M. Begun, and R. L. Moody, *Anal. Chem.* **56**, 1667 (1984).
112. W. Krasser and A. J. Renouprez, *Solid State Commun.* **41**, 231 (1982).
113. B. H. Loo, *J. Chem. Phys.* **75**, 5955 (1981).
114. A. Otto, *Appl. Surf. Sci.* **6**, 309 (1980).
115. L. A. Sanchez, R. C. Birke, and J. R. Lombardi, *Chem. Phys. Lett.* **79**, 218 (1981).
116. H. Yamada, Y. Yamamoto, and N. Tani, *Chem. Phys. Lett.* **86**, 397 (1982).
117. C. Y. Chen, E. Burstein, and S. Lundquist, *Solid State Commun.* **32**, 63 (1979).
118. S. L. McCall, P. M. Platzman, and P. A. Wolff, *Phys. Lett.* **77A**, 381 (1981).
119. V. Laor and G. C. Schatz, *J. Chem. Phys.* **76**, 2888 (1982).
120. G. C. Schatz, *Acc. Chem. Res.* **17**, 370 (1984).
121. F. W. King, R. P. Van Duyne, and G. C. Schatz, *J. Chem. Phys.* **69**, 4472 (1978).

122. R. P. Van Duyne and G. C. Schatz, *Surf. Sci.* **101**, 425 (1980).
123. A. Otto, *Surf. Sci.* **75**, 1392 (1978).
124. J. I. Gersten, *J. Chem. Phys.* **72**, 5779 (1980).
125. J. I. Gersten and A. Nitzan, *J. Chem. Phys.* **73**, 3023 (1980).
126. C. K. Chen, A. R. B. De Castro, and Y. R. Shen, *Phys. Rev. Lett.* **46**, 145 (1981).
127. T. Maniv and H. Metiu, *Chem. Phys. Lett.* **79**, 79 (1981).
128. P. K. K. Pandey and G. C. Schatz, *J. Chem. Phys.* **80**, 2959 (1984).
129. J. I. Gersten, R. L. Birke, and J. R. Lombardi, *Phys. Rev. Lett.* **43**, 147 (1979).
130. E. Burstein, Y. J. Chen, C. Y. Chen, S. Lundquist, and E. Tossatti, *Solid State Commun.* **29** (1979).
131. A. Otto, J. Timper, J. Billman, and I. Pockrand, *Phys. Rev. Lett.* **45**, 46 (1980).
132. A. Otto, J. Timper, J. Billman, G. Kovaco, and I. Pockrand, *Surf. Sci.* **92**, 155 (1980).
133. R. P. Cooney, M. W. Howard, M. R. Mahoney, and T. P. Mernogh, *Chem. Phys. Lett.* **79**, 459 (1981).
134. M. W. Howard, R. P. Cooney, and A. J. McQuillan, *J. Raman Spectrosc.* **9**, 273 (1980).
135. R. E. Kunz, J. G. Gordon, M. R. Philpott, and A. Girlando, *J. Electroanal. Chem.* **112**, 391 (1980).
136. M. Fleischmann, P. J. Hendra, and A. J. McQuillan, *J. Electroanal. Chem.* **65**, 933 (1975).
137. D. A. Zwemer, J. E. Rowe, C. V. Shank, and C. A. Murray, *C. R.—Conf. Int. Raman Spectrosc., 7th, 1980*, p. 414 (1980).
138. M. W. Howard and R. P. Cooney, *Chem. Phys. Lett.* **87**, 299 (1982).
139. C. S. Allen and R. P. Van Duyne, *Chem. Phys. Lett.* **63**, 455 (1979).
140. B. Pettinger and H. Wetzel, *Chem. Phys. Lett.* **78**, 398 (1981).
141. V. V. Marinjick, *J. Electroanal. Chem.* **110**, 111 (1980).
142. J. E. Pemberton and R. P. Buck, *J. Phys. Chem.* **85**, 248 (1981).
143. T. J. Sinclair, *C. R.—Conf. Int. Raman Spectrosc., 7th, 1980*, p. 408 (1980).
144. C. G. Blatchford, J. R. Campbell, and J. A. Creighton, *Surf. Sci.* **108**, 411 (1981).
145. H. Wetzel, B. Pettinger, and U. Wenning, *Chem. Phys. Lett.* **75**, 173 (1980).
146. J. R. Lombardi and R. C. Birke, *Surf. Sci.* **95**, L259 (1980).
147. M. E. Lippitsch, *Chem. Phys. Lett.* **79**, 224 (1981).
148. T. M. Cotton, S. G. Shultz, and R. P. Van Duyne, *J. Am. Chem. Soc.* **102**, 7960 (1980).
149. A. V. Bobrov, Ya. M. Kimel'fel'd, and M. Mostovaya, *J. Mol. Struct.* **60**, 431 (1980).
150. K. Manzel, W. Schulze, and M. Moskovits, *Chem. Phys. Lett.* **85**, 183 (1982).
151. M. Moskovits and D. Dilella, *C. R.—Conf. Int. Raman Spectrosc., 7th, 1980*, p. 394 (1980).

152. M. Kerker, *Pure Appl. Chem.* **56**, 1429 (1984).
153. P. C. Lee and D. Melsel, *J. Phys. Chem.* **86**, 3391 (1982).
154. J. A. Creighton, R. K. Chang, and T. E. Furtak, eds., *Surface Enhanced Raman Spectroscopy.* Plenum, New York, 1982.
155. C. Jennings, R. Aroca, A.-M. Hor, and R. O. Loutfy, *Anal. Chem.* **56**, 2033 (1984).
156. R. Aroca and R. O. Loutfy, *J. Raman Spectrosc.* **12**, 262 (1982).
157. C. J. Sandroff and D. R. Herschbach, *J. Phys. Chem.* **86**, 3277 (1982).
158. D. A. Weitz, S. Garoff, J. I. Gersten, and A. Nitzan, *J. Chem. Phys.* **78**, 5324 (1983).
159. L. Van Haverbecke and H. A. Herman, *Pergamon Ser. Environ. Sci.* **7**, 127 (1982).
160. S. K. Brahma, P. E. Hargraves, W. F. Howard, and W. H. Nelson, *Appl. Spectrosc.* **37**, 55 (1983).
161. L. Tosi and A. Garnier-Suilkrot, *J. Chem. Soc., Dalton Trans.* p. 103 (1982).
162. M. Z. Zgierski and M. Pawlikowski, *Chem. Phys.* **65**, 335 (1982).
163. L. V. Haley and J. A. Koningstein, *J. Phys. Chem.* **87**, 621 (1983).
164. G. H. Atkinson, ed., *Time-Resolved Vibration Spectroscopy.* Wiley, New York, 1983.
165. K. B. Eisenthal, R. M. Hochstrasser, W. Kaiser, and A. Laubereau, eds., *Picosecond Phenomena*, Part III. Springer-Verlag, Berlin, 1982.
166. D. H. Auston and K. B. Eisenthal, eds., *Ultrafast Phenomena*, Vol. 4. Springer-Verlag, Berlin, 1984.
167. J. Birman, H. Cummins, and K. Rebane eds., *Light Scattering in Solids.* Plenum, New York, 1979.
168. K. B. Eisenthal, ed., *Applications of Picosecond-Spectroscopy to Chemistry.* Reidel, Dordrecht, The Netherlands, 1984.
169. A. Lauberau and M. Stockburger, eds., *Time-Resolved Vibrational Spectroscopy.* Springer-Verlag, Berlin, 1985.
170. H. J. Bowley, S. M. H. Collin, D. L. Gerrard, D. I. James, W. F. Maddams, P. B. Tooke, and I. D. Wyatt, *Appl. Spectrosc.* **39**, 1004 (1985).
171. F. Ni and H. A. Scheraga, *J. Raman Spectrosc.* **16**, 337 (1985).
172. D. L. Gerrard and K. P. L. Williams, *Opt. Laser Technol.* **17**, 245 (1985).
173. T. Hirschfeld and B. Chase, *Appl. Spectrosc.* **40**, 133 (1986).
174. M. Kasha, *J. Chem. Phys.* **20**, 71 (1962).
175. D. L. Gerrard and W. F. Maddams, *Appl. Spectrosc.* **30**, 554 (1976).
176. C. H. J. Wells, *Introduction to Molecular Photochemistry*, p. 75. Chapman & Hall, London, 1972.
177. S. K. Freeman, *Applications of Laser Raman Spectroscopy*, p. 45. Wiley, New York, 1975.
178. H. J. Bowley and D. L. Gerrard, *Opt. Laser Technol.* **18**, 33 (1986).
179. M. Nichols, *Spectra-Phys. Laser Rev.* **4**, 1 (1977).

180. J. M. Harris, R. W. Chrisman, F. E. Lytle, and R. S. Tobias, *Anal. Chem.* **48**, 1937 (1976).

181. T. L. Gustafron and F. E. Lytle, *Anal. Chem.* **54**, 634 (1982).

182. A. B. Harvey, ed., *Chemical Applications of Non-Linear Raman Spectroscopy.* Academic Press, New York, 1982.

183. W. Kiefer and D. A. Long, eds., *Non-Linear Raman Spectroscopy and Its Chemical Applications.* Kluwer, Boston Inc., Hingham, Massachusetts, 1982.

CHAPTER

10

RAMAN SPECTROSCOPY OF CATALYSTS

MEHMED MEHICIC

BP Research
Warrensville Research & Environmental Science Center
Cleveland, Ohio

JEANETTE G. GRASSELLI

Department of Chemistry
Ohio University
Athens, Ohio

10.1. INTRODUCTION

10.1.1. Advantages of Raman Spectroscopy in Catalytic Studies

Raman spectroscopy is a very powerful technique for studying catalytic systems, including those used in the heterogeneous processes important in the chemical and petroleum/energy production industries. Many other spectroscopies are applicable to this area as well. However, Raman and infrared (IR) spectroscopies, which have complementary aspects, are among the most popularly used. The majority of the more than 400 papers published on the use of optical techniques to characterize heterogeneous catalysts during the period 1985–1988 focused on Raman and IR applications (1).

There are numerous reasons for the importance of Raman spectroscopy in characterizing catalytic materials and systems. In the design of a good catalyst, choice of support material and preparation conditions strongly affect the stability, activity, and selectivity of catalysts, and must be considered. Raman spectroscopy can play an important role in studying the surface layers of catalysts fabricated using different supports and conditions or in studying *in situ* preparations of catalysts, and thereby aid in finding optimum

Analytical Raman Spectroscopy, Edited by Jeanette G. Grasselli and Bernard J. Bulkin. Chemical Analysis Series, Vol. 114.
ISBN 0-471-51955-3

structures. A brief summary of these special advantages of Raman spectroscopy in heterogeneous catalysis follows. A number of important books contain excellent overviews of this area (2–10).

1. *Raman spectroscopy provides a molecular-level scrutiny of a catalyst surface (a key factor controlling chemical selectivity) and of the bulk structure of a supported catalyst.* In heterogeneous processes, supported metal oxides are widely used as catalysts (11, 12). Raman enables the identification of the structure, composition, and quantity of these metal oxides or other active metal catalysts as a function of surface coverage and preparation conditions (Fig. 10.1). It can provide information on catalyst/support interactions and on the nature and dispersion of active surface sites. Pyridine is often used to obtain information on the types of sites that are available on a surface; its frequency shifts according to whether it is adsorbed at a Lewis acid site or a Brönsted acid site, or whether it is hydrogen bonded to a surface OH group of an oxide (13). In addition to surface characterization, Raman can provide information on the bulk unsupported catalyst, impregnated bulk species, or on the support material in a supported catalyst provided the material is not highly opaque or refractive. Fresh catalysts can be studied as well as used, deactivated, and regenerated catalysts. Raman has a particular

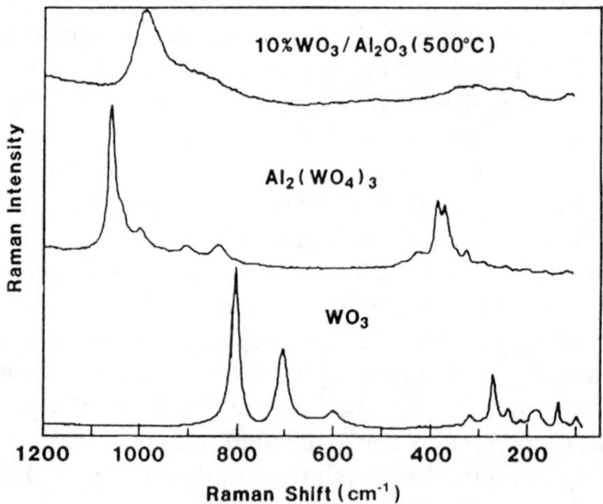

Figure 10.1. Comparison of surface oxides and crystalline oxides. Raman spectra of cystalline WO_3, crystalline $Al_2(WO_4)_3$, and 10% WO_3/Al_2O_3. Reprinted from Wachs et al. (11), with permission.

advantage over X-ray diffraction (XRD) because it can be used for amorphous as well as crystalline materials.

In profiling the entire solid catalyst, numerous techniques can be combined. For example, X-ray photoelectron spectroscopy (XPS), which characterizes the top few *atomic* layers, can be combined with XRD and IR spectroscopy for analysis of the *bulk*. Raman fits in between these for characterization of the top *molecular* layers. All of the available surface spectroscopies have been used to examine the structure of the surface and adsorbates; see Table 10.1 (14). Such knowledge about catalytic systems can provide a scientific basis for the design and improvement of tailored catalysts with long-term stability and activity.

Because Raman allows easy access to the low-frequency region ($< 1000\,cm^{-1}$), the vibratory motions of adsorbed or chemisorbed molecules have also been studied (15). These molecules may be reactants, products, or intermediates that exist only as surface species during activation of the catalyst or during a reaction. The structural changes and distribution of these species can be probed as conditions change, and a knowledge of reaction mechanisms can be obtained. Spectroscopic characterization of surface-adsorbed molecules on typical heterogeneous catalysts, which are generally strong scatterers, is a more difficult problem. But examples such as a study of the interaction of acetylene exposed to alumina and alumina-supported rhodium catalysts show the promise of the Raman technique (16).

Infrared spectroscopy pioneered the study of adsorbates and is still widely used (5, 9, 17, 18). Electron energy loss spectroscopy (EELS) and inelastic electron tunneling spectroscopy (IETS) are also now being used; the latter, however, is very sample limited. Raman has also had application as a resonance-enhanced (RRS) or surface-enhanced (SERS) process (19). (In the enhanced process, spectral interpretation may be more difficult than in the unenhanced process). Some of the newer, powerful methods emerging over the last decade or so for the study of adsorbed species include surface spectroscopies such as angle-resolved photoemission spectroscopy (ARPES), molecular beam surface scattering, and surface extended X-ray absorption fine-structure (SEXAFS), and stimulated desorption techniques such as temperature-programmed desorption using thermal gravimetric analysis (TGA) coupled with Fourier transform–infrared (FT-IR) spectroscopy.

2. *Inorganic materials active as catalysts or promoters are easily studied by Raman, whereas support materials (typically* SiO_2, Al_2O_3, etc.) *are poor Raman scatterers* and do not interfere with the study of the active phase. This is not the case in IR spectroscopy, where support materials absorb very strongly and the spectrum of the bulk support material may overwhelm the catalyst spectrum. However, routine FT-IR transmission spectra of

Table 10.1. Some of the Surface Characterization Techniques Frequently Utilized to Determine the Structure and Composition of Solid Surfaces[a]

Surface Analysis Method	Acronym	Physical Basis	Type of Information Obtained
Low-energy electron diffraction	LEED	Elastic back-scattering of low-energy electrons	Atomic surface structure of surfaces and adsorbates
Auger electron spectroscopy	AES	Electron emission from surface atoms excited by electron, X-ray, or ion bombardment	Surface compostion
High-resolution electron energy-loss spectroscopy	HREELS	Vibrational excitation of surface atoms by inelastic reflection of low-energy electrons	Structure and bonding of surface atoms and adsorbates
Infrared spectroscopy	IRS	Vibrational excitation of surface atoms by absorption of IR radiation	Structure and bonding of adsorbates
X-rays and UV photoelectron spectroscopy	XPS, UPS (ESCA)	Electron emission from atoms excited by X-rays or UV light	Electronic structure and oxidation state of surface atoms and adsorbates
Ion scattering spectroscopy	ISS	Elastic reflection of inert-gas ions	Atomic structure and composition of solid surfaces
Secondary ion mass spectroscopy	SIMS	Ion-beam-induced ejection of surface atoms as positive and negative ions	Surface composition
Extended X-ray absorption fine-structure analysis	EXAFS	Interference effects in photoemitted electron wavefunction in X-ray absorption	Atomic structure of surfaces and adsorbates
Thermal desorption spectroscopy	TDS	Thermally induced desorption or decomposition of adsorbates	Adsorption energetics and composition of adsorbates

Source: From Spencer and Somorjai (14).

[a] Adsorbed species present at concentrations of 1% of a monolayer can be readily detected.

reasonable quality can be obtained of supported catalysts using a high-pressure diamond cell even when transmission is extremely low (20). By use of Raman, supported catalysts can be studied in the form of powders or opaque pellets and do not require much preparation. Overall, Raman is the technique of choice for studying the molecular surface layers of supported catalysts, and especially for elucidating interactions between the catalyst and the support.

3. *Raman can be used in the diagnostics of mass transport phenomena* (*of reactants and products*) *in heterogeneous processes* by providing simultaneous determination of local values of the temperature and molecular concentration of gases at various distances from a gas–solid catalyst interface. There are no other noninvasive experimental methods for obtaining this type of information at present. Transport mechanisms influence the kinetics of a reaction and the nature and role of intermediate species, and thus it is important that they be understood. This was described for the catalytic hydrogenation of acetylene to ethylene over platinum (21). Oxygen isotopes were used to study mechanisms and growth of scale on chromia and iron–chromium spinels at high temperature (22).

4. *Aqueous systems such as impregnation solutions, slurries, and other catalyst preparation systems can be studied by Raman* because water is a poor scatterer and will not interfere with the species of interest.

In the preparation of heterogeneous catalysts, metal salts/oxides are usually dispersed on inorganic oxide supports. The metal species can be impregnated and/or adsorbed onto the supports as crystalline or, less likely, amorphous oxide phases, both of which are readily studied by Raman (12, 23). The preparation can call for an ion-exchange reaction or precipitation from an aqueous solution. *In situ* Raman can be readily applied to the study of the mechanism of precipitate formation and the development of crystallinity. The colloidal nature of the precipitate can cause problems for many other techniques. Schrader et al. discuss the formation of Bi/Mo catalysts as precipitates from aqueous solutions as a function of pH and other preparation conditions (Fig. 10.2) (24).

After preparation, the catalyst may be dried (in air), calcined, or heat treated (in O_2 at elevated temperature) and reduced (25, 26). Raman can be used as a probe during each of these steps.

Systems involving H_2 and other adsorbates can also be probed (27). Hydrogen is strongly chemisorbed on most transition metals, and it has been successfully investigated by Raman in a study of silica-supported Ni.

5. *An* in situ, *real-time, noninvasive Raman mode is easily executed, thus enabling one to study catalytic materials under reaction conditions*, including a broad range of temperatures and pressures and full flow of gas reactants. Corrosive environments can also be studied since *in situ* Raman cells are

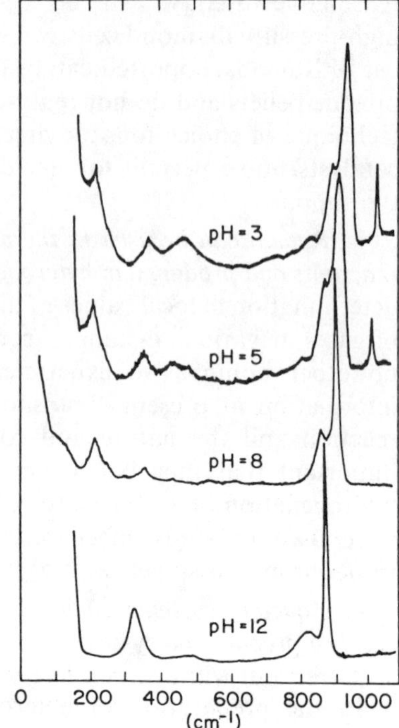

Figure 10.2. Raman spectra of molybdenum species in aqueous solution as a function of pH. Reprinted from Schrader et al. (24), with permission.

usually of quartz. Catalyst performance may be altered during *in situ* IR experiments, as catalysts must be compressed into thin disks so that one can obtain transmission spectra. This alters pore structure and the transport of gases to interior surfaces. *In situ* Raman offers advantages by avoiding this problem (28). The presence of reactant and product gases above the catalyst surface generally does not interfere with analysis of surface species (28, 29). Compared with Raman scattering from solids, scattering from gases is weak because the number of molecules per scattering volume is small.

In situ Raman spectroscopy of catalysts is very useful in obtaining reliable information about catalyst structure. Obviously, much structural information can be obtained *ex situ* also. However, when a catalyst is removed from a reactor for spectroscopic study its surface may be altered owing to exposure to new conditions. Raman scattering is very surface sensitive, and this must be taken into account for the typically opaque heterogeneous catalysts with high refractive indexes. *Ex situ* spectroscopy is suitable for bulk composition analysis and is used routinely to follow catalyst performance in commercial plants. *In situ* study offers a more detailed and realistic approach where the immediate response of the catalyst to changing conditions (temperature, feed composition, etc.) can be studied (30). Kinetic studies are also easily implemented using Raman data.

The design of an *in situ* cell is relatively simple because visible laser beams, typically used as the source in Raman spectrometers, are easy to manipulate. There are static cells, such as the one described by Mehicic et al. (31) for a backscattering experiment. This type of cell can be easily adapted for 90° scattering. However, backscattering is particularly convenient for work with opaque solids, especially when the laser power is properly adjusted and the beam is defocused. The price to pay is the longer acquisition time required.

When catalyst absorption and heating is a problem during an *in situ* experiment, one can do several things to minimize the problem. The deep red lines of a Kr^+ laser can be used (647.1, 676.4, 752.5, and 799.3 nm). They are generally not as damaging to the sample as the Ar^+ green/blue lines and can also be beneficial in reducing fluorescence. The disadvantage of using longer wavelengths for the source is loss of scattering intensity (which is the inverse fourth power of excitation wavelength) and reduced detector response. Another approach is to rotate the sample so that a fresh area is constantly exposed to the laser beam. Rotating sample cells have been reported (26, 32–34). One can also keep the sample stationary and move the laser beam. In a rotating lens design, the laser beam moves around the sample in a circle (35; 36). We should also mention here some similar designs using oscillating mirrors or prisms, with consideration given to slit orientation with respect to the cylindrical image of the scattering area. A high-pressure cell for use in both IR and Raman has been described (37). Commercial ones are also available (Aries, Inc., United States). A novel combination of Raman spectroscopy and a rotating analyzer ellipsometer was applied to elucidate states of surface carbon species and their hydrogenation on cobalt (38).

Two relatively recent advances provide additional capability for *in situ* real-time analysis. The first is spectral detection using an optical multichannel detection (OMA). This offers speed and multiplexing for a good signal-to-noise ratio and is especially useful for kinetics experiments. The second advance is Raman microspectroscopy using a microprobe (see item 6, below). *In situ* Raman microspectroscopy is also benefiting from the introduction of commercial hot stage cells (Linkham, Ltd., United Kingdom), which can be adapted for such spectroscopy.

A combination of analytical techniques can be used to elucidate in much detail the catalytic processes occurring in the *in situ* cell. The reaction products (gases) coming out of the cell can be analyzed by gas chromatography (GC), FT-IR, mass spectrometry (MS), etc., or by combined techniques (such as GC/MS, GC/IR, TGA/IR/MS, and MS/MS) at given intervals or continuously during on-line analysis. *In situ* Raman analysis of the catalyst combined with product analysis is a very powerful combination of techniques. It allows one to correlate input parameters (temperature, feed composition, etc.) with structure/phase composition of the solid catalyst and with product composition, and thus to obtain information on reaction parameters such as catalyst selectivity, conversion, specific rate, and activation energy. Another level of complexity can be unraveled with the study of molecular species (such as intermediate reactants) adsorbed on the catalyst surface. Knowledge of these species should allow complete understanding of a catalytic process. See Table 10.1 for a list of techniques for studying adsorbed species.

6. *The Raman microprobe is available for high spatial resolution experiments that provide important and detailed information on heterogeneous materials such as typical catalyst particles.* It uses an optical microscope and low laser power and has a spatial resolution of 1 μm. Although reasonably well established, Raman microspectroscopy is not yet widely employed in analytical laboratories. It should, however, be considered an important technique by the Raman spectroscopist. The normal Raman experiment yields a wealth of information that can be correlated to catalyst activation mechanisms, catalyst aging, structural changes related to selectivity and conversion, etc. The microprobe experiment, on the other hand, provides data on phase segregation having far-reaching implications for catalyst performance and analysis of "hot spots" that may degrade performance. Examples where a microprobe has been used include the segregation and identification of Co/Mo catalyst phases on alumina supports (6, 39, 40) and the identification of the types and distribution of carbon deposited on steam-reforming catalysts (41). Both the defocused Raman experiment and the microprobe should be used in studying heterogeneous systems. However, caution should be exercised so that samples are representative when using the microprobe for the chemical analysis of such systems.

In general, the power of Raman spectroscopy lies in the fact that it is a universal molecular structure tool. It is not limited to certain structures, states of matter, crystallinity, etc. There are some limitations, however, owing to the selection rules, weakness of the Raman effect,, and sample fluorescence when present. Fortunately, the effects of fluorescence can often be minimized, as described below.

Sample purification, surface treatment, or some form of quenching should be first considered, but in practice these are often impossible or impractical. The simplest method is to "burn off" fluorescing contamination by using as intense a laser beam as possible without damaging the sample. Frequency modulation and derivative spectra have also been successfully used to deal with the fluorescence of surface hydroxide ions (42). The most effective, way, however, is to use the newly developed near-IR FT-Raman technique (43, 44) where the energy of the excitation wavelength is outside the region in which fluorescence is excited. In addition, fluorescence and Raman processes can be time resolved based on the large difference of their time scales (microseconds for fluorescence versus femtoseconds for Raman). This however, is not an easy experiment.

Finally, it should be mentioned that fluorescence is usually less of a problem when Raman microprobes are used, for two reasons: one is that the focusing results in a more efficient "burn off," and the other is that almost always one can find a nonfluorescing spot on the sample.

Later sections of this chapter will focus on Raman studies of specific catalysts of industrial importance, particularly zeolites, hydrodesulfurization catalysts, and oxidation catalysts.

10.1.2. Miscellaneous Applications

Raman has been used in many areas where catalysts are important. In a study of acetylene polymerization on oxide surfaces, Raman has shown the formation of a range of different length polymers (45, 46).

Raman, along with ^{29}Si NMR (nuclear magnetic resonance), has provided information on the concentrations and structures of various hydrolyzed species during the gelation of a silicon alkoxide (the gelation technique and use of silica gels is important in the production of high-quality glasses and ceramics). The progress of reactions with and without chemical additives (used to change the rate of gelation, the release of solvent molecules, and final product homogeneity) can be monitored. Raman is able to provide information even after gelation when highly condensed oligomeric species exist; unlike NMR, which is limited to the early stages of the reaction (47).

Raman has been used to characterize the structures resulting during the preparation of Pd catalysts supported on silica and La oxide (these catalysts are active and highly selective in the synthesis of methanol) (48). Also triosmium catalysts were characterized; Os clusters, rather than dispersed Os, were found to be of catalytic importance in the isomerization of 1-hexene (49).

Raman is uniquely suited to supply information about metal–metal bonds in supported metal catalysts and can thus indicate when metals clusters have broken up into mononuclear metal complexes or aggregated to form crystallites (49).

Raman and other techniques were used to obtain information on crystal-field–related properties of Y_2O_3. In one test, the characteristics of rare earth ions (Eu^{3+}, Dy^{3+}, and Er^{3+}) doped into a C_2 site of the crystal Y_2O_3, were studied and used to predict the behavior of ions in the C_{3i} sites (50).

Layered solids, such as vermiculite (a layered silicate), some of which have advantages over zeolites in certain catalytic systems, have been studied by SERS Raman (51). Raman (SERS) has also been used in studying oxygen electrocatalysts (52), and it has been proposed that it can be used in studying *in situ* corrosion processes where metal is in contact with water or an electrolyte solution (53).

10.2. ZEOLITES

Catalytic cracking, i.e., the conversion of large petroleum molecules to smaller hydrocarbons, is one of the most important catalytic operations in the petroleum industry. Ninety percent of the world's catalytic cracking (as well as numerous other petrochemical processes) is done using zeolite catalysts (48). Zeolites are unique among the many commercial oxide catalysts in that they are crystalline. They are much more active than their amorphous predecessors used for catalytic cracking before the 1960s. They are composed of an aluminosilicate framework of the type $M_{x/n}[(AlO_2)_x(SiO_2)_y]\cdot zH_2O$ (where n is the charge of the metal cation, M^{n+}, which is usually Na^+, K^+, or Ca^{2+}). This framework can be used to support metals or metal complexes such as heavy metals used in heterogeneous catalysis. Of the 34 natural zeolites and some 100 synthetic ones, only a few are industrially important: zeolites A, X, and Y (which differ in their characteristic Si/Al ratios and therefore differ in their crystal lattice parameters), erionite, and synthetic mordenite (3). Therefore, most spectroscopic studies are confined to these structures.

Raman spectroscopy can be used to follow the synthesis of a zeolite, such as ZSM-5 (55, 56). The strength of Raman is in its applicability to aqueous solutions, solid gels, and crystallized products. In the experiment described by Baranska et al. (55), Raman spectra were taken of a starting aluminosilicate gel, then during crystallization, and then when the ZSM-5 catalyst was 100% crystallized (Fig. 10.3). A pH of >9 is required for nucleation, and a pH of <6 for crystallization (55). The alkalinity of the gel mixture is a function of the SiO_2/Al_2O_3 ratio and is a crucial synthesis parameter. The Raman spectrum of ZSM-5 product is also given by Chao et al. (56), where various bands are interpreted in terms of the vibrational modes of the skeletal framework. The kinetics and mechanisms of zeolite formation are of obvious

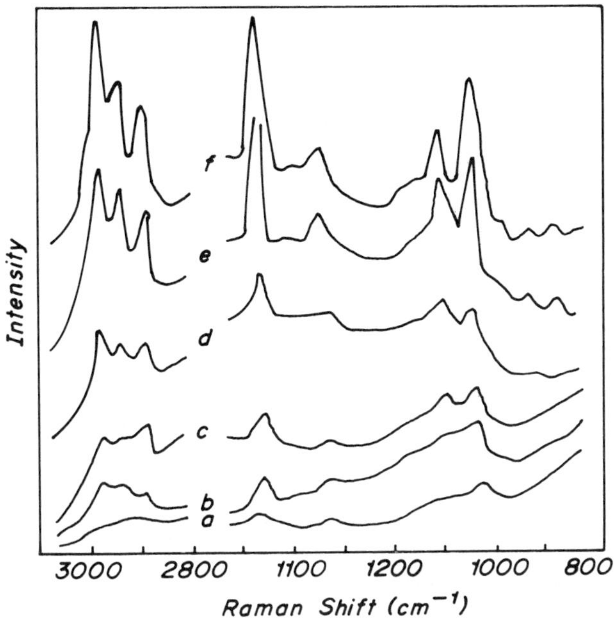

Figure 10.3. Raman spectra of (a) gel and ZSM-5 after crystallization with increasing degree of crystallinity; (b) 2%; (c) 18%; (d) 50%, (e) 77%; (f) 100%. Reprinted from Baranska et al. (55), with permission.

interest. Dutta and Puri (56a) also applied Raman spectroscopy to examine solution and solid phases present during various stages of ZSM-5 synthesis. They were able to characterize the alumino silicate zeolite framework which is distinguished from A, X, and Y zeolites by a Raman band characteristic of the five-member building units of ZSM-5. Other workers used Raman spectroscopy as one of the major analytical tools applied (57–63). An excellent illustration is the study of the crystallization of zeolites A, X, and Y (57). Raman spectroscopy of a solution formed during zeolite crystallization indicates the disappearance of $Al(OH)_4^-$ species at 621 cm^{-1} (Fig. 10.4) before the appearance of any zeolite structure (Fig. 10.5). This suggests the formation of soluble polymeric aluminosilicate complex aggregates, which provide the nuclei for zeolite crystal growth. A solution transport mechanism for the formation of four- and six-member-ring versus five-member-ring systems is proposed. More recent work (64) has determined that $Al(OH)_4^-$ ions are necessary in order for the transformation of aluminosilicate gel to zeolite A

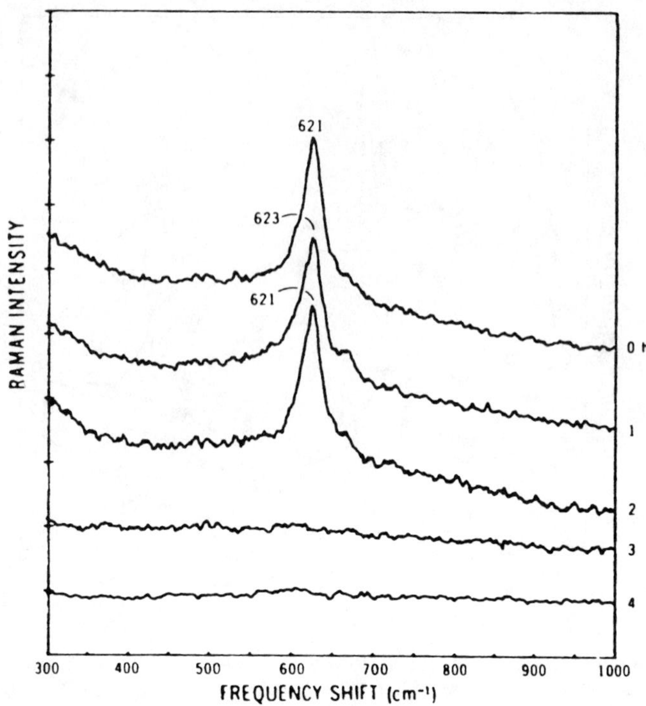

Figure 10.4. Raman spectra of liquid phase in zeolite A synthesis. From Roozeboom et al. (57).

to proceed. During nucleation, the gel (consisting of randomly connected four-member rings) reorganizes its structure by interaction with these ions. A systematic study of the Raman spectra of the gels and of the solid phase during crystallization (Fig. 10.6) allowed band assignments to be made. Results were confirmed by X-ray diffraction and ^{27}Al NMR (64). A similar Raman spectroscopic study, combined with ^{29}Si NMR (65), of the synthesis of zeolite Y indicates the presence of six-member aluminosilicate rings during the nucleation period. These are formed either upon aging or in the presence of a high concentration of Si in solution. For zeolite Y, six-member aluminosilicate rings connect to each other via four-member silicate rings, and two such units connect to form a sodalite cage (65). For zeolite X, the crystallization process favors the formation of four-member aluminosilicate rings as intermediate species.

Organic cations, which are often used in the synthesis of zeolites, can be trapped inside various parts of the framework of the zeolite, thus having a chemical effect as well as a template effect (shaping the structure). Raman

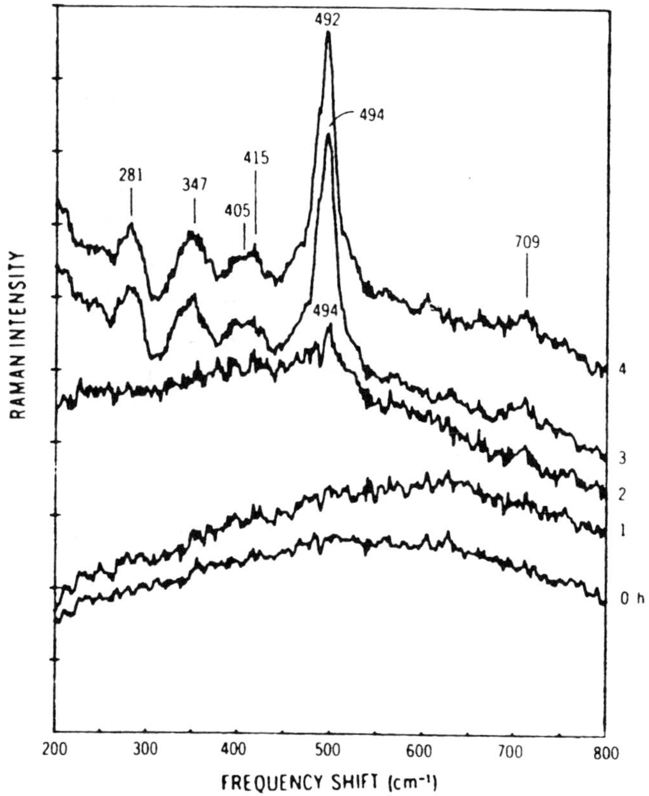

Figure 10.5. Raman spectra of solid phase in zeolite A synthesis. From Roozeboom et al. (57).

spectroscopy can be an effective tool for their study. A good example of this is the study of tetramethylammonium ion and its role in the synthesis of ZK-4 (66). Aluminosilicate bonding in zeolites and the effect of cation exchange is also described by Dutta and Del Barco (67). Hydrated zeolite A was completely exchanged with Li^+, Na^+, K^+, Tl^+, and NH_4^+. There is evidence that Li^+ distorts the zeolite framework whereas the others do not. There is strong hydrogen bond formation between the cation and the lattice oxygens in the case of NH_4^+.

The aspects of zeolite structure that seem important in catalysis are acidity, pore size, and electrostatic fields (54). Raman is useful in studying molecules adsorbed on zeolite surfaces which were treated to reduce fluorescence (68). The electrostatic fields in the supercage of alkali metal–exchanged X and Y zeolites were probed using adsorbed benzene (see Ref. 54, additional

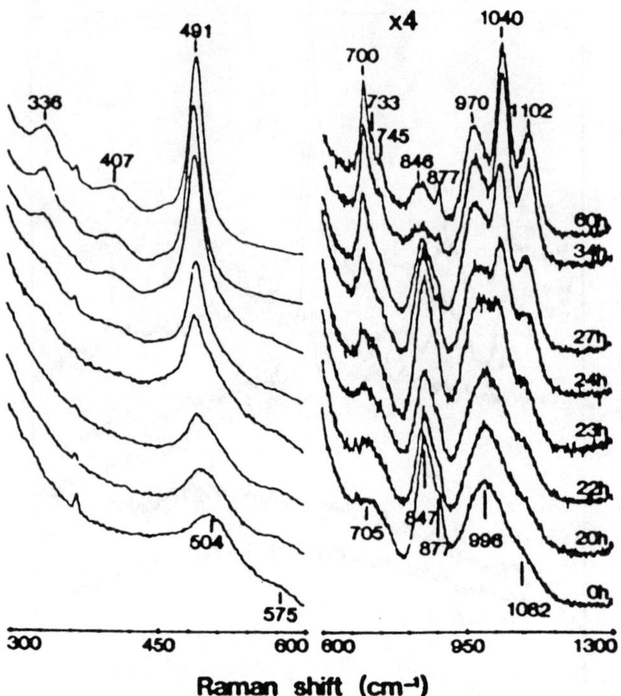

Figure 10.6. Raman spectra of the solid phase during zeolite A synthesis at 50°C at various times (starting composition $8.6Na_2O1Al_2O_31SiO_2556H_2O$). Reprinted from Dutta and Shieh (64), with permission.

references cited therein). Both the electronic and the vibrational energy levels of benzene are perturbed during adsorption as a function of cation size and benzene loading (Table 10.2). Higher field strengths at the active adsorption sites result from crowding of larger cations into the fixed geometry of the zeolite supercage. This helps explain the unique activity of X-type zeolites for the alkylation of toluene to styrene with methanol. The frequency shifts of the Raman active breathing mode of benzene were related to selectivity.

The study of adsorbed species also contributes to elucidation of the active sites in zeolites. Raman spectroscopy is well suited for studying adsorption, activation, and reactions of molecules on zeolite surfaces (46, 69, 70). Charge-transfer interactions between the oxide framework and the adsorbed species, and shielding of sites by the oxide framework were explored in a study of bromine adsorption on zeolites X, Y, and A (71). The aromatic bromination of benzene is reported by Krasser et al. (72). Detailed analysis of adsorption sites and the bromination reaction was possible.

Zeolites containing transition metals are important in hydrogenation,

Table 10.2. Comparison of Trends in Electronic and Vibrational Spectra as a Function of Changes in Cation Size, SiO_2/Al_2O_3 Ratio, and Benzene Loading

System perturbation	Result (at one C_6H_6/supercage) Electronic ($\pi - \pi^*$)	Vibrational (ν_1)
Increase cation radius ($Na^+ \rightarrow Cs^+$)	Red shift[a]	Red shift
Increase SiO_2/Al_2O_3(X → Y)	(Blue shift)[b]	Blue shift
Increase C_6H_6 loading	Red shift	Blue shift
Increase electrostatic field at C_6H_6 (Y site II → X site III)	(Red shift)[b]	Red shift

Source: From Freeman and Unland (54).

[a] A red shift is a shift to lower wavenumber or energy.

[b] Regarded as tentative because diffuse reflectance data were extrapolated back to one molecule per supercage.

hydrocracking, and hydroisomerization (73, 74). The introduction of heavy metals into zeolites can, in principle, happen in several different ways. There can be interstitial (intracavity) substitution or direct lattice substitution. Raman spectroscopy should be a good technique for the study of these processes. However, spectra of metal-substituted zeolites were no different than those of unsubstituted zeolites (75). This could be either due to low loading of heavy metals (below detection limits) or because the metal ion is held by adsorption or interstitial substitution with little perturbation of the basic zeolite structure. This latter interpretation agrees well with recent EXAFS (extended X-ray absorption fine-structure) work (73, 76).

A major problem often encountered by workers in zeolite Raman spectroscopy is strong fluorescence. It has been suggested that heavy-metal impurities such as Fe^{3+}, Cr^{3+}, and Mn^{2+} are the culprit (59).

Finally, theoretical studies such as a normal coordinate analysis with different force fields for pentasil structure zeolites (77) are of great interest for the spectroscopic understanding of zeolite structures. An additional review of zeolite spectroscopy is included in Moeller et al. (78).

10.3. HYDRODESULFURIZATION CATALYSTS

Hydrodesulfurization (HDS) is probably the most important of the hydroprocessing methods because of current emphasis on environmental issues such as acid rain. In this process sulfur is removed from petroleum feedstocks or coal products. A typical catalyst is Mo or W oxide supported on γ-alumina (γ-Al_2O_3) for petroleum processing or on SiO_2–Al_2O_3 for coal liquefaction processes. Cobalt or nickel act as promoters (3). The HDS catalyst is activated by sulfiding, which is usually accomplished by exposure to an H_2S/H_2 mixture at elevated temperatures. Similar catalysts are also used for hydrodenitrogenation (HDN).

Experimental procedures in the preparation of the catalysts are extremely important, and Raman spectroscopy provides a tool for studying these in relation to catalyst activity. The level of metal loading, pH of the metal

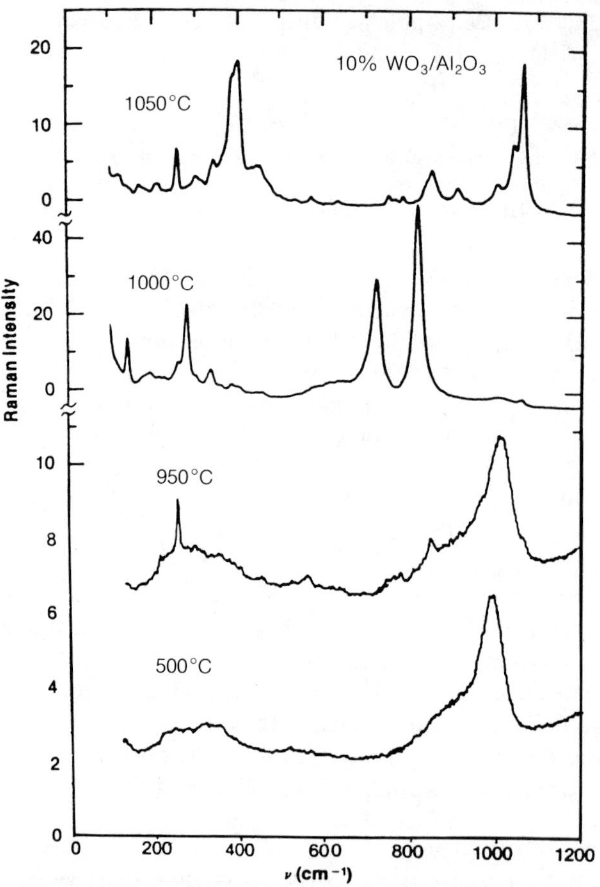

Figure 10.7. Raman spectra of 10% WO_3/Al_2O_3 as a function of calcination temperature. Reprinted from Wachs et al. (11), with permission.

solutions, type and source of the support, temperature and method of calcination, percentage of promoter, degree of hydration, etc., all influence the structure of the resulting catalyst. The changes in the Raman spectra as each of these parameters varied, especially in experiments *in situ*, have resulted in a large body of literature on the assignments of the various bands in the spectra. Good reviews are found in a number of sources (2, 12, 79–82).

Although a complete consensus on the structures that are present still does not exist, the Raman spectrum can definitely discriminate between different oxide states. This is shown in Fig. 10.7 for 10% WO_3/Al_2O_3 as a function of calcination temperature (11). The sample at 500°C exhibits a strong 986 cm^{-1} band (W=O) characteristic of the supported tungsten oxide as distinguished from crystalline WO_3 (see Fig. 10.1). As the sample is heated to 950°C, bands in the low-frequency region appear which are due to θ-alumina. This indicates a phase change of γ-Al_2O_3 to the θ-Al_2O_3. The high-frequency shift of the 986 cm^{-1} band is attributed to an increase in surface density of WO_x species. At 1000°C crystalline WO_3 and crystalline $Al_2(WO_4)_3$ are observed (see Fig. 10.1), and at 1050°C all phases convert to

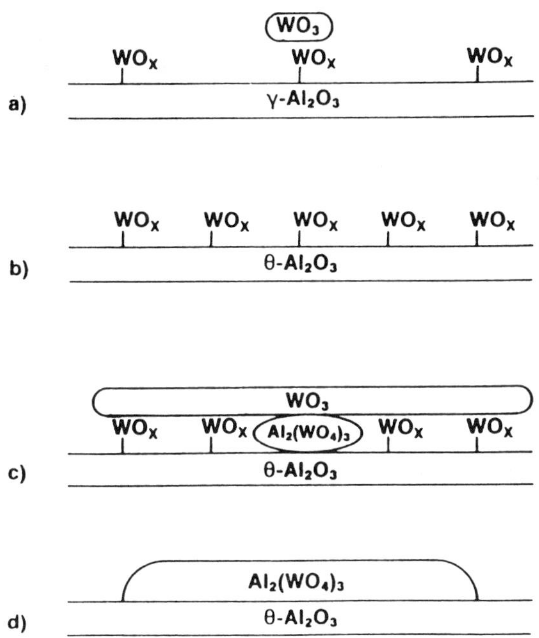

Figure 10.8. Structural transformations of 10% WO_3/Al_2O_3 system with calcination temperature. Structure after heating to (a) 500°C, (b) 950°C, (c) 1000°C, and (d) 1050°C. Reprinted from Wachs et al. (11), with permission.

the thermally stable $Al_2(WO_4)_3$. These transformations are schematically shown in Fig. 10.8 (11) and are typical of adsorbed surface metal oxides on an oxide support. These systems are susceptible to hydration and dehydration, which also affect the order in the surface oxide layer and cause perturbation in the Raman spectrum. Similar studies with 8% NiO on Al_2O_3 revealed that the surface nickel oxide shows only Ni—O bands at $\sim 550\,cm^{-1}$ and no sensitivity to water. It is concluded that the surface nickel oxide is absorbed into defects in the surface of the alumina support by diffusion. This is discussed more fully in the section on promoted Mo catalysts.

In the following subsections, some historical perspective and the current state of understanding for several specific HDS catalyst systems will be reviewed.

10.3.1. MoO_3-Based Catalysts

10.3.1.1. Background and Preparation Variables

MoO_3-based catalysts are typically promoted by CoO and sometimes by NiO. The support is typically γ-Al_2O_3 and sometimes SiO_2/Al_2O_3.

The earliest Raman investigators (83) of a 10% MoO_3/γ-Al_2O_3 catalyst (the percentage of Mo oxide loading is based on weight percent of the metal oxide) were unable to obtain a Raman spectrum of any active phase/support interaction species, and they postulated the formation of a new, Raman-inactive compound.

Brown and Makovsky (84) followed with a study of a commercial CoO–MoO_3/Al_2O_3–SiO_2 catalyst used in the PERC/ERDA coal liquefaction SYNTHOIL process. Good spectra were obtained by rotating the samples during analysis. For the unused, unsulfided catalyst, they found no evidence for bulk (molecular) MoO_3, but there were bands that they assigned to bridged Mo—O—Mo (877 cm^{-1}) and to terminal Mo—O (948 cm^{-1}) structures. An unused, sulfided catalyst showed molecular MoS_2 (409 and 384 cm^{-1}), which was also evident in a used catalyst from a coal liquefaction reactor.

In subsequent work, Brown and co-workers (85) studied a series of commerical Ni-promoted catalysts with varying MoO_3 loading on Al_2O_3 supports. A Co-promoted coal liquefaction SYNTHOIL process catalyst was also included. The goal was structural characterization of the catalyst surface and identification of the metal oxide–support "interaction species" postulated by numerous workers (86–89). In this work Brown et al. paid no attention to the role of the promoter; its type and concentration were not taken into consideration in interpreting spectral changes. Because commercially available catalysts were used, preparation details were not known (yet these are variables that influence catalyst activity and spectral character). At high

MoO_3 loading (30%) only bulk MoO_3 was observed. At intermediate MoO_3 loading (18%) a sample promoted with 3.5% NiO showed spectral characteristics of bulk MoO_3 (but these were less intense than those of the 30% loaded sample discussed above) and a broad band centered at approximately 950 cm^{-1}. This feature was interpreted as an octahedrally coordinated Mo—O fundamental stretching frequency. It was presumed to be due to an interaction species between the MoO_3 and the support.

Catalysts with low MoO_3 loading (10% and 15%) showed no bulk MoO_3, suggesting that the surface structure consists of a monolayer of MoO_3 that retains the properties of bulk MoO_3 (90–92). Unpromoted 10% MoO_3 catalyst and Co-promoted 15% MoO_3 SYNTHOIL catalyst show, in addition to the 950 cm^{-1} band, a tetrahedrally coordinated molybdate ion, MoO_4^{2-}.

Further structural work in this area combining Raman spectroscopy with other surface and bulk spectroscopies [XPS, XRD, ISS (ion scattering spectroscopy)] was carried out (93, 94). In a series of MoO_3/γ-Al_2O_3 catalysts, several distinct molybdenum species were observed depending on molybdena loading (93). The results were in general agreement with the aforementioned earlier work by Brown and colleagues (84, 85). As a result of these and other publications (95–103) the structures presumed to be present in MoO_3–Al_2O_3 systems are (1) at low Mo loading (less than 10%)—tetrahedral Mo species (890–920 cm^{-1}, depending on degree of distortion); (2) at medium Mo loading—a polymolybdate phase (strong 950 cm^{-1}); and (3) at high Mo loading—free MoO_3 (996 and 820 cm^{-1}) and/or (4) $Al_2(MoO_4)_3$ ($\sim$ 1000 cm^{-1}). The dispersion of these species on the surface of the catalyst is a function of the preparation conditions (96–104), as mentioned previously.

The extreme importance of calcination conditions and exposure to water vapor for the structures present on a catalyst surface was recently discussed by Stencel et al. (105, 106) and Payen et al. (107) for Mo catalysts on SiO_2 and on Al_2O_3, by Chan et al. (108) for WO_3, MoO_3, and V_2O_5 on TiO_2 and Al_2O_3, and by Horsley et al. (109) for WO_3/Al_2O_3. Band shifts as large as 60 cm^{-1} for the Mo=O stretching mode (950–1000 cm^{-1}) indicate changes in structure of surface species with calcination temperature and with exposure to moisture (Fig. 10.9) (105). The changes are reversible (Fig. 10.10) (108). The effect of laser power at the sample produces similar changes owing to local calcination and dehydration. This was effectively demonstrated by Payen et al. using the Raman MOLE (Molecular Optics Laser Examiner) microprobe (Fig. 10.11) (107). Stencel et al. (106) point out the extreme care which must be exercised in interpreting the Raman spectrum and the necessity of carefully controlled experimental conditions (Fig. 10.12) (106). A global interpretation of the Raman bands of supported molybdenum oxide and reduced species has been presented by Payen et al. (110). Changes in the spectra are attributed to chemical effects such as ligand heterogeneity,

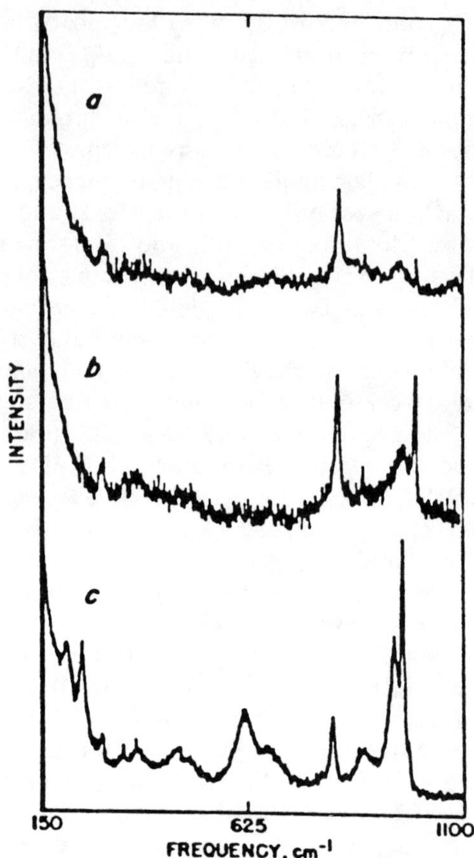

Figure 10.9. Raman spectra of impregnated 8% Mo/SiO_2 catalyst: (a) air stabilized; (b) *in situ* O_2 calcined at 500°C; (c) H_2O exposed. Reprinted from Stencel et al. (105), with permission.

coordinative heterogeneity, and oxidation number of Mo ions. The terminal Mo—O band wavenumber shifts are related to these effects.

Note that γ-Al_2O_3 is used exclusively as the support in commercial petroleum HDS processes. Silica-supported catalysts are inactive for this reaction, although SiO_2 and TiO_2 supports are used for oxidation catalysis (see below) and SiO_2–Al_2O_3 is used for coal products. This emphasizes the critical importance of studying the structures that form at support/catalyst interfaces under various conditions in order to obtain a better understanding of their relationship to catalytic activity. Raman spectroscopy has been a unique tool for such studies.

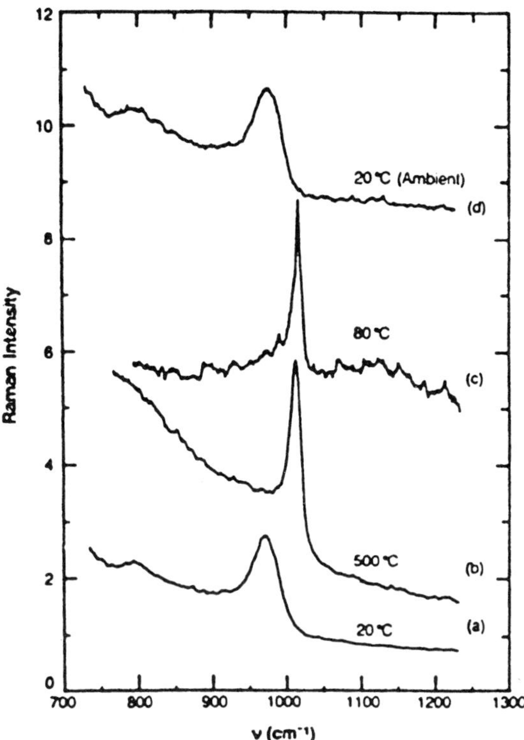

Figure 10.10. Raman spectra of 7 wt.% tungsta on titania (Degussa): (a) *in situ* initial conditions at room temperature; (b) *in situ* calcination at 500°C in dry air; (c) *in situ* postcalcination at 80°C in dry air; (d) postcalcination with exposure to ambient air for 12 h at room temperature. Reprinted from Chan et al. (108), with permission. Copyright 1984, American Chemical Society.

10.3.1.2. Promoter (Cobalt-Containing) Structures

HDS, HDN, and hydrogenation reactions all benefit from promotion of the metal oxide catalysts with Co (or Ni, discussed in the next subsection) (3). Promoter chemistry is very complex and much less understood than that of the basic Mo catalyst. Numerous theories about the role of the promoter have been proposed. Massoth (79) summarizes these theories and says that Co promotion increases Mo monolayer area, improves the ease of Mo reduction, increases H_2 mobility, prevents catalyst deactivation, enhances surface segregation, and prevents MoS_2 crystallization; moreover, its activity is due to Co intercalation of MoS_2, MoS_2/Co_9S_8 synergism, or specific kinetic effects. It is clear that the chemical state of Co (or Ni) and its dispersion in the catalyst structure are critical factors.

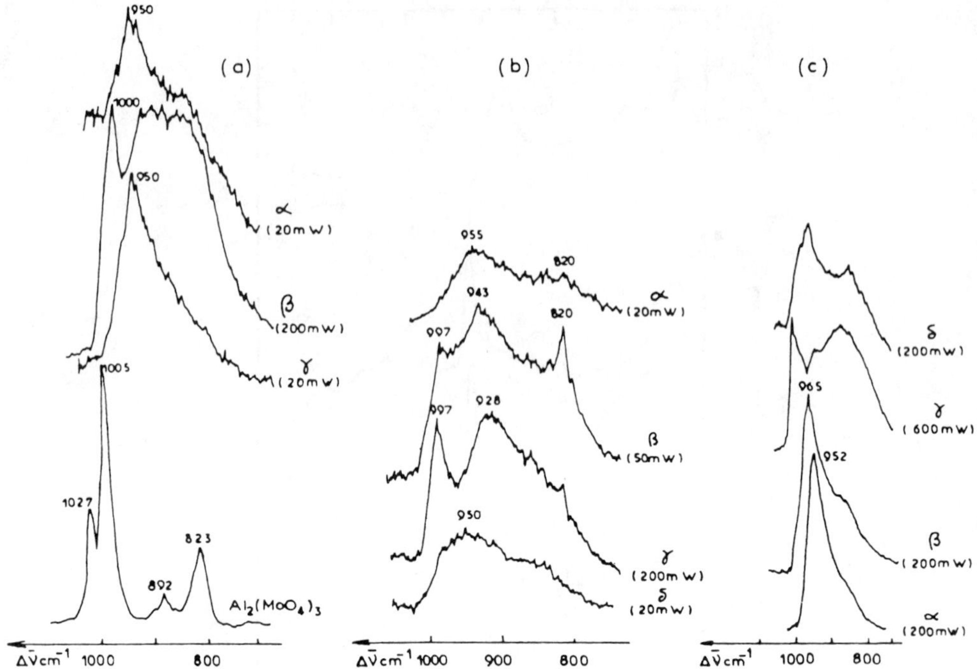

Figure 10.11. Effect of the laser power on the Raman spectra of different catalysts: (a) 3.6 Ni 14 Mo catalyst; (b) 3 Co 14 Mo catalyst; (c) 14 Mo catalyst. In each case, the laser power at the sample is given in parentheses. To check the reversibility of the phenomenon, the Greek letters refer to the order of the successive records. In (a), the spectrum of $Al_2(MoO_4)_3$ is given for comparison. From Payen et al. (107).

Makovsky et al. (94) used Raman spectroscopy in addition to several other techniques to study promoter structures in a $CoO-MoO_3/\gamma-Al_2O_3$ catalyst series with fixed MoO_3 loading (15%) and varying concentrations of CoO (0–8%). They found that at CoO concentrations below 6%, the catalyst contained $CoMoO_4$ and irreducible Co^{2+} ions in a tetrahedral coordination. With a further increase in the CoO concentration (7–8%), crystalline Co_3O_4 appeared, coinciding with an increase in Mo reducibility and a decrease in surface area. Payen and co-workers (95) reported similar findings in their study of structure–concentration relationships. At low Co loading, a tetrahedral Co^{2+} was present; and at higher concentrations, Co_3O_4 clusters were observed.

Makovsky et al. (94) summarized the state of understanding of a $CoO-MoO_3/\gamma-Al_2O_3$ catalyst. Molecular three-dimensional species were observed, including $CoMoO_4$, Co_3O_4, MoO_3, $CoAl_2O_4$, and $Al_2(MoO_4)_3$

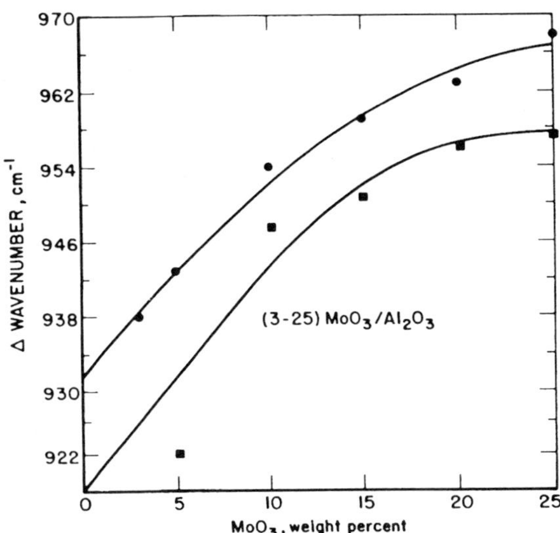

Figure 10.12. Plot of the band position of the molybdenum–oxygen stretching frequency as a function of molybdenum concentration. *Key*: (●) spectra acquired in air within 3 days of catalyst synthesis; (■) spectra acquired 3 months after catalyst synthesis. From Stencel et al. (106).

(93, 97, 111). In addition, two-dimensional species were reported that involved MoO_3–Al_2O_3 and Co–Mo interactions (85, 112, 113). Also there were dispersed cobalt and molybdenum ions with coordination symmetry dependent on the available γ-Al_2O_3 surface sites (93, 114, 115). The appearance of these species is a function of concentration as discussed above. It is also influenced by preparation parameters, as has been discussed, such as calcination conditions (93, 111), acidity of impregnating solutions, and the sequence of impregnation (116–118), and the condition of the γ-Al_2O_3 support prior to impregnation (94).

10.3.1.3. Promoter (Nickel-Containing) Structures

The Ni-promoted catalyst is similar to the Co-promoted catalyst in its performance, but much less research has been reported on it. Jeziorowski and Knoezinger (99, 100) reported on the oxide form of NiO–MoO_3/γ-Al_2O_3. The calcined catalyst, without the NiO promoter, was characterized by a two-dimensional polymolybdate (97, 101) and molecular MoO_3. After impregnation with $Ni(NO_3)_2$, there was a reaction of these two Mo phases to form Ni^{2+} heteropolymolybdates. Drying and calcination of the impregnated sample at elevated temperatures created conditions for

MoO_3–Al_2O_3 interaction (101), which distorted the Ni^{2+} species, resulting in broadened Raman spectra. The interaction between Ni^{2+} and molybdenum phases is strong since Ni^{2+} is pulled out from the support surface spinel lattice where it is located after $Ni(MoO_3)_2$ impregnation. The aforementioned authors postulated that the role of the Ni promoter is to increase the formation of a two-dimensional polymolybdate structure and to decrease the tendency to form molecular MoO_3. In subsequent work (119), the same authors proposed that the sequence of impregnation is important and influences the distribution of Ni^{2+} between the support spinel matrix and molybdate monolayers, which are proposed to be forming on the support surface. Ion scattering studies (119, 120) indicate that Ni is in the topmost layer of the catalyst. Ni-promoted and -unpromoted MoO_3/Al_2O_3 catalysts and the numerous analytical methods used in their structural characterization have been reviewed by Knoezinger (121).

Raman was used to study the impregnation solutions and surface structures of Mo–Ni–P hydrogenation catalysts. (The petroleum-processing industry has long used Mo–Ni/γ-alumina as a hydrogenation catalyst. It was recently improved by the addition of P.) Up to this point little has been understood about the chemical nature of these catalysts. The composition of the impregnation solutions is an important factor in determining the resulting catalyst activity. To prepare P–Mo catalysts, an impregnation solution consisting of a P–Mo–O anion (aqueous solution) can be used. The Raman spectra of three anions ($PMo_{12}O_{40}^{3-}$, $P_2Mo_{18}O_{62}^{6-}$, and $P_2Mo_5O_{23}^{6-}$) were obtained (this work can be used to identify the anions present at various preparation conditions, specifically pH or Mo/P atomic ratio). The catalysts prepared from these anionic solutions respectively were found to have decreasing activity in thiophene desulfurization. The catalysts all had Raman peaks at 950 cm^{-1}; however, the peak strength varied and was found to correlate with catalytic activity. It was proposed by Wang et al. (122) that this peak is related to surface-active structures on the catalyst. When Ni is added to P–Mo structures it is "entrapped" in the P–Mo–O network and the catalyst containing Ni has lower activity than the P–Mo catalysts. The Mo–Ni catalyst system was also studied. Raman spectra indicate that the formation of the surface-active structure is stimulated by the addition of P. When P is added, the structure is maintained at a range of temperatures (122).

Chan and Wachs (123) studied NiO supported on γ-Al_2O_3. They conclude that Ni is *absorbed* into the alumina support forming a NiO–Al_2O_3 surface spinel structure. It is characterized by a Raman band in the 300–800 cm^{-1} region that undergoes large thermal broadening, indicating very strong interactions, and by relative insensitivity to moisture. Low-oxidation-state metal oxides (Ni^{2+}, Cu^{2+}, Co^{2+}) can be accommodated by the γ-Al_2O_3 lattice to form surface spinels. The authors suggest that oxides with high oxidation states (Re^{7+}, W^{6+}, Mo^{6+}, Cr^{6+}, and V^{5+}) cannot be accommodated by γ-Al_2O_3 and are *adsorbed* onto the support surface. These systems are characterized by the Raman metal–oxygen (M=O) stretching band in the 900–1100 cm^{-1} region as well as by moisture sensitivity.

10.3.1.4. Activated (Sulfided) Molybdenum-Containing Catalysts

A sulfided form of the CoO–MoO_3/Al_2O_3 catalyst is the active form for hydrodesulfurization and hydroconversion of coal. In addition to some of the aforementioned references, several deal specifically with the sulfided form (84, 114–126). Sulfiding generally involves heating the catalyst (400–450°C) under H_2S/H_2 gas for several hours. There is no agreement as to which oxide form of the catalyst provides the most active sulfided form. There is also disagreement as to which sulfided species is catalytically active. Raman spectra are weak and difficult to obtain. Brown et al. (84, 124) reported the formation of MoS_2 in sulfided samples and no oxygenated molybdenum species, in particular, tetrahedral MoO_4^{2-}. This supports a model proposing that molecular MoS_2 is the active form of sulfided catalysts (90, 127, 128). Other proposed models are the monolayer model (129) and the intercalation model (130). After mild sulfiding, Payen et al. (131) find a $WS_3(MoS_3)$-like phase and oxysulfide. After complete sulfiding, $WS_2(MoS_2)$ crystallites were observed. Brown et al. also reported the presence of Mo—S species in used catalysts, the spectroscopic observation being very difficult owing to large amounts of an ordered high-carbon material. Regeneration of used and presulfided catalysts (600°C under air) did not yield MoO_4^{2-} species found in fresh catalysts. Resulfiding of regenerated catalysts yielded species whose Raman spectra were characteristic of the presulfided form.

Schrader and Cheng (125) reported very careful studies of species sulfided at different temperatures. For samples sulfided by 10% H_2S in H_2 at 400°C, MoS_2 structures were reported, as they were by Brown and colleagues in the aforementioned work (84, 124). However, when stepwise heating (150°, 250°, and 350°C) was employed, different species were observed at each step: Mo-oxysulfide, reduced molybdate, and a surface MoS_2 compound (Table 10.3) (126). Reoxidation modified the oxysulfide and sulfide structures. As a result of this work, the authors proposed an interesting model that combines anion vacancies with terminal and bridging Mo—S linkages (Fig. 10.13) (125). Follow-up *in situ* work by the same authors (126) was particularly concerned with the behavior of individual oxide phases under reduction and sulfiding conditions. These phases were $CoMoO_4$, Co_3O_4, MoO_3, and aggregated or polymeric molybdate phases, which react at different rates, with $CoMoO_4$ being the least reactive.

Reduction is also important in the activation of the catalyst. Payen et al.

Table 10.3. Band Positions for Stepwise Sulfiding

Sample	Region (cm^{-1})				
	0–200	201–400	401–600	601–800	801–1000
Co(3)Mo(5)					
150°C		325 w, br	480 w, br	690 s	820 w 940 s 950 sh
250°C		330 w, br	467 w 525 vw		820 w 940 w
350°C		350 m	407 w		
Co(3)Mo(7.5)					
150°C		354 w, br	468 w, br	700 m	820 w 940 s
250°C	135 m	335 w, br	466 m		940 s
350°C		360 m, br	405 m 465 w		940 m
Co(3)Mo(10)					
150°C		290 m, br 345 m, br	445 w 530 w	700 m	821 w 940 s 950 sh
250°C		325 m, br	460 w, br		940 s
350°C		222 w, br	407 br		940 m
Co(3)Mo(15)					
150°C		325 w, br	445 m 530 m, br		819 m 940 s 998 vw
250°C		325 m, br	445 m	700 w	940 s
350°C		370 m, br	405 m	700 w	940 m

Source: From Schrader and Cheng (126).

(131) report on various pretreatments of the oxide catalysts, followed by H_2 reduction. They find dehydration to be the most important reaction in producing an active catalyst. Stencel et al. (106) and others (93, 102, 105–107, 132) found that water content of the catalysts greatly influenced the Raman spectra (see previous discussion). At low Mo loading, samples were very difficult to reduce. At higher Mo concentrations, two consecutive reduction steps were identified: formation of reduced molybdenum species as Mo(V)—OH and the formation of a bridge structure Mo(X)—O—Mo(IV), where X = V or IV, the latter being characterized by the 760 cm^{-1} stretching mode.

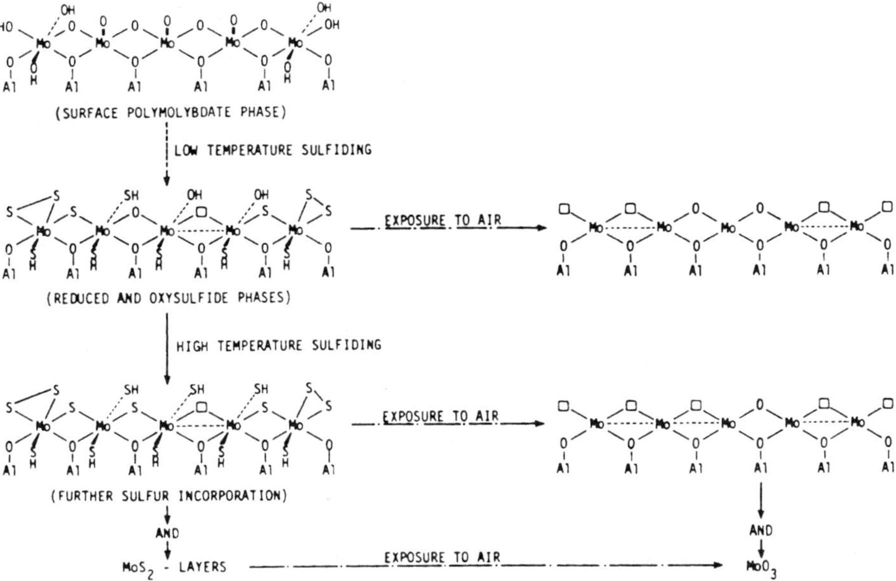

Figure 10.13. Model for partially sulfided and reduced Mo/γ-Al_2O_3. From Schrader and Cheng (125).

Abdo et al. (132a) used HCl and HBr for reduction and noted a complex situation resulting on the catalyst surface. Based mostly on electron paramagnetic resonance (EPR) data, they postulated the existence of isolated Mo(V) species and Mo(V) species coupled via bridging oxide ligands.

Cheng et al. (133) studied *in situ* a series of alumina-supported MoO_3 catalysts containing 1.25–20% molybdates on γ-Al_2O_3, showing the evidence of aggregated molybdate species in impregnated dried samples. Upon calcination, MoO_3 is clearly produced in samples with higher molybdenum loadings. Also there was an indication of aggregated or polymolybdate species.

10.3.1.5 Conclusion

The reader is referred to numerous reports in the literature dealing with preparation parameters and structural characterization: (a) different oxide supports (e.g., TiO_2, CeO_2, or ZrO_2) (98, 116, 134–137); (b) the effect of impregnation sequence for Co-promoted (119) and Ni-promoted MoO_3/γ-Al_2O_3 (117); (c) different preparation methods (104, 138); (d) hydration/dehydration of the surface and calcination (93, 102, 105–107, 132); (e) structural comparison of NiO–MoO_3/Al_2O_3 and CoO–MoO_3/Al_2O_3 (103), and comparison of MoO_3/Al_2O_3 and WO_3/Al_2O_3 (139); (f) oxospecies on Al_2O_3 (140); (g) surface sites using pyridine as a molecular probe for acidity (39, 102, 141); and (h) model hydrodesulfurization reactions on thiophene (39, 142) and hydrogenation of *o*-xylene (143).

A Raman microprobe has been used with success in the study of HDS catalysts (39, 95, 143–145). Littlejohn and Chang (146) describe a study of sulfur- and nitrogen-containing species in a wet flue gas desulfurization and denitrification system, a type of analysis that could be useful in monitoring HDS/HDN processes. A rapid and elegant method for quantitative determination of crystalline MoO_3 and WO_3 over a wide range of concentrations using Raman has been developed (147). It uses a KNO_3 internal standard, thereby eliminating many potentially troublesome experimental parameters (e.g., laser power, focusing, or sampling volume).

10.3.2. WO_3-Based Catalysts

WO_3-based catalysts are important not only in hydrodesulfurization but in olefin metathesis (148) as well (see Section 10.5.1, below). There is a considerable volume of published literature on the WO_3/Al_2O_3 catalyst (11, 149) and the WO_3/SiO_2 catalyst (139, 149).

Salvati et al. (150) found WO_3/Al_2O_3 catalysts to be structurally similar to MoO_3/Al_2O_3 catalysts. At low WO_3 loading, the catalyst surface contained tetrahedral WO_4^{2-} interaction species, whereas loadings over 15% led to octahedral WO_3-like interaction structures. Above monolayer coverage (approximately 24% WO_3), bulk WO_3 was detected, rather than $Al_2(WO_4)_3$; which some investigators believed existed (151–153) but others did not (149, 154, 155). The question of $Al_2(WO_4)_3$ formation is not resolved. Thomas et al. (155) showed that, in coprecipitation of aluminum nitrate and ammonium metatungstate, $Al_2(WO_4)_3$ is formed only at a high calcination temperature of 1100°C. At lower calcination temperatures, 550° and 900°C, Raman signals for WO_3 dominate. Reduction studies were also carried out by Salvati and co-workers in an attempt to estimate the degree of metal–support interaction by comparing the reducibility of the catalyst with that of bulk metal oxides. Bulk WO_3 could be reduced to metallic tungsten with H_2 at 550°C, whereas the interacted tungsten species (at submonolayer WO_3 loading) could not be reduced. In the Ni-promoted catalyst, Ni species were incorporated into the tungsten monolayer, displacing some of the tungsten interaction species. These species could be almost completely reduced to Ni metal.

Work by Chan and co-workers [108, 156–158] was based on estimating Raman scattering cross sections for different tungsten species. Crystalline

WO_3 is a very strong scatterer (approximately a factor of 32 better than $Al_2(WO_4)_3$ and 160 times that of tungsten oxide surface species on alumina), which could account for some of Thomas and co-workers' observations on the effect of calcination temperature (155). A high scattering cross section for crystalline WO_3 supported on SiO_2 was also found by Thomas et al. (139) using aluminum nitrate as an internal standard. As a result of the above findings, Chan et al. disagreed with Salvati's (150) proposal of WO_3-like octahedral structures. They concluded that there are crystalline and amorphous structural transformations occurring in the WO_3/Al_2O_3 system and that these are influenced by calcination temperature and the surface density of the tungsten oxide species on the Al_2O_3 surface. At lower WO_3 loading (below monolayer), a highly dispersed and amorphous surface complex is formed. With increasing temperature, the alumina surface area decreases, leading to a close-packed monolayer surface WO_3 complex. At still higher temperatures, distinct oxide phases are formed, namely, WO_3 and $Al_2(WO_4)_3$. The $Al_2(WO_4)_3$ phase is formed from the reaction of WO_3 crystallites and the support. The approximate detection limits are 0.1 wt.% for WO_3 on Al_2O_3 and 1 wt.% for $Al_2(WO_4)_3$ on Al_2O_3. The high-dispersion model is supported by previous work by Murrell et al. (159) and Thomas et al. (149) on WO_3/SiO_2 and WO_3/Al_2O_3 systems, indicating small clusters of polymeric units.

Stencel and co-workers (160) extensively studied the effects of O_2 calcination and H_2O–D_2O exposure on the WO_3/Al_2O_3 catalyst. These studies, along with $^{18}O/^{16}O$ isotope substitution studies, enabled them to propose the assignment of the fundamental vibrations of surface tungstate and the mechanism of the interaction of H_2O and surface tungstate. The effects of H_2O adsorption on the surface structure were also reported by Chan et al. for WO_3/Al_2O_3 and WO_3/TiO_2 systems (108) and by Horsley et al. for WO_3/Al_2O_3 (109). The work by Horsley et al. included X-ray absorption near-edge spectroscopy (XANES) as well as Raman spectroscopy.

This powerful combination of techniques provided information that allowed the latter investigators to conclude that the catalyst contained a surface complex where the supported WO_3 has formed a polymeric structure on the alumina support composed of WO_4 and WO_6 units joined in infinite chains. The XANES spectra can distinguish between tetrahedral and octahedral surface tungsten oxide complexes and can monitor the role of coordinated water. The Raman spectra provide detailed molecular structure information and were interpreted from classical group theory assignments of bands for tetrahedral, octahedral, and bridged M—O—M structures: ν_{asym} at 860 cm^{-1} and ν_{sym} at 250 cm^{-1} for linear bridged structures and at 750 and 500 cm^{-1}, respectively, for the bent structure. A bending mode near 200 cm^{-1} is also observed. In general, the position of the highest Raman

Table 10.4. Highest Wavenumber Raman Band of Tungsten Oxide Compounds

Compound	Wavenumber (cm^{-1})	Compound	Wavenumber (cm^{-1})
Octahedra		Tetrahedra	
Li_6WO_6	740	$CaWO_4$	913
WO_3	805	Cs_2WO_4	920
$FeWO_4$	856	Na_2WO_4	928
$CoWO_4$	886	$[WO_4]^{2-}_{aq}$	931
$MnWO_4$	886	$BaWO_4$	940
$NiWO_4$	893	$Fe_2(WO_4)_3$	1026
$CuWO_4$	908	$Al_2(WO_4)_3$	1060
H_2WO_4	951		
$(NH_4)_6H_2(W_3O_{10})_4$	980		

Source: Reprinted from Horsley et al. (109) with permission. Copyright 1987, American Chemical Society.

band reflects the shortest W—O bond (highest bond order). Thus, regular tetrahedral groups (WO_4) will show Raman bands at higher frequencies than octahedral (WO_6) groups. However, distortions in the structures will change the bond orders and shift the Raman bands, resulting in much overlap. Table 10.4 gives the highest wavenumber Raman band of some WO_3 compounds (109). Note that only the octahedra have bands below 910 cm^{-1} and only the tetrahedra have bands above 980 cm^{-1}. Again, the above discussion points to the caution that must be exercised when interpreting Raman spectra, while recognizing the unique capability of Raman for spectra–structure correlations.

Iannibello et al. (161) reported that the addition of aqueous ammonium paratungstate solutions to γ-Al_2O_3 resulted in a large pH increase in the liquid phase (suggesting strong interactions between W(VI) and the basic sites of γ-Al_2O_3) and the formation of aluminum tungstate–like surface species. Electronic spectra indicated tetrahedrally coordinated chemisorbed W(VI), which is in agreement with Raman spectra suggesting isolated W(VI) oxo species in tetrahedral coordination. Tetrahedral coordination was increasingly distorted as the tungsten content was increased. The addition of Ni(II) in the solution enhanced the chemisorption of W(VI). Otherwise chemisorption was limited to 18–20 wt.% WO_3 on commercial γ-Al_2O_3 without the addition of Ni. There was no evidence of a $NiWO_4$ phase at NiO concentrations of 2.5 wt.% on a 25.5 wt.% WO_3/γ-Al_2O_3 catalyst.

The nature of the surface species generated during the sulfurization of tungsten-based catalysts was studied (162). The formation of oxide, oxysulfide, WS_3, and WS_2 was sensitive to the presence of water or Ni.

10.4. OXIDATION CATALYSIS

Oxidation catalysis is used in a variety of processes to produce a range of industrially important chemicals (3, 7). The processes range from the oxidation of simple molecules like SO_2 and CO, to the selective or partial oxidation of hydrocarbons in which aldehyde, carboxyl, epoxy, or nitrile groups can be added, and to energy conversions in living organisms that are catalyzed by metal-containing enzymes. Industrial catalysts are typically complex transition metal oxides on a variety of supports.

10.4.1. Vanadium Oxide Catalysts

Vanadium oxide is a widely used catalyst in the oxidation of SO_2, CO, and hydrocarbons (163, 164)). Structural studies involving Raman spectroscopy have been reported in several publications (165–175).

Roozeboom and co-workers described the structure of V ions in the V(V) oxide/γ-Al_2O_3 system (165) as a function of preparation method (wet impregnation versus ion exchange of hydroxyl groups on the support surface), monomeric/polymeric vanadate ions (as described in Ref. 165a), and V surface coverage, i.e., concentration (up to about 5 wt.%). Interpretation of spectra was based on comparison with known V compounds. In Fig. 10.14 the spectra of V_2O_5 and V_2O_4 are shown (172). Bulk V_2O_5 exhibits a sharp Raman band at 997 cm^{-1} that is due to the symmetrical stretching mode of the terminal V=O. Structures present on the wet impregnated catalyst surface were found to be concentration dependent (Fig. 10.15) (165). At low V coverage there were isolated tetrahedral VO_4 structures (830 cm^{-1}) and a two-dimensional polymeric network of distorted vanadate octahedra (970 cm^{-1}). At higher coverage V_2O_5 crystallites (997 cm^{-1}) were found. An advantage of Raman spectroscopy over XRD for such systems is that Raman detects small crystallites as well as amorphous structures. The monolayer catalyst prepared by ion exchange had a more uniform surface and consisted mainly of an octahedral polyvanadate network along with a small amount of isolated monomeric tetrahedral units. There was no evidence of $AlVO_4$ in either preparation. The investigators recognized the difficulties in both qualitative and quantitative interpretations owing to large differences in the scattering cross sections; V_2O_5 is a much better scatterer than other V structures. In this respect the situation is very similar to that of the WO_3/γ-Al_2O_3 HDS catalyst, described earlier in this chapter.

Roozeboom and co-workers also studied and compared V_2O_5 supported on a variety of oxides: γ-Al_2O_3, CeO_2, Cr_2O_3, SiO_2, TiO_2, and ZrO_2 (166). As in the first study, the effect of preparation method and V concentration (1–40 wt.%) was investigated, as well as the effect of heat treatment. Formation

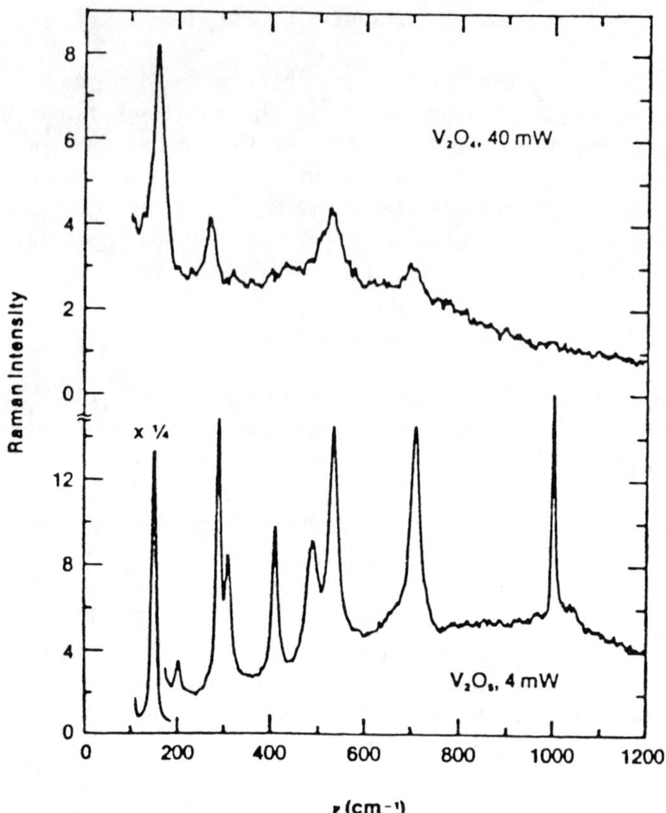

Figure 10.14. Laser Raman spectra of V_2O_5 and V_2O_4. Reprinted from Wachs and Chan (172), with permission.

of V surface structures was found to be dependent on V concentration, preparation method, and support type. At low concentrations the two-dimensional polymeric vanadate phase was observed for all supports and for both preparation methods. At medium and high concentrations differences showed up: wet impregnated samples had V_2O_5 crystallites, whereas ion-exchanged samples did not. An exception was the SiO_2-supported samples, where V_2O_5 crystallites were found in both preparations. Heat treatment (373–1073 K) was generally found to change the nature of the surface vanadate monolayer, however the γ-Al_2O_3 support provided good monolayer stability. Stability decreased in the order γ-$Al_2O_3 > TiO_2 > CeO_2$. Also, with heat treatment, there was an indication of ZrV_2O_7 formation on the ZrO_2 support, solid solution formation on TiO_2,

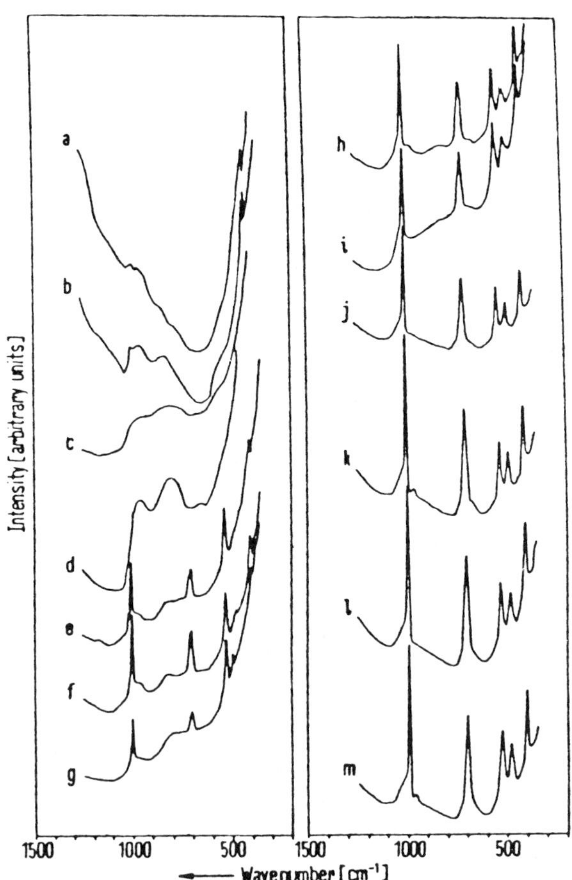

Figure 10.15. Raman spectra of impregnated vanadium (V) oxide/γ-Al_2O_3 catalysts and V_2O_5. (a) 0.5% V (by weight); (b) 1.1% V; (c) 1.7% V; (d) 2.1% V (e) 3.7% V; (f) 4.0% V; (g) 4.7% V; (h) 5.2% V; (i) 7.0% V; (j) 15.0% V; (k) 33.3% V; (l) 35.6% V; (m) V_2O_5. (All percentages are percentages by weight.) Reprinted from Roozeboom et al. (165), with permission.

and dispersion of crystalline V_2O_5 in the ion-exchanged SiO_2-supported sample.

Van Hengstum et al. (167) addressed preparation methods, supports, and Raman characterization of bulk V_2O_5 in considerable detail. An impressive set of papers by Wachs and co-workers (168–175) correlates the structural changes in V_2O_5/TiO_2 (anatase) catalysts with catalytic activity for the oxidation of *o*-xylene to phthalic anhydride. They point out that V_2O_5 species coordinated to the TiO_2 support constitute the active site, and complete monolayer surface coverage is necessary for optimum conversion and selectivity. The Raman spectra can be used to follow the development of such surface vanadia species for different loadings, preparation conditions, and calcination temperatures. The sharp feature at 997 cm^{-1} due to the V_2O_5 crystallites is replaced by a broad Raman band between 850 and 1000 cm^{-1} for the polymeric vanadia species. For good examples of thermal effects on Raman spectra, see Murrell et al. (159). With increasing temperature, reversible thermal broadening of bands from crystalline materials was noted. Band sharpening and shifting occurred for amorphous materials. This was attributed to the desorption of water from the surface.

Oyama et al. (175a) have shown that the determination of the active surface area of supported and unsupported V_2O_5 by oxygen chemisorption is sensitive to the temperatures used for sample reduction and oxygen adsorption. Raman spectroscopy confirms that a portion of the vanadia dispersed on SiO_2 is present as isolated vanadyl species bonded to the support. The remainder of the vanadia is present as crystallites of V_2O_5 that are too small to be detected by X-ray diffraction. $^{18}O/^{16}O$-exchange experiments demonstrate that at 640 K oxygen chemisorption probes the isolated vanadyl groups but that vanadyl groups in the interior of V_2O_5 crystallites are not affected.

Hanuza and co-workers (176) reported on the structural determination of V_2O_5/MgO catalysts used for oxidative dehydrogenation of ethylbenzene to styrene. Complete assignments of IR and Raman bands were made. The position of the terminal V=O band was related to bond order, band length, and vanadium oxidation state. Bands due to V—O—V bridge structures were also assigned.

10.4.2 Manganese Oxide Catalysts

Mn/Al_2O_3 catalysts are active in oxidation reactions, particularly for CO oxidation, where they are usually promoted by CuO or CoO. Strohmeier and Hercules (177) reported the only Raman study on this catalyst (in combination with ISS, XRD, and XPS), although numerous analytical techniques were previously employed (see references cited in 178). The authors encountered difficulties with reference compounds MnO, α-Mn_2O_3, β-MnO_2, α-Mn_3O_4, and $MnAl_2O_4$ since only α-Mn_3O_4 and $MnAl_2O_4$ have reasonable Raman spectra while the others have only broad features or no features at all. Computerized signal averaging or multiplexing, sometimes useful in analyzing poor scatterers, was used successfully with MnO_2 (178). Strohmeier and Hercules showed that Raman did not identify α-Mn_3O_4 and $MnAl_2O_4$ in the catalyst, therefore supporting XRD findings. Combination of data from all the techniques indicated very weak interaction of manganese with the support since there was aggregation to bulk-like manganese oxide even at low metal loadings. Calcination temperature had a strong effect on catalytic structures: at 400°C β-MnO_2 was found on the catalyst, at 600°C α-Mn_2O_3 was found, while at intermediate temperatures a mixture of the two was observed. Strohmeier and Hercules also showed that the Mn/Al_2O_3 catalyst (i.e., supported manganese oxides on the surface) had reduction and sulfidation properties similar to those of the bulk oxides. H_2/500°C reduction yielded MnO, whereas H_2S/H_2/500°C sulfidation yielded α-MnS.

10.4.3. Bismuth Molybdate Catalysts

The most important catalysts in selective oxidation/ammoxidation of olefins are based on bismuth molybdates. This goes back to the original discovery of $Bi_9PMo_{12}O_{52}$ supported on SiO_2 by Veatch, Callahan, and co-workers (179–181). This catalyst was used for the one-step synthesis of acrylonitrile from propylene, ammonia, and air in a fluidized-bed reactor. The technology was revolutionary and is still the dominant one for this industrially very important reaction. Today's catalysts are very complex multicomponent compounds containing a variety of metals (like Co, Ni, and Fe) and promoters such as K or P. But they are "bismuth molybdate" in nature, are redox catalysts, and exhibit high selectivity and conversion.

Numerous mechanisms have been proposed by various authors: Matsuura and Schuit (182–184), Haber and co-workers (185, 186), Sleight and co-workers (187, 188), Keulks and co-workers (189), and Grasselli and co-workers (190–195). Several good reviews on mechanistic aspects are available (3, 189, 196–202).

There are some generally accepted features of the mechanism. A catalytic redox cycle involving two metal centers was first suggested by Mars and van Krevelen (163). A hydrocarbon molecule is oxidized by a lattice oxide ion, O^{2-}, on one metal cation site, M_1^{n+}. Hydrocarbon leaves, partially oxidized, while the metal is reduced by two electrons. The second metal cation, M_2^{m+}, accepts these two electrons, thus oxidizing the first cation back to the original state. Electrons are further used to reduce molecular oxygen (which is part of the reactant feed) to oxide ions, thus replenishing lattice oxygen. In this step, the second cation is oxidized back to the original state. Therefore, the cycle is complete. Mechanisms suggested by Grasselli, Haber, Sleight, and Keulks suggest that Mo is the M_1^{n+} center while Bi is M_2^{m+}; Matsuura reverses these roles.

In the case of propylene oxidation, in the first step, propylene is chemisorbed, an α-H is abstracted, and a symmetrical, resonantly stabilized allyl radical is formed. To produce a product, such as acrolein (or acrylonitrile), an oxygen (or nitrogen) insertion takes place from another catalyst surface site. Finally, the lattice oxygen is regenerated by dissociative chemisorption of molecular oxygen and reduction to oxide ions.

The role of lattice defects on reaction kinetics was studied by Brazdil and co-workers (203) among others. Matsuura and co-workers (204) studied the acidic properties of molybdate-based catalysts and the influence on selective propylene oxidation.

Much of the analytical work, including Raman, involves identification of different bismuth molybdate phases (α-$Bi_2Mo_3O_{12}$, β-$Bi_2Mo_2O_9$, γ-Bi_2MoO_6, γ'-Bi_2MoO_6) and molybdates formed with promoters and metal cations. Studies of sites responsible for activity or selectivity have used isotope labeling to support some mechanistic hypothesis. There are numerous works reporting vibrational spectra (in particular, Raman) of these compounds and the correlation of structural features with their activity and selectivity (205–218). Structures were also studied using X-ray diffraction (219–222), neutron diffraction (223–226), and the X-ray absorption methods of EXAFS and XANES (226, 227). Although the structures are well understood, there are difficulties with the detailed interpretation of the vibrational spectra because of the complexity of the structures. Lattices are distorted in a sense that there is no clear definition of six-fold coordination (octahedral) and four-fold coordination (tetrahedral), although these designations are commonly used in the discussions. In addition, the spectroscopic and structural data on many oxo-molybdenum and some oxo-bismuth cations (3, 228–247) are not as useful as one may expect. One of the reasons lies in symmetry. Free complex cations in solution have relatively high symmetry and therefore degenerate normal modes of vibration and simpler spectra. The site symmetry of the cation in the crystalline lattice is typically much lower than the molecular point group symmetry of free cations, therefore lifting degeneracy (site splitting) and resulting in more complex spectra. On top of that, the number of atoms in the unit cell is considerable, adding to the complexity. Other effects and interactions change frequencies, intensities, and shapes of the spectral bands. There is overlap between frequency regions typical of particular molecular groups, creating more difficulties in sorting out observed bands. The most thorough vibrational analysis is given by Matsuura and co-workers (207). It includes normal coordinate analysis and models involving different types of distorted tetrahedra. Some general useful groupings of vibrational bands can be assigned. Figure 10.16 and Tables 10.5 and 10.6 show IR, Raman, and X-ray data for the bismuth molybdate phases.

Kovats (202) presents specific structural groups of interest in the vibrational assignment of bismuth molybdates and an overview of this area. The highest frequency bands are associated with Mo=O stretching (990 cm^{-1} in MoO_3) and symmetric and antisymmetric stretching of O=

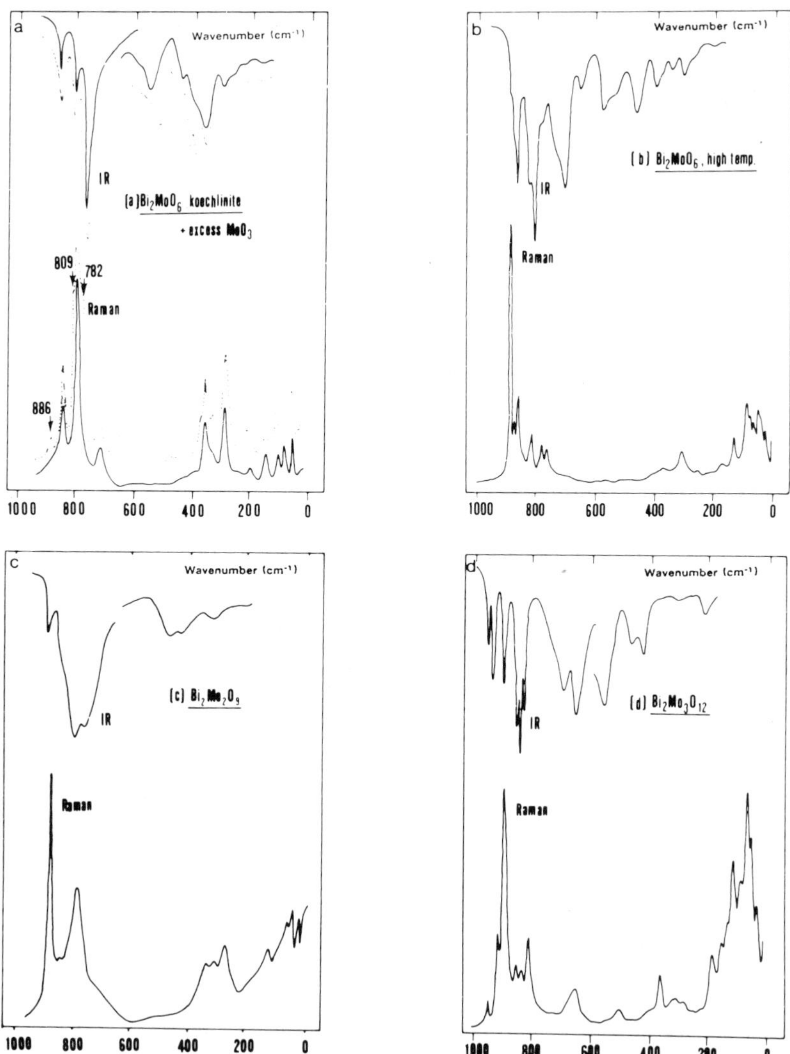

Figure 10.16. Infrared and Raman spectra for Bi-molybdates: (a) γ-Bi_2MoO_6; (b) γ'-Bi_2MoO_6; (c) β-$Bi_2Mo_2O_9$; (d) α-$Bi_2Mo_3O_{12}$. Reprinted from Matsuura et al. (207), with permission.

MO=O structures in α-$Bi_2Mo_3O_{12}$ and β-$Bi_2Mo_2O_9$. For α phase, the 900–950 cm^{-1} region was assigned to ν_{sym} (955, 919, 897 cm^{-1}), and the 800–850 cm^{-1} region to ν_{asym} (853, 836, 811 cm^{-1}) by comparison with tetrahedral MoO_4^{2-} anion. The multiplicity of bands, three for each type of symmetry, comes from the fact that there are three crystallographically different

Table 10.5. Infared and Raman Spectra of Bismuth Molybdates

Bi_2MoO_6 (γ)		$Bi_2MoO_6 + 4\%\ MoO_3$ (extra bands)		Bi_2MoO_6 (γ')		$Bi_2Mo_2O_9$ (β)		$Bi_2Mo_3O_{12}$ (α)	
IR	Raman	IR	Raman	IR	Raman	IR	Raman	IR	Raman
842 m	843_{30}		886_{16}	890 sh	892_{100}	890 m	892_{100}	948 m	955_{10}
798 w	799_{100}	832 sh		880 sh	883_{30}	840 sh	845_{30}	931 m	919_{30}
735 vs	715_{15}		809 sh	868 vs	864_{41}	800 s	793_{60}	902 m	897_{82}
600 w		760 sh	782 sh		851_{23}	770 s		860 s	853_{20}
		710 w		825 m	822_{37}	740 m	715_{10}	847 s	836_{18}
				810 vs		710 sh		830 m	811_{28}
				780 w	787_{25}			715 m	
					770_{29}			670 s	654_{12}
				740 m					
				710 s	695_{25}				
				645 m					
570 m		545 shifted		562 m		500 sh		570 m	
		from 570		527 w				550 sh	
450 m				467 w		473 m		470 m	
						446 m		440 m	
								415 sh	
408 m		408 st		391 w	399_{9}	409 m		392 m	
390 sh	399_{5}	390 st		380 w	375_{13}	392 w		350 w	360_{18}
352 s	354_{9}	358 st		335 w	359_{10}	362 m	356_{15}	314 m	317_{9}
304 m	326_{15}			300 w	315_{25}	325 vw	323_{16}	303 vw	302_{3}
287 w	295_{13}			290 vw	303_{27}	275 m	286_{21}	276 vw	282_{8}
277 vw	278_{18}				258_{16}	266 w	260_{5}	262 vw	248_{5}
268 vw					226_{9}		180_{2}		
248 vw					206_{9}				
	194_{5}								

Source: Reprinted from Matsuura et al. (207), with permission.

Table 10.6. Crystallographic Data for Bismuth Molybdates

	Bi_2MoO_6 Koechlinite (γ)	Blasse (γ')	$Bi_2Mo_2O_9$(β)	$Bi_2Mo_3O_{12}$(α)
Crystal system	Orthorhombic	Orthorhombic	Orthorhombic	Orthorhombic
Monoclinic space group	$Pca\ 1_1$	$P\ 2_1/c$	$P\ 2_1/c$	$P\ 2_1/c$
Coordination of Mo^{6+}	Distorted octahedral	Tetrahedral	Trigonal-bipyramidal or tetrahedral	Trigonal-bipyramidal or distorted tetrahedral
Mo—O bond length (Å)	Bi_2MoO_6 (γ)	$Bi_2Mo_3O_{12}$ (α)		
	γ-	α_1-	α_2-	α_3-
	Mo—O_1 1.86	Mo_1—O_4 1.69	Mo_2—O_1 1.72	Mo_3—O_{12} 1.68
	—O_4 2.24	—O_5 1.72	—O_9 1.73	—O_{11} 1.78
	—O_4 1.76	—O_2 1.85	—O_6 1.86	—O_8 1.86
	—O_5 2.24	—O_{10} 1.92	—O_3 1.89	—O_7 1.87
	—O_5 1.75	—O_{10} 2.25	—O_8 2.13	—O_6 2.30
	—O_6 1.93			

Source: Reprinted from Matsuura et al. (207) with permission.

molybdenum atoms in the unit cell. In the β phase, there is one type of O=Mo=O structure and therefore much simpler spectra where the band at around 890 cm^{-1} can be assigned to ν_{sym} and that at around 845 cm^{-1} to ν_{asym}. Lower-order bonds, such as the Mo—O—Mo bridging structure, are expected to vibrate at the lower frequency. In MoO_3, the higher frequency band at 885 cm^{-1} is assigned to ν_{asym} and the lower frequency band at 815 cm^{-1} to ν_{sym} (239). In γ-Bi_2MoO_6, the assignment for stretching of the Mo—O—Mo bridge is 845 cm^{-1}, ν_{asym} [Mitchell and Trifiro (205) assign this mode to the terminal Mo=O structure, but the crystallographic bond distances are more in agreement with the lower-order bond], and 790 cm^{-1}, ν_{sym}. In the β phase, this structure is indicated in the 700–800 cm^{-1} region. In the α phase, this bridging structure has bismuth cation close to the oxygen anion, which could lower its frequency to the observed 600–750 cm^{-1} region. There are at least three bonds, maybe more, in this region. This could be an indication of other bonding, such as Bi—O—Mo. Note that the frequency of Mo—O—Mo stretching is lower in bismuth molybdates than in MoO_3. Obviously, structures are "looser," perturbed by the addition of bismuth cation, and distorted. The general range of the Mo—O—Mo bridge vibration can be given as 840–900 cm^{-1} for asymmetric stretching and 700–820 cm^{-1} for symmetric stretching.

The situation in the deformation range, below 500 cm^{-1}, is far more complex and far less understood. Kovats (202) indicates the 320 cm^{-1} deformation of the MoO_4^{2+} tetrahedron as a central point, with O=Mo=O deformations at higher frequency and Mo—O—Mo deformations at the lower frequency. All the compounds have very complex spectra in this region with many bands. Taking the most prominent bands and limiting the "typical" Mo—O deformation region to 200–400 cm^{-1}, Kovats (202) described a reasonable assignment: (1) the Mo—O—Mo deformation is at 285 cm^{-1} for MoO_3 and the β and γ phases of bismuth molybdate; (2) the O=Mo=O deformation is at 370 cm^{-1} for the α phase and between 360 and 375 cm^{-1} for the β phase; (3) a strong band at 355 cm^{-1} in the γ phase was tentatively explained as partially bridging, where the bond order would be somewhere in between the two limiting structures, Mo—O—Mo and O=Mo=O.

In addition to the Mo—O deformations, there are many other vibrations possible in this region, such as lattice vibrations, deformations, and possibly the stretching of mixed linkages (Bi—O—Mo), and deformations and stretching motions associated with Bi—O structures as, for example, indicated by the very rich, low-frequency spectra of Bi_2O_3 (248).

Two additional polymorphs of γ-Bi_2MoO_6 are recognized (249, 250), and their Raman spectra recorded (Ref. 207 for γ'-Bi_2MoO_6, and Ref. 250 for both γ' and γ'' forms; see Fig. 10.17). γ-Bi_2MoO_6 is very well characterized (219) as an orthorhombic *Pca* 2_1 (C_{2v}^5) unit cell (koechlinite structure) with

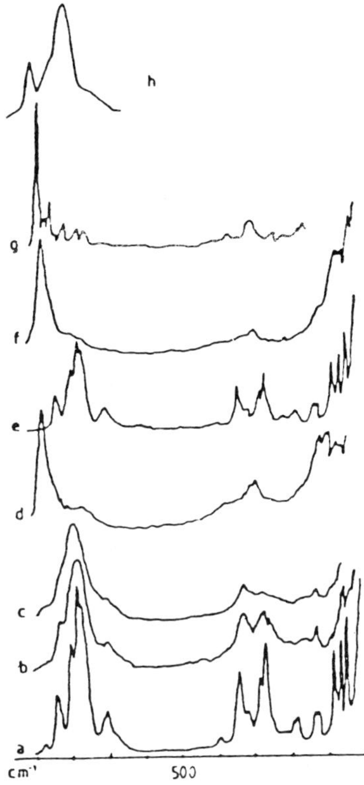

Figure 10.17. Raman spectra: (a) ambient; (b) 300°C; (c) 550°C; (d) 600°C (γ''); (e) cooling back to ambient (γ); (f) 750°C (γ'); (g) cooling back to ambient (γ'); (h) 1 h near 900°C ($\gamma_{HT} + \gamma'$). Reprinted from Gaucher et al. (250), with permission.

Mo in a highly distorted octahedral environment (site symmetry C_1). The compound is highly lamellar along (010). The γ'-Bi_2MoO_6, often referred to as the "Blasse structure," is monoclinic $P2_1/c$ (C^5_{2h}) space group symmetry. The third, γ''-Bi_2MoO_6, referred to as the "Erman phase," is not well understood. Another, high-temperature structure $\bar{\gamma}_{HT}$ (obtained between 850°C and its melting point) was indicated by Erman and co-workers in their studies of different γ-bismuth molybdate structures (251–253).

There is another catalytically very important phase, described first by Sleight and Jeitscho (254). It is structurally and spectroscopically characterized in several publications (210, 218, 255–259). Its composition is $Bi_3FeMo_2O_{12}$ and its structure is that of a scheelite, so it is often represented as $Bi_3(FeO_4)(MoO_4)_2$ to emphasize tetrahedral structures. Another bismuth–iron–molybdenum compound with the formula, $Bi_2Fe_2Mo_2O_{12}$, was reported (210); an alternate formula, $Bi_2O_3 \cdot Fe_2O_3$–$2MoO_3$, emphasizes individual oxides. Some authors disagreed (255), indicating it is not a compound but rather a mixture of known phases. The question of

$Bi_2Fe_2Mo_2O_{12}$ is an open question in view of almost identical XRD data (259) and identical IR and Raman spectra (257). Grzybowska and co-workers (257) find some differences in the surface composition by X-ray photoelectron spectroscopy, suggesting a possibility of different surfaces but the same bulk structures. This would not be unusual since the surface composition can vary with the method of preparation for the same bulk structure [see, for example, Linn and Sleight (255, Note 2)] . Linn and Sleight (255) also suggest that the Mossbauer spectra of $Bi_2Fe_2Mo_2O_{12}$ (210) are better interpreted as arriving from the mixture of $Bi_3FeMo_2O_{12}$ and α-Fe_2O_3.

Brazdil and co-workers (260) reported on the activation of $Bi_3FeMo_2O_{12}$ for propylene oxidation and ammoxidation. It is not much better than bismuth molybdate in terms of conversion and selectivity (193, 261). It was found that $Bi_3FeMo_2O_{12}$ undergoes a reduction-induced restructuring during the redox cycle, resulting in a considerable increase in activity (acrylonitrile yield improves from 50.9% to 73.0%) and selectivity (acrylonitrile selectivity improves from 78.4% to 83.3%). Using *in situ* Raman spectroscopy in conjunction with X-ray photoelectron spectroscopy and X-ray diffraction, these investigators identified several phases after restructuring: $Bi_3FeMo_2O_{12}$, α-$Bi_2Mo_3O_{12}$, β-$FeMoO_4$ (high-temperature phase), and some γ-Bi_2MoO_6. The increase in performance can be traced to catalytically very active α-$Bi_2Mo_3O_{12}$, which is further enhanced by β-$FeMoO_4$.

The activation energies for the reoxidation of α, β, and γ forms of bismuth molybdate were measured by Brazdil et al. (193) using a pulsed microreactor. The results suggested some unique redox behavior of the β-$Bi_2Mo_2O_9$ compound; the restructuring of the catalyst surface of disproportionation of β phase into α and γ phases was postulated. The X-ray diffraction study of β-$Bi_2Mo_2O_9$, after the compound went through reduction and a reoxidation cycle, did not indicate any changes in bulk composition, so the study was continued using *in situ* Raman spectroscopy (194, 213, 260) (Fig. 10.18). First, it was shown that the γ phase was stable at 500°C for an extended period of time. This was expected based on earlier work by Egashira and co-workers (262), who indicated it as metastable up to around that temperature. Work by Kumar and Ruckenstein (263) on very thin thermally evaporated β-$Bi_2Mo_2O_4$ films (300–500 Å), using electron diffraction, suggests the onset of decomposition into γ-Bi_2MoO_6 and MoO_2 at lower temperatures, i.e., 400°C in air and 350°C *in vacuo*. The same occurs on heating in reducing atmosphere (264, 265) where X-ray diffraction was employed. In the work by Brazdil and co-workers described here, XRD showed nothing (however, it is not comparable to electron diffraction in looking at a true surface), and since MoO_2 is an extremely poor Raman scatterer it would be very difficult to detect (178). After one redox cycle—reduction with propylene and

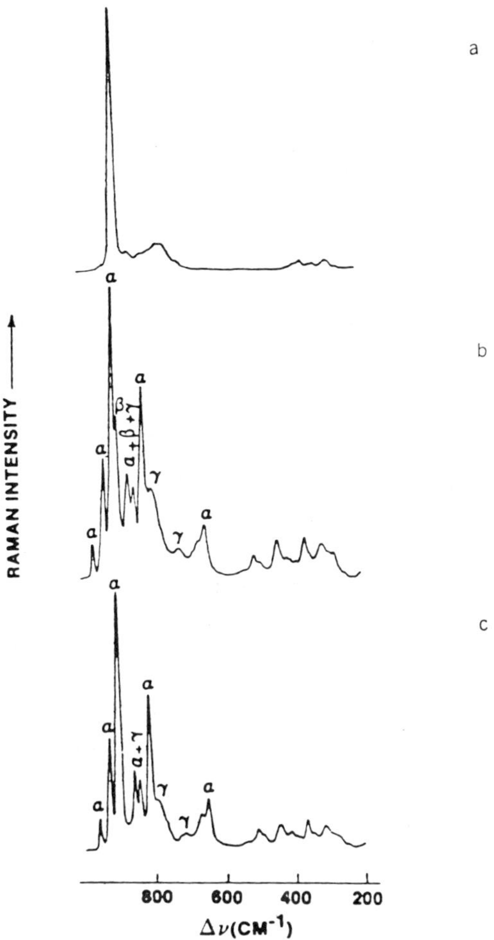

Figure 10.18. Raman spectra of: (a) $Bi_2Mo_2O_9$ after 600°C in air; (b) $Bi_2Mo_2O_9$ after reduction with propylene and reoxidation with air at 430°C; (c) after 480°C in flowing air. Here α refers to $Bi_2Mo_3O_{12}$, β to $Bi_2Mo_2O_9$, and γ to Bi_2MoO_6. From Brazdil et al. (260).

reoxidation with air at 430°C for 30 min—the spectral evidence for all three phases is clear. The initial disproportionation has taken place. By further reoxidation under air at 430°C for 24 h, a complete disproportionation of β-$Bi_2Mo_2O_9$ into γ-$Bi_2Mo_3O_{12}$ and γ-Bi_2MoO_9 takes place. This suggests that the β phase becomes unstable even under 500°C once the nucleation centers of the α and β phases are present. Similar experiments on the α and γ compounds indicated complete stability under the same conditions. The mechanism (A) for this disproportionation was suggested by Brazdil (260).

$$Bi_2Mo_2O_9 \xrightarrow{\text{Red.}} Bi_2MoO_{6-x} + (x+y)[O] + MoO_{3-y}$$

$$Bi_2Mo_2O_9 + MoO_{3-y} \longrightarrow Bi_2Mo_3O_{12-y} \tag{A}$$

$$Bi_2MoO_{6-x} + \frac{x}{2}O_2 \xrightarrow{\text{Reoxid.}} Bi_2MoO_6$$

$$Bi_2Mo_3O_{12-y} + \frac{y}{2}O_2 \xrightarrow{\text{Reoxid.}} Bi_2Mo_3O_{12}$$

The kinetic measurements also suggest that after restructuring, the surface is γ-Bi_2MoO_6-rich with α-$Bi_2Mo_3O_{12}$ in subsurface layers. An attempt to study this by X-ray photoelectron spectroscopy was not conclusive (266). In principle, it should be possible since for different phases there is variation in the Bi/Mo concentration as 2:3, 2:2, and 2:1. At these electron kinetic energies, the escape depths are no more than 20–30 Å; and, after restructuring, the surface on that scale is likely to be crystallographically very poorly defined. Therefore, the difficulties with interpretation are understandable.

Isotope tracer studies (^{18}O) have been very fruitful in mechanistic studies cited earlier and in studying the functioning of the catalyst structure. Work by several authors (261, 267–273) has clearly indicated participation of lattice oxygen in the oxidation process and its incorporation into the reaction products. Hoefs and co-workers (208) showed that incorporation of ^{18}O into the lattice of α, β, and γ bismuth molybdates produced isotope shifts and measurable spectral changes that could be used to study different sites and their function in a multifunctional catalyst. There are two recent works following this reasoning by Glaeser and co-workers (274) and by Matsuura and co-workers (275).

Glaeser and co-workers (274) selected γ-Bi_2MoO_6 as a model compound for their study, following an earlier kinetic study by Ueda and co-workers (276) that showed it was possible to use the ^{18}O tracer to independently probe oxide ions responsible for the α-H abstraction and oxygen insertion into an allylic intermediate during the oxidation of propylene. However, site identification was not possible from kinetic results alone; Glaeser and co-workers used *in situ* Raman spectroscopy to study spectral changes of the catalyst cycled through reduction with different reactants and reoxidation with ^{18}O. Pulse kinetic experiments were also included, indicating that the rate-determining step in catalyst reoxidation is the diffusion of oxide ions through the catalyst bulk. Advantage was taken of the following facts:

Table 10.7. Raman Spectra of Partially Reduced[a] γ-Bi_2MoO_6 Reoxidized with $^{18}O_2$

		Major Band Positions[b] After:					
Reductant	Initial	First Cycle	Second Cycle	Third Cycle	Fourth Cycle	Fifth Cycle	Sixth Cycle
Butene[c]	844	841 (−3)	841 (−3)	841 (−3)	840 (−4)	840 (−4)	840 (−4)
	803	801 (−2)	802 (−1)	802 (−1)	798 (−5)	798 (−5)	797 (−6)
	725	724 (−1)	724 (−1)	725 (0)	722 (−3)	722 (−3)	723 (−2)
Methanol[d]	844	839 (−5)	839 (−5)	838 (−6)	836 (−8)	832 (−12)	831 (−13)
	803	803 (0)	801 (−2)	802 (−1)	799 (−4)	787 (−16)	787 (−16)
	725	721 (−4)	720 (−5)	719 (−6)	717 (−8)	709 (−16)	709 (−16)
Propene[c]	844	835 (−9)	832 (−12)	830 (−14)	830 (−14)	830 (−14)	830 (−14)
	803	792 (−11)	792 (−11)	790 (−13)	790 (−13)	786 (−17)	785 (−18)
	725	715 (−10)	715 (−10)	713 (−12)	712 (−13)	710 (−15)	708 (−17)
Ammonia[c]	844	838 (−6)	836 (−8)	834 (−10)	832 (−12)	830 (−14)	830 (−14)
	803	802 (−1)	799 (−4)	795 (−8)	790 (−13)	790 (−13)	789 (−14)
	725	722 (−3)	719 (−6)	716 (−9)	713 (−12)	710 (−15)	710 (−15)

Source: Reprinted from Glaeser et al. (274), with permission.

[a]Degree of reduction = 50 μmol [O] vacancies per m^2 per cycle: Raman spectra taken at 430°C.

[b]Numbers in parentheses indicate the magnitude and direction of the shift; Units, cm^{-1}.

[c]Reduced and reoxidized at 430°C.

[d]Reduced at 400°C; reoxidized at 430°C.

- Oxidation of butene to butadiene involves only α-H abstraction (276).
- Methanol oxidation to formaldehyde occurs on Mo—O centers in molybdates (277).
- Propylene oxidation is a two-step process of α-H abstraction followed by oxygen insertion (194, 196).

Using the highest frequency bands in the 700–800 cm^{-1} region that are associated with stretching vibrations of Mo—O polyhedra (207), Glaeser and co-workers (274) showed that there is no isotope shift when butane is a reducing agent whereas there is a shift with methanol and propylene (Table 10.7). This clearly indicates that oxygen involved in α-H abstraction is not associated with molybdenum but rather with the bismuth cations, as suggested earlier by Grasselli and co-workers (194). Oxygen, associated with molybdenum, inserts into an allylic intermediate. Specific oxygens in the structure are suggested in an assignment based on structural studies by neutron diffraction (224). Finally, the role of dissociative chemisorption of O_2 is assigned to the Bi—O—Bi structures, which contain two lone electron peaks on bismuth. This function of bismuth was established by electrochemical studies (278). Figure 10.19 shows a schematic representation of this mechanism.

Matsuura and co-workers (275) selected α-$CoMoO_4$ and α-$MnMoO_4$ (with 1% Bi and 3% Fe) as model compounds in following the earlier work (279), which indicated that lattice oxygen used in oxidation was supplied by $M^{2+}MoO_4$ structures in the bulk of the catalyst. The ^{18}O tracer was substituted into the lattice by replacing oxygen ions used in the oxidation of propylene. IR and Raman spectral changes indicated the location of the isotope tracer. (Fig. 10.20 and 10.21). In this very fine set of spectra, one can observe all the effects that may occur:

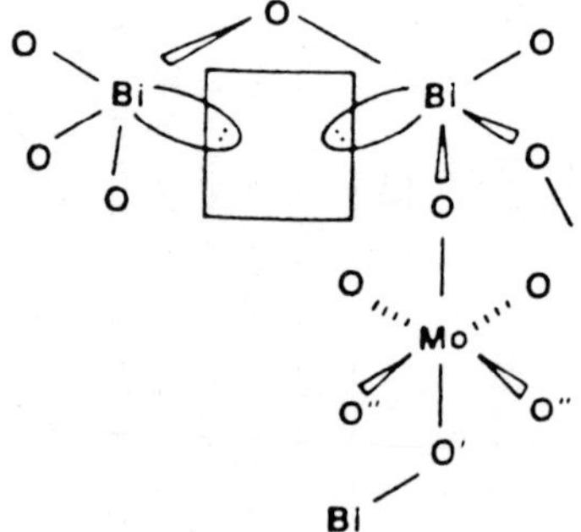

Figure 10.19. Schematic representation of the active-site structure on the surface of Bi_2MoO_6. *Key*: O′, Oxygen responsible for α-H abstraction; O″, oxygen associated with Mo responsible for oxygen insertion; *square*, proposed center for O_2 reduction and dissociative chemisorption. Reprinted from Glaeser et al. (274), with permission.

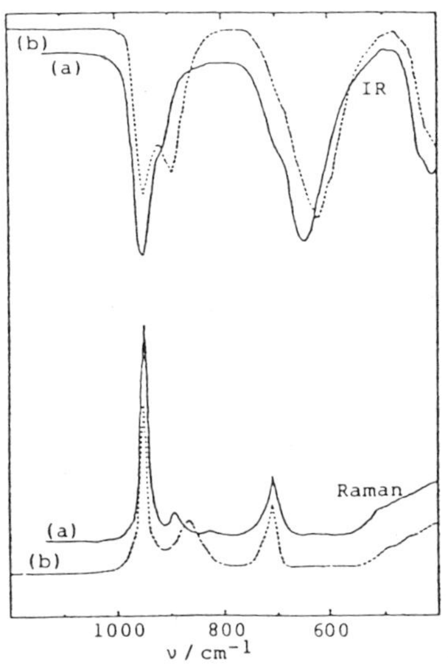

Figure 10.20.

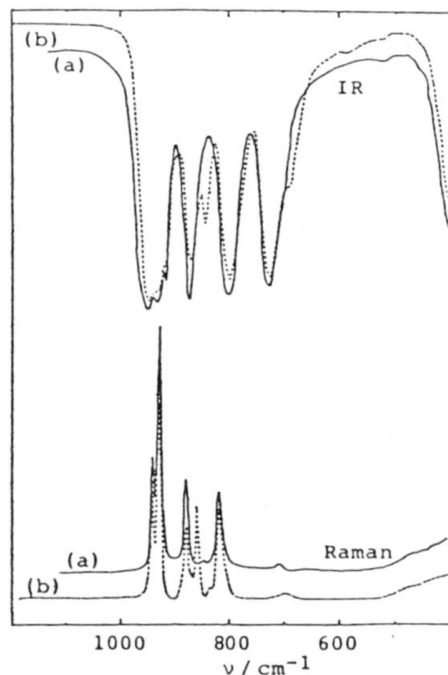

Figure 10.21.

Figure 10.20. (Left) IR and Raman spectra of α-$CoMoO_4$ (a) before and (b) after $C_3H_6 + {}^{18}O_2$ reaction. From Matsuura et al. (275).

Figure 10.21. (Right) IR and Raman spectra of α-$MnMoO_4$ (a) before and (b) after $C_3H_6 + {}^{18}O_2$ reaction. Reprinted from Matsuura et al. (275), with permission.

- The reduction of integrated peak intensities for the ${}^{16}O$ structure
- Band shifts
- Appearance of new bands

Having good crystallographic data (280, 281) and a precise vibrational assignment based on normal coordinate analysis, the authors were able to identify two specific oxygens in the $M^{2+}MoO_4$ lattice that participate in selective olefin oxidation. Furthermore, they suggest a mechanism for the transport of oxide ions through the lattice. They also suggest that the bismuth-rich (010) and (100) planes of $M^{2+}MoO_4$ are olefin oxidation sites and that the (001) plane, with no bismuth coverage, is the oxygen adsorption site.

During spectroscopic analysis of more complex catalytic systems, one is likely to encounter a variety of metal molybdates, depending upon the promoters and other metals added to the composition. Many reference Raman spectra (and IR) of various well-defined model compounds can be found in the hydrodesulfurization literature (for Co, Cr, Al, and Fe) (97) or

Table 10.8. IR and Raman Band Positions of Simple Molybdates

Compound	IR Bands (cm^{-1})[a,b]
α-$NiMoO_4$–20% SiO_2	*953, 932, 928, 875, 795, 700, 675, 640, 470, 445, 420, 390, 362, 336, 295
α-$CoMoO_4$–20% SiO_2 (green)	*944, 920 sh, 915 sh, 868, 815, 790, 650, 570, 475, 435, 410, 330, 300
β-$CoMoO_4$ (purple) (ATR)	*930, 840, 780, 710, 410
$FeMoO_4$	*920, 860, 790, 650
$Fe_2(MoO_4)_3$–20% SiO_2	985, 965 sh, 957, 890 sh, *830, 815 sh, 785 sh, 470, 410, 370, 330
$Fe_2(MoO_4)_3$	990, 960, *830, (v. br.), 785 sh, 620 sh, 400, 370, 330, 280
$Cr_2(MoO_4)_3$	970, 930, *850 (v. br.), 770, 610, 555, 380, 310, 250
α-$Bi_2(MoO_4)_3$	950, 930, 900, *860, *855, 830, 745, 720–650 br, *550, 465, 440, 420, 380, 350, 315
β-$Bi_2Mo_2O_9$	890 (v. br.), *845, *810, *770, *710, 470, 440, 380, 340, 290
γ-Bi_2MoO_6 (koechlinite)	845, 800, *740, 550, 455, 410, 360
γ'-Bi_2MoO_6 (high temp.)	880, 870, 825 sh, *810, 780 sh, 720, 650, 570, 550, 520, 490, 460, 390, 380, 330, 300
$In_2(MoO_4)_3$	935, 920 sh, *884, 857, 843, 834, 818, 805, 724, 450–400 multiple bands
$Sc_2(MoO_4)_3$	984, 972, *940, 890, 850, 828, 820, 500–400 multiple bands
MoO_3–20% SiO_2	980, *850, 820, 690, 570, 480, 370
SiO_2	1190, 1160 sh, *1090, 790, 625, 480, 380

Compound	Raman Bands (cm^{-1})[c]
β-$NiMoO_4$	**955, **945, 915, 890, 830
α-$NiMoO_4$	**970, 920, 715
β-$CoMoO_4$ (purple)	935, **928, 868, 810
α-$CoMoO_4$ (green)	**940, 880, 820, 700
$Fe_2(MoO_4)_3$	990, 960, 938, 820, **780
α-$Bi_2(MoO_4)_3$	990, 950, 920, **900, 850, 835, 810, 650
β-$Bi_2Mo_2O_9$	**885, 750
γ-Bi_2MoO_6 (koechlinite)	852, 810 sh, **795, 785 sh, 715, 402
γ'-Bi_2MoO_6 (high temp.)	**900, 880, 870, 820, 790, 770
MoO_3	995, **818, 665
$Cr_2(MoO_4)_3$	**970, 830, 360
$FeMoO_4$	**937, **929
$In_2(MoO_4)_3$	997, **981, 954, 869, 849, 843, **820, 400–300 br.
$Sc_2(MoO_4)_3$	**989, 958, 950, 840 sh, **822, 400–300 br.
CrO_3	1008, **969, 902, 497

Source: Reprinted from Hazle and Mehicic (286), with permission.

[a]The strong SiO_2 bands near 1100 cm^{-1} were not listed for any of the supported samples.

[b]* = Most intense band in the active phase.

[c]** = Most intense band in spectrum.

the selective oxidation literature (for Mn, Cd, Co, Fe^{2+}, Fe^{3+}, Pb, Ca, and Ni) (203, 247, 275, 282–285). The IR and Raman data for many compounds of interest are listed in Table 10.8 (286).

Real multicomponent commercial catalysts are very complex and contain many phases. They present a formidable challenge to the analytical spectroscopist. The selective oxidation catalyst 50% $Ni_3Co_5Fe_3BiPK_{0.1}Mo_{12}O_{52.5}$/50% SiO_2, described by Rao and Menon (287), illustrates this point very well. It also illustrates the need for a multitechnique analytical approach for a more thorough understanding of such structures.

Raman spectroscopy is a powerful technique for studying aqueous solutions (288, 289), and it has been used to study the preparation of selective oxidation/ammoxidation catalysts. A good example of such work is the study by Schrader et al. (290) of the formation of α-$Bi_2Mo_3O_{12}$, γ-Bi_2MoO_6, and α-$NiMoO_4$, including the incorporation of excess MoO_3. They show the influence of solution pH on aggregation and the formation of complex ions, and ultimately on a particular bismuth molybdate. The importance of the degree of aggregation of both Bi and Mo was also indicated in Schrader and co-workers' earlier publication (243). For $NiMoO_4$, solution pH is a key factor influencing the Ni/Mo ratio, i.e., the final composition of the catalyst. Incorporation of excess molybdenum, as MoO_3, in these catalysts is of interest because it affects the final catalytic properties. In γ-Bi_2MoO_6, MoO_3 is easily incorporated. The molybdenum oxide species are dispersed throughout the lattice, probably along bismuth–oxygen planes. At low pH levels, however, there is indication of MoO_3 segregation. In α-$Bi_2Mo_3O_{12}$, MoO_3 is not readily incorporated because of the formation of large molybdenum aggregates early in the precipitation process. α-$NiMoO_4$ also cannot incorporate excess molybdenum by dispersion throughout the lattice; it tends to segregate MoO_3 and retain its original structure. The overall stoichiometry of the precipitated catalyst is governed by segregation of the MoO_3 phase.

The question of support/active phase interaction is always of interest because catalyst performance is greatly influenced by the type and the physical properties of support. In selective oxidation catalysis, there are no published reports dealing with this, although a considerable body of work exists in hydrodesulfurization catalysis (see Section 10.3 above), and it has bearing on oxidation catalysis. Some of the publications dealing with the SiO_2 support and model molybdate compounds were reviewed by Kovats (202).

10.4.4. Iron Oxide Catalysts

The Fe_2O_3–MoO_3 catalyst, alone or Te promoted, is used in the oxidation of methanol to formaldehyde, the oxidation and the isomerization of 1-butene, and the epoxidation of olefins (218, 291–293). Very little Raman work has been reported in this area. Villa and co-workers (218) reported a study of a ternary Fe–Mo–Bi oxides system in an Fe-rich region, where Bi can be considered a promoter. Various amounts of Bi_2O_3 were added to $Fe_2(MoO_4)_3$. Only at high concentration of Bi_2O_3 (3.5 wt.%), where catalytic activity was constant, were they able to detect a new phase, γ-Bi_2MoO_6. They suggest that the active and selective phase when Bi dopant is added is Fe(III)-molybdate with Bi in interstitial or substitutional sites which may give rise to [Bi(V)=O]—[Mo(VI)=O] groups.

Earlier Raman work on the Te-promoted system (291) reported identification of $Fe_2(MoO_4)_3$ and MoO_3 phases. It was shown that the presence of Te promoter is essential to selectivity in oxidation of 1-butene to butadiene, since it decreases the mobility of the lattice oxygen in the ferric molybdate lattice.

Cairati and co-workers (294) reported on the study of an Fe_2O_3–MoO_3 fluid bed catalyst with SiO_2 and Al_2O supports. A low surface area favors $Fe_2(MoO_4)_3$. An increase in the surface area decreases activity and $Fe_2(MoO_4)_3$ concentration. The authors propose the formation of Fe(III)MoAl or Fe(III)MoSi. The possible Fe_2O_3 structure was excluded, since in the oxidation of methanol it would principally produce CO_2 (295).

10.5. OTHER SYSTEMS

10.5.1. Metathesis Catalysis

Metathesis or disproportionation of olefins (7) is industrially a very important type of process. It involves the exchange of alkylidene groups of two olefins via double bond cleavage. An example is the conversion of propylene into equimolar amounts of ethylene and 2-butene. Several metal oxides are active: Re_2O_7, CoO–MoO_3, and WO_3, in order of decreasing activity. Typical supports are SiO_2 and Al_2O_3. Tungsten and molybdenum are well known in other catalytic processes, such as hydrodesulfurization and oxidation. Rhenium is a very active metathesis catalyst, even at room temperature.

Kerkhof et al. (296) reported on the Raman spectroscopy investigation of the Re_2O_7/Al_2O_3 system. They present the spectrum of aqueous perrhenate ion, ReO_4^-: 970 cm^{-1} [ν_{sym}(ReO)], 916 cm^{-1} [ν_{asym}(ReO)], and 332 cm^{-1} [δ(OReO)], which is in good agreement with other data found in Wachs et al. (11) and Woodward and Roberts (297). From the spectral similarity of the Re_2O_7/Al_2O_3 catalyst and the ReO_4^- anion, they conclude that the catalyst surface contains distorted ReO_4^- tetrahedra and that this distortion is dynamic rather than static. Band broadening of the 916 cm^{-1} feature, but no shifting, indicates a dynamic rather than a static situation. The distortion of the tetrahedra is tentatively attributed to its interaction with the surface hydroxyl groups. This is in agreement with the fact that the Re_2O_7/Al_2O_3 system is easier to reduce than the WO_3/Al_2O_3 catalyst, where interactions are stronger. These investigators observe no other rhenium species, even at high loading, and compare this to previous molybdenum-on-alumina work by Medema (97) and Giordano (298), where tetrahedral species are present at low concentrations but, at higher loading, octahedral species, aluminum molybdate, and MoO_3 are suggested. They explain this by the fact that rhenium is much heavier than molybdenum, giving less surface coverage at the same weight percent loading. Subsequent work by Nakamura and co-workers (299) using IR spectroscopy shows excess Re_2O_7, in addition to ReO_4^- ion at high loadings of 17%. Also, earlier work by Yao and Shelef (300), using electron spin resonance and temperature-programmed reduction, suggests the existence of a two-dimensional phase in addition to three-dimensional crystallites.

More recent work on the rhenia–alumina catalyst by Wang and Hall (301) employed Raman, IR, and visible reflectance spectroscopy. Their results are in very good agreement with those of Kerkhof et al. (296), although the two works use different preparation methods, i.e., the incipient wetness method by Kerkhof et al. and equilibrium adsorption by Wang and Hall. The latter authors report the existence of only a monomeric rhenia species—a distorted tetrahedral ReO_4^- anion. They also show that, *in vacuo*, there is a reversible and temperature-dependent disorder of tetrahedra (which increases with temperature) owing to the increasing strength of the interaction with the support. This is manifested by the shifting and splitting of vibrational bands associated with ReO_4^- tetrahedron. The sensitivity to moisture was also shown, where the alumina-bound rhenia could be easily hydrolyzed to $HReO_4$. This sensitivity is in agreement with earlier findings by Olsthoorn and Boelhouwer (302) using IR spectroscopy.

The most recent work by Wachs and co-workers (11) is in complete agreement with Kerkhof et al. (296) and Wang and Hall (301), also indicating tetrahedral ReO_4^- species on the aluminum support. Wachs et al. do not indicate a method of preparation of their catalyst but do provide a good summary of vibrational spectra of several rhenium compounds. Additional spectra of some other relevant compounds can be found in Kerkhof et al. (296), Wang and Hall (301), and Salvati et al. (303).

MoO_3 and WO_3 supported on SiO_2 were reported in a detailed study by Thomas and co-workers (304).

A very detailed study by Raman and XPS of the support/active phase interactions in a number of tungsten and rhenium catalysts, prepared by a variety of methods on SiO_2 and Al_2O_3, was published by Kerkhof and co-workers (305). Previously discussed tungsten surface compounds are formed on both SiO_2 and γ-Al_2O_3. Interactions are stronger on the γ-Al_2O_3 support, resulting in better stability and better dispersion; more crystallites are formed on the SiO_2 support. Similar results were found for rhenia.

10.5.2. Olefin Hydrogenation

Hydrogenation is used extensively in industrial processes and is important in petroleum refining and in the petrochemical industry. Particular examples include fatty oil refining, methanol synthesis, and alcohol synthesis from aldehydes prepared by the aldol reaction. It involves the reaction of gaseous hydrogen with, generally, an unsaturated organic substance, at varying temperatures and pressures and in the presence of various catalysts. (In its broadest sense hydrogenation can include the reaction of hydrogen with any substance; therefore, the reaction of H_2 and CO syngas in the Fischer–Tropsch synthesis, the reaction of H_2, CO, and alkenes in hydroformylation reactions, considered under the general classification of Fischer–Tropsch reactions, and hydrodesulfurization can be considered hydrogenations.) In this section we shall discuss hydrogenation of olefins, followed in Section 10.5.3 by syngas processes.

Ward et al. (306) found silica-supported zirconium hydrides to be moderately active for olefin hydrogenation and, in further studies, Rh hydride species to be particularly effective even in catalyzing hindered olefins. Raman and IR were used to identify the catalytic species SiO_2-supported RhH_2. The investigators note that since the support is not transparent in IR spectra, the combination of IR and Raman is effective in studying a wide range of frequencies (306).

Raman and IR were used to study the intermediate Ru-based clusters formed on silica during the preparation of a catalyst containing small Ru particles encapsulated in the silica surface. This catalyst is used in α-olefin hydrogenation, isomerization, or CO reduction to hydrocarbons (307).

A fused Fe catalyst with V_2O_5 promoter was studied during activation and during the hydrogenation of acetophenone. Raman spectra were obtained of the reduced catalyst surface during activation by H_2, then D_2; the spectra indicate the formation of H bonds with the surface (a hydride-like structure is believed to exist). In this process active surface sites are formed. During hydrogenation at industrial catalysis conditions of high temperature and pressure, the Raman spectrum is assigned to one or two intermediates, possibly having C=O and C—C bonds (308).

10.5.3. Syngas Processes

Synthesis gas, CO/H_2, is used as a feedstock in some of the most important industrial catalytic processes. It is used, for example, in methanol synthesis, hydroformylation, and Fischer–Tropsch catalysis. Some Raman spectroscopic work has been published in these areas.

10.5.3.1. Fischer–Tropsch Catalysis

The synthetic fuel industry relies on the catalytic hydrogenation of CO to form methane and long-chain hydrocarbons. There is renewed interest in this technology (first developed in the early part of this century). Coal, tar sands, and oil shale are all sources of CO and H_2 (the most convenient source being the gasification of coal by steam).

In Fischer–Tropsch catalysis, syngas, CO/H_2, is converted into high-boiling hydrocarbons such as diesel fuel and wax (over a fixed-bed catalyst) or into lower boilers—gasoline-grade hydrocarbons (over a fluid-bed catalyst). This chemistry is widely utilized in South Africa by Sasol. Two forms of the process exist: one uses a Co catalyst at about 200°C and 7–10 bar; the other uses an Fe catalyst in a fluid bed at about 200°C and 15–35 bar (Hydrocol process). An early catalyst was composed of Fe and Co (later ThO_2 was added) promoted with K_2CO_3 and Cu. Selective Ni catalysts exist for methane synthesis and Zn–Cr–Cu catalysts for methanol synthesis. Shape-selective zeolites are sometimes added to Fischer–Tropsch catalysts to provide selectivity. With the availability of surface science techniques, the mechanisms for this process are finally being elucidated (14).

An excellent Raman study in this area is by Zhang and Schrader (309). They performed *in situ* studies of an alkali-promoted, fused iron catalyst. Again, it should be stressed that, of all spectroscopies, Raman is probably the easiest to do *in situ*, i.e., under the conditions of high temperature and high pressure employed in this process. It is considerably easier to do than the IR experiment since these catalysts have very poor transmittance. Other spectroscopic work [IR, XPS, AES (Auger electron spectroscopy), Mossbauer] is also listed in this publication.

In a steady-state condition, the catalyst is a mixture of bulk oxides and carbides that are formed during an activation or induction period (310) in contact with syngas. Zhang and Schrader (309) discuss the active components, i.e., iron, iron carbide, and iron oxides, Fe_2O_3 and Fe_3O_4. They used Raman to study the composition of a reduced commercial alkali-promoted fused iron catalyst under an H_2 atmosphere. Their study indicates the formation of Fe—H bonds because of Raman bands at 1902 and 1951 cm^{-1} (terminal stretching) and 1625 cm^{-1} (stretching of bridging species). Assignment was based on the available literature for these types of structures (229, 311–313). In an attempt to identify adsorbed species that exist during Fischer–Tropsch catalysis, they next studied CO/H_2 adsorption. In addition to the Fe—H bond already discussed, there are numerous bands associated with CO adsorption: C≡O terminally adsorbed on different adsorption sites (1975, 2080, 2160, and 2175 cm^{-1}), and chemisorbed CO at 1820 cm^{-1}. Multinuclear activation of the C≡O bond, where electron back-donation from the metal occurs, results in significant weakening of the triple bond (note that gaseous CO vibrates at 2143 cm^{-1}). The authors suggest that this chemisorbed CO may be a precursor of dissociated CO. A probable oxygenated, formate-like intermediate is characterized by the 1556 cm^{-1} band. It is discussed by several authors (314–319). Formation of Fe—C bonds is indicated by the 360, 400, and 450 cm^{-1} vibrations, which have also been discussed by other authors (320–323). With the foregoing assignments, it is possible to study the actual working of fused iron catalysts *in situ* and provide reasonable assignments for surface adsorbed structures; see Table 10.9 (309). The chemisorption and activation of CO takes place on Fe^0 as well as on Fe^{n+} sites.

Previous work by Loktev and co-workers (324, 325) shows that initially the bulk Fe catalyst is 85% α-Fe but the final composition of a working catalyst is 65% Fe_3O_4, 25% Fe carbides, and 8% α-Fe. They conclude that bulk oxides and carbides are not important but a monoxidized surface structure with a defect iron carbide structure is the catalytically active phase. Zhang and Schrader (309) find the surface more complex; CO adsorption studies confirm surface iron oxide(s). They suggest two active phases on the surface: Fe^0/Fe-carbide and an Fe-oxide (possibly Fe_3O_4). It is suggested that the carbide phase is responsible for hydrocarbon formation (via adsorption, activation, and dissociation of CO); hydrocarbons are a dominant product. The Fe-oxide phase may be responsible for the formation of oxygenated products such as alcohols. The oxygen-containing, formate-like

Table 10.9. Assignment of Raman Bands for Adsorbed Species Associated with Fischer–Tropsch Synthesis on an Alkali-Promoted, Fused Iron Catalyst

Band Position (cm^{-1})	Assignment
1951 (1948), 1902	Fe—H stretch of terminal H species
1625	Fe—H stretch of bridging H species
1970	C—O stretching of CO adsorbed on Fe^0 sites
2070	C—O stretching of CO adsorbed on Fe^{2+} sites
2130, 2160, 2175	C—O stretching of CO adsorbed on Fe^{3+} sites
1820, 1850, 1890	C—O stretching of CO adsorbed to form multinuclear Fe^0 clusters
1556, 1586–1600 region	Vibrations due to formate-like species
360, 400, 450	Fe—C stretch of adsorbed CO species

Source: Reprinted from Zhang and Schrader (309), with permission.

intermediate, $C_1H_xO_y$, is indicated in their *in situ* studies of working catalysts and is given an assignment in Table 10.9.

10.5.3.2. Hydroformylation Catalysis

The hydroformylation (or oxo) process involves the reaction of syngas, CO/H_2, and an olefin to produce straight or branched aldehydes. Chains are longer than the original olefin by one carbon atom. Aldehydes can be hydrogenated into corresponding alcohols, which are valuable in the chemical industry. The process typically is a homogeneously catalyzed reaction. The catalyst can be a Co, Ru, or Rh metal center with CO ligands. The addition of amine- or phosphine-containing ligands may improve activity and selectivity. Cobalt carbonyls are preferred catalysts because of their higher activity and lower cost.

Woo and Hill (326) used *in situ* Raman spectroscopy to study a $Co_2(CO)_6(PPh_3)_2$ catalyst supported on SiO_2 and γ-Al_2O_3, for the hydroformylation of propylene. They also studied the same catalyst in a homogeneous liquid-phase process in order to compare reaction intermediates. Apparently, intermediates are different on the two different supports. Also, it appears that intermediates in the liquid-phase, homogeneously catalyzed system are different from those in the supported catalyst. The authors follow earlier *in situ* IR work (327–329) and provide a good discussion of the ease of the *in situ* Raman experiment compared to the IR.

The use of Raman spectroscopy with FT-IR and photoacoustic FT-IR spectroscopy to study cobalt and rhodium carbonyls covalently anchored to inorganic oxides by phosphines or amines was demonstrated in a seies of papers by Woo and Hill (25, 330, 331). Raman bands between 950 and 1600 cm^{-1} were used to follow the intraction of the phenyl groups in the phosphine with hydroxyl groups on the surface of the γ-alumina support. A new Raman band at 1797 cm^{-1} was assigned to a carbonyl species hydrogen bonded to hydroxyl groups on the surface of the γ-Al_2O_3. These species disappeared at elevated temperatures and a partially decarbonylated structure is formed. Under hydroformylation conditions, the spectroscopic evidence indicated a conversion to a tetrameric cobalt cluster compound. An understanding of the relationship between surface intemediates and catalytic activity for these and other hydroformylation catalysts (332–334) is greatly enhanced by *in situ* Raman studies.

Ruthenium was a subject of a study by Theolier and co-workers (335). They report IR and Raman studies of a $Ru_3(CO)_{12}$ cluster supported on silica, in a follow-up of previous IR studies of the same system (336–338) and other metal carbonyls supported on inorganic oxides (339). Theolier and co-workers (335) were able to study this Ru system with the exclusion of oxygen and water, and they report the formation of the grafted cluster $HRu_3(CO)_{10}(O—Si\lessdot)$. This can be an intermediate for small Ru metal clusters (14 Å), in addition to some Ru(II) carbonyl species encapsulated in the silica surface. This reference is also a good source for the IR stretching frequencies, ν(CO), of numerous ruthenium carbonyl complexes.

10.5.3.3. Methanation

Methanation is the conversion of CO and H_2 to methane. Stencel and co-workers (340) report an IR and Raman study of a sulfur-resistant methanation catalyst, $NiO/Cr_2O_3/MgSiO_3$, which has 70% CO-to-CH_4 conversion (compared to 20% for the catalyst without chromium, i.e., $NiO/MgSiO_3$). The resistance to sulfur poisoning is important since otherwise an additional catalytic step of hydrodesulfurization of feed is necessary. The as-prepared catalysts are found to contain a CrO_4^{2-} structure both on the surface and in the bulk. Further data also indicate SiO_4- and/or Si_2O_7-type structures. The authors suggest there is cross-linking of these Si and Cr structures to produce Cr—O—Si linkages. There is tentative evidence for MgO and NiO. It is proposed that the molecular structure of the catalyst is represented by MgO and NiO dispersed in a quasi-tridimensional Cr—O—Si network. Reducing and sulfiding the catalyst leads to reduction of the CrO_4^{2-} concentration, reordering of the Cr—O—Si network, changes in SiO_n coordination, and reduction of Cr^{6+} to Cr^{3+} or Cr^{2+}. Adsorbed sulfur is suggested to be in a sulfide form that, with subsequent oxidation, leads to SO_4^{2-}. Data also suggest the formation of SO_4^{2-} during methanation.

10.5.4. Olefin Dehydrogenation

A chromium oxide catalyst is of great industrial importance in the reactions of olefins, where it is used in dehydrogenation, polymerization, and dehydrocyclization processes (7, 341). For polymerization a SiO_2 support is typically used, and for redox processes an Al_2O_3 support is employed (208).

Zaki and co-workers (342) studied a chromium catalyst on SiO_2 and γ-Al_2O_3 in order to characterize Cr—O surface species and the catalytically active Cr oxidation state, as a function of support type and Cr content, which is known to have an influence. Raman spectroscopy is one of the numerous techniques used to structurally characterize this catalyst. From this combined technique approach, several conclusions are reached. When the SiO_2 support is used, the catalyst surface contains three chromium species: polychromates, chromium chromate, and α-Cr_2O_3. With increasing chromium content, there is an increase in α-Cr_2O_3. The relative chromium chromate concentration increases with a decrease in polychromate concentration. With the γ-Al_2O_3

support, there is no α-Cr_2O_3 detected at low Cr concentrations (below 5.2 wt.% Cr_2O_3); only chromate species are present. The α-Cr_2O_3 phase is identified when concentrations are higher than 5.2 wt.%. Additional results, not based on Raman spectroscopy include the suggestion of coupled Cr^{3+}–Cr^{6+} species, which may be responsible for catalytic activity for H_2O_2 decomposition (a model reaction used for the study).

The interaction of chromium oxide with Al_2O_3, TiO_2, and silica supports was investigated by Hardcastle and Wachs (343). The influence of the nature of the oxide support, calcination temperature, and chromium oxide loading upon the molecular state of the supported chromium oxide was determined. Different structures resulted with differing surface-hydroxyl chemistries of Al_2O_3, TiO_2, and SiO_2 supports.

Zaki et al. (342) present good reviews of previous work, using techniques other than Raman spectroscopy, on chromium surface structures and on oxidation states that may be responsible for catalytic activity. There also are several good references giving vibrational assignments and IR/Raman spectra of chromium reference compounds (11, 344–350).

REFERENCES

1. R. L. Austermann, D. R. Denley, D. W. Hart, P. B. Himelfarb, R. M. Irwin, M. Narayana, R. Szentirmay, S. C. Tang, and R. C. Yeates, *Anal. Chem.* **59**, 68R (1987); D. L. Gerrard and H. J. Bowley, *ibid.* **60**, 368R (1988); D. L. Gerrad and J. Birnie, *ibid.* **62**, 140R (1990).
2. W. N. Delgass, G. L. Haller, R. Kellerman, and J. H. Lunsford, *Spectroscopy in Heterogeneous Catalysis.* Academic Press, New York, 1979.
3. B. C. Gates, J. R. Katzer, and G. C. A. Schuit, *Chemistry of Catalytic Processes.* McGraw-Hill, New York, 1979.
4. J. M. Thomas and R. M. Lambert, *Characterization of Catalysts.* Wiley (Interscience), New York, 1980.
5. M. L. Deviney and J. L. Gland, eds., *Catalyst Characterization Science. Surface and Solid State Chemistry,* ACS Symp. Ser. No. 288. Am. Chem. Soc., Washington, D.C., 1985.
6. J. G. Grasselli, M. K. Snavely, and B. J. Bulkin, *Chemical Applications of Raman Spectroscopy.* Wiley (Interscience), New York, 1981.
7. K. Weissermel and H. J. Arpe, *Industrial Organic Chemistry.* Verlag Chemie, New York, 1978.
8. R. J. H. Clark and R. E. Hester, eds. *Advances in Infrared and Raman Spectroscopy,* Vols. 1–18. J. Wiley, New York, 1975–1989.
9. R. F. Willis, *Vibrational Spectroscopy of Solids.* Springer-Verlag, New York, 1980.

10. *Proc. 10th Int. Conf. Raman Spectrosc.* (W. L. Peticolas and B. Hudson, eds.) Eugene, Oregon (1986); *11th* (R. J. H. Clark and D. A. Long, eds.) London, U.K. (1988); *12th* (J. R. Durig and J. F. Sullivan, eds.) Columbia, South Carolina (1990).
11. I. E. Wachs, F. D. Hardcastle, and S. S. Chan, *Spectroscopy* **1**, 30 (1986).
12. L. Dixit, D. L. Gerrard, and H. J. Bowley, *Appl. Spectrosc. Rev.* **22**(2–3), 189 (1986).
13. H. M. Ismail, C. R. Theocharis, D. N. Waters, M. J. Zaki, and R. B. Fahim, *J. Chem. Soc., Faraday Trans. 1* **83**, 1601 (1987).
14. N. D. Spencer and G. A. Somorjai, *Rep. Prog. Phys.* **46**, 1 (1983).
15. A. Wokaun, A. Baiker, W. Fluhr, M. Meier, and S. Miller, *J. Vac. Sci. Technol.* [2]**B3**, 1397 (1985).
16. D. M. Hanson, Y. Udagawa, and K. Tohji, *J. Am. Chem. Soc.* **108**, 3884 (1986).
17. L. H. Little, *Infrared Spectra of Adsorbed Species.* Academic Press, London, 1966.
18. M. L. Hair, *Infrared Spectroscopy in Surface Chemistry.* Dekker, New York, 1967.
19. R. L. Garrell, *Anal. Chem.* **61**, 401A (1989).
20. M. Mehicic and R. Barbour, BP America R & D, Cleveland, Ohio, private communication.
21. J. Doppelbauer, G. Leylendecker, and D. Baeuerle, *Appl. Phys* **B33**, 141 (1984).
22. J. C. Hamilton and R. J. Anderson, *J. Electrochem. Soc.: Solid State Sci. Technol.* **132**, 1753 (1985).
23. Z. Wang, W. Lu, and E. Min, *China-Jpn.-U.S. Symp. Heterog. Catal. Relat. Energy Probl., 1982* **A25C** (1982).
24. G. L. Schrader, M. S. Basista, and C. B. Bergman, *Chem. Eng. Commun.* **12**, 121 (1981).
25. S. Woo and C. Hill, *J. Mol. Catal.* **24**, 165 (1984).
26. C. Cheng, J. Ludowise, and G. Schrader, *Appl. Spectrosc.* **34**, 146 (1980).
27. G. Haller, *Catal. Rev.—Sci. Eng.* **23**, 477 (1981).
28. S. Woo and C. Hill, *J. Mol. Catal.* **29**, 231 (1985).
29. See Ref. 15, p. 1398.
30. J. F. Brazdil, M. Mehicic, L. C. Glaeser, M. A. S. Hazle, and R. K. Grasselli, in *Catalyst Characterization Science. Surface and Solid State Chemistry* (M. L. Deviney and J. D. Gland, eds.), ACS Symp. Ser. No. 288. Am. Chem. Soc., Washington, D.C., 1985.
31. M. Mehicic, M. A. Hazle, J. R. Mooney, and J. G. Grasselli, in *Chemical, Biological, and Industrial Applications of Infrared Spectroscopy* (J. R. Durig, ed.). Wiley, New York, 1985.
32. W. Kiefer and H. J. Bernstein, *Appl. Spectrosc.* **25**, 609 (1971).

33. F. R. Brown, L. E. Makovsky, and K. H. Rhee, *Appl. Spectrosc.* **31**, 563 (1977).
34. S. S. Chan and A. T. Bell, *J. Catal.* **89**, 433 (1984).
35. N. Zimmerer and W. Kiefer, *Appl. Spectrosc.* **28**, 279 (1974).
36. W. Kiefer, *Adv. Infrared Raman Spectrosc.* **3**, 1 (1977).
37. R. O. Kagel, R. A. Koster, and W. T. Allen, *Appl. Spectrosc.* **30**, 350 (1976).
38. M. Watanabe, *J. Catal.* **110**, 37 (1988).
39. E. Payen, M. C. Dhamelincourt, R. Dhamelincourt, J. Grimblot, and J. P. Bonnelle, *Appl. Spectrosc.* **36**, 30 (1982).
40. D. Espinat, H. Dexpert, E. Freund, G. Martino, C. Couzi, P. Lespade, and F. Cruege, *Appl. Catal.* **16**(3), 343 (1985).
41. C. Johnson and K. Thomas, *Fuel* **63**, 1073 (1984).
42. H. Knoezinger and E. Taglauer, *Symp. Chem. Phys. Catal., Pet. Div., Am. Chem. Soc., Atlanta* p. 357 (1981).
43. H. Buijs, *Spectroscopy* **1**, 14 (1986).
44. D. B. Chase, *Anal. Chem.* **59**, 881A (1987).
45. V. Rives-Arnau and N. Sheppard, *J. Chem. Soc., Faraday Trans. 1* **76**, 394 (1980).
46. J. Heaviside, P. J. Hendra, P. Tsai, and R. P. Cooney, *J. Chem. Soc., Faraday Trans. 1* **74**, 2542 (1978).
47. I. Artaki, M. Bradley, T. Zerda, and J. Jonas, *J. Phys. Chem.* **89**, 4399 (1985).
48. J. A. Rabo, ed., *Zeolite Chemistry and Catalysis.* Am. Chem. Soc., Washington, D.C., 1976.
49. M. Deeba, B. Streusand, G. L. Schrader, and B. C. Gates, *J. Catal.* **69**, 218 (1981).
50. J. Gruber, R. Leavitt, C. Morrison, and N. Chang, *J. Chem. Phys.* **82**, 5373 (1985).
51. B. R. York, S. A. Solin, N. Wada, R. H. Raythatha, I. D. Johnson, and T. J. Pinnavaia, *Solid State Commun.* **54**, 475 (1985).
52. E. Yeager, D. A. Scherson, and C. A. Fierro, in *Catalyst Characterization Science. Surface and Solid State Chemistry* (M. L. Deviney and J. D. Gland, eds.), ACS Symp. Ser. No. 288. Am. Chem. Soc., Washington, D.C., 1985.
53. G. Nauer, P. Strecha, N. Brinda-Konopik, and G. Liptay, *J. Therm. Anal.* **30**, 813 (1985).
54. J. J. Freeman and M. L. Unland, *J. Catal.* **54**, 183 (1978).
55. H. Baranska, B. Czerwinska, and A. Labudzinska, *J. Mol. Struct.* **143**, 485 (1986).
56. K.-J. Chao, T. C. Tasi, M.-S. Chen, and I. Wang, *J. Chem. Soc., Faraday Trans. 1* **77**, 547 (1981).
56a. P. K. Dutta and M. Puri, *J. Phys. Chem.* **91**, 4329 (1987).
57. F. Roozeboom, H. E. Robson, and S. S. Chan, *Zeolites* **3**, 321 (1983).
58. C. L. Angell and W. H. Flank, in *Molecular Sieves II* (J. R. Katzer, ed.), ACS Chem. Ser. No. 40. Am. Chem. Soc., Washington, D.C., 1977.
59. B. D. McNicol, G. T. Pott, K. R. Loos, and N. Mulder, *Adv. Chem. Ser.* **121**, 152 (1973).

60. B. D. McNicol, G. T. Pott, and K. R. Loos, *J. Phys. Chem.* **76**, 3388 (1972).
61. J. L. Guth, P. Gaullet, P. Jacques, and R. Wey, *Bull. Soc. Chim. Fr.* **3-4**, 121 (1980).
62. J. L. Guth, P. Gaullet, and R. Wey, in *Zeolites* (L. V. C. Rees, ed.) Proc. 5th Int. Conf., p. 30. Heyden, London 1980.
63. C. L. Angell, *J. Phys. Chem.* **77**, 222 (1973).
64. P. K. Dutta and D. C. Shieh, *J. Phys. Chem.* **90**, 2331 (1986).
65. P. K. Dutta, D. C. Shieh, and M. Puri, *J. Phys. Chem.* **91**, 2332 (1987).
66. P. K. Dutta, B. Del Barco, and D. C. Shieh, *Chem. Phys. Lett.* **127**, 200 (1986).
67. P. K. Dutta and B. Del Barco, *J. Phys. Chem.* **89**, 1861 (1985).
68. T. A. Egerton, A. H. Hardin, Y. Kozirovski, and N. Sheppard, *J. Catal.* **32**, 343 (1974).
69. T. A. Egerton, A. H. Hardin, and N. Sheppard, *Can. J. Chem.* **54**, 586 (1976).
70. N. T. Tam, P. Tasi, and R. P. Cooney, *Aust. J. Chem.* **31**, 255 (1978).
71. R. P. Cooney and P. Tsai, *J. Raman Spectrosc.* **8**, 195 (1979).
72. W. Krasser, A. Fadini, and A. Renouprez, *J. Mol. Struct.* **60**, 427 (1980).
73. M. Sano, T. Maruo, H. Yamatera, M. Suzuki, and Y. Saito, *J. Am. Chem. Soc.* **109**, 52 (1987).
74. T. I. Morrison, P. J. Viccaro, and G. K. Shenoy, in *Intrazeolite Chemistry* (G. C. Stucky and F. G. Dwyer, eds.), ACS Symp. Ser. No. 218. Am. Chem. Soc., Washington, D.C., 1980.
75. M. Mehicic, H. Fochler, and M. A. Hazle, BP America R&D, Cleveland, Ohio (unpublished results).
76. M. R. Antonio, BP America R&D, Cleveland, Ohio (private communication).
77. P. Walther, *Z. Chem.* **26**, 189 (1986).
78. K. Moeller, C. Penker, W. Pilz, and W. Schirmer, *Wiss. Z. Humboldt-Univ. Berlin, Math.-Naturwiss. Reihe* **34**, 22 (1985).
79. F. E. Massoth, *Adv. Catal.* **27**, 265 (1978).
80. F. E. Massoth and G. MuraliDhar, in *Chemistry and Uses of Molybdenum* (H. F. Barry and P. C. H. Mitchell, eds.), Proc. 4th Int. Conf., Golden, Colorado. Climax Molybdenum Co., Ann Arbor, Michigan, 1982.
81. H. P. Boehm and H. Knoezinger, in *Catalysis: Science and Technology* (J. R. Anderson and M. Boudart, eds.), Vol. 4. Springer-Verlag, Berlin, 1983.
82. Y. I. Yermakov, B. N. Kuznetsov, and V. A. Zakharov, *Catalysis by Supported Complexes.* Elsevier, Amsterdam, 1981.
83. P. L. Villa, F. Trifiro, and I. Pasquon, *React. Kinet. Catal. Lett.* **1**, 341 (1974).
84. F. R. Brown and L. E. Makovsky, *Appl. Spectrosc.* **31**, 44 (1977).
85. F. R. Brown, L. E. Makovsky, and K. H. Rhee, *J. Catal.* **50**, 162 (1977).
86. J. H. Ashley and P. C. H. Mitchell, *J. Chem. Soc. A* p. 2730 (1969).
87. G. N. Asmolov and O. V. Krylov, *Kinet. Catal.* **11**, 847 (1970).

88. W. H. J. Stork, J. G. F. Coolegem, and G. T. Pott, *J. Catal.* **32**, 497 (1974).
89. F. Massoth, *J. Catal.* **36**, 164, (1975).
90. J. T. Richardson, *Ind. Eng. Chem. Fundam.* **3**, 154 (1964).
91. J. M. J. G. Lipsch and G. C. A. Schuit, *J. Catal.* **15**, 174 (1969).
92. J. Sonnemans and P. Mars, *J. Catal.* **31**, 209 (1973).
93. D. S. Zingg, L. E. Makovsky, R. E. Tischer, F. R. Brown, and D. M. Hercules, *J. Phys. Chem.* **84**, 2898 (1980).
94. L. E. Makovsky, J. M. Stencel, F. R. Brown, R. E. Tischer, and S. S. Pollack, *J. Catal.* **89**, 334 (1984).
95. E. Payen, J. Barbillat, J. Grimblot, and J. P. Bonnelle, *Spectrosc. Lett.* **11**, 997 (1978).
96. L. Wang and W. K. Hall, *J. Catal.* **66**(1), 251 (1980).
97. J. Medema, C. van Stam, V. H. J. de Beer, A. J. A. Konings, and D. C. Koningsberger, *J. Catal.* **53**, 386 (1978).
98. H. Knoezinger and H. Jeziorowski, *J. Phys. Chem.* **82**, 2002 (1978).
99. H. Jeziorowski and H. Knoezinger, *C.R.—Conf. Int. Spectrosc. Raman, 7th, 1980* p. 434 (1980).
100. H. Jeziorowksi and H. Knoezinger, *Appl. Surf. Sci.* **5**, 35 (1980).
101. H. Jeziorowski and H. Knoezinger, *J. Phys. Chem.* **83**, 1166 (1979).
102. F. R. Brown, R. Tischer, L. E. Makovsky, and K. H. Rhee, *Prepr., Div. Pet. Chem., Am. Chem. Soc., Anaheim Meeting* **23**, 65 (1978).
103. P. Dufresne, E. Payen, J. Grimblot, and J. P. Bonnelle, *J. Phys. Chem.* **85**, 2344 (1981).
104. L. Wang and W. K. Hall, *J. Catal.* **83**, 242 (1983).
105. J. M. Stencel, J. R. Diehl, J. R. D'Este, L. E. Makovsky, L. Rodrigo, K. Marcinkowska, A. Adnot, P. C. Roberge, and S. Kaliaguine, *J. Phys. Chem.* **90**, 4739 (1986).
106. J. M. Stencel, L. E. Makovsky, T. A. Sarkus, J. de Vries, R. Thomas, and J. A. Moulijn, *J. Catal.* **90**, 314 (1984).
107. E. Payen, S. Kasztelan, J. Grimblot, and J. P. Bonnelle, *J. Raman Spectrosc.* **17**, 233 (1986).
108. S. S. Chan, I. E. Wachs, L. L. Murrell, L. Wang, and W. K. Hall, *J. Phys. Chem.* **88**, 5831 (1984).
109. J. A. Horsley, I. E. Wachs, J. M. Brown, G. H. Via, and F. D. Hardcastle, *J. Phys. Chem.* **91**, 4014 (1987).
110. E. Payen, J. Grimblot, and S. Kasztelan, *J. Phys. Chem.* **91**, 6642 (1987); E. Payen, S. Kasztelan, S. Houssenbay, R. Szymanski, and J. Grimblot, *ibid.* **93**, 6501 (1989).
111. M. LoJacono, A. Cimino, and G. C. A. Schuit, *Gazz. Chim. Ital.* **103**, 1281 (1973).
112. F. R. Brown, R. E. Tischer, L. E. Makovsky, and K. H. Rhee, *Prepr., Div. Pet. Chem., Am. Chem. Soc., Anaheim Meeting* **23**, 65 (1978).

113. R. L. Chin and D. M. Hercules, *J. Phys. Chem.* **86**, 360 (1982).
114. P. Gajardo, P. Grange, and B. Delmon, *J. Catal.* **63**, 201 (1980).
115. R. L. Chin and D. M. Hercules, *J. Phys. Chem.* **86**, 3079 (1982).
116. A. Nishijima, H. Shimada, T. Sato, Y. Yoshimura, and J. Hiraishi, *Polyhedron* **5**, 243 (1986).
117. M. A. Apecetche, M. Houalla, and B. Delmon, *SIA, Surf. Interface Anal.* **3**, 90 (1981).
118. K. S. Chung and F. E. Massoth, *J. Catal.* **64**, 320 (1980).
119. H. Knoezinger, H. Jeziorowski, and E. Taglauer, *Stud. Surf. Sci. Catal.* **7** (Pt. A, New Horiz. Catal.), 604 (1981).
120. H. Knoezinger, J. Jeziorowski, and E. Taglauer, *Proc. Int. Congr. Catal., 7th, Tokyo, 1980* Paper A42 (1981).
121. H. Knoezinger, *Erdoel, Erdgas, Kohle* **102**, 300 (1986).
122. Z. Wang, L. Wanzhen, and M. Enze, *China-Jpn.-U.S. Symp. Heterog. Catal. Relat. Energy Probl., 1982* **A25C**, 6 (1982).
123. S. S Chan and I. E. Wachs, *J. Catal.* **103**, 224 (1987).
124. F. R. Brown, L. E. Makovsky, and K. H. Rhee, *J. Catal.* **50**, 385 (1977).
125. G. L. Schrader and C. P. Cheng, *J. Catal.* **80**, 369 (1983).
126. G. L. Schrader and C. P. Cheng, *J. Catal.* **85**, 488 (1984).
127. G. Hagenbach, P. Courty, and B. Delmon, *J. Catal.* **31**, 264 (1973).
128. V. H. J. de Beer, C. Bevelander, T. H. M. van Sint Fiet, and P. G. A. J. Werter, C. H. Amberg, *J. Catal.* **43**, 68 (1976).
129. G. C. A. Schuit and B. C. Gates, *AIChE J.* **19**, 417 (1973).
130. R. J. H. Voorhoeve, *J. Catal.* **23**, 236 (1971).
131. E. Payen, S. Kasztelan, J. Grimblot, and J. P. Bonnelle, *J. Mol. Struct.* **143**, 259 (1986); **174**, 71 (1988).
132. L. Rodrigo, K. Marcinkowska, A. Adnot, P. C. Roberge, S. Kaliaguine, J. M. Stencel, L. E. Makovsky, and J. R. Diehl, *J. Phys. Chem.* **90**, 2690 (1986).
132a. S. Abdo, A. Kazusaka, and R. F. Howe, *J. Phys. Chem.* **85**, 1380 (1981).
133. C. P. Cheng, J. D. Ludovise, and G. L. Schrader, *Appl. Spectrosc.* **34**, 146 (1980).
134. C. P. Cheng and G. L. Schrader, *J. Catal.* **60**, 276 (1979).
135. K. Y. S. Ng and E. Gulari, *J. Catal.* **92**, 340 (1985).
136. J. Leyrer, B. Vielhaber, M. I. Zaki, Z. Shuxian, J. Weitkamp, and H. Knoezinger, *Mater. Chem. Phys.* **13**, 301 (1985).
137. Jeziorowski, H. Knoezinger, P. Grange, and P. Gajardo, *J. Phys. Chem.* **84**, 1825 (1980).
138. A. Lopez Agudo, F. J. Gil, J. M. Calleja, and V. Fernandez, *J. Raman Spectrosc.* **11**, 454 (1981).
139. R. Thomas, M. C. Mittelmeijer-Hazeleger, F. P. J. M. Kerkhof, J. A. Moulijn, J. Medema, and V. H. J. de Beer, in *Chemistry and Uses of Molybdenum* (H. F.

Barry and P. C. H. Mitchell, eds.), Proc. 3rd Int. Conf., Ann Arbor. Climax Molybdenum Co., Ann Arbor, Michigan, 1979.

140. A. Iannibello, S. Margeno, P. Tittarelli, G. Morelli, and A. Zecchina, *J. Chem. Soc., Faraday Trans. 1* **80**, 2209 (1984).

141. G. L. Schrader and C. P. Cheng, *J. Phys. Chem.* **87**, 3675 (1983).

142. A. Spoyakina, B. Gigov, and D. Shopov, *React. Kinet. Catal. Lett.* **19**, 11 (1982).

143. M. V. Landau, L. N. Alekseenko, B. K. Nefedov, M. A. Kipnis, D. A. Agievskii, G.D. Chukin, S. A. Sergienko, and V. I. Kvashonkin, *Kinet. Katal.* **25**, 934 (1984).

144. E. Payen and J. Barbillat, *Actual. Chim.* **4**, 60 (1980).

145. B. Sombret, P. Dhamelincourt, F. Wallart, A. C. Muller, M. Bouquet, and J. Grosmangin, *J. Raman Spectrosc.* **9**, 291 (1980).

146. D. Littlejohn and S.-G. Chang, *Environ. Sci. Technol.* **18**, 305 (1984).

147. J. P. Baltrus, L. E. Makovsky, J. M. Stencel, and D. M. Hercules, *Anal. Chem.* **57**, 2500 (1985).

148. K. Weissermel and H. J. Arpe, *Industrial Organic Chemistry*. Verlag Chemie, New York, 1978.

149. R. Thomas, J. A. Moulijn, and F. P. J. M. Kerkhof, *Recl. Trav. Chim. Pays-Bas* **96**, 134 (1977).

150. L. Salvati, Jr., L. E. Makovsky, J. M. Stencel, F. R. Brown, and D. M. Hercules, *J. Phys. Chem.* **85**, 3700 (1981).

151. P. Biloen and G. T. Pott, *J. Catal.* **30**, 169 (1973).

152. W. H. J. Stork and G. T. Pott, *Recl. Travl. Chim. Pays-Bas* **96**, 104 (1977).

153. G. T. Pott and W. H. J. Stork, in *Preparation of Catalysts* (B. Delmon, P. A. Yacobs, and G. Poncelet, eds.), p. 537. Elsevier, Amsterdam, 1976.

154. K. T. Ng and D. M. Hercules, *J. Phys. Chem.* **80**, 2094 (1976).

155. R. Thomas, F. P. M. J. Kerkhof, J. A. Moulijn, J. Medema, and V. H. J. de Beer, *J. Catal.* **61**, 559 (1980).

156. S. S. Chan, I. E. Wachs, and L .L. Murrell, *J. Catal.* **90**, 150 (1984).

157. S. S. Chan, I. E. Wachs, L. L. Murrell, and N. C. Dispenziere, Jr., *Stud. Surf. Sci. Catal.* **19** (Catal. Energy Scene), 259 (1984).

158. S. S. Chan, I. E. Wachs, L. L. Murrell, and N. C. Dispenziere, Jr., *J. Catal.* **92**, 1 (1985).

159. L. L. Murrell, D. G. Grenoble, R. T. K. Baker, E. B. Prestridge, S. C. Fung, R. R. Chianelli, and S. P. Cramer, *J. Catal.* **79**, 203 (1983).

160. J. M. Stencel, L. E. Makovsky, J. R. Diehl, and T. A. Sarkus, *J. Raman Spectrosc.* **15**, 282 (1984).

161. A. Iannibello, P. L. Villa, and S. Marengo, *Gazz. Chim. Ital.* **109**, 521 (1979).

162. E. Payen, S. Kasztelan, J. Grimblot, and J. P. Bonnelle, *Catal. Today* **4**, 57 (1988).

163. P. Mars and D.W. van Krevelen, *Chem. Eng.*, Spec. Suppl. **3**, 41 (1954).

164. W. M. H. Sachtler, *Catal. Rev.* **4**, 27 (1970).

165. F. Roozeboom, J. Medema, and P. J. Gellings, *Z. Phys. Chem.* (Wiesbaden) **111**(2), 215 (1978).

165a. F. Roozeboom, T. Fransen, P. Mars, P. J. Gellings, *Z. Anorg. Allg. Chem.* **449**, 25 (1979).

166. F. Roozeboom, M. C. Mittelmeijer-Hazeleger, J. A. Moulijn, J. Medema, V. H. J. De Beer, and P. J. Gellings, *J. Phys. Chem.* **84**(21), 2783 (1980).

167. A. J. Van Hengstum, J. G. Van Ommen, H. Bosch, and P. J. Gellings, *Appl. Catal.* **5**, 205 (1983).

168. I. E. Wachs, R. Y. Saleh, S. S. Chan, and C. C. Chersich, CHEMTECH **15**, 756 (1985).

169. I. E. Wachs, R. Y. Saleh, S. S. Chan, and C. C. Chersich, *Appl. Catal.* **15**, 339 (1985).

170. I. E. Wachs, S. S. Chan, and R. Y. Saleh, *J. Catal.* **91**, 366 (1985).

171. R. Y. Saleh, I. E. Wachs, S. S. Chan, and C. C. Chersich, *J. Catal.* **98**, 102 (1986).

172. I. E. Wachs and S. S. Chan, *Appl. Surf. Sci.* **20**, 181 (1984).

173. I. E. Wachs, S. S. Chan, C. C. Chersich, and R. Y. Sazeh, in *Catalysis on the Energy Scene* (S. Kaliaguine and A. Mahay, eds.), p. 275. Elsevier, Amsterdam, 1984.

174. I. E. Wachs, S. S. Chan, and C. C. Chersich, in *Reactivity of Solids* (P. Barret and L. C. Dufour, eds.), p. 1047. Elsevier, Amsterdam, 1985.

175. R. Y. Saleh, I. E. Wachs, S. S. Chan, and C. C. Chersich, *Prepr., Div. Pet. Chem., Am. Chem. Soc.* **31**, 272 (1986).

175a. S. T. Oyama, G. T. Went, K. B. Lewis, A. T. Bell, and G. A. Somorjai, *J. Phys. Chem.* **93**, 6786 (1989).

176. J. Hanuza, K. Hermanowicz, W. Oganowski, and B. Jezowska-Trzebiatowska, *Bull. Pol. Acad. Sci., Chem.* **31**, (3–7), 1984.

177. B. Strohmeier and D. M. Hercules, *J. Phys. Chem.* **88**, 4922 (1984).

178. J. G. Grasselli, M. A. S. Hazle, J. R. Mooney, and M. Mehicic, *Keynote Lect.—Colloq. Spectrosc. Int., 21st, Cambridge England* (G. F. Kirkbright, ed.). Heyden, London, 1979.

179. F. Veatch, J. L. Callahan, J. D. Idol, and E. C. Milberger, *Chem. Eng. Prog.* **56**, 65 (1960).

180. Sohio, Dutch Patent 7,006,454 (1970); Offenleungsschrift 2,203,710 (1972).

181. J. L. Callahan, R. K. Grasselli, E. C. Milberger, and H. A. Strecker, *Ind. Eng. Chem. Prod. Res. Dev.* **9**, 134 (1970).

182. I. Matsuura and G. C. A. Schuit, *J. Catal.* **25**, 314 (1972).

183. I. Matsuura, *J. Catal.* **33**, 420 (1974).

184. I. Matsuura, *J. Catal.* **35**, 452 (1974).

185. J. Haber and B. Grzybowska, *J. Catal.* **29**, 489 (1973).

186. A. Bielanski and J. Haber, *Catal. Rev. Sci. Eng.* **19**, 1 (1979).

187. A. W. Sleight and W. J. Linn, *Ann. N.Y. Acad. Soc. Sci.*, **272**, 22 (1976).
188. A. W. Sleight, W. J. Linn, and K. Aykan, CHEMTECH p. 235 (1978).
189. G. W. Keulks, L. D. Krenzke, and T. M. Notermann, *Adv. Catal.* **27**, 183 (1978).
190. J. D. Burringston, C. T. Kartisek, and R. K. Grasselli, *J. Catal.* **63**, 235 (1980).
191. J. D. Burrington and R. K. Grasselli, *J. Catal.* **59**, 79 (1979).
192. J. D. Burrington, C. T. Kartisek, and R. K. Grasselli, *J. Catal.* **75**, 225 (1982).
193. J. F. Brazdil, D. D. Suresh, and R. K. Grasselli, *J. Catal.* **66**, 347 (1980).
194. R. K. Grasselli, J. D. Burrington, and J. F. Brazdil, *Faraday Discuss. Chem. Soc.* **72**, 203 (1981).
195. R. K. Grasselli, J. F. Brazdil, and J. D. Burrington, *Proc. Int. Cong. Catal., 8th, Berlin (West), 1983*, Vol. 5, p. 369 (1984).
196. R. K. Grasselli and J. D. Burrington, *Adv. Catal.* **30**, 133 (1981).
197. D. B. Dadyburjor, S. S. Jewur, and E. Ruckenstein, *Catal. Rev. Sci. Eng.* **19**, 293 (1979).
198. K. van der Wiele and P. J. van den Berg, in *Comprehensive Chemical Kinetics*, (C. H. Bamford and C. F. H. Tipper, eds.), Vol. 20. Elsevier, Amsterdam, 1978.
199. R. Higgins and P. Hayden, *The Chem. Soc., Catal Spec. Period. Rep.* **1**, 168 (1977).
200. D. J. Hucknall, *Selective Oxidation of Hydrocarbons.* Academic Press, New York, 1974.
201. G. C. A. Schuit, *J. Less-Common Met.* **36**, 329 (1974).
202. W. D. Kovats, Ph.D. Thesis, University of Wisconsin-Madison (1984).
203. J. F. Brazdil, L. C. Glaeser, and R. K. Grasselli, *J. Catal.* **81**, 142 (1983).
204. I. Matsuura, S. Mizuno, and H. Hashiba, *Polyhedron* **5**, 111 (1986).
205. P. C. H. Mitchell and F. Trifiro, *J. Chem. Soc. A* p. 3183 (1970).
206. F. Trifiro, P. Centola, and I. Pasquon, *J. Catal.* **10**, 86 (1968).
207. I. Matsuura, R. Schut, and K. Hirakawa, *J. Catal.* **63**, 152 (1980).
208. E. V. Hoefs, J. R. Monnier, and G. W. Keulks, *J. Catal.* **57**, 331 (1979).
209. T. G. Alkhazov, K. Yu Adzhamov, L. Ya. Margolis, O. V. Krylov, and G. W. Keulks, *Kinet. Katal.* **18**, 591 (1977).
210. T. Notermann, G. W. Keulks, A. Skliarov, Yu. Maximov, L. Ya. Margolis, and O. V. Krylov, *J. Catal.* **39**, 286 (1975).
211. F. Trifiro, L. Kubelkova, and I. Pasquon, *J. Catal.* **19**, 121 (1970).
212. F. Trifiro, H. Hoser, and R. D. Scarle, *J. Catal.* **25**, 12 (1972).
213. See Ref. 31, pp. 94–96.
214. W. D. Kovats and C. G. Hill, Jr., *Appl. Spectrosc.* **40**, 1215 (1986).
215. M. Akimoto and E. Echigoya, *J. Chem. Soc., Faraday Trans. I* p. 1757 (1979).
216. M. Akimoto and E. Echigoya, *J. Catal.* **35**, 278 (1974).
217. T. G. Alkhazov and K. Yu. Adzhamov, *Kinet. Katal.* **15**, 201 (1974).
218. P. L. Villa, A. Szabo, F. Trifiro, and M. Carbucicchio, *J. Catal.* **47**, 122 (1977).

219. A. F. van den Elzen and G. D. Rieck, *Acta Crystallogr.*, **B29**, 2436 (1973).
220. A. F. van den Elzen and G. D. Rieck, *Acta Crystallogr.* **B29**, 2433 (1973).
221. A. F. van den Elzen and G. D. Rieck, *Mater. Res. Bull.* **10**, 1163 (1975).
222. H. Y. Chen and A. W. Sleight, *J. Solid State Chem.* **63**, 70 (1986).
223. F. Theobald, A. Laarif, and A. W. Hewat, *Ferroelectrics* **56**, 219 (1984).
224. R. G. Teller, J. F. Brazdil, and R. K. Grasselli, *Acta Crystallogr.* **C40**, 2001 (1984).
225. F. Theobald, A. Laarif, and A. W. Hewat, *Mater. Res. Bull.* **20**, 653 (1985).
226. M. R. Antonio, R. G. Teller, D. R. Sandstrom, M. Mehicic, and J. F. Brazdil, *J. Phys. Chem.* **92**, 2939 (1988).
227. M. R. Antonio, J. F. Brazdil, L. C. Glaeser, M. Mehicic, and R. G. Teller, *J. Phys. Chem.* **92**, 2338 (1988).
228. K. Nakamoto, *Infrared and Raman Spectra of Inorganic and Coordination Compounds.* Wiley, New York, 1986.
229. J. R. Ferraro, *Low Frequency Vibrations of Inorganic and Coordination Compounds.* Plenum, New York, 1971.
230. F. A. Cotton and G. Wilkinson, *Advanced Inorganic Chemistry: A Comprehensive Text*, 4th ed. Wiley, New York, 1980.
231. A. F. Wells, *Structural Inorganic Chemistry*, 5th ed. Oxford University Press, New York, 1986.
232. N. Weinstock, H. Schulze, and A. Muller, *J. Chem. Phys.* **59**, 5063 (1973).
233. J. Aveston, E. W. Anacker, and J. S. Johnson, *Inorg. Chem.* **3**, 735 (1964).
234. D. S. Honig and K. Kustin, *Inorg. Chem.* **11**, 65 (1972).
235. J. F. Ojo, R. S. Taylor, and A. G. Sykes, *J. Chem. Soc., Dalton Trans.* p. 500 (1975).
236. F. Trifiro, P. Forzatti, and P. L. Villa, in *Preparation of Catalysts* (B. Delmon, F. A. Jacobs, and G. Poncelet, eds.). Elsevier, Amsterdam, 1976.
237. H. T. Evans, B. M. Gatehouse, and P. Leverett, *J. Chem. Soc., Dalton Trans.* p. 505 (1975).
238. K. H. Tytko and O. Glemser, *Adv. Inorg. Chem. Radiochem.* **19**, 239 (1976).
239. F. A. Cotton and R. M. Wing, *Inorg. Chem.* **4**, 867 (1965).
240. R. H. Busey and O. L. Keller, Jr., *J. Chem. Phys.* **41**, 215 (1964).
241. W. P. Griffith and P. J. B. Lesniak, *J. Chem. Soc. A* p. 1066 (1969).
242. C. P. Cheng and G. L. Schrader, *J. Catal.* **60**, 276 (1979).
243. G. L. Schrader, M. S. Basista, and C. B. Bergman, *Chem. Eng. Commun.* **12**, 121 (1981).
244. V. A. Maroni and T. G. Spiro, *J. Am. Chem. Soc.* **88**, 1410 (1966).
245. V. A. Maroni and T. G. Spiro, *Inorg. Chem.* **7**, 183 (1968).
246. H. Jeziorowski and H. Knoezinger, *J. Phys. Chem.* **83**, 1166 (1979).
247. W. Oganowski, J. Hanuza, B. Jezowska-Trzebiatowska, and J. Wrzyszcz, *J. Catal.* **39**, 161 (1975).

248. R. J. Betsch and W. B. White, *Spectrochim. Acta* **34A**, 505 (1977).
249. A. Watanabe and H. Kodama, *J. Solid State Chem.* **35**, 240 (1980).
250. P. Gaucher, V. Ernst, and P. Courtine, *J. Solid State Chem.* **47**, 47 (1983).
251. L. Ya. Erman and E. L. Galperin, *Russ. J. Inorg. Chem.* (*Engl. Transl.*) **13**, 487 (1968).
252. L. Ya. Erman and E. L. Galperin, *Russ. J. Inorg. Chem.* (*Engl. Transl.*) **15**, 441 (1970).
253. L. Ya. Erman, E. L. Galperin, and B. P. Sobelev, *Russ. J. Inorg. Chem.* (*Engl. Transl.*) **16**, 258 (1971).
254. A. W. Sleight and W. Jeitscho, *Mater. Res. Bull.* **9**, 951 (1974).
255. W. J. Linn and A. W. Sleight, *J. Catal.* **41**, 134 (1976).
256. W. Jeitschko, A. W. Sleight, W. R. McClellan, and J. F. Weiher, *Acta Crystallogr., Sect. B* **B32**, 1163 (1976).
257. B. Grzybowska, E. Payen, L. Gengembre, and J. P. Bonnelle, *Bull. Pol. Acad. Sci., Chem.* **31**, 245 (1984).
258. P. Forzatti, P. L. Villa, N. Ferlazzo, and D. Jones, *J. Catal.* **76**, 188 (1982).
259. M. Lo Jacono, T. Notermann, and G. W. Keulks, *J. Catal.* **40**, 19 (1975).
260. See Ref. 30, pp. 28–33.
261. L. D. Krenzke and G. W. Keulks, *J. Catal.* **64**, 295 (1980).
262. M. Egashira, K. Matsuo, S. Kagawa, and T. Seiyama, *J. Catal.* **58**, 409 (1979).
263. J. Kumar and E. Ruckenstein, *J. Solid State Chem.* **31**, 41 (1980).
264. J. Haber, *J. Less-Common Met.* **54**, 243 (1977).
265. P. A. Batist, personal communication to J. Kumar and E. Ruckenstein.
266. M. Mehicic, BP America R&D, Cleveland, Ohio (unpublished results).
267. G. W. Keulks, *J. Catal.* **19**, 232 (1970).
268. G. W. Keulks and L. D. Krenzke, *Proc. Int. Congr. Catal., 6th, 1976*, Vol. 2, p. 806 (1977).
269. R. D. Wragg, P. R. Ashmore, and J. A. Hockey, *J. Catal.* **22**, 49 (1971).
270. T. Otsubo, H. Mirura, Y. Morikawa, and T. Shirasaki, *J. Catal.* **36**, 240 (1975).
271. Y. Moro-Oka, W. Ueda, S. Tanaka, and T. Ikawa, *Proc. Int. Congr. Catal., 7th, 1980*, Part B, p. 1086 (1981).
272. K. M. Sancier, P. R. Wentrcek, and H. Wise, *J. Catal.*, **39**, 141 (1975).
273. L. D. Krenzke and G. W. Keulks, *J. Catal.* **61**, 316 (1980).
274. L. C. Glaeser, J. F. Brazdil, M. A. S. Hazle, M. Mehicic, and R. K. Grasselli, *J. Chem. Soc., Faraday Trans. 1* **81**, 2903 (1985).
275. I. Matsuura, H. Hashiba, and I. Kanesaka, *Chem. Lett. Chem. Soc. Jpn., 1986*, pp. 533–6.
276. W. Ueda, Y. Moro-oka, and T. Ikawa, *J. Chem. Soc., Farady Trans. 1* **78**, 495 (1982).
277. C. Machiels and A. W. Sleight, in *Chemistry and Uses of Molybdenum* (H. F.

Barry and P. C. H. Mitchell, eds.), Proc. 4th Int. Conf., Golden, Colorado, p. 411. Climax Molybdenum Co., Ann Arbor, Michigan, 1982.

278. M. J. Verkerk, M. W. J. Hammink, and A. J. Burggraff, *J. Electrochem. Soc.* **130** 70 (1983).
279. I. Matsuura, *Proc. Int. Congr. Catal. 7th, Tokyo, 1980*, Part B, p. 1099 (1981).
280. G. W. Smith and J. A. Ibers, *Acta Crystallogr.*, **19**, 269 (1965).
281. S. C. Abrahams and J. M. Reddy, *J. Chem. Phys.* **43**, 2533 (1965).
282. P. Tarte and M. Liegeois-Duyckaerts, *Spectrochim. Acta* **28A**, 2029 (1972).
283. P. Tarte and M. Liegeois-Duyckaerts, *Spectrochim. Acta* **28A**, 2037 (1972).
284. S. S. Saleem, *Infrared Phys.* **27**, 309 (1987).
285. P. Forzatti, P. L. Villa, D. Ercoli, and F. Gasparini, *Ind. Eng. Chem. Prod. Res. Div.* **16**, 26 (1977).
286. M. A. Hazle and M. Mehicic, BP America R&D, Cleveland, Ohio (unpublished spectra).
287. T. S. R. Prasada Rao and P. G. Menon, *J. Catal.* **51**, 64 (1978).
288. D. E. Irish, *Adv. Infrared Raman Spectrosc.* **2** (1976).
289. R. S. Tobias, in *The Raman Effect* (A. Anderson, ed.), Vol. 2. Dekker, New York, 1973.
290. G. L. Schrader, U. Sivrioglu, and M. A. Basista, in *Chemistry and Uses of Molybdenum* (H. F. Barry and P. C. H. Mitchell, eds.), Proc. 4th Int. Conf., Golden, Colorado. Climax Molybdenum Co., Ann Arbor, Michigan, 1982.
291. I. Pasquon, F. Trifiro, and G. Caputo, *Chim. Ind.* (*Milan*) **55**, 168 (1973).
292. F. Trifiro, V. DeVecchi, and I. Pasquon, *J. Catal.* **15**, 8 (1969).
293. F. Trifiro, S. Notarbartolo, and I. Pasquon, *J. Catal.* **22**, 324 (1971).
294. L. Cairati, M. Carbucicchio, D. Ruggeri, and F. Trifiro, in *Preparation of Catalysts II. Scientific Bases for the Preparation of Heterogeneous Catalysts* (B. Delmon, P. A. Jacobs, and G. Poncelet, eds.). Elsevier, New York, 1975.
295. F. Trifiro and I. Pasquon, *J. Catal.* **12**, 412 (1968).
296 F. P. J. M. Kerkhof, J. A. Moulijn, and R. Thomas, *J. Catal.* **56**(2), 279 (1979).
297. L. A. Woodward and H. L. Roberts, *Trans. Faraday Soc.* **52**, 615 (1956).
298. N. Giordano, S. C. J. Bart, A. Vachi, A. Castellan, and G. Martinotti, *J. Catal.* **36**, 81 (1975).
299. R. Nakamura, F. Abe, and E. Echigoya, *Chem. Lett., Chem. Soc., Jpn., 1981*, pp. 51–54.
300. H. C. Yao and M. Shelef, *J. Catal.* **44**, 392 (1976).
301. L. Wang and W. K. Hall, *J. Catal.* **82**, 177 (1983).
302. A. A. Olsthoorn and C. Boelhouwer, *J. Catal.* **44**, 197 (1976).
303. L. Salvati, Jr., G. L. Jones, and D. M. Hercules, *Appl. Spectrosc.* **34**, 624 (1980).
304. R. Thomas, J. A. Moulijn, V. H. J. de Beeg, and J. Medema, *J. Mol. Catal.* **8**, 161 (1980).

305. F. P. J. M. Kerkhof, J. A. Moulijn, R. Thomas, and J. C. Oudejans, *Stud. Surf. Sci. Catal.* **3** (Prep. Catal. II), 77 (1979).
306. M. Ward, T. Harris, and J. Schwartz, *J. Chem. Soc., Chem. Commun.* **8**, 357 (1980).
307. A. Theolier, A. Choplin, L. D'Ornelas, and J. Basset, *Polyhedron* **2**, 119 (1983).
308. A. Bobrov, Ya. Kimel'fel'd, L. Glebov, G. Kliger, and S. Loktev, *Chem. Phys. Lett.* **118**, 156 (1985).
309. H. Zhang and G. L. Schrader, *J. Catal.* **95**, 325 (1985).
310. K. M. Sancier, W. E. Isakson, and H. Wise, *Adv. Chem. Ser.* **178**, 129 (1979).
311. G. Wedler and D. Borgmann, *Ber. Bunsenges. Phys. Chem.* **78**, 67 (1974).
312. H. D. Kaesz and R. B. Saillant, *Chem. Rev.* **72**, 231 (1972).
313. B. F. G. Johnson and J. Lewis, *Adv. Inorg. Chem. Radiochem.* **24**, 261 (1981).
314. G. Blyholder and L. D. Neff, *J. Phys. Chem.* **66**, 1664 (1962).
315. N. A. Rubene, A. A. Davydov, A. V. Kravstsov, N. V. Usheva, and S. I. Smolyaninov, *Kinet. Catal.* (*Engl. Transl.*) **17**, 400 (1976).
316. V. G. Amerikov and L. A. Kasatkina, *Kinet. Catal.* (*Engl. Transl.*) **12**, 137 (1971).
317. G. Busca and V. Lorenzelli, *J. Catal.* **66**, 155 (1980).
318. C. H. Rotchester and S. A. Topham, *J. Chem. Soc., Faraday Trans. 1* **75**, 872 (1979).
319. J. A. Amelse, L. H. Schwartz, and J. B. Butt, *J. Catal.* **72**, 95 (1981).
320. S. J. Teichner, F. Blanchard, B. Pommier, and J. P. Raymond, *Abstr. 8th North Am. Meet. Catal. Soc., Philadelphia, May 1–4* (1983).
321. M. Tanaka and G. Blyholder, *J. Phys. Chem.* **76**, 3180 (1972).
322. D. Dwyer and G. A. Samorjai, *J. Catal.* **52**, 291 (1978).
323. D. Dwyer and G. A. Samorjai, *J. Catal.* **56**, 249 (1979).
324. S. M. Loktev, L. I. Markarenkova, E. V. Slivinskii, and S. D. Entin, *Kinet. Catal.* (*Engl. Transl.*) **13**, 933 (1972).
325. S. M. Loktev, A. N. Bashkirov, E. V. Slivinskii, L. I. Zvezdkina, and Yu. B. Kagan, *Kinet. Catal.* (*Engl. Transl.*) **14**, 175 (1973).
326. S. I. Woo and C. G. Hill, Jr., *J. Mol. Catal.* **29**, 231 (1985).
327. R. Whyman, *J. Organomet. Chem.* **81**, 97 (1974).
328. R. Whyman, *J. Organomet. Chem.* **94**, 303 (1975).
329. D. E. Morris and H. B. Tinker, CHEMTECH, p. 554 (1975).
330. S. I. Woo and C. G. Hill, Jr., *J. Mol. Catal.* **29**, 209 (1985).
331. S. I. Woo and C. G. Hill, Jr., *J. Mol. Catal.* **15**, 309 (1982).
332. A. Kazusaka and R. F. Rowe, *J. Mol. Catal.* **9**, 183 (1980).
333. A. Brenner and R. L. Burwell, Jr., *J. Catal.* **52**, 353 (1978).
334. J. S. Kristoff and D. F. Shriver, *Inorg. Chem.* **13**, 499 (1974).
335. A. Theolier, A. Choplin, L. D'Ornelas, J. M. Basset, G. Zanderighi, and C. Sourisseau, *Polyhedron* **2**, 119 (1983).

336. V. L. Kutznetsov, A. T. Bell, and Y. Yermakov, *J. Catal.* **65**, 374 (1980).
337. J. Goodwin and C. Naccache, *J. Mol. Catal.* **14**, 259 (1981).
338. A. Zecchina, E. Guglielminotti, A. Bossi, and M. Camia, *J. Catal.* **74**, 225 (1982).
339. D. C. Bailey and S. H. Langer, *Chem. Rev.* **81**, 110 (1981).
340. J. M. Stencel, E. B. Bradley, and F. R. Brown, *Appl. Spectrosc.* **34**, 319 (1980).
341. C. Groeneveld, P. P. M. M. Wittgen, A. M. Van Kersbergen, P. L. M. Mestrom, C. E. Nuijten, and G. C. A. Schutt, *J. Catal.* **59**, 153 (1979).
342. M. I. Zaki, N. E. Fouad, J. Leyrer, and H. Knoezinger, *Appl. Catal.* **21**, 359 (1986).
343. J. D. Hardcastle and I. E. Wachs, *J. Mol. Catal.* **46**, 173 (1988).
344. J. A. Campbell, *Spectrochim. Acta* **21**, 1333 (1965).
345. G. Michel and R. Cahay, *J. Raman Spectrosc.* **17**, 79 (1986).
346. W. P. Doyle and P. Eddy, *Spectrochim. Acta* **23A**, 1903 (1967).
347. W. B. Grant and S. Rodhokrishna, *Solid State Commun.* **13**, 109 (1973).
348. W. Scheuerman and G. J. Ritter, *J. Mol. Struct.* **6**, 240 (1970).
349. R. Srivastova and L. L. Chase, *Solid State Commun.* **11**, 349 (1972).
350. T. R. Hart, R. L. Aggarwal, and B. Lax, in *Proc. Int. Conf. Light Scattering Solids, 2nd, 1971* p. 174 (1971).

CHAPTER

11

RAMAN SPECTROSCOPY FOR BIOLOGICAL APPLICATIONS

PHAM V. HUONG

Laboratoire de Spectroscopie
Moleculaire et Cristalline
Universite de Bordeaux I
33405 Talence, France

11.1. INTRODUCTION

Biological systems are complicated, both chemically and physically. Their molecular constituents range from very simple species, like ions or molecules containing one or a few atoms, to very large polymers or macromolecules. These constituents can be in any physical state: solid, liquid, or gas, or be in aqueous or organic solutions, depending on the specific part of the organism being studied. Samples can contain many of these constituents and therefore are generally heterogeneous.

Moreover, biological samples are often living systems and undergo reactions or interactions. Thus knowledge of the dynamics of the systems is necessary to understand the mechanisms involved.

Raman spectroscopy and its many variations, (such as resonance Raman spectroscopy) (1–4), offer a suitable method for the study of these molecular constituents and their reactions.

We shall first select the most important problems concerning biological structures (such as identification of functional groups, conformations, and detection of interaction sites) and mechanisms (such as hydrogen bonding, proton transfer, isomerization, or disulfide conversion). We shall then present appropriate methods for elucidating them. Finally we shall illustrate these various aspects with some examples.

Analytical Raman Spectroscopy, Edited by Jeanette G. Grasselli and Bernard J. Bulkin. Chemical Analysis Series, Vol. 114.
ISBN 0-471-51955-3

11.2. SPECIFIC PROBLEMS IN BIOLOGICAL STUDIES

11.2.1. Identification of Biomolecules

The identification of entire biomolecules is not always easy. But with Raman spectroscopy, identification of chemical groups, such as the C=O and NH groups in peptides, or in amino acids, the purine and pyrimidine bases [adenine (A), guanine (G), cytosine (C), thymine (T), and uracil (U)], and protein S—S bonds is trivial. Characterization of the arrangement of these groups in hydrogen-bond pairing or in stacking is also possible. The identification method is based on comparison of spectral features of unknown components with the spectral characteristics of known model molecules and known structures. The model molecules are monomers or oligomers, and structural information is often gained initially by X-ray crystallography. Such is the case, for instance, for structural determination of oligopeptides (5–7) and nucleotides (8, 9) as well as the determination of conformational arrangements in α-helix and random-coil proteins (10, 11).

11.2.2. Structure in Aqueous Solutions

The structures of some large molecules of biological origin and interest have already been determined by X-ray crystallography. But in biological systems most molecules are surrounded by water or an organic environment. So, an important question is whether the properties in solution will differ from those in the solid state. Raman spectroscopy, which is able to record molecular spectra of compounds both in the solid state and in aqueous solution, offers the answer (12, 13).

11.2.3. Dynamics and Interactions

In living systems, molecular and conformational dynamics occur, and they are responsible for physiological functions. Also molecular and ionic interactions are present in all living systems. Raman, resonance Raman, and time-resolved Raman spectroscopies are helpful in their study.

11.3. INSTRUMENTATION AND PROCEDURES

All the Raman spectra given in this chapter were recorded on normal Raman spectrometers: Jobin-Yvon RAMANOR 2G double monochromator, and Coderg T 800 triple monochromator. Raman microscopy was performed on a single-channel detection Jobin-Yvon MOLE and a DILOR OMARS 89 with multichannel detection. Let us remember that in Raman microscopy, instead of directly illuminating the sample, the incident laser is sent through an optical microscope (2) and scattered light is collected by backscattering from an area as small as 1 μm^2 (Fig. 11.1).

These instruments are frequently equipped with lasers emitting in the visible range with fixed or tunable wavelengths, but altraviolet (UV) lasers and attachments can also be used. Raman spectroscopy using UV excitation has the advantage of avoiding fluorescence that often appears in biological systems at visible wavelengths (14, 15). Nevertheless, although the quality of Raman spectra using UV lasers can be good, they are not as easy to interpret as Raman spectra in the visible range because of the poor selectivity of chromophore wavelengths in the UV. Nearly all important chromophores absorb in so narrow a range that the selective resonance enhancement is not easy to obtain.

Sometimes fluorescence can be suppressed in the Raman simply by changing the wavelength of the exciting line because fluorescence appears at fixed frequencies and the Raman frequency shifts are independent of the wavelength of the exciting line.

Use of excitation in the near-infrared (IR) is also helpful in avoiding fluorescence especially in Fourier transform Raman spectroscopy (16).

The resonance Raman spectra, obtained when the excitation energy approachs or reaches a nonstationary level of a molecule (Fig. 11.2), were recorded with several devices and equipment already described (2).

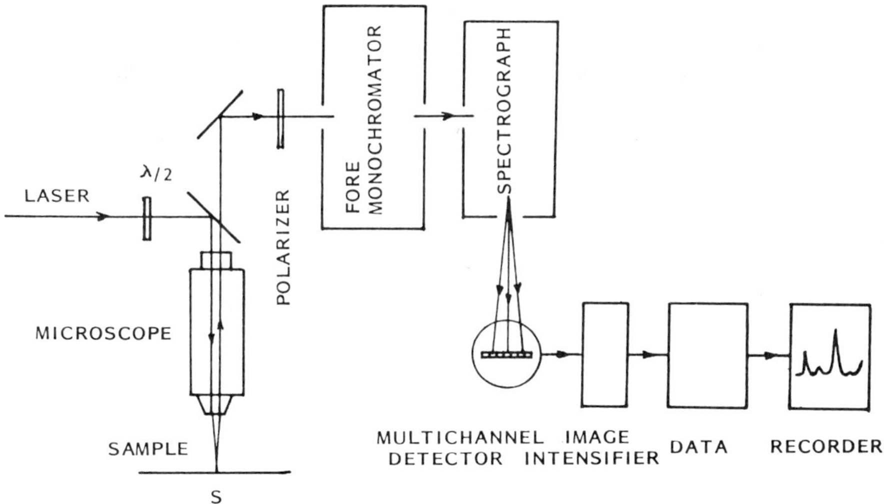

Figure 11.1. Scheme of a Raman microspectrometer equipped with a multichannel detector.

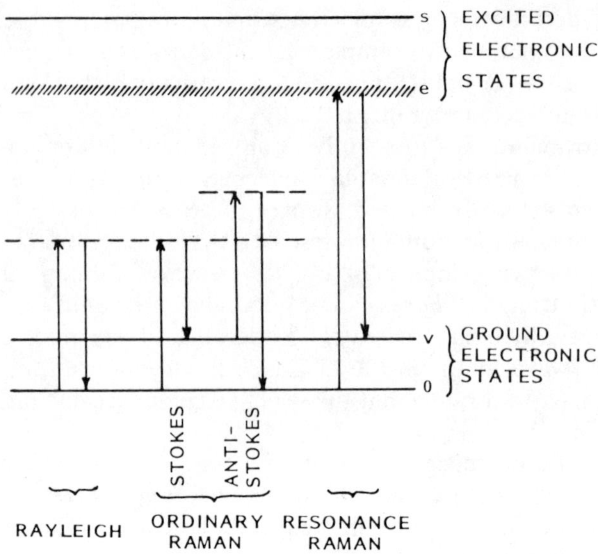

Figure 11.2. Energy levels involved in Raman scattering.

Besides these common techniques for biological studies, coherent anti-Stokes Raman spectroscopy (CARS) can be used, especially to avoid fluorescence (17, 18). In CARS, the sample is illuminated by two lasers with frequencies, ω_1 and ω_2, one of which, ω_1, is fixed and the other, ω_2, is scanned. When $2\omega_1 - \omega_2$ coincides with the frequency of an anti-Stokes Raman vibration or phonon, ω_3, an enhanced Raman line appears at this frequency (Fig. 11.3). The advantage of CARS resides in the fact that the Raman effect is coherently enhanced while fluorescence is not enhanced by the same magnitude.

Fluorescence can sometimes be removed via time discrimination by the

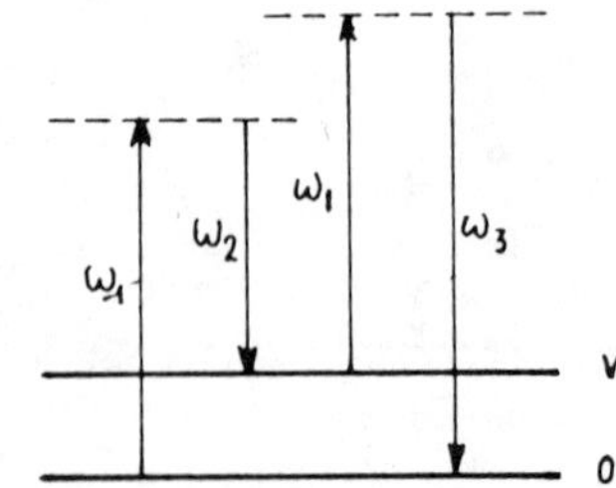

Figure 11.3. Energy level diagram in CARS.

simple principle that the Raman effect is instantaneous while fluorescence has a much longer lifetime in its emission process.

For the study of intermediates, labile compounds, or short-lived substances, rapid-scanning Raman spectrometers offer the possibility of recording molecular spectra in 10^{-4} s, and time-resolved Raman spectrometers allow Raman spectra to be recorded in a nanosecond time scale (19, 20).

All of these instruments can be used to study biological samples at variable temperatures ranging from liquid helium temperatures up to 100°C (13, 21).

11.4. STRUCTURES

11.4.1. Raman Spectra of Living Organisms/Tissues/Cells

Raman spectroscopy is a nondestructive method. Living organisms/tissues/cells can be studied without further preparation. A difficulty comes from the fact that, in biological samples, several constituents can coexist in a small area. So Raman microscopy must be used, as it offers a spatial resolution on the order of 1 μm^2.

By illuminating the wall of a living cell by Raman microscopy, its molecular spectrum can be recorded (2). Of course, moving the *XY* plate supporting the sample to present another area to the laser spot allows the Raman spectrum characteristic of this area to be recorded. This technique is helpful

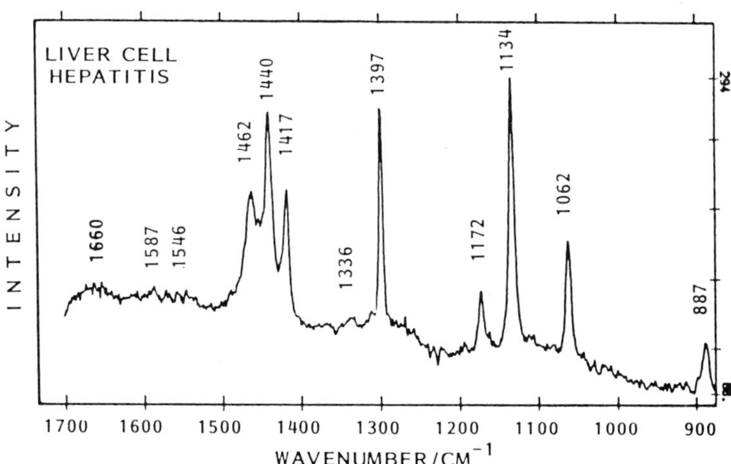

Figure 11.4. Raman spectrum of a liver cell (from a patient with hepatitis treated with the drug methotrexate).

in characterizing liver cells (Fig. 11.4), muscles (22, 23), gallstones (24), eye lenses (25), etc., which are heterogeneous samples.

The following two examples show how Raman spectroscopy aids in understanding the conformations of large structures such as proteins and the dynamics of lipid-chain disordering; in both cases the spectra of entire pieces of tissue have been obtained.

11.4.1.1. Eye Lens

In the Raman spectrum of a rabbit eye lens (Fig. 11.5) the absence of an amide I band at 1650–1660 cm^{-1} indicates that the α-helical conformation (the most stable of the several possible conformations of proteins) is absent. A band at 1675 cm^{-1}, however, indicates the presence of β-sheet or random-coil structure. Disulfide bonds, S—S (500–550 cm^{-1}), which serve to stabilize the coiled arrangement or secondary structure of a protein, are not detected either. On the contrary, a strong S—H band at 2569 cm^{-1} is present, indicative of broken S—S bonds by reduction. The relative intensity of this sulfhydryl Raman band in lens protein with respect to the 2728 cm^{-1} C—H band is often shown as a function of age and cataract progession (10).

11.4.1.2. Biological Membranes

Some nerve cells contain one-layer membrane composed mainly of proteins and long-chain phospholipids. Micro-Raman can be used to study these membranes (2). Raman spectra of nerve fibers can be recorded at several temperatures, in the C—H stretching region. As the three Raman bands at 2850, 2880, and 2930 cm^{-1} are sensitive to chain conformation and molecular interactions, their intensities change upon heating, reflecting a change in the intermolecular ordering of the phospholipids (26). Complementary information can be obtained by following correlative changes in the C—C stretching region; in particular, the intensity of the 1080 cm^{-1} band increases with heating, indicating an increase in intramolecular chain disorder. One can deduce that the disordering of lipid chains must play an important role in the permeability of a membrane.

11.4.2. Raman Spectra of Isolated Biomolecules

11.4.2.1. Structures and Conformations of t-RNA

11.4.2.1.1. Structures in Solutions and Solids. Some biologically active substances can be isolated and their structures deduced by X-ray diffraction. The structures determined by this technique correspond to the

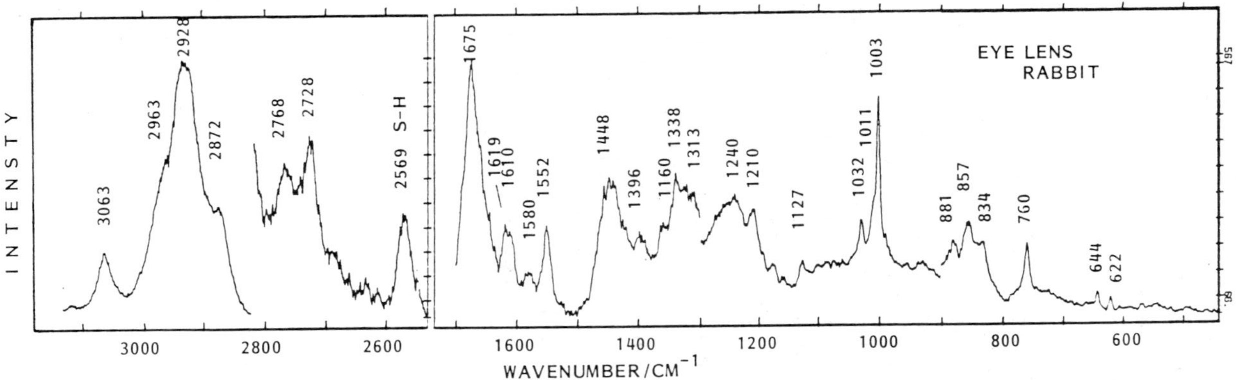

Figure 11.5. Raman spectrum of a rabbit eye lens.

solid state. However, nearly all bioactive substances are in the liquid state and generally exist in a form that is viable in aqueous solutions. Thus, it is interesting to know if the structure determined in the solid state is valuable in understanding the structure in its active state. Raman spectroscopy is helpful in the elucidation of such a problem, as this technique can be applied to both physical states.

The structures of transfer ribonucleic acids (t-RNA) for aspartic acid (t-RNA–Asp) (Fig. 11.6) and phenylalanine (t-RNA–Phe) have been determined by X-ray diffraction (27, 28). These compounds play an important role in protein synthesis. Their specific interactions with synthetases lead to the correct attachment of amino acids, which are joined by peptide bonds to form long protein chains. In another function, their interactions with messenger RNA codons allow the deciphering of the genetic message.

The Raman spectrum of t-RNA–Asp in aqueous solution is displayed in Fig. 11.7. The bands represent the bases G, A, C, and U, and the PO_2^- and —OPO— groups (29) (note that RNA is a polymer of purine and pyrimidine ribonucleotides linked together by 3, 5′-phosphodiester bridges). The band at 814 cm^{-1} gives information about the geometry of the RNA backbone. When the ratio of intensities of this band and the 1100 cm^{-1} band approaches 1.8, the phosphodiester groups in t-RNA are found to be in ordered configuration. The band at 814 cm^{-1} disappears when the ribophosphate backbone is completely disordered. It is also found that the structure of the backbones of t-RNA–Asp and t-RNA–Phe are very similar and superimpose well in the core of the structure, although the conformations of nucleotides differ markedly (28, 30).

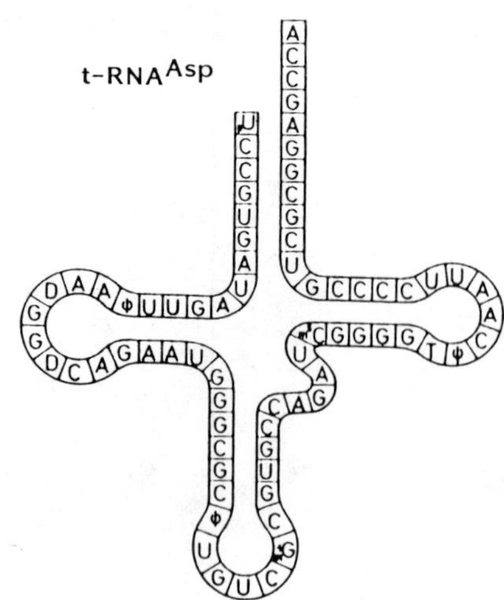

Figure 11.6. Cloverleaf representation of the t-RNA–Asp structure.

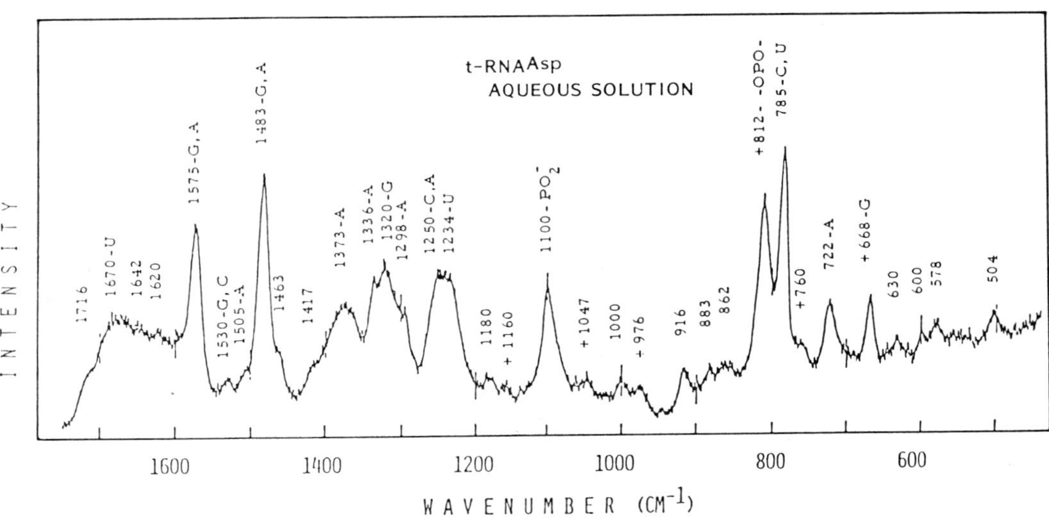

Figure 11.7. Raman spectrum of t-RNA–Asp in aqueous solution.

The Raman intensity of each band depends also on base stacking. In particular, the intensity of the 668 cm^{-1} line decreases three times when passing from the all-stacked t-RNA to the unstacked form. In t-RNA–Asp, it is found that G-base stacking is more effective than in t-RNA–Phe. This can be explained by the greater proportion of G bases localized in the helical regions of the structure.

When comparing the Raman spectrum of t-RNA–Asp in aqueous solution to that of the solid sample (Fig. 11.8), quasi-identical features are recorded, suggesting great similarities between the t-RNA molecule in the crystal form and in solution. The slight difference observed for bands corresponding to G and C bases suggests a slightly more effective stacking of these bases in the solid.

11.4.2.1.2. Endo-melting (Ordered to Disordered Structure). Biomolecules are mostly macromolecules. Some of them have regions of local structure stabilized by hydrogen bonds. These hydrogen bonds are not intermolecular. They are a special type of intramolecular interaction that bind different segments such as nucleotides. We call them *endo-hydrogen bonds*. These bonds stabilize the helical structure in many biomolecules and the stem in the cloverleaf of t-RNAs (see Fig. 11.6). When heated, these endo-hydrogen bonds can

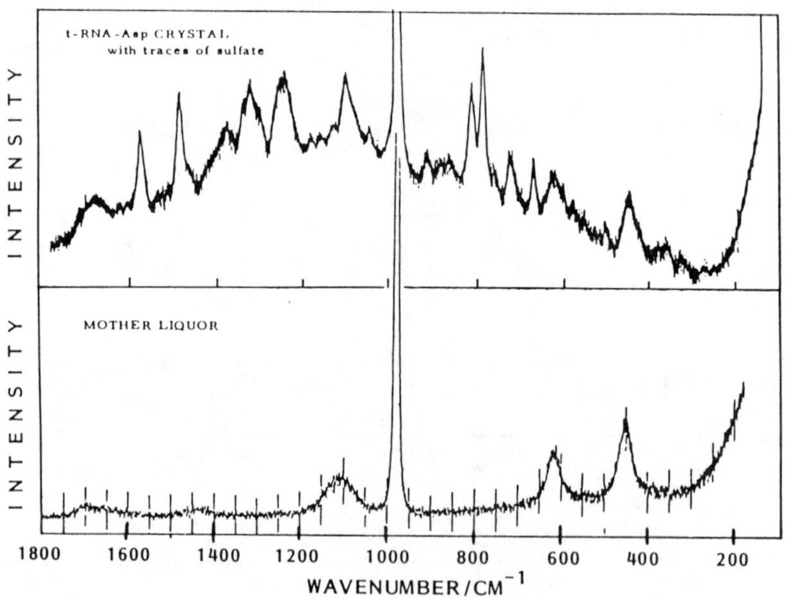

Figure 11.8. Raman spectrum of solid t-RNA–Asp, compared to that of the mother liquor.

be broken; therefore a large change in the conformation of the biomolecules can be observed. We call it *endo-melting*. During endo-melting, a biomolecule goes from an ordered to a completely disordered conformation.

When t-RNA–Asp is heated for example, its Raman spectrum smoothly changes up to 79°C, but above this temperature distinct spectral changes occur in both the ribose–phosphate backbone and base stacking (Fig. 11.9). In particular the band at 812 cm^{-1} completely disappears. This is the proof of a conformation change from an ordered to a totally disordered structure.

11.4.2.1.3. Conformation Transitions Between Ordered States. Raman spectroscopy can also be used to observe the transition from an ordered conformation to another ordered conformation. We call this transformation a *conformation transition* (13, 21). Taking t-RNA–Asp again as an example, a careful plot of the relative intensity of characteristic Raman bands with

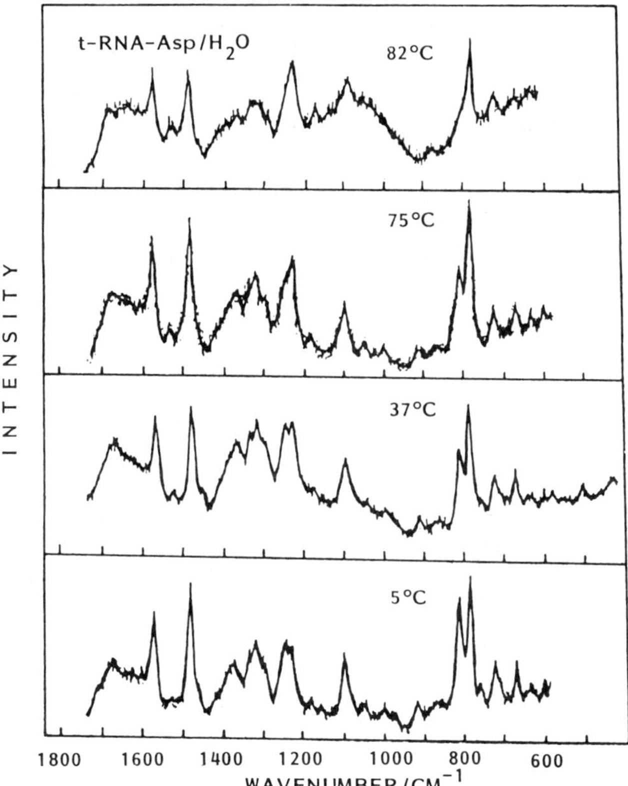

Figure 11.9. Raman spectrum of aqueous t-RNA–Asp at various temperatures.

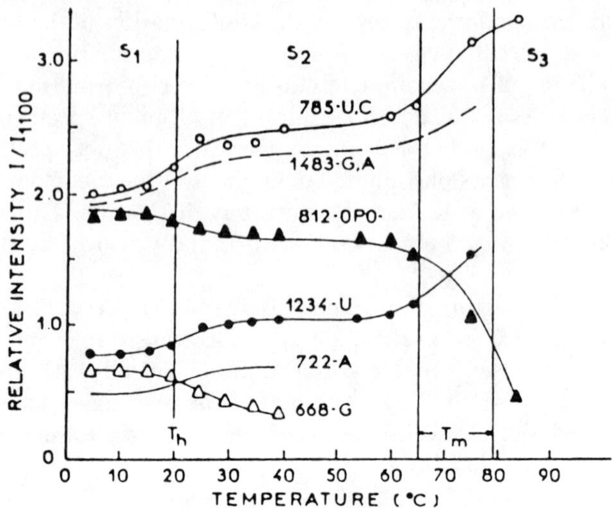

Figure 11.10. Relative Raman intensity of t-RNA–Asp bands versus temperature.

respect to the $1100\,cm^{-1}$ band (taken as reference) versus temperature (Fig. 11.10) reveals a conformation transition temperature, T_H, at 20°C. In the low-temperature structure, G-base stacking is shown to be more effective than in the high-temperature structure (13). We should mention that this analytical technique is currently used for studies of phase transitions in solids.

11.5. MOLECULAR DYNAMICS

Some biomolecules play a dynamic role in living systems. Transfer RNAs constantly move. Each t-RNA recognizes and attaches to a complementary molecule (charged with a specific amino acid) and transfers its amino acid, then detaches, allowing another t-RNA molecule to continue the process.

Raman spectroscopy can help to study the dynamics of these nucleic molecules. For t-RNA–Asp, the intensity of its Raman bands exhibits marked changes near 70°C. This temperature, or the *endo-melting temperature*, T_m, varies with the nature of the base pairing and base stacking. We have found that the U–A base pairs are the earliest to be affected by heating, and G–C pairs are the last to be affected, as indicated by broadened Raman bands.

It is shown that T_m is much higher for t-RNA–Asp than for t-RNA–Phe, which is ~ 50°C (29). This fact was interpreted (2, 31) as being due to a larger number of G–C base pairs in t-RNA–Asp, since these base pairs are known to be more stable than other base pairs.

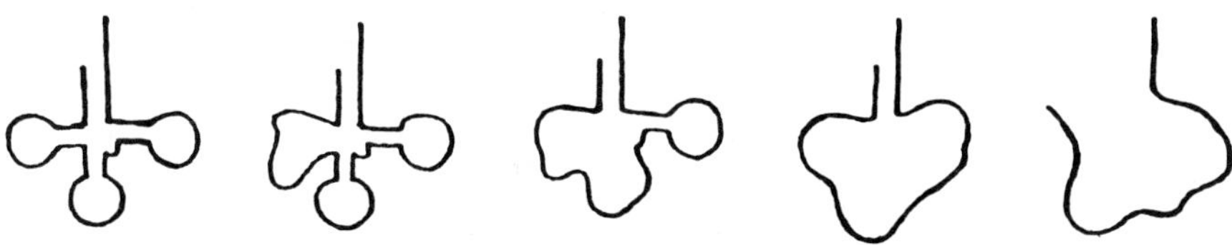

Figure 11.11. Model for cloverleaf opening in t-RNA–Asp.

Melting of the arms of t-RNA–Asp can be explained by the following (Fig. 11.11): the T_Ψ arm contains the maximum number of G–C hydrogen bonds, therefore it is the last opened in the cloverleaf conformation during the dynamic disordering of t-RNA–Asp (21). We recall that at 95°C a complete endo-melting and disordering is observed. We have also shown that all these conformational changes are, with some irregular hystereses, reversible.

11.6. INTERACTIONS

11.6.1. Selective Enhancement of Raman Bands Corresponding to the Interaction Site of Hemoglobin

Biomolecules are generally macromolecules containing a great number of atoms. Therefore numerous vibrations are possible ($3N - 6$, where N represents the number of atoms). However, in a biomolecule, many sequences of units like peptides or nucleotides are repeated and thus the number of effective vibrations is reduced. Nevertheless, in living systems, interactions do not affect all parts of a biomolecule but are often localized in active sites. If one uses selective resonance Raman enhancement, only Raman bands corresponding to the chromophore of these sites will be recorded.

Recall that the resonance Raman effect occurs when the excitation energy approaches or coincides with a nonstationary electronic excited state of a molecule (Fig. 11.2). For molecules having many chromophores with different excited electronic levels, one can choose an appropriate excitation wavelength and thus selectively enhance the Raman bands associated with a given chromophore.

Hemoglobin is a large biomolecule containing the protein globin and a heme (metalloporphyrin) site. As the catalytic reactions and interactions are localized mainly at the heme site and precisely on its central metal atom (Fe), attention must be focused on this part of the biomolecule. In the resonance Raman spectrum of a red blood corpuscle, only bands corresponding to the metallophorphyrin are observed (1). All less interesting vibrations corresponding to the remaining part of protein and side chains are absent. Similar results can be obtained with cytochrome-*c* (also an iron porphyrin) (2).

11.6.2. Changes in the Active Site of Hemoglobin with Interactions

Knowledge of the characteristic Raman spectrum of an active group in a biomolecule makes the study of interactions on this site easy. Good spectra have been obtained of oxyhemoglobin and reduced hemoglobin, thus providing information on O_2 uptake and displacement. The interaction of the heme group with carbon monoxide has also been proved (32).

Among the changes that can occur in a metalloporphyrin ring, size of the ring (core) is particularly marked. Resonance Raman spectroscopy offers an easy means to follow modification of core size. A linear relationship proposed by Huong (33) allows a rapid evaluation of the radius Ct–N (where Ct represents the central metal atom and N the nitrogens that surround the metal) of the core from resonance Raman frequencies:

$$d = a - b\bar{\nu},$$

where d is the Ct–N distance and the parameters a and b are respectively 4.86 and 0.0018 for band IV and 5.87 and 0.00236 for band V of metalloporphyrins.

This relationship is well verified for a large number of metalloporphyrins, and it has been rationalized and extended by many authors (34–36). It is applicable for solid state as well as for liquid state.

11.6.3. α-Helical and Random-Coil Structures in Proteins

Among the specific arrangements of the peptide backbone in proteins, α-helix and random-coil structures are of great interest.

Peptide carbonyl stretching (amide I vibration) as well as (amide III vibrations) NH bending coupled with N—C stretching are very sensitive to the conformation of the peptide backbone. In an α-helical structure, the amide I band is located between 1640 and 1660 cm^{-1} while the amide III band appears between 1265 and 1300 cm^{-1} (37).

In the Raman spectrum of insulin (bovine pancreas) (Fig. 11.12) the amide I band at 1661 cm^{-1} indicates that the structure is highly α-helical (38, 39). Nevertheless, a shoulder at 1678 cm^{-1} indicates also the presence of some random-coil structure. In contrast, the Raman spectrum of γ-globulin (bovine Cohn fraction) (Fig. 11.12) with its intense band at 1676 cm^{-1}, shows that this protein has mostly random-coil structure.

Another vibration, mainly connected with the C—C stretching vibration, around 940 cm^{-1}, is also sensitive to the conformation of proteins. In insulin,

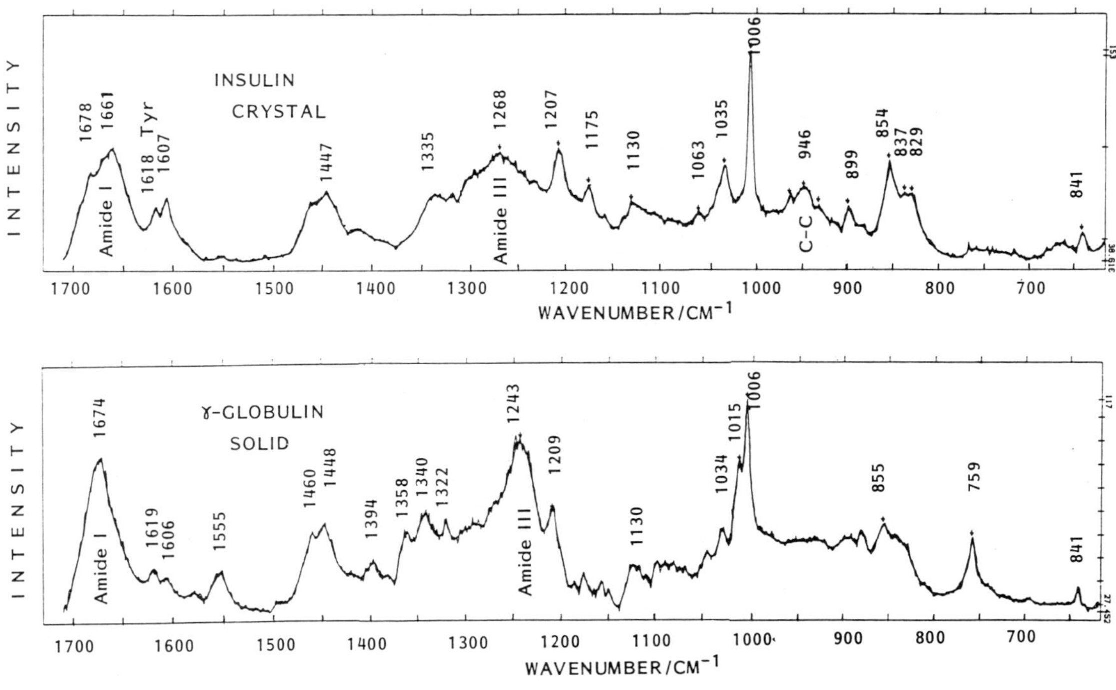

Figure 11.12. Raman spectra of insulin (bovine pancreas) and γ-globulin (bovine).

this vibration, appearing at 945 cm^{-1}, is characteristic of α-helical structure. This band disappears in a random-coil structure, as it does in the Raman spectrum of γ-globulins such as that described above.

11.7. MECHANISMS

11.7.1. Vision

The visual pigment in the rod cells of the retina has been identified as rhodopsin, which has a glycoprotein component, opsin, and a chromophore (the associated nonprotein carotenoid, 11-*cis*-retinal). By resonance Raman spectroscopy, if retinal can be easily identified (40–44) at concentrations as low as 1 ppm.

This chromophore is connected to the glycoprotein opsin by Schiff-base bonding. The Raman spectrum of a model Schiff base of retinal compared with the spectrum of retinal shows a slight change in the ν(C=C) frequency of the polyene side chain and a large modification in the enhancement of the scattered light (the same laser line in the visible is used for excitation). This resonance enhancement is connected with the shift of the absorption maximum of the two chemical species: retinal absorbs near 370 nm, while its Schiff-base absorbs at 490 nm. In addition, even at low concentrations, the ν(C=O) and ν(C=N) vibrations are clearly identified (1, 40).

Polyene chains (found in Carotenoids exist in living systems in several forms, and Raman spectroscopy can be used to easily distinguish them. Two similar carotenoids, which differ only by one double bond in a long polyene chain, have quite different resonance Raman spectra (45). Even retinal can exist in many isomeric forms. The all-*trans*-retinal has a ν(C=C) frequency that differs from that of 11-*cis*-retinal, enabling these forms to be distinguished.

Cis-trans-Retinal Isomerizations. The visual pigment in the retina converts light energy into nervous impulses for transmission to the brain. The mechanism of cis-trans isomerization of retinal, a photochemical conversion, is considered by many authors to play an important role in vision (46). As mentioned above, the Raman spectra of the two isomers are different, allowing their identification. Time-resolved Raman spectroscopy offers rapid scanning whereby one can follow the picosecond isomerization conversion using artificial photoinduction (44, 47). Even though the mechanism is not entirely known, this technique of time-resolved spectroscopy can be applied to the study of labile intermediary species and used to follow the rapid evolution of molecular species, step by step.

Protonated Schiff Bases. When the rhodopsin schiff base is protonated , the nitrogen atom is positively charged (46) and can be characterized by the Raman vibration

$$\nu\left(\mathrm{C{=}N^{+}}\!\!\begin{smallmatrix}/\\ \diagdown_{\mathrm{H}}\end{smallmatrix}\right)$$

which is different in the case of unprotonated Schiff base (40). Huong showed, in 1975, that the Raman spectrum of the protonated Schiff base of retinal depends on the nature of the associated anion (designated as A^- below), linked through hydrogen bonding with the N^{+}—H group (40)

$$\gt N^{+}\text{—}H\cdots A^{-}$$

The corresponding absorption maximum is shifted toward the middle of the visible when A^- is a phenolate anion (47).

When the phenolate anion is associated with the base and a laser line in the middle of the visible is used for excitation, the resonance Raman scattering intensity is greatly increased. Among the phenolate anions that could occur in protein, tyrosine phenolate is generally associated with the N^+H group of rhodopsin.

An important role of the associated anion is to assist in reversible proton transfer

$$\gt N^{+}\text{—}H\cdots O\text{—}C_6H_4\text{—} \rightleftharpoons \gt N\cdots HO\text{—}C_6H_4\text{—}$$

and this anion could play the role of signal transmission.

Resonance Raman spectroscopic studies reveal a large enhancement of light-scattering intensity (48) in the middle of the visible range when the phenolate anion is linked with the NH group of the protonated Schiff base of retinal. This leads us to speculate that chemicals in this family could be used to improve vision.

11.7.2. Transmission in Nervous Systems

γ-Aminobutyric acid (GABA) is well known for its important role as inhibitory transmitter in the central nervous system. In the solid or condensed state, GABA (Fig. 11.13) exists as a zwitterion:

$$^{+}H_3N\text{—}(CH_2)_3CO_2^{-}$$

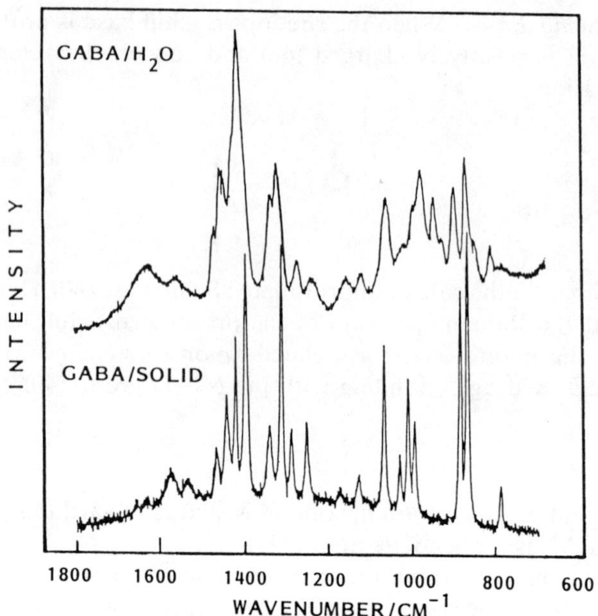

Figure 11.13. Raman spectrum of GABA in aqueous solution and as a solid.

Its neutral form, $H_2N(CH_2)_3COOH$, can also be isolated in an inert medium (49). When the concentration of this acid increases, several interaction steps occur, in particular hydrogen bonding and proton transfer from the carboxylic group to the amine group. We have shown that the proton transfer in GABA involves a trimolecular mechanism:

$$\overbrace{-CO_2^- \cdots}^{1}\overbrace{H_3^+N-(CH_2)_3-CO_2^-}^{2}\overbrace{\cdots H_3N^+}^{3}-$$

This double proton transfer could explain the role of GABA as a transmitter in the central nervous system.

11.7.3. Protonation in Biomolecules

Hydrogen bonds are responsible for stabilizing conformations (in nucleotide chains, peptide chains, and other structures; therefore proton transfer is involved in reactivity and various mechanisms.

Raman spectroscopy is an appropriate tool for characterizing such proton transfer reactions. Amino acids like methionine

$$CH_3SCH_2CH_2CH(NH_2)CO_2H$$

actually do not exist in this neutral form. Its Raman spectrum reveals no C=O stretching or $\nu(NH_2)$ stretching vibrations (Fig. 11.14). On the contrary, for methionylglycine, the observatiion of $\nu(C{=}O)$ at 1648 cm^{-1} and $\nu(NH_2)$ at 3340 cm^{-1} attests to the presence of C=O and NH_2 in this compound.

11.7.4. S—H/S—S Conversion

The stretching vibration $\nu(S{-}S)$ is very easily identified in Raman spectroscopy, as it is located in a narrow frequency range, 510–550 cm^{-1} (Fig. 11.15). Disulfide bonds in protein are structure-stabilizing bonds (which crosslink the long peptide chains). These disulfide bonds result from "dimerization" (by oxidation) of "monomers" containing the thiol group S—H. Such is the case for the conversion of cysteine $HSCH_2CH(NH_2)CO_2H$ to cystine

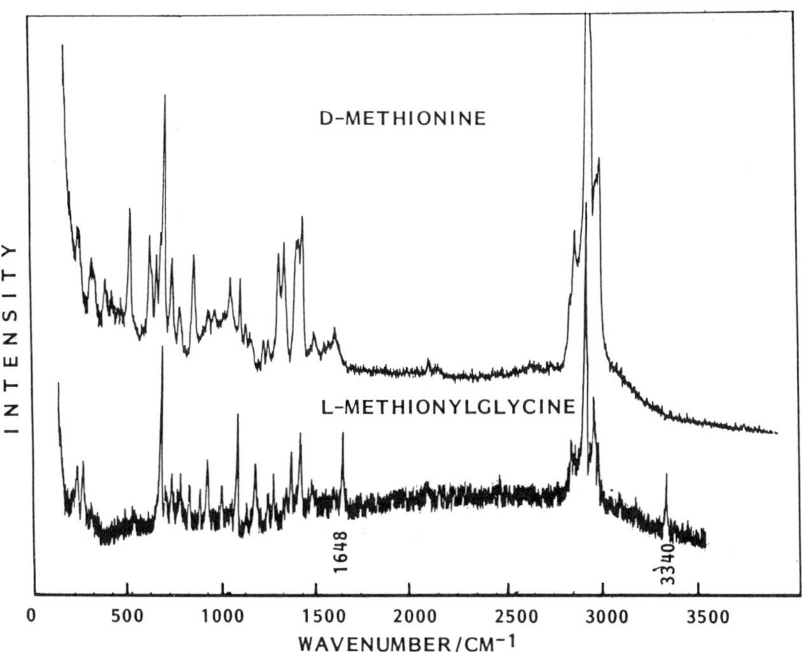

Figure 11.14. Raman spectra of methionine and methionylglycine.

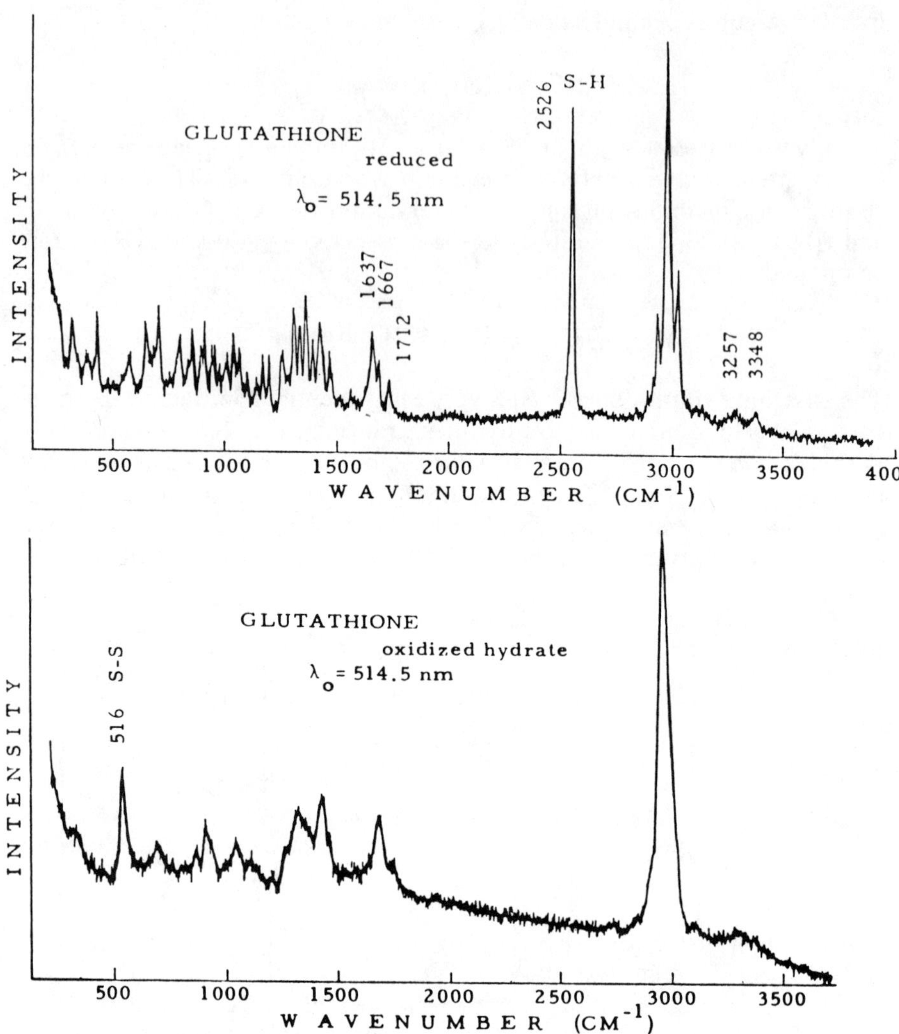

Figure 11.15. Raman spectra of glutathione, oxidized and reduced.

$[—SCH_2CH(NH_2)CO_2H]_2$ and for the conversion of glutathione to oxidized glutathione (4, 50, 50a).

The Raman characteristics of these two species are quite obvious. The S—H stretching vibration is located in the 2500–2600 cm^{-1} region and completely disappears in the oxidized form. In the oxidized form, S—S stretching appears. Of course, this conversion also affects the adjacent S—C vibrations.

11.7.5. Carcinogenesis

Recently, blue and blue-green particles were observed in cancer tumors and blood of leukemia patients (51, 52). All of these particles do not give good Raman spectra, but those located in leukemic blood, pleural fluid of lung cancer, or generalized cancer blood are easy to investigate. In such cases, the blue particles appeared as crystal-like aggregates and give good Raman spectra with no fluorescence. The same Raman spectra were obtained with blue particles in cancerous samples cultivated on glucose containing pre-warmed blood.

In the micro-Raman spectra of such blue particles (Fig. 11.16) (which are basically the same for all blue particles independent of the source of cancer) spectral features from 500 to 1750 cm^{-1}, corresponding to vibrations of organic components, can be observed. Below 500 cm^{-1}, bands at 210, 265, and 440 cm^{-1} appear in the spectral range indicative of metal–ligand structures like metal–oxygen, metal–nitrogen, and metal–sulfur groups.

When laser lines are used that approach the electronic levels corresponding to metallic transitions in the near-UV the Raman bands in the 200–450 cm^{-1} range are greatly enhanced (53). This specific resonance effect confirms that the observed Raman bands are due to metal–ligand interactions.

For the blue-green particles, the Raman spectrum is quite different (Fig. 11.17), but it is identical for all blue-green corpuscles spotted in any cancerous sample, again independently of their origin. If strong bands are observed in the spectral range where the nucleic base vibrations as well as the organic component vibrations are usually located, a strong line is also observed at 670 cm^{-1}. On the lower-frequency side, bands at 250 and 350 cm^{-1} can correspond to metal–nitrogen and metal–sulfur vibrations by comparison to known complexes.

The metal has recently been identified by fluorescence X-ray (54, 55) as copper, and it is observed in the bulk of both blue and blue-green particles. Therefore, we assign the observed metal–ligand Raman frequencies to Cu—O, Cu—N, and Cu—S vibrations.

The assignment of the Raman bands is based on previous studies of model complexes. We have assigned copper–nucleotide complexes (8, 9), copper–tyrosine complexes (56, 57), copper–peptide complexes, and copper–amino acid complexes (55) in such previous studies.

The Raman spectra of blue and blue-green cancer particles, when compared to those of the above model copper-organic complexes, show that

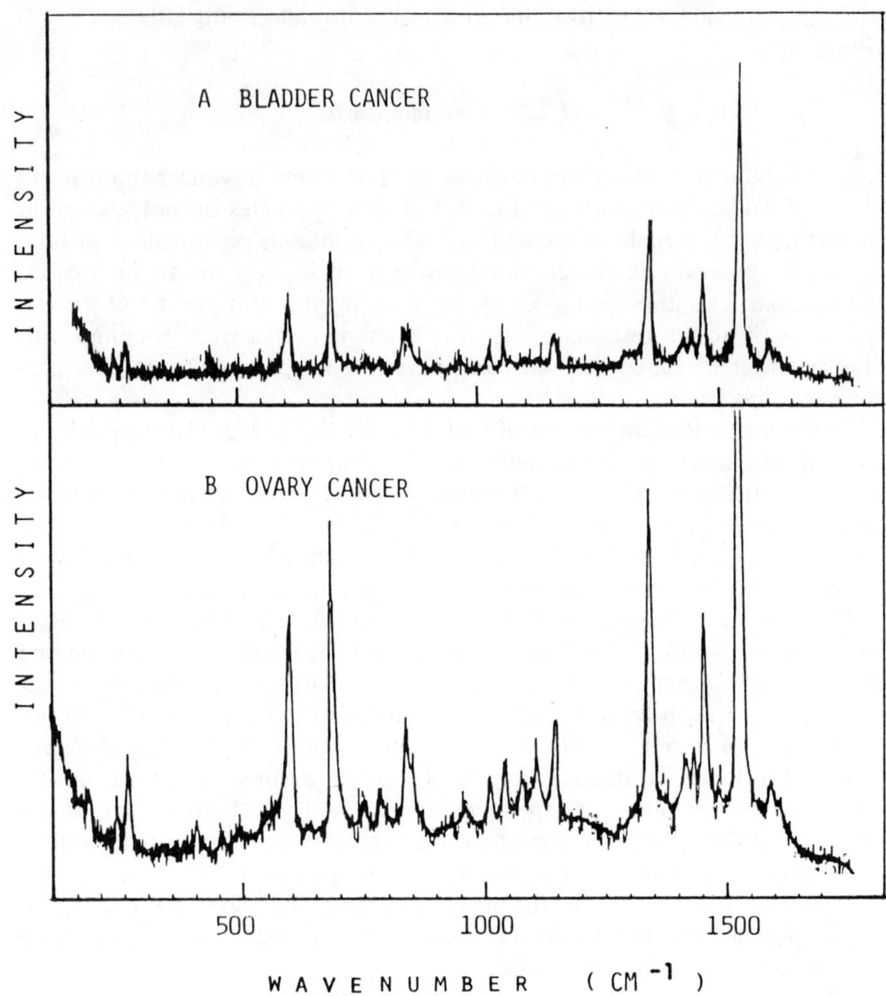

Figure 11.16. Raman spectra of blue particles in cancer tumors.

copper–nucleotide complexes and copper–cystine complexes are likely to be present.

At this stage of our study, the complete assignment of the Raman bands of the blue cancer particles is not yet achieved. Nevertheless, we can already state that: (a) the blue complexes contain more than one ligand, (b) the ligands contain both nucleic and protein bases, and sulfur-containing bases are present.

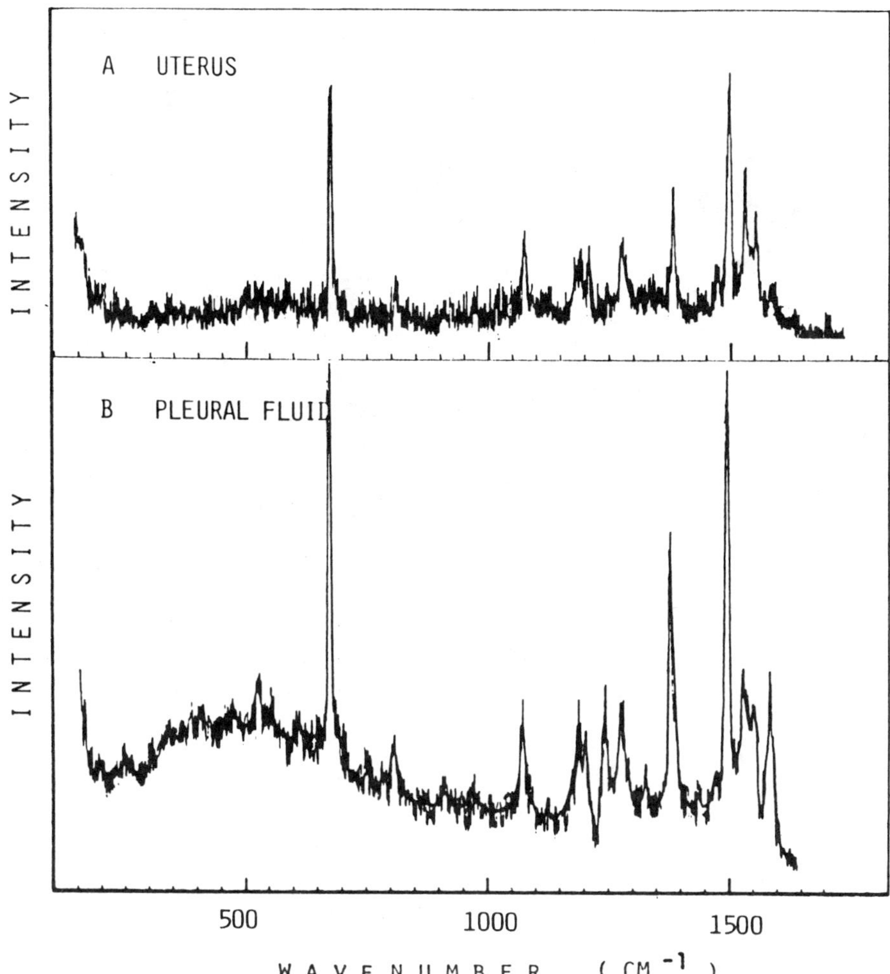

Figure 11.17. Raman spectra of blue-green particles in cancer.

In addition, electron microscopy reveals, in the neighborhood of the colored particles, the presence of spherules of copper and other metals (53). The fact that such organometallic complexes exist suggests that they play a role in inducing disordered cell growth resulting in cancer.

We suggest (55) that in organisms, the blue partices create a local disorder that is the triggering factor initiating the missing of codes and messages; this may be the origin of cancer because cell division is no longer regulated (see Fig. 11.18).

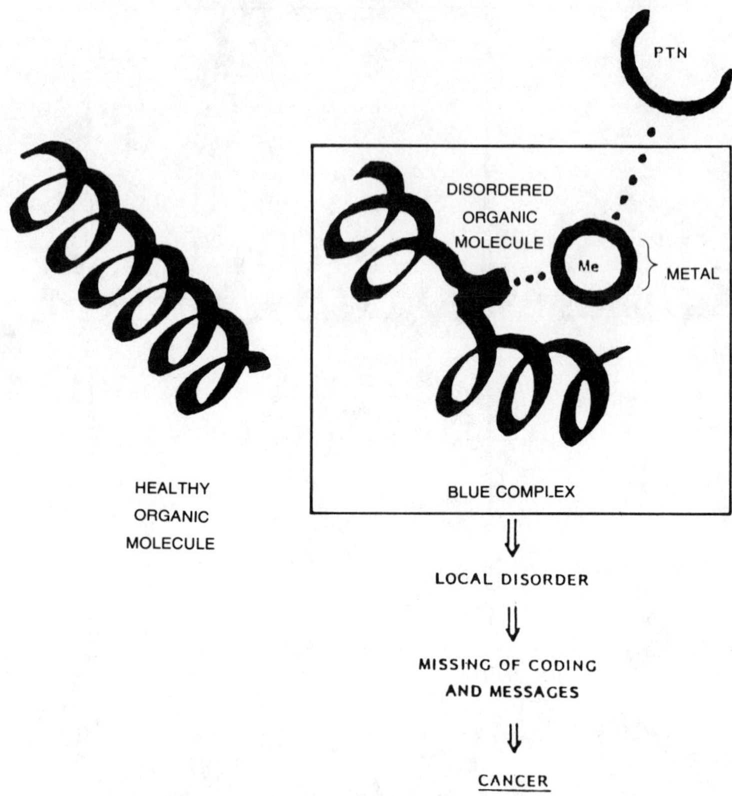

Figure 11.18. A mechanism of carcinogenesis.

This carcinogenesis model could give a satisfying explanation for the selective reactivity of a cancer drug, i.e., how a drug kills cancerous cells and not healthy cells.

In healthy cells/tissue there is a low level of metal, therefore no blue complex exists. In cancerous cells, organometallic complexes are present that can be destroyed by completing equilibria induced by anticancer drugs. This model could help direct therapeutic research to consider the thermodynamic point of view: a competitive equilibrium might be created to remove the metal involved in the blue disordering complexes (55).

Pharmaceutical research could also consider conformations of the ligands involved in blue complexes. Anticancer agents could have similar conformations to nucleotides and sulfur-containing bases but with stronger basic properties, acquired with suitable substituents (4, 55).

11.8. CONCLUSIONS

The foregoing examples show Raman spectroscopy, with its numerous variances such as resonance Raman, rapid scanning Raman, and Raman microscopy, to be a versatile and nondestructive method very suitable for biological studies; for the identification of delicate structures and conformations of molecules in living cells/tissue/organisms. The Raman technique is also helpful for the investigation of conformational dynamics that often play important roles in biological systems, such as the action of t-RNA or the permeability of membranes. Raman spectroscopy is a precious tool for following the interaction modes and reactivity of molecules involved in mechanisms like vision, nervous transmission, and carcinogenesis.

This powerful technique merits use by biologists and scientists working in related scientific and medical areas.

ACKNOWLEDGMENTS

The author wishes to express his warm thanks to J. C. Cornut, R. Cavagnat, S. P. Plouvier, D. Lespiaux, P. Lambert, R. Giege, J. Lorosch, P. Carmona, and J. F. Goussot for their helpful collaboration.

REFERENCES

1. P. V. Huong, in *Vibrational Spectra and Structure* (J. R. Durig, ed.), Vol. 9, p. 143. Elsevier, New York and Amsterdam, 1980.
2. P. V. Huong, in *Reviews on Analytical Chemistry* (L. Niinisto, ed.), p. 209. Kiadó, Budapest, 1982.
3. J. Lascombe and P. V. Huong, eds., *Raman Spectroscopy, Linear and Non-Linear*, Wiley, New York and Chichester, 1982.
4. P. V. Huong, *J. Pharm. Biomed. Anal.* **4**, 811 (1986).
5. M. Avignon, P. V. Huong, J. Lascombe, M. Marraud, and J. Neel, *Biopolymers* **8**, 69 (1969).
6. M. Avignon and P. V. Huong, *Biopolymers* **9**, 427 (1970).
7. M. Marraud, J. Neel, M. Avignon, and P. V. Huong, *J. Chim. Phys.* **67**, 959 (1970).
8. P. Carmona, P. V. Huong, and E. Gredilla, *J. Raman Spectrosc.* **19**, 315 (1988).
9. P. Carmona, P. V. Huong, and E. Gredilla, *Raman Spectrosc., Proc. Int. Conf., 10th, 1986* p. 2/15 (1986).
10. A. T. Tu, *Raman Spectroscopy in Biology*. Wiley, New York and Chichester, 1982.
11. T. G. Spiro, ed., *Biological Applications of Raman Spectroscopy*. Wiley, New York, 1985, 1986.

12. P. V. Huong and R. Giege, *Biochimie* **63**, 921 (1981).
13. R. Giege, P. V. Huong, and D. Moras, *Spectrochim. Acta* **42A**, 378 (1986).
14. M. Stockburger, T. Alshuth, D. Desterhelt, and W. Gartner, in *Spectroscopy of Biological Systems* (R. J. H. Clark and R. E. Hester, eds.), p. 483. Wiley, Chichester, 1986.
15. I. Harada and H. Takeuchi, in *Spectroscopy Biological Systems* (R. J. H. Clark and R. E. Hester, eds.), p. 113. Wiley, Chichester, 1986.
16. C. G. Zimba, V. M. Hallmark, J. D. Swalen, and J. F. Rabolt, *Appl. Spectrosc.* **41**, 721 (1981).
17. L. A. Carreira, L. P. Goss, and T. B. Malloy, *J. Chem. Phys.* **66**, 4360 (1977).
18. C. B. Moore, ed., *Chemical and Biochemical Applications of Lasers*. Academic Press, New York, 1979.
19. R. Mathies, in *Chemical and Biochemical Applications of Lasers* (C. B. Moore, ed.), p. 55. Academic Press, New York, 1979.
20. A. Campion, A. Terner, and M. A. El-Sayed, *Nature* (*London*) **265**, 659 (1971).
21. P. V. Huong, R. Giege, D. Moras, M. B. Commarmond, and J. C. Thierry, in *Raman Spectroscopy, Linear and Non-Linear* (J. Lascombe and P. V. Huong, eds.), p. 705. Wiley, New York and Chichester, 1982.
22. I. M. Asher, E. B. Carew, and H. E. Stanley, in *Physiology of Smooth Muscle* (E. Bulbring and M. F. Shuba, eds.), p. 229. Raven Press, New York, 1976.
23. M. Pezolet, M. Pigeon-Gosselin, J. Nadeau, and J. P. Caille, *Biophys. J.* **31**, 1 (1980).
24. H. Ishida, R. Kamoto, S. Uchida, A. Ishitani, Y. Ozaki, K. Iriyama, E. Tsukie, H. Shibata, F. Ishihara, and H. Kameda, *Appl. Spectrosc.* **41**, 407 (1987).
25. N. T. Yu and E. J. East, *J. Mol. Chem.* **250**, 2196 (1975).
26. M. Pezolet, D. Georgescault, and H. Richard, in *Raman Spectroscopy, Linear and Non-Linear* (J. Lascombe and P. V. Huong, eds.), p. 773. Wiley, Chichester, 1982.
27. S. M. Kim, J. L. Sussman, F. L. Suddath, G. F. Quigley, A. McPherson, A. H. J. Wang, N. C. Seeman, and A. Rich, *Proc. Nat. Acad. Sci. U.S.A.* **71**, 4970 (1974).
28. D. Moras, M. B. Comarmond, J. Fisher, R. Weiss, J. C. Thierry, J. P. Ebel, and R. Giege, *Nature* (*London*) **288**, 669 (1980).
29. M. C. Chen, R. Giege, R. C. Lord, and A. Rich, *Biochemistry* **14**, 4385 (1975).
30. E. Westhof, P. Dumas, and D. Moras, *J. Mol. Biol.*, **184**, 119 (1985).
31. P. V. Huong, E. Audry, R. Giege, D. Moras, J. C. Thierry, and M. B. Comarmond, *Biopolymers* **23**, 71 (1984).
32. P. V. Huong and J. C. Cornut, *Proc. Int. Conf. Raman Spectrosc., 9th, Tokyo, 1984* p. 50 (1984).
33. P. V. Huong and J. C. Pommier, *C. R. Hebd. Seances Acad. Sci., Ser. C* **285**, 519 (1977).
34. T. G. Spiro, J. D. Stong, and P. Stein, *J. Am. Chem. Soc.* **101**, 2648 (1979).
35. K. N. Solovkov, L. L. Gladkov, A. T. Gradyushko, N. M. Ksenofontova, A. M. Shulga, and A. S. Starukhin, *J. Mol. Struct.* **45**, 267 (1978).

36. N. Blom, D. P. Strommen, and K. Nakamoto, in *Spectroscopy of Biomolecules* (A. J. P. Alix, L. Bernard, and M. Manfait, eds.), p. 363. Wiley, New York and Chichester, 1985.
37. J. L. Koenig and B. Fruschour, *Biopolymers* **11**, 1871 (1972).
38. N. T. Yu, C. S. Liu, and D. C. O'Shea, *J. Mol. Biol.* **70**, 117 (1972).
39. W. L. Peticolas, W. L. Kubaser, G. A. Thomas, and M. Tsuboi, in *Biological Applications of Raman Spectroscopy* (T. G. Spiro, ed.), Vol. 1, Chapter 4. Wiley, New York, 1985.
40. P. V. Huong, R. Cavagnat, and F. Cruege, in *Lasers in Physical Chemistry and Biophysics* (J. Joussot-Dubien, ed.), p. 425. Elsevier, Amsterdam, 1975.
41. M. E. Heyde, D. Gill, R. G. Kilponen, and L. Rimai, *J. Am. Chem. Soc.* **93**, 6776 (1971).
42. C. C. Askren, N. T. Yu, and J. F. R. Kuck, Jr., *Exp. Eye Res.* **29**, 647 (1979).
43. Y. Ozaki, A. Mizuno, and O. Iriyama, *J. Biol. Chem.* **262**, 15545 (1987).
44. G. Hayward, W. Carlsen, A. Siegman, and L. Stryer, *Science* **211**, 942 (1980).
45. P. V. Huong, *C. R. Hebd. Seances Acad. Sci., Ser. C* **286**, 25 (1978).
46. M. Sulkes, A. Lewis, and A. Marcus, *Biochemistry* **17**, 4712 (1978).
47. M. A. El-Sayed, *Springer Ser. Opt. Sci.* **26**, 295 (1981).
48. P. V. Thoai, *J. Mol. Struct.* **143**, 423 (1986).
49. P. V. Huong and J. C. Cornut, *J. Chem. Phys.* **65**, 4748 (1976).
50. P. V. Huong and S. P. Plouvier. in *Molecular Spectra and Structure* (J. R. Durig, ed.), Vol. 17A, p. 497. Elsevier, New York and Amsterdam, 1989.
50a. P. V. Huong and D. Lespiaux, in *Spectroscopy of Biological Molecules* (A. Bertoluzzi, ed.), p. 66. Wiley, New York and Chichester, 1989.
51. P. V. Huong and S. R. Plouvier, *J. Mol. Struct.* **115**, 489 (1984).
52. S. R. Plouvier and P. V. Huong, *Biorheology, Suppl.* **1**, 345 (1984).
53. P. V. Huong, *J. Mol. Struct.* **141**, 203 (1983).
54. P. V. Huong, S. R. Plouvier, and P. Lambert, in *Spectroscopy of Biomolecules* (A. J. P. Alix, L. Bernard, and M. Manfair, eds.), p. 231. Wiley, New York and Chichester, 1985.
55. P. V. Huong, in *Molecules in Physics, Chemistry and Biology* (J. Maruani, ed.). Vol. 4, p. 87. Kluwer, Amsterdam, 1988.
56. P. V. Huong, P. Carmona, and J. Lorosch, *Raman Spectrosco., Proc. Int. Conf., 10th, 1986* p. 4/3 (1986).
57. J. Lorosch, W. Haase, and P. V. Huong, *J. Inorg. Biochem.* **27**, 53 (1986).

CHAPTER

12

CHEMICAL APPLICATIONS OF GAS-PHASE RAMAN SPECTROSCOPY*

WILLIAM F. MURPHY

Steacie Institute for Molecular Sciences
National Research Council
Ottawa, Canada

12.1. INTRODUCTION

In this chapter, a summary of current and past analytical applications of Raman scattering (RS) studies of gas-phase samples will be presented. Indeed, such applications were not common in the prelaser era, and are less than widespread even today. The chief reason for this, to be discussed more fully below, is that the RS process is an energy-limited phenomenon, and this fact, coupled with the low densities encountered in gaseous samples, complicates the experimental problem of observing the Raman spectrum of such samples. In spite of this fact, Raman spectral measurements can provide unique information about selected systems, in which cases the use of this technique can prove to be well worth the effort required to make such measurements.

In the explosion of activity which followed the discovery of the Raman effect in 1928, many spectra of gas-phase samples were observed and reported. [See Weber's review (1) for a summary of this period, and references to the original work.] However, from that time and extending into the era of laser excitation, such studies were largely confined to specialized measurements of high-resolution spectra, to obtain information on molecular vibrational and rotational energy levels (2, 3), and, more recently, to perform measurements of absolute scattering cross sections in order to gain insight into the nature of scattering intensities. [See Murphy et al. (4) for a summary of prelaser results, and Schrötter and Klöckner (5) for a summary of more recent

*NRCC No. 32813.

Analytical Raman Spectroscopy, Edited by Jeanette G. Grasselli and Bernard J. Bulkin. Chemical Analysis Series, Vol. 114.
ISBN 0-471-51955-3

work.] Throughout this period, there were relatively few reports of analytical studies of gaseous systems by Raman spectroscopy.

The value of such applications was formally recognized in the publication of the proceedings of a 1973 workshop on the measurement of gas properties using RS techniques (6). In that proceedings, the capabilities of RS for measuring gas temperatures and species concentrations were emphasized by many contributors, and later applications of Raman spectroscopy were clearly foretold in such areas as remote sensing, e.g., atmospheric pollution studies, and the monitoring of combustion systems.

It should be emphasized that for many applications, the various nonlinear techniques such as CARS (coherent anti-Stokes RS) have proven to be better suited for obtaining required information from hostile environments. Owing to my lack of experience in this area, I will not be discussing it in detail, and the reader is referred to recent texts and reviews (7–10) for a more comprehensive discussion of such techniques. However, it should be noted that there is a close relationship between a spontaneous RS cross section and the electric susceptibility that governs the intensity of various coherent Raman signals [see Nibler (10), for example, for a discussion of this point]. Thus, Raman cross-section measurements under controlled laboratory conditions can be of value in the interpretation of various nonlinear spectra [e.g., see Frunder et al. (11)], and classical Raman studies will continue to have application, albeit indirect, in areas served by nonlinear effects.

In other, more chemically oriented, applications, RS has been used to measure reaction products in laboratory and field situations. It is especially suited for detecting homonuclear diatomic molecules, whose infrared (IR) absorption spectrum is symmetry forbidden. Examples of various applications will be discussed in more detail below.

12.2. EXPERIMENTAL ASPECTS

For an indication of the considerations involved in making viable quantitative measurements of gas-phase Raman spectra, the instrumentation and procedures used in our laboratory will be presented in some detail. In addition, other approaches will be mentioned in order to provide a basis for selection between potential methods in a given application.

12.2.1. Instrumentation

12.2.1.1. Source

The overriding consideration in obtaining Raman spectra of gaseous samples is that the detection of molecularly scattered light is an energy-limited process.

Therefore, a paramount concern is to maximize the effect of the available input energy. The light source used to excite the spectrum is often a high-powered (1–10 W) rare-gas-ion laser (Ar^+ or Kr^+). For high-resolution applications, the laser output is normally limited to a single mode, so the observed spectrum will not be significantly broadened by the width of the excitation source. However, for most analytical applications, such high resolution is not required; the multimode output with a line width of $\sim$0.1–0.3 cm^{-1} for a typical Ar^+ laser is acceptable. Since a continuous-wave rare-gas laser is a convenient source of high average power illumination, pulsed lasers have not, in general, had widespread application for spontaneous RS applications. Since gaseous samples are, for the most part, optically "clean," it is usually not necessary to filter out the background emission spectrum from the laser discharge.

12.2.1.2. Sample Irradiation

In order to maximize the sample illumination, it is helpful to multipass the incident laser through the sample volume. Several variations of systems to accomplish this have been described (3). In our own work, a double spherical mirror arrangement (12) is used outside of the laser cavity, and signal enhancement factors of 10–20 are routinely achieved. Various intracavity multiple reflection devices have also been used (3); the disadvantage of such devices is that special care must be taken in their design and use in order not to reduce the gain of the laser cavity and thus limit or halt the laser action. The observation of many weak overtone and combination transitions of allene (13) and torsional overtone transitions of ethane (14), for example, has been accomplished with the use of such a device.

12.2.1.3. Sample Cells

Various sample cell designs have been presented that involve mounting windows of an appropriate material on a metal body (1, 15). However, we have found cells constructed from fused silica to be convenient to work with, and these have been used routinely in our laboratory (12). The cell body is cylindrical, and the windows for transmission of the laser beam are mounted at Brewster's angle to minimize transmission losses. Even so, through a "light pipe" action, some stray laser light can leak into the cell body, and a background fused silica spectrum can often be observed, extending to $\sim$500 cm^{-1}. For example, this background is prominent in the weak rotational spectrum of water vapor (16). This occurs because the density of the fused silica wall material is so much higher than that of the gaseous sample, in spite of the fact that the transfer optics do not efficiently image the cell walls onto the monochromator entrance slit. Another disadvantage of such cells is that the internal surfaces of the Brewster windows cannot be cleaned easily and any contamination of these surfaces will lead to a loss of enhancement efficiency of the multipass arrangement. Despite these inconveniences, this arrangement has proved to be quite satisfactory in our laboratory.

In other situations, it may be possible to observe the Raman spectrum of the sample *in situ*, without need for a cell. For example, combustion systems may be studied by multipassing the incident beam through the flame produced with an appropriately designed burner (17) without the need to confine the sample. The advantage of eliminating the windows is appreciable, as can be seen by comparing the signal obtained from the air spectrum with and without a sample cell under otherwise comparable optical conditions.

12.2.1.4. Transfer Optics

The scattered light is normally imaged onto the entrance slit of the monochromator using a simple lens. Reflection optics, or a low f number collection lens combined with a suitable transfer lens, have also been used. The important point to check here is that the limiting aperture of the monochromator (normally the first collimator mirror) is completely filled, i.e., that the monochromator f number is appropriately matched to that of the transfer lens. Otherwise, a mismatch will result in a loss in collection efficiency (overfilling) or a reduction in monochromator resolving power (underfilling). A secondary factor is that the transfer system should be achromatic through the spectral region of the Raman spectrum; otherwise a loss in efficiency will result as the focus of the sample image moves away from the slit plane as the frequency is varied. Such an effect can be especially insidious if the transfer optics are optimized at different strong bands in the spectrum at various times, causing a variation in relative band intensities in different scans.

Included as part of the transfer optics is an analyzer for determining the polarization properties of the scattered light. We have had equal success with a high-efficiency Polaroid material (type HN-22) and a high-quality photographic polarization filter. The latter is more convenient to use, since it is mounted in a standard photographic mount.

A final item in the transfer optics is the "pseudo" polarization scrambler, a wedge of birefringent material that mixes the polarization of the scattered light to avoid problems due to the difference in monochromator transmission efficiency for the different polarizations (18). In our own case, this consists of a 25 mm square section of calcite, having a 1° wedge angle, oriented so that the slit height is parallel to the plane mutually perpendicular to the two surfaces of the scrambler. Although this device does not function correctly in the limit of small slit heights (18), the operation of our scrambler is satisfactory down to a height of 2 mm, as seen by the independence of the spectral sensitivity calibration (see below) to the polarization of the incident light.

12.2.1.5. Monochromator and Detector

In the detection of RS, the requirement to discriminate the Raman signal from the strong scattered and stray light at the excitation frequency is a prime consideration. Although a gaseous sample is optically "cleaner" than many other types of samples and thus does not generate an excessive amount of stray light, most studies of gas-phase spectra in the modern, laser, era have been carried out with a double monochromator, designed to discriminate against stray light interference. Such a monochromator often consists of a Czerny–Turner double holographic grating instrument having an aperture of $f/6$ to $f/8$ and a dispersion such that a mechanical slit width of 100 μm corresponds to a spectral slit width of $\sim 1\ \text{cm}^{-1}$. Thus, the use of a mechanical slit width of 100–300 μm permits the efficient transfer of the image of the focused laser beam into the monochromator, while maintaining a reasonable spectral resolution capability. The mechanical scanning drive is of the cosecant type to achieve an abcissa readout in wavenumber units. The usual detector is a photomultiplier having a GaAs photocathode, chosen for its low thermal signal and high efficiency, especially in the red region of the visible spectrum. The photon-counting method of measuring the detector response is used not because the light signals are exceedingly small but because of the more direct compatibility with digital data-acquisition systems.

Other methods of measuring Raman spectra have rarely been used for gaseous samples. The multichannel Raman spectrograph (19) has been applied in the study of many types of condensed-phase systems, but such a system has been used only rarely (20) for gas-phase work. Workers may be discouraged by the lack of resolution available in typical applications of this technique, or by the potential lack of sensitivity to the signal from the low-density sample in the presence of the background signals generated by this type of detector. The availability of charge-coupled devices for use in Raman studies (21) may make such applications more attractive.

The other recent innovation in the observation of Raman spectra has been the application of Fourier transform (FT) spectrometer (22–24). Again, the major interest in these devices has been in the measurement of condensed-phase spectra. For gas-phase applications, the chief advantage of this technique over a dispersion instrument is its throughput under high resolution (spectral slit: $< 1\ \text{cm}^{-1}$) conditions (25). Under such conditions, the area of the resolution-defining aperture is larger for the interferometer than the dispersive spectrometer, so the transmitted energy and thus the detected signal is greater. Spectra of hydrogen and deuterium gases, measured using this technique, have been reported (26), and more recently hydrogen spectra in a flame were measured in order to observe transitions from high rotational levels (27).

A further observation method employs narrow band filters to pass only scattered radiation in the frequency range of interest. This could involve, for example, a pair of such filters—one to monitor a strong band from a reference molecule, and the other to observe a band from the sample molecule. Such single-purpose devices are used, for example, in the remote detection of atmospheric pollutants (28).

The final requirement in detecting RS is a means to record the observed spectrum. A digital spectrometer-control and data-recording system should be used so that the spectral data can be available for subsequent manipulation by computer-based numerical techniques. The many advantages of such a capability are widely recognized and need no further elaboration here. One point that may not be generally appreciated is the large dynamic range realizable in a digitally recorded photon-counting detection system (29), which can extend from the thermal background signal of less than 10 cps to 10^6 cps or greater, if proper allowance for the pulse resolution capabilities of the system electronics has been made.

12.2.2. Instrumental Calibration

In analytical applications of spectroscopy, it is important that the response of the spectrometer be appropriately calibrated. This involves calibration of two factors: the wavenumber readout and the spectral sensitivity as a function of the wavenumber. It is worthwhile to describe such calibration procedures in some detail. I will do so in terms of the scanning monochromator system I am most familiar with; many of the considerations may be adapted to other situations in a straightforward manner.

12.2.2.1. Frequency Calibration

In the calibration and measurement of frequency values, there are two considerations: the accuracy and the reproducibility (precision) of the measurement. Accuracy may be achieved by a suitable calibration procedure; however, the success of such a procedure rests on its reproducibility. In discussions of the Raman difference spectroscopy technique (30, 31), in which two or more spectra are observed alternately at each wavenumber position, one of the disadvantages of rescanning is often attributed to mechanical scanning errors related to the $\sim 2\,\text{cm}^{-1}$ accuracy specifications given for the typical commercial instrument. However, the *reproducibility* of the monochromators I am familiar with is much better, with band peak positions under minimal spectral slit width conditions usually being repeatable within two or three steps of the drive, corresponding to 0.04–0.06 cm^{-1}. Further, in a multiple scan of a strong, narrow band, I have found little evidence of band broadening when compared with a single scan. Thus, in practice I have found no reason to fault such systems on the basis of drive reproducibility. Indeed, the wavenumber calibration technique we have developed in our laboratory is critically dependent on the reproducibility of the monochromator drive system and has been found to be quite satisfactory in practice.

The procedure we use entails the measurement of a calibration curve that is stored for access by the calibration program. The mechanical frequency display of our commercial monochromator is given in units of wavenumbers in air, so that these are the working abscissa units for instrument operation. However, the frequency calibration function is designed to convert these units to vacuum wavenumbers or vacuum wavenumber shifts, to provide accurate values for the differences in molecular energy levels, directly comparable to values determined by other techniques. The difference in the two scales (vacuum versus air wavenumber shift) amounts to 0.83 cm^{-1} at a 3000 cm^{-1} shift from the 514.5 nm line of the Ar^+ laser. We begin operation of the spectrometer by synchronizing the digital data-acquisition system with the monochromator drive. Taking care to avoid backlash, the monochromator is positioned at the peak of a known emission line and the data-acquisition-system wavenumber value is initialized at that point. This value is also provided to the calibration program, where the calibration curve is offset suitably to zero it at this point. Thus, the wavenumber-calibration procedure depends critically on the operator initialization of the monochromator drive, but in practice the error in this procedure is rarely found to be more than one or two steps ($<0.05\ cm^{-1}$). Furthermore, the initialization emission line is chosen to lie within the spectral region under investigation in order to minimize the magnitude of the correction required over the range of the spectrum. For 514.5 nm Ar^+ excitation, it is convenient to use the 546.1 nm Hg emission line from the fluorescent room lights; this wavelength corresponds to a Raman shift of 1123 cm^{-1}.

The correction function itself is derived in two parts. The overall, absolute correction is obtained from a scan of the Ne emission spectrum throughout the spectral range used. The correction function relating the observed to the actual wavenumber values is derived from a polynomial fit of the differences as a function of the observed values. When we investigated the reproducibility of this procedure, we found that the residuals in the fit, varying up to 0.1 cm^{-1} or more, were reproducible from trial to trial, indicating that there was a further non-random error that possibly could be included in the calibration. Thus, as a second step, the comb spectrum obtained by passing a "white light" continuum through an etalon was recorded. A commercial etalon was chosen with a reflector spacing which gave fringes at $\sim 6\,\text{cm}^{-1}$ intervals. An accurate value of the fringe spacing is not required for the calibration procedure, which only makes use of the fact that the fringe spacing is constant in wavenumbers. From the difference between the average observed fringe spacing and the actual value as a function of observed wavenumber, it was found that a periodic variation existed having a peak-to-peak amplitude of $\sim 0.2\,\text{cm}^{-1}$ and a period of $100\,\text{cm}^{-1}$. In our instrument, the lead screw undergoes one revolution every $100\,\text{cm}^{-1}$, which allowed us to identify this residual periodic error as an error involving the lead-screw system. Thus, when this component was added to the drive calibration routine, the mean error of the fit of the Ne emission spectrum was $0.040\,\text{cm}^{-1}$, and subsequent Ne spectra could be measured with a mean error of $0.052\,\text{cm}^{-1}$.

12.2.2.2. Intensity Response Calibration

The spectral sensitivity calibration has been carried out using two types of procedures. One is based on the response of the monochromator system to a calibrated continuum source (32–34) (e.g., a tungsten filament lamp whose color temperature and emissivity as a function of wavelength have been determined). The other is to measure the relative Raman cross sections of reference spectra, such as H_2 and D_2 rotational and vibrational transitions, with a series of exciting lines (34–37) and to use the relation between the measured values and the known values to derive the response calibration at discrete points throughout the desired frequency range. The calibration function is then obtained by a suitable interpolation between these points.

We have chosen to use the former procedure; it is experimentally more convenient, in that a single spectrum of the calibration source provides the calibration data, and there is little chance of missing short-range variations in the response curve, as could happen with the second method. Some problems have been noted for this first method, the continuum source method (38), but they have not occurred in our work.

The expression for the Raman cross section differs by a factor of ν_0/ν_s, where ν_0 and ν_s are the wavenumbers of the incident and scattered radiation, depending on whether the quantity measured is the power of the scattered light or the number of scattered photons per unit time. However, this difference must not be confused with the experimental detection method used; it is accounted for by the sensitivity calibration procedure. Thus, if the quantity to be measured is the power of the scattered radiation, one uses the cross-section expression given in Eq. (2), in Section 12.3.1, below, and calibrates monochromator sensitivity in terms of the power output of the standard lamp. On the other hand, if the scattered photon rate is chosen, the cross-section expression in Eq. (2) is scaled by the factor ν_0/ν_s and the calibration is carried out in terms of the photon output of the standard lamp. The detection method used, photon counting or photocurrent measurement, is not relevant in this consideration; the important factor is the calibration of the wavenumber dependence of the instrumental response in terms of the quantity to be measured, the power of, or the rate of photons in, the scattered light.

A second consideration that arises is related to the fact that most monochromator systems are operated using mechanically constant slit widths when measuring a spectrum. For the commonly used Czerny–Turner design, the spectral slit width is approximately constant in wavelength throughout the visible region of the spectrum, which means that it is not constant when expressed in wavenumbers. Thus, the question arises as to whether the sensitivity calibration should be adjusted to compensate for the variation in spectral slit width, when expressed in wavenumbers. To measure the total scattered power for a given transition, we integrate the signal observed for that transition over a suitable wavenumber range, since the monochromator system used for Raman measurements is normally equipped with a cosecant, wavenumber drive. Thus, the signal to be integrated, the spectrum, should be expressed in units of power per unit wavenumber, in order that the resulting integrated intensity have the correct units. This means that the calibration data for the standard lamp must be expressed in units of power per unit wavenumber, which could entail a recalculation of the calibration data furnished for the lamp. However, the variation of the spectral slit as a function of wavenumber requires no additional compensation; again, as for the photons/power case, we are interested only in the response of the detection system to the standard source under the experimental conditions to be used to measure the cross sections of interest. Since we are interested only in the instrumental response at a given wavenumber position compared to that at the reference wavenumber, the effect of wavenumber variation in the spectral slit is cancelled in the final result obtained from the ratio of the two measurements.

For the experimental determination of the spectral sensitivity calibration function, a sample of Eastman white reflectance standard (39) is mounted at the usual sample position. The image of the reference source, a ribbon filament microscope illuminator lamp, is focused onto the reflectance standard by an aluminized, silica-overcoated mirror. The lamp calibration data are furnished in units of $\mu W \cdot cm^{-2} \cdot sr^{-1} \cdot nm^{-1}$; they are converted to a wavenumber basis by multiplying by the square of the wavelength (40). The transfer optics used are the same ones used in the Raman experiment; separate calibration function determinations are made with and without the polarization analyzer in the beam. Correct polarization scrambler operation is verified by observing that the calibration result is independent of the analyzer orientation throughout the spectral range. The calibration function is generated by taking the ratio of the lamp emissivity to the observed signal at the lamp calibration points; these data are collected into tabular form, and an appropriate interpolation routine is used to generate calibration values to apply to an experimental spectrum. It is estimated that the major error in the procedure is due to the wavelength variation of the reflectance of the mirror used to image the filament onto the reflectance standard. For aluminum, this has a value of $\pm 2.4\%$ over the 450–750 nm calibration region. [The total variation of the reflectance of the white reflectance standard over this range is 0.2% (39).] A correction for this effect is not made, however, since the actual reflectance dependence for the mirror used is not known, but the effect of this variation will be negligible over the maximum 50 nm or so separation between a reference band and the Raman band to be measured.

The calibration results have been checked by determining the cross sections for the rotational and vibrational lines in H_2 relative to each other and to the N_2 vibrational Q branch, with several of the Ar^+ laser excitation lines. The largest deviation from literature values (5) was less than 5%, much better than the 10% error limits quoted for these values.

12.3. APPLICATIONS

12.3.1. Theoretical Background

In a Raman-scattering experiment, the power of the light scattered in a given direction due to a given (vibrational) transition of the molecule is given by

$$I_{T,i} = N\left(\frac{\partial \sigma}{\partial \Omega}\right)_i I_0, \tag{1}$$

where I_0 is the power density of the incident radiation, $(\partial\sigma/\partial\Omega)_i$ is the differential scattering cross section for the ith vibrational band, and N is the number of scattering molecules (5). If the assumptions of Placzek's polarizability model (41) are satisfied, the differential cross section may be expressed in terms of the properties of the molecular polarizability. For the conventional rectilinear experimental arrangement (observation direction orthogonal to both the direction and polarization vector of the linearly polarized incident light), the differential cross section is given by (15)

$$\left(\frac{\partial \sigma}{\partial \Omega}\right)_i = \left(\frac{\pi^2}{90\epsilon_0^2}\right)\nu_s^4\left[1 - \exp\left(\frac{-hc\nu_i}{kT}\right)\right]^{-1} g_i(45\bar{\alpha}'^2 + 7\gamma'^2), \tag{2}$$

where ν_s is the wavenumber of the scattered radiation; ν_i is the band origin and g_i is the degeneracy of the vibrational mode; $\bar{\alpha}'$ and γ'^2 (units: Cm^2/V, where C is coulombs and V volts) are the invariants (41) of the derivative of the molecular polarizability with respect to the *dimensionless* normal coordinate; ϵ_0 is the permittivity of vacuum ($8.8542 \times 10^{-12}\,CV^{-1}\,m^{-1}$); and h, c, k, and T have their usual definitions (Planck's constant, velocity of light, Boltzmann's constant, and temperature, respectively). As mentioned above, if the intensities are expressed in terms of the scattered photon rate, the cross-section expression should be multiplied by the factor ν_0/ν_s, where ν_0 is the wavenumber of the incident radiation.

The polarizability invariants $\bar{\alpha}'$ and γ'^2 give rise to "trace" and "quadrupole" spectra, respectively, to use Placzek's terminology (41). These are often called isotropic and anisotropic spectra, terms that incorrectly describe their spatial distribution characteristics. By taking into account the polarization properties of the scattered light, these two spectra may be separated (42). The two orthogonally polarized components of the scattered light are given by

$$I_{\parallel} \sim 45\bar{\alpha}'^2 + 4\gamma'^2$$

and

$$I_{\perp} \sim 3\gamma'^2, \tag{3}$$

where $I_{\parallel}$ denotes the component of the scattered light whose polarization vector is parallel, and $I_{\perp}$ perpendicular, to that of the incident light. Thus, from the two polarized components, measured by a suitable polarization analysis of the scattered light, one can obtain the trace and quadrupole spectra from the relationships

$$I_t = I_{\parallel} - \tfrac{4}{3}I_{\perp}$$

and

$$I_q = \tfrac{7}{3}I_{\perp}. \tag{4}$$

Note that these are normalized such that their sum will give the total scattering spectrum (42).

The advantage of performing this separation is that the trace and quadrupole spectra have different characteristics, so that it is preferable to consider them separately. The trace spectrum is governed by a more limited set of selection rules. It is observed only for totally symmetric vibrational modes (i.e., those vibrational motions maintaining the equilibrium symmetry of the molecule) and only for transitions between rotational levels having the same total angular momentum. The selection rules governing the quadrupole spectrum are much less restrictive as to both vibrational and rotational transitions. Thus, compared to its trace spectrum counterpart, the quadrupole spectrum of a gaseous sample will usually consist of more bands, each having a more complicated rotational structure. The characteristics of these two spectra can be compared in Fig. 12.1, where the Raman spectra of *n*-propane are displayed.

Although the integrated intensities of the two spectra are comparable, the few well-resolved, narrow rotational Q branches in the trace spectrum are by far the prominent features in the total spectrum. The occurrence of perturbations is easily discerned in the trace spectrum. For example, the effect of the ubiquitous Fermi resonances involving the C—H stretching modes of hydrocarbons is obvious in this region of the propane trace spectrum, while the large number of allowed transitions, with their accompanying unresolved rotational structure, complicates the quadrupole spectrum so that little can be concluded about such perturbations from this spectrum without a detailed high-resolution analysis. Of course, as with any spectroscopic selection rule, the fact that a transition is allowed does not guarantee its observation. In the case of propane, the symmetric CH_2 deformation modes at 1391, 1462, and 1474 cm^{-1} in the trace spectrum are not observed. Similar behavior is found for the symmetric hydrogenic deformation modes of ethane (43) and water (44), but note that the

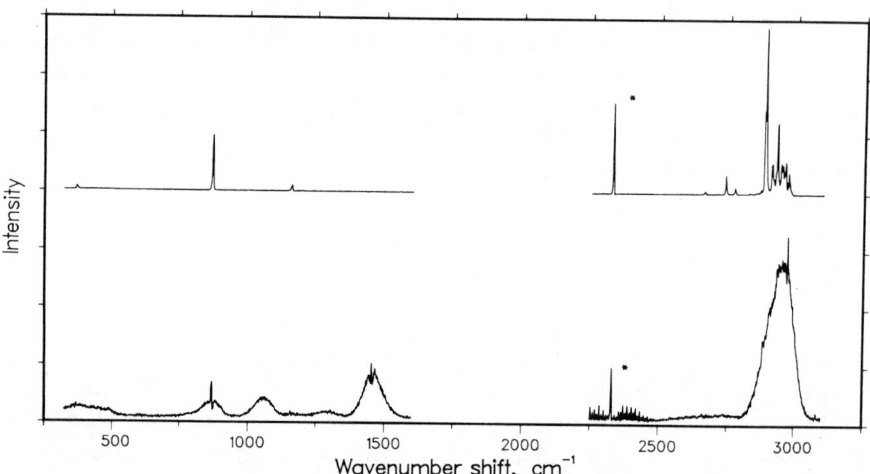

Figure 12.1. The trace (top) and quadrupole (bottom) Raman spectra of gaseous *n*-propane. The quadrupole spectrum has been scaled up by a factor of 10 for presentation purposes. The sample pressures were 200 mm Hg of *n*-propane and 500 mm Hg of nitrogen added for reference purposes. The nitrogen vibrational band is indicated by an asterisk. Excitation and spectrometer conditions: Ar^+ laser; 5 W at 514.5 nm; multipass optics; 2 cm^{-1} spectral slit; 1 cm^{-1} step size; 1.5-s sample period; two scans of each polarized component. From Gough et al. (45).

corresponding modes of deuterium-substituted isotopomers are easily observed (43, 45).

For quantitative applications, the advantages of working with the trace spectrum are obvious. The well-resolved, narrow bands permit the baseline to be determined in a straightforward manner, so that an unambiguous and precise band area may be measured. The contribution of the quadrupole spectrum to the total spectrum can compromise the determination of an accurate baseline, and affect the accuracy of the band-area determination. Thus, if it is at all feasible to perform a polarization analysis of the scattered light, the advantages of using the trace scattering spectrum for obtaining quantitative results are well worth the extra effort required to obtain it.

12.3.2. Methodology

12.3.2.1. Measurement of Species Abundance

For quantitative measurements, once the scattering cross section for a given transition is known, the abundance of scattering molecules can be determined from the scattered intensity using the relation in Eq. (1). Since it is difficult to measure the absolute power of the scattered radiation, such measurements are normally made on a relative basis, in terms of a suitable reference. In many systems, including atmospheric ones, the nitrogen vibrational transition at 2331 cm^{-1} provides a useful reference; the Q-branch cross section for this band has been critically reviewed and tabulated for several standard excitation sources (5). Note that, for the trace scattering cross section, a small correction is required for the contribution of the quadrupole cross section to the reported value (43).

In other cases, any strong, well-resolved band may be used, possibly after performing an appropriate calibration. To give an indication of the relative cross sections that may be expected, the compilation of literature results (5) is summarized in Fig. 12.2. Two other popular reference bands have been the totally symmetric band of methane, and the S_1 rotational line in hydrogen at 587 cm^{-1}. In the latter case, the cross section is well established on both an experimental and theoretical basis, and this line has served as a primary standard for the determination of other RS cross sections (5).

Another approach for obtaining the necessary cross-section data for a given application has been to estimate them by means of some intensity model. Various models have been proposed (46), including the bond polarizability, or electro-optical parameter, model (47, 48), and the Cartesian polarizability derivative model (49). However, recent results from our laboratory cast some doubt on the degree of transferability of such parameters (50), and such estimates should be used only with caution. An alternative

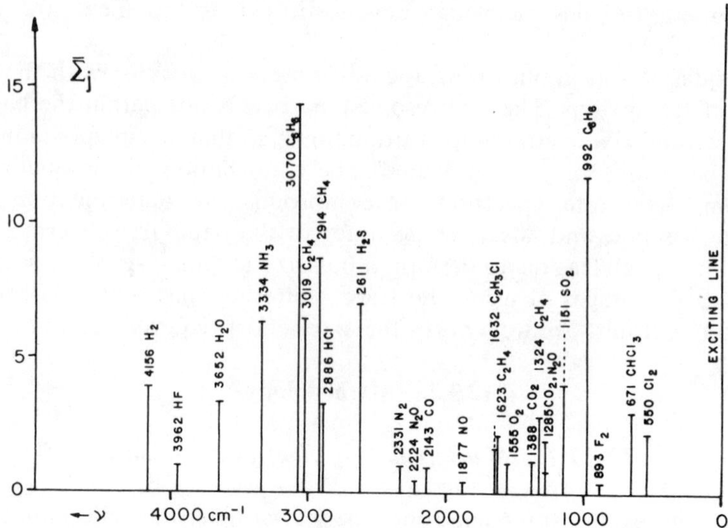

Figure 12.2. RS cross sections for vibrational bands of selected gases. The cross sections have been divided by the ν_s^4 and temperature factors in Eq. (2) and have been normalized by the corresponding nitrogen value. From Schrötter and Klöckner (5).

source for cross sections might be *ab initio* quantum mechanical calculations (51–53); recent results for smaller molecules correspond quite well with the corresponding experimental values. However, it is best, if at all possible, to calibrate the measurement directly, or at least to use a reference having an experimental cross section that is well established in the literature.

12.3.2.2. *Measurement of Temperature*

The determination of the temperature of a sample, or of one species in a system, is based on considerations similar to those required for species density measurements. In the latter, according to Eq. (1), the integrated intensity of a band is related to the total number of scattering molecules and to the total differential cross section for the transition. Specifically, the temperature factor in Eq. (2) originates from the assumption that the population of the various levels of the ith normal mode is given by a Boltzmann distribution and that the vibration is harmonic, to account for the contribution from each of the transitions $v \rightarrow v + 1$, where $v = 0, 1, 2, \ldots$. On the other hand, if transitions from individual levels can be resolved and the relative cross section for each transition is known, the relative population of the initial levels can be determined from their relative intensity. If the relative populations obey a

Boltzmann distribution, a temperature may be derived for the scattering species.

In situations near ambient temperatures, the relative intensities of pure rotational transitions have normally been used to derive temperature information. For nitrogen, for example, some 20 rotational transitions, spaced at $\sim 8\,cm^{-1}$, may be observed and are easily resolved with a conventional monochromator. Using established relations for the relative intensities of the transitions, temperature values may be determined (54) to an accuracy of $\pm 2°C$. A similar analysis for the symmetric top case has been described and applied to the case of ammonia (55).

Under high-temperature conditions, the relative intensities of the rotational spectrum can still provide a sensitive measure of temperature (56). However, in addition, the structure of vibrational bands may be used for temperature determination (57). If the vibration is sufficiently anharmonic, the vibrational "hot bands" ($v \rightarrow v+1$, where $v > 0$) are displaced with respect to the fundamental ($v = 0 \rightarrow 1$) so that the individual transitions may be resolved. The relations between the intensities of the fundamental and hot bands may then be used to measure the population distribution between the levels, and thus obtain a (vibrational) temperature of the species in a manner similar to the rotational case.

It has been shown that the rigid rotor and harmonic oscillator intensity expressions, which are useful for an initial estimate, must be corrected for higher-order effects in order to obtain accurate estimates of the sample temperature (58). Centrifugal distortion couples molecular vibration with rotation and must be taken into account for accurate rotational temperature measurements (58, 59). The mechanism that allows vibrational temperature measurements is the mechanical anharmonicity, which shifts the hot bands with respect to the fundamental so that they may be observed. In addition, for accurate temperature measurements, the effects of electrical anharmonicity, the higher-order derivatives of the polarizability, which have been neglected in deriving Eq. (2), must be taken into account (60). These corrections are largest for hydride molecules but are still nonnegligible for molecules such as nitrogen at high temperatures. With such corrections, temperature determinations from the Raman spectrum can be quite accurate; the aforementioned references may be consulted for details.

There are other molecular species that cannot be treated in such a straightforward manner. The *Q* branch of the symmetric stretching vibration of the water molecule has been used to monitor the temperature of that species, for example, but its complicated rotational structure does not permit resolution of individual hot-band transitions, so that the above outlined method cannot be used. In this case, the band contour for a range of temperatures has been calculated from the well-determined molecular constants for this molecule to simulate observed spectra and thus determine the water-vapor temperature in the experimental system (61, 62).

In any attempt to model the temperature behavior of a complicated contour such as that of water, the variation in widths of lines due to the various rotation–vibration transitions, as well as the temperature dependence of the line widths, must be taken into account. This has become a very active area of investigation, especially for application in nonlinear scattering processes (63).

12.3.3. Survey of Applications

The following discussion is intended to provide a flavor of what may be accomplished in the way of analysis of gas-phase systems using spontaneous Raman spectroscopy. It is not intended to be an exhaustive compilation of available results. It will, I hope, be useful to the interested reader as an introduction to the literature.

12.3.3.1. Combustion Diagnostics

As mentioned above, the 1973 workshop in Raman gas diagnostics (6) foretold the ensuing activity in this area. Valuable initial studies were carried out using spontaneous RS. If acceptable results can be achieved, this technique is recommended, since it is experimentally simpler than alternative ones. This approach continues to be recommended in a recent survey of laser-based combustion diagnostic techniques (64); a section of the latter publication provides a good summary of applications of RS in this area.

Owing to such problems as the interference from the luminous background emission present in many systems of practical interest, the various nonlinear scattering processes have been found to be better suited in applications such as studies of gas turbines and internal combustion engines. A comprehensive consideration of these techniques is beyond the scope of the present article; the interested reader should refer to the many reviews concerned with this work (see, e.g., 10, 65, 66).

There are two properties of the RS processes, both spontaneous and nonlinear, that make them especially well suited for studies of combustion systems. The first is the spatial definition that can be achieved. The intersection of the narrow laser beam and the optic axis of the observation system defines the sample volume, which can be scanned throughout the combustion system with suitable optics. This property may be contrasted with an absorption technique, for example, where the sample volume is defined by the length of the transmitted beam through the system, so the observed spectrum is an average response along the beam.

The second property is that the technique is noninvasive, in contrast to mechanical temperature probes or sampling devices that must be inserted into the combustion system and can thus perturb the properties to be measured. The scattering techniques do require optically transparent windows for passing the incident and scattered light; however, these need not intrude into the reaction volume itself and should not seriously affect the process being studied.

Initial spontaneous Raman studies were carried out with flat-flame laboratory burners using hydrogen and small hydrocarbon molecules as fuels. Such flames have low luminosity, so that classical Raman spectra can be observed without serious interference. Temperature measurements of several systems have been summarized (64). Notable among them are studies of hydrogen/air flames (17, 67), which provide good examples of spectra and their analysis. Species profiles have been generated for CH_4, O_2, H_2, CO, H_2O, and CO_2 in a methane/air flame (68). Figure 12.3 summarizes the

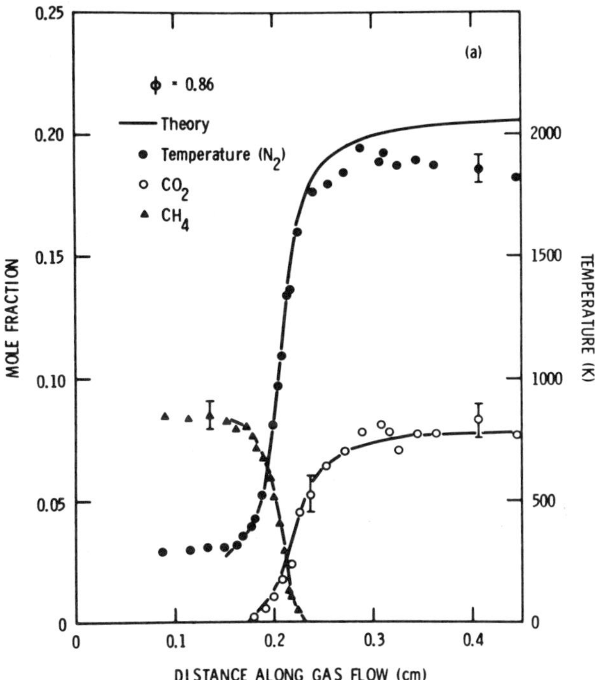

Figure 12.3. Spontaneous Raman spectroscopic measurements of concentration and temperature profiles in an atmospheric methane/air flame. The theoretical estimates are obtained from a model calculation involving 24 reactions. From Bechtel et al. (68).

results of this work, showing the spatial precision and accuracy obtainable in species concentration and temperature measurements. By means of time-resolved techniques, instantaneous mapping of fuel abundances has been carried out in turbulent diffusion flames (69). A Raman system for measuring temperatures in hydrogen/oxygen flames induced by oblique shocks has been described (69a). Also, the characteristics of a flame in the wake of a sphere in an air stream has been studied by Raman spectroscopy (69b).

Species detection is not limited to observation of the vibrational and rotational transitions discussed here so far. The presence of oxygen atoms in a H_2/O_2 flame was detected by Dasch and Bechtel, who observed transitions between the levels of the ^3P manifold of the electronic ground state (70). Other examples of atomic Raman spectroscopy are cited in this report.

As mentioned above, spontaneous Raman scattering is finding continuing application in the area of combustion diagnostics. Recent work (71) involves the use of a polychromator system for simultaneously monitoring the frequency shifts of eight selected species in methane flames; the observed results are corrected for background fluorescence that can occur under some flame conditions. Another proposed application is the measurement of temperatures in the space shuttle main engine preburner (72) from the observation of H_2 rotational and vibrational structure. Because of physical constraints, the various nonlinear and fluorescence techniques are not feasible, but spontaneous Raman scattering, excited and observed remotely via fiber optics, is predicted to have the required sensitivity and accuracy.

12.3.3.2. Remote Sensing

Laser-based techniques for remote sensing have had a wide application since the time of the development of the laser itself (73). A recent review of this field (74) discusses and compares the various techniques available and is a convenient reference to the previous literature. Raman-scattering applications in this area have been limited because of the low available sensitivity, but the specificity achievable by monitoring at the Raman-shifted frequencies of the species of interest has provided valuable results in selected cases.

The general remote-sensing experiment consists of firing a pulsed laser and collecting the return radiation by means of a collimator (typically a 0.5- to 1.0-m-diameter mirror) aligned with the laser direction. In the case of the Raman experiment, the collected radiation is dispersed or filtered to isolate and measure the bands of interest. Of course, the nitrogen vibrational band provides a convenient reference in atmospheric studies. Temperature measurements may be made from the nitrogen rotational intensities (28, 75) by means of the procedure outlined above. Range measurements can be made by correlating the time-resolved response with the time-of-flight of the transmitted and returned radiation.

Early results, including observations of pollutants, such as automobile exhausts and SO_2, and water-vapor profiles, have been reviewed by Inaba and Kobayashi (76). More recently, several groups have continued to develop systems for monitoring atmospheric water vapor (77–79, 79a). RS is felt to be especially useful over shorter ranges (up to 1–2 km), and it has the advantage of providing temperature information in addition to species detection.

The development and realization of a mobile LIDAR (light detection and ranging) system (a field remote Raman system) for the measurement of methane plumes in the atmosphere has been described in some detail (80). The operation of this system in monitoring gas vented from a high-pressure transmission system has also been described (81); its specifications of 10% concentration accuracy and a spatial resolution of 0.5×1.4 m at 1 km indicate the continuing utility of RS in this type of application.

12.3.3.3. Analytical Applications

There have been a variety of applications that demonstrate the capability of the Raman technique to discriminate the presence and properties of chemically similar species. In a chemical equilibrium, it is often possible to find a group of bands due to the various species involved. For example, in a study of the monomer/dimer equilibrium of acetic acid vapor (82), the pressure dependence of the C—C stretching bands of the two components over the range of ~0.25–1.5 atm could be used with gas-density measurements to obtain the equilibrium constant for the system. Determinations were carried out at 100° and 155°C. It was found that deviations from ideal gas behavior were appreciable and had to be taken into account to obtain a satisfactory result.

In a similar application, the energy differences between rotational conformers may be determined from the temperature dependence of the relative intensities of bands attributed to the species involved. The Raman spectrum is in general better suited for such a study because it is dominated by narrow Q-branch trace scattering components, whereby bands due to the two species should be more easily resolved than in IR absorption, where the broader rotational structure can complicate the assignment of the contributions from the two species. A problem in such studies is to attribute correctly the bands used and to eliminate contributions from extraneous features in the more complicated spectra. The measurement of the trace scattering spectrum in such systems could possibly improve this situation. Such an approach, making use of the total Raman spectrum, was used to derive the enthalpy difference between the trans and gauche conformers of *n*-butane (83). More recently, two measurements of ΔH for the trans-trans and trans-gauche forms of *n*-pentane have been reported (84, 85). From the temperature dependence of the intensity ratio of the same pair of bands, enthalpy differences of 560 ± 100 (84) and 430 ± 28 (85) cal/mol were determined for this molecule. The latter value was corrected for small contributions from the gauche-gauche conformer, to give a result of 465 ± 30 cal/mol. The difference in the two results was ascribed to differences in instrumental resolution (85).

An alternative approach to the determination of the energy difference between rotational conformers is to measure and analyze the structure of bands due to the torsional mode of vibration associated with the transition from one form to the other. One of the parameters of the analysis is the energy difference between the ground state of the two conformers, which can be obtained from the fit of the observed spectrum. Such measurements have been reported, for example, for *n*-butane (86, 87), and for halogen-substituted propanes (88).

If the differential cross sections for the chosen bands were known, the relative populations could be determined directly, using the relation given in Eq. (1). This was accomplished in recent studies in this laboratory, in which all of the observed trace scattering cross sections for a series of isotopomers were fit in terms of a set of isotopically independent intensity parameters. When two conformers of a given isotopomer could be present, their relative population was included in the fit. In each case investigated (43, 45, 50), the relative populations were consistent with a random distribution of the proton among the available sites. This method has also been used in a study of *n*-butane (88a). The relative trans-gauche population so measured is consistent with other recent determinations; reasons for the discrepancies between the earlier results (83, 86, 87) and the current ones were discussed (88a).

Raman spectroscopy has been used in the analysis of several types of gas mixtures. The components of natural gas samples have been determined to an accuracy of 0.002 mole fraction (89). A polychromator system was constructed and used in the analysis of automobile exhaust gases (20). Gases produced in the combustion of wood and various plastics were monitored with a sensitivity of ~ 1000 ppm for most components (90). From an analysis of their rotational Raman spectra, mixtures of CO_2, O_2, N_2, and CO were analyzed by fitting the observed spectra to a linear combination of spectra of the pure components involved (91). An accuracy of 1% relative concentration was achieved. Chemical reactions occurring within tungsten/halogen light bulbs were monitored by correlating the concentration of $CBrF_3$ and Br_2 with the filament burn time (92). Several trace compounds were also identified in the bulb. A. Raman system designed to measure air contamination in argon-filled insulating glass windows has been described (92a). A RS-based instrument has been developed to monitor patient respiration in hospital operating rooms (92b). It can measure respiratory gases to 0.43 vol% and volatile anesthetics to 0.03 vol%. In each of these applications, the ability of RS to monitor major components is well demonstrated; however, in addition, if the spectrum of a species includes one or more narrow, strong Q branches, its detection under lower concentration conditions can be achieved in favorable circumstances.

The strong rotational lines of the hydrogen molecule make it especially well suited for detection by RS, as shown in the earlier discussion of combustion diagnostics. In addition, its production in a reaction may be monitored quantitatively (93). Since the para nuclear spin species can only exist in even (odd if ortho) angular momentum states, the Raman spectrum provides an excellent means of monitoring the conversion of natural to para hydrogen (94). The rotational Raman transitions of H_2 and D_2 have been proposed for use in the calibration of Thomson scattering in tokomak fusion plasma experiments (95). Also, the contribution of Raman scattering, mainly from H_2, to the ultraviolet (UV) continuum observed from the atmospheres of the large planets has been discussed (96).

An application that poses some experimental challenges is the analysis of fluid inclusions in mineral samples. In one experiment (97), the presence of gaseous CO_2 and liquid water in inclusions in quartz was observed. In another (98), gaseous H_2S, CH_4, CO_2, H_2O, and solid carbonaceous material were observed in inclusions in calcite cements. While H_2S and CH_4 were found in all proportions in the different inclusions studied, CO_2, if present at all, was observed only in the 10–20% concentration range. It was estimated that CO_2 concentrations down to 0.5% should be observable for these samples (98). Detailed consideration of the experimental limitations to the quantitative analysis of mineral inclusions by Raman spectroscopy has led to questioning whether literature values of scattering cross sections apply to the case of fluid inclusions in minerals (99). In later work, spectra of known gas mixtures were used to derive response factors for each species in order to combine the instrumental sensitivity correction and the cross sections for the various species (100). It is unclear how this approach can better deal with uncertainties in the wavelength variation of the transmission of the incident and Raman-scattered radiation within an actual mineral sample.

Several other examples of Raman measurements under difficult experimental conditions have been reported. The temperature of hot water vapor (400–600 K) has been determined in a two-phase flow, which included droplets of liquid water that gave rise to a large stray-light background (101). Plasma diagnostic measurements have included the study of chemical deposition systems. Local temperatures and species concentrations in the heated zone of a chemical-vapor-deposition reacter have been measured during the deposition of boron on a boron fiber from a H_2/BCl_3 reactive mixture (102). In other studies, the temperature and species abundance for such molecules as SiH_4 and WF_6 have been studied as a function of the distance above the heated susceptor in the reactor (103, 103a). High temperature, gas-phase syntheses used in the manufacture of fiber optics involve producing high purity silica particles by introducing suitable silicon compounds into a methane/oxygen or hydrogen/oxygen flame; the particles are then deposited to form a glass which is later drawn into a fiber. The low luminosity of these particles allows spontaneous RS to be used to characterize these systems (103b, 103c). In a shock-tube diagnostic application, the temperature and abundance of N_2 and N_2^+ have been measured (104). In molecular beams, the rotational Raman spectra of N_2 and CO_2 have been used to determine the thermal distributions over their rotational states, as a function of distance from the beam nozzle (105). In the same study, jet cooling was observed for the different nuclear spin species of CH_4, and for vibrational and rotational cooling of SF_6; in the latter case the temperature parameters measured for the two degrees of freedom differed significantly, with the rotational cooling process being more efficient (105). In another application, the time-resolved response of low-pressure gases to IR multiphoton excitation has been observed with an apparatus optimized for such work (106).

12.4. CONCLUSIONS

From the foregoing discussion, it is clear that gas-phase, spontaneous Raman spectroscopy is not a routine technique that can be easily applied with reasonable chance of success in all situations. However, despite the limitations imposed by the fact that it is an energy-limited process, it is capable of producing worthwhile results in situations that can exploit its advantages. The relative experimental simplicity of the technique provides an advantage over higher-order, nonlinear scattering processes, if experimental problems, such as interfering background radiation from the scattering source, can be avoided. Compared with IR absorption techniques, it is capable of a much higher degree of spatial discrimination, and the strong, narrow Q branches are more convenient for quantitative measurements, especially if effort has been taken to separate the trace scattering spectrum. Thus, the potential of gas-phase Raman spectroscopy should not be forgotten; it can serve as a valuable technique in those cases able to take advantage of its unique properties.

REFERENCES

1. A. Weber, in *The Raman Effect* (A. Anderson, ed.), Vol. 2, pp. 543–757. Dekker, New York, 1973.
2. S. Brodersen, in *Raman Spectroscopy of Gases and Liquids* (A. Weber, ed.), pp. 7–69. Springer-Verlag, Berlin, 1979.
3. A. Weber, in *Raman Spectroscopy of Gases and Liquids* (A Weber, ed.), pp. 71–121. Springer-Verlag, Berlin, 1979.
4. W. F. Murphy, W. Holzer, and H. J. Bernstein, *Appl. Spectrosc.* **23**, 211–218 (1969).
5. H. W. Schrötter and H. W. Klöckner, in *Raman Spectroscopy of Gases and Liquids* (A. Weber, ed.), pp. 123–166. Springer-Verlag, Berlin, 1979.
6. M. Lapp and C. M. Penney, eds., *Laser Raman Gas Diagnostics.* Plenum, New York, 1974.
7. A. B. Harvey, ed., *Chemical Applications of Nonlinear Raman Spectroscopy.* Academic Press, New York, 1981.
8. Y. R. Shen, *The Principles of Nonlinear Optics.* Wiley, New York, 1984.
9. J. J. Valentini, in *Laser Spectroscopy and Its Applications* (L. J. Radziemski, R. W. Solarz, and J. A. Paisner, eds.), pp. 507–564. Dekker, New York, 1987.
10. J. W. Nibler and J. J. Yang, *Annu. Rev. Phys. Chem.* **38**, 349–381 (1987).
11. H. Frunder, R. Angstl, D. Illig, H. W. Schrötter, L. Lechuga-Fossat, J.-M. Flaud, C. Camy-Peyret, and W. F. Murphy, *Can. J. Phys.* **63**, 1189–1194 (1985).
12. W. Kiefer, H. J. Bernstein, H. Wieser, and M. Danyluk, *J. Mol. Spectrosc.* **43**, 393–400 (1972).
13. W. Knippers, K. van Helvoort, M. De Felici, J. Reuss, and S. Stolte, *Chem. Phys.* **105**, 27–39 (1986).
14. R. Fantoni, K. van Helvoort, W. Knippers, and J. Reuss, *Chem. Phys.* **110**, 1–16 (1986).
15. J. Martín and S. Montero, *J. Chem. Phys.* **80**, 4610–4619 (1984).
16. W. F. Murphy, *J. Chem. Phys.* **67**, 5877–5882 (1978).
17. D. P. Aeschliman, J. C. Cummings, and R. A. Hill, *J. Quant. Spectrosc. Radiat. Transfer* **21**, 293–307 (1979).
18. G. F. Bailey and J. R. Scherer, *Spectrosc. Lett.* **2**, 261–265 (1969).
19. A Campion and W. H. Woodruff, *Anal. Chem.* **59**, 1299A–1308A (1987).
20. M. D'Orazio and R. Hirschberger, *Opt. Eng.* **22**, 308–313 (1983).
21. C. A. Murray and S. B. Dierker, *J. Opt. Soc. Am. A* **3**, 2151–2159 (1986).
22. T. Hirschfeld and B. Chase, *Appl. Spectrosc.* **40**, 133–137 (1986).
23. D. B. Chase, *J. Am. Chem. Soc.* **108**, 7485–7488 (1986).
24. C. G. Zimba, V. M. Hallmark, J. D. Swalen, and J. F. Rabolt, *Appl. Spectrosc.* **41**, 721–726 (1987).

25. G. N. Zhizhin and M. N. Popova, *J. Appl. Spectrosc.* **32**, 662–665 (1980).
26. D. E. Jennings, A. Weber, and J. W. Brault, *Appl. Opt.* **25**, 284–290 (1986).
27. D. E. Jennings, A. Weber, and J. W. Brault, *J. Mol. Spectrosc.* **126**, 19–28 (1987).
28. J. A. Cooney, *Opt. Eng.* **22**, 292–301 (1983).
29. S. Kint, R. H. Elsken, and J. R. Scherer, *Appl. Spectrosc.* **30**, 281–287 (1976).
30. J. Laane, in *Vibrational Spectra and Structure* (J. R. Durig, ed.), Vol. 12, pp. 405–467. Elsevier, New York, 1983.
31. R. Savoie, M. Langlais, and P. Beauchesne, *Appl. Spectrosc.* **41**, 671–674 (1987).
32. M. D'Orazio and B. Schrader, *J. Raman Spectrosc.* **2**, 585–592 (1974).
33. F. J. Purcell, R. Kaminski, and E. Russavage, *Appl. Spectrosc.* **34**, 323–326 (1980).
34. R. Ouillon and S. Adam, *J. Raman Spectrosc.* **12**, 281–286 (1982).
35. H. Hamaguchi, I. Harada, and T. Shimanouchi, *Chem. Lett.* pp. 1405–1410 (1974).
36. R. A. Hill, A. J. Mulac, and D. R. Smith, *Appl. Spectrosc.* **30**, 183–186 (1976).
37. S. Montero, D. Bermejo, and M. A. Lopez, *Appl. Spectrosc.* **30**, 628–630 (1976).
38. M. Malyj and J. E. Griffiths, *Appl. Spectrosc.* **37**, 315–333 (1983).
39. F. Grum and T. E. Wightman, *Appl. Opt.* **16**, 2775–2776 (1977).
40. J. R. Scherer and S. Kint, *Appl. Opt.* **9**, 1615–1622 (1970).
41. G. Placzek, in *Handbuch der Radiologie* (E. Marx, ed.), 2nd ed., Vol. 6, Part II, pp. 205–374. Akad. Verlagsges. Leipzig, 1934. [Translation: UCRL-Trans-524(L), U. S. At. Energy Comm., 1961. Available from Office of Technical Services, Dept. of Commerce, Washington, D.C.].
42. J. R. Scherer, S. Kint, and G. F. Bailey, *J. Mol. Spectrosc.* **39**, 146–148 (1971).
43. K. M. Gough and W. F. Murphy, *J. Chem. Phys.* **85**, 4290–4296 (1986).
44. W. F. Murphy, *Mol. Phys.* **33**, 1701–1714 (1977).
45. K. M. Gough, W. F. Murphy, T. Stroyer-Hansen, and E. N. Svendsen, *J. Chem. Phys.* **87**, 3341–3346 (1987).
46. A. Rupprecht, *J. Mol. Spectrosc.* **108**, 165–205 (1984).
47. S. Montero and G. del Rio, *Mol. Phys.* **31**, 357–363 (1976).
48. M. Gussoni, in *Vibrational Intensities in Infrared and Raman Spectroscopy* (W. B. Person and G. Zerbi, eds.), pp. 221–238. Elsevier, Amsterdam, 1982.
49. M. P. Bogaard and R. Haines, *Mol. Phys.* **41**, 1281–1289 (1980).
50. K. M. Gough and W. F. Murphy, *J. Chem. Phys.* **87**, 1509–1519 (1987).
51. G. B. Bacskay, S. Sæbø, and P. R. Taylor, *Chem. Phys.* **90**, 215–224 (1984).
52. E. N. Svendsen and T. Stroyer-Hansen, *Mol. Phys.* **56**, 1025–1032 (1985).
53. M. J. Frisch, Y. Yamaguchi, J. F. Gaw, H. F. Schaefer, III, and J. S. Binkley, *J. Chem. Phys.* **84**, 531–532 (1986).
54. A. S. Gilbert and H. J. Bernstein, in *Laser Raman Gas Diagnostics* (M. Lapp and C. M. Penney, eds.), pp. 161–169. Plenum, New York, 1974.

55. G. H. Miller and A. J. Mulac, *J. Quant. Spectrosc. Radiat. Transfer* **25**, 53–58 (1981).
56. M. C. Drake, C. Asawaroengchai, D. L. Drapcho, K. D. Veirs, and G. M. Rosenblatt, *Temp.: Its Meas. Control Sci. Ind.* **5**, 621–629 (1982).
57. M. C. Drake, M. Lapp, and C. M. Penney, *Temp.: Its Meas. Control Sci. Ind.* **5**, 631–638 (1982).
58. M. C. Drake, C. Asawaroengchai, and G. M. Rosenblatt, *ACS Symp. Ser.* **134**, 231–237 (1980).
59. L. M. Cheung, D. M. Bishop, D. L. Drapcho, and G. M. Rosenblatt, *Chem. Phys. Lett.* **80**, 445–450 (1981).
60. M. C. Drake, *Opt. Lett.* **7**, 440–441 (1982).
61. J. L. Bribes, R. Gaufrès, M. Monan, M. Lapp, and C. M. Penney, *Appl. Phys. Lett.* **28**, 336–337 (1976).
62. R. J. Hall and J. A. Shirley, *Appl. Spectrosc.* **37**, 196–202 (1983).
63. D. A. Greenhalgh, R. J. Hall, F. M. Porter, and W. A. England, *J. Raman Spectrosc.* **15**, 71–79 (1984); D. A. Greenhalgh and L. A. Rahn, *J. Raman Spectrosc.* **21**, 847–855 (1990).
64. J. H. Bechtel, C. J. Dasch, and R. E. Teets, in *Laser Applications* (J. F. Ready and R. K. Erf, eds.), Vol. 5, pp. 129–212. Academic Press, Orlando, Florida, 1984.
65. R. J. Hall and A. C. Eckbreth, in *Laser Applications* (J. F. Ready and R. K. Erf, eds.), Vol. 5, pp. 213–309. Academic Press, Orlando, Florida, 1984.
66. G. M. Dobbs and A. C. Eckbreth, *Proc. SPIE—Int. Soc. Opt. Eng.* **644**, 20–27 (1986).
67. M. C. Drake and J. W. Hastie, *Combust. Flame* **40**, 201–211 (1981).
68. J. H. Bechtel, R. J. Blint, C. J. Dasch, and D. A. Weinberger, *Combust. Flame* **42**, 197–200 (1981).
69. M. B. Long, D. C. Fourguette, M. C. Escoda, and C. M. Layne, *Opt. Lett.* **8**, 244–246 (1983).
69a. P. Kasal and J. Algermissen, *Z. Flugwiss. Weltraumforsch.* **13**, 399–404 (1989).
69b. T. Kadota, K. Sumida, and F. Hiasa, *JSME Int. J. [2]* **33**, 355–361 (1990).
70. C. J. Dasch and J. H. Bechtel, *Opt. Lett.* **6**, 36–38 (1981).
71. A. R. Masri, R. W. Bilger, and R. W. Dibble, *Combust. Flame* **68**, 109–119 (1987).
72. J. A. Shirley, *NASA Conf. Publ.* **2372**, 603–617 (1985).
73. E. D. Hinkley, ed., *Laser Monitoring of the Atmosphere*. Springer-Verlag, New York, 1976.
74. W. B. Grant, in *Laser Spectroscopy and Its Applications* (L J. Radziemski, R. W. Solarz, and J. A. Paisner, eds.), pp. 565–621. Dekker, New York, 1987.
75. Yu. F. Arshinov, S. M. Bobrovnikov, V. E. Zuev, and V. M. Mitev *Appl. Opt.* **22**, 2984–2990 (1983).
76. H. Inaba and T. Kobayashi, in *Laser Monitoring of the Atmosphere* (E. D. Hinkley, ed.), pp. 153–200. Springer-Verlag, New York, 1976.

77. J. A. Cooney, K. Petri, and A Salik, *Appl. Opt.* **24**, 104–108 (1985).

78. R. J. Schwiesow, *J. Appl. Meteorol.* **22**, 881–890 (1983).

79. S. H. Melfi and D. Whiteman, *Bull. Am. Meteorol. Soc.* **66**, 1288–1292 (1985); S. H. Melfi, D. Whiteman, and R. Ferrare, *J. Appl. Meteorol.* **28**, 789–806 (1989).

79a. A. Ansmann, M. Riebesell, and C. Weitkamp, *Opt. Lett.* **15**, 746–748 (1990).

80. J. D. Houston, S. Sizgoric, A Ulitsky, and J. Banic, *Appl. Opt.* **25**, 2115–2121 (1986); R. M. Bilbe, S. J. Bullman, and F. Swaffield, *Meas. Sci. Technol.* **1**, 495–499 (1990).

81. J. D. Houston and D. R. Brown, *NASA Conf. Publ.* **2431**, 295–297 (1986).

82. R. Gaufrès, J. Maillols, and V. Tabacik, *J. Raman Spectrosc.* **11**, 442–448 (1981).

83. A. L. Verma, W. F. Murphy, and H. J. Bernstein, *J. Chem. Phys.* **60**, 1540–1544 (1974).

84. M. Maissara, J. C. Cornut, J. Devaure, and J. Lascombe, *Spectrosc. Int. J.* **2**, 104–119 (1983).

85. I. Kanesaka, R. G. Snyder, and H. L. Strauss, *J. Chem. Phys.* **84**, 395–397 (1985).

86. J. R. Durig and D. A. C. Compton, *J. Phys. Chem.* **83**, 265–268 (1979).

87. D. A. C. Compton, S. Montero, and W. F. Murphy, *J. Phys. Chem.* **84**, 3587–3591 (1980).

88. J. R. Durig, S. E. Godbey, and J. F. Sullivan, *J. Chem. Phys.* **80**, 5983–5993 (1984).

88a. W. F. Murphy, J. M. Fernández-Sánchez, and K. Raghavachari, *J. Phys. Chem.* **95**, 1124–1139 (1991).

89. D. E. Diller and R. F. Chang, *Appl. Spectrosc.* **34**, 411–414 (1980).

90. M. Aldén, J. Blomqvist, H. Edner, and H. Lundberg, *Fire Mater.* **7**, 32–37 (1983).

91. D. Schiel and W. Richter, *Fresenius' Z. Anal. Chem.* **327**, 335–337 (1987).

92. J. A. Sell, G. L. Eesley, and D. L. Simon, *Appl. Spectrosc.* **39**, 137–142 (1985).

92a. A. C. R. Pipino, *Appl. Spectrosc.* **44**, 1163–1169 (1990).

92b. R. A. Van Wagenen, D. R. Westenskow, R. E. Benner, D. E. Gregonis, and D. L. Coleman, *J. Clin. Monitoring* **2**, 215–222 (1986); D. R. Westenskow, K. W. Smith, D. L. Coleman, D. E. Gregonis, and R. A. Van Wagenen, *Anesthesiology* **70**, 350–355 (1989); D. R. Westenskow and D. L. Coleman, *Biomed. Instrum. Technol.* **23**, 485–489 (1989).

93. L. J. Johnston, D. J. Lougnot, V. Wintgens, and J. C. Scaiano, *J. Am. Chem. Soc.* **110**, 518–524 (1988); J. C. Netto-Ferreira, W. F. Murphy, R. W. Redmond, and J. C. Scaiano, *J. Am. Chem. Soc.* **112**, 4472–4476 (1990).

94. W. F. Murphy, unpublished work.

95. H. Röhr, *Phys. Lett.* **81A**, 451–453 (1981).

96. W. D. Cochran, *Adv. Space Res.* **1**, 143–153 (1981).

97. E. V. Guseva, F. P. Mel'nikov, R. Y. Orlov, and M. E. Uspenskaya, *Dokl. Akad. Nauk SSSR* **272**, 197–200 (1983).

98. N. Guilhaumou, B. Velde, and C. Beny, *Bull. Minéral.* **107**, 193–202 (1984).

99. B. Wopenka and J. D. Pasteris, *Appl. Spectrosc.* **40**, 144–151 (1986).

100. B. Wopenka and J. D. Pasteris, *Anal. Chem.* **59**, 2165–2170 (1987).

101. S. Neti, C. M. Anastasia, W. R. Smith, and J. C. Chen, *J. Raman Spectrosc.* **18**, 333–337 (1987).

102. J. Bouix, M. P. Berthet, M. Boubehira, J. Dazord, and H. Vincent, *J. Electrochem. Soc.* **129**, 2338–2343 (1982).

103. W. G. Breiland, M. E. Coltrin, and P. Ho, *Proc. SPIE—Int. Soc. Opt. Eng.* **385**, 146–151 (1983).

103a. R. Gaufrès, P. Huguet, D. Boya, and J. Lafforet, *Mater. Res. Soc. Symp. Proc.* **168**, 131–136 (1990).

103b. M. D. Allendorf, J. R. Bautista, and E. Potkay, *J. Appl. Phys.* **66**, 5046–5051 (1989); M. D. Allendorf and R. E. Palmer, *High Temp. Sci.* **26**, 45–58 (1989).

103c. I. Souche, *J. Chim. Phys.* **87**, 301–312 (1990).

104. J. W. Glaser and S. Lederman, *AIAA J.* **21**, 85–91 (1983).

105. G. Luijks, S. Stolte, and J. Reuss, *Chem. Phys.* **62**, 217–229 (1981).

106. E. Mazur, *Rev. Sci. Instrum.* **57**, 2507–2511 (1986).

INDEX

Abscissa correction, 116
Abscissa errors, in quantitative analysis by Raman spectroscopy, 115–116, 123
Absorption, correction for, 114
Absorption techniques, emission techniques versus, 128–129
Acetic acid, bandfitting in Raman spectrum of, 82–83
Acetone, Raman determination of, 110, 111
AlInAs/InP alloys, 204
Alloy semiconductors:
- alloy composition of, 200
- carrier concentration of, 206–207
- ion implantation and annealing of, 206
- potential fluctuations, of, 200–206
- Raman spectroscopy of, 200–211

Aluminum oxide, as catalyst support, 340, 344–345
γ-Aminobutyric acid (GABA), in nervous system transmission, 413–414
Analyzer, 56
Angle-resolved photoemission spectroscopy (ARPES), of catalysts, 327
Anisotropic Raman spectra, 45, 61
- in gas-phase spectroscopy, 435
- of uniaxially oriented polymers, 253–274

Anti-Stokes Raman spectroscopy, 5, 7, 8, 9, 140, 316
Antitumor agents:
- carcinogenesis model for, 419–420
- Raman determination of, 109

Aqueous systems:
- of biological molecules, 398
- Raman spectroscopy of, 279–282, 329

Area, in quantifying spectroscopic peaks, 118
Argon lasers, 22
Aromatic amines, Raman determination of, 109
Aromatic compounds, Raman spectra of, 287
Arsenic-implanted silicon, 169
Atomic displacement polarization selection rules, 147–150
Auger electron spectroscopy (AES), surface analysis by, 328
Average ligand number, of mononuclear complexes, 88
Azo dyes, Raman determination of, 109, 110, 112

Bandfitting programs, spectral analysis with, 80–83
Baseline:
- curvature, correction for, 119–121
- error due to, 80
- for quantitative analysis, 76

Beer's Law, 83
Biological mechanisms, Raman studies on, 412–421
Biological systems:
- Raman spectroscopy of, 90, 397–423
  - instrumentation, 398–401
  - problems in, 398
  - structures, 401
- resonance Raman scattering of, 13, 397

Birefringent wedge, 55
Bismuth molybdate catalysts, Raman spectroscopy of, 359–374
Bisulfate ion, symmetry-lowering effect of, 69
Blanks, in quantitative analysis, 80
Block diagram, for Raman scattering experimental setup, 154
Blue particles, in leukemia and cancer fluids, 417–420
BNDFIT computer program, Gaussian–Lorentzian bandshapes from, 81–82
Boron-implanted silicon, 169

Cadmium bromide ion, Raman spectrum of, 68
Cadmium nitrite complexes, Raman spectra of, 74, 89
Calcination, effect on catalyst surfaces, 343, 353
Cancer, drugs for therapy of, *see* Antitumor agents
Carcinogenesis, Raman studies on, 417–420
Carpenter's square, 46
CARS, 15, 17, 425
  of biological systems, 400
Catalyst Raman spectroscopy, 279, 325–395
  advantages of, 325–333
  bismuth molybdate catalysts, 359–374
  Fischer–Tropsch catalysis, 378–380
  fluorescence problem in, 339
  hydrodesulfurization catalysts, 339–342
  of hydroformylation catalysis, 380–381
  iron oxide catalysts, 374–375
  metathesis catalysis, 375–377
  methanation, 381
  of olefin dehydrogenation, 381–382
  in olefin hydrogenation, 377
  oxidation catalysis, 355–375
  in syngas processes, 377–381
  tungsten oxide-based, 340–342, 352–354
  vanadium oxide catalysts, 355–358
  zeolites, 334–339
Catalytic cracking, zeolite catalyst use in, 334
Catecholamines, Raman determination of, 109
CdZnTe alloys, optic phonons of, 203
Cell(s):
  division, aberrant, in cancer, 419
  growth, Raman studies on, 419
  Raman spectroscopy of, 401–402
CF_4/H_2 plasma, reactive ion etching with, 170
Charge coupled device (CCD) detector, 29, 32, 40, 124
Charge injection device (CID) detector, 29
Chemical deposition systems, gas-phase Raman studies on, 445–446
Chlorate ion, Raman spectrum of, 65
Chlorite ion, Raman spectrum of, 65
Cobalt-containing catalyst promoter structures, 345–347
Coherent anti-Stokes Raman spectroscopy, *see* CARS
Collection geometry, 60–61
Collection optics, 26–28
Colloids, surface-enhanced Raman spectroscopy of, 303
Combustion diagnostics, using gas-phase Raman spectroscopy, 441–442
Compton effect, 1
Computer(s):
  program, for bandfitting, 81
  Raman instrument control by, 90
  use in Raman detection systems, 30
  use in spectral data storage, 76, 80–81
Concentration matrix, 83
Configurations, experimental, 49
Conformation:
  Raman sensitivity to changes in, 230–231
  transition in biomolecules, Raman studies on, 407–408
Containers, for samples, *see* Sample containers and cells
Continuous wave lasers, *see* CW lasers
Copolymerizations, Raman spectroscopy of, 311–312
Copper-ligand complexes, Raman frequencies of, 417
Coupled LO-plasmon modes, in polarizability theory of Raman scattering, 146, 181
Cross correlation, in quantifying spectroscopic peaks, 118, 119
CSRS, 17
Cumulative formation constants, of mononuclear complexes, 87
CW lasers, 22–23
Cysteine, Raman spectroscopy of, 415

DABCO (1,4-diazabicyclo[2,2,2]octane), spectral isolation factor analysis of, 84
Data treatment, in quantitative analysis by Raman spectroscopy, 116–123
Degree of depolarization, *see* Depolarization ratio(s)
Degree of formation, of mononuclear complexes, 87
Depolarization ratio(s), 14, 61
  accurate values for, 45
Depolarization ratio(s), measurement of, 12–13
Detectors and detection systems, 29–30, 40–41
  diode array type, 124
  dynamic range of, 124
  for gas-phase Raman spectroscopy, 429–430

Diamond, atomic displacement effects in, 147–150
Dielectric constant tensor, 139
Digital data recording, in gas-phase Raman spectroscopy, 430
Digital filtering, correlation combined with, 119
Diode lasers, 39
Dispersion relations, for lattice vibrations, 152–154
Disulfide bonds, 402
  in biomolecules, Raman spectroscopy of, 415–417
Dry-processing procedures, effects on semiconductors, 199
Dye lasers, 24
Dyes, Raman determination of, 109

Effective ionic charge, in polarizability theory of Raman scattering, 142
Electrochemistry:
  Raman spectroscopy use in, 59, 85
  surface-enhanced Raman spectroscopy use in, 303
Electron energy loss spectroscopy (EELS), of catalysts, 327
Emission techniques:
  versus absorption techniques, 128–129
  experimental flexibility of, 129–130
  linear dynamic range in, 128
Endo-melting of biomolecules, Raman spectroscopy of, 406–407
Equilibrium constants, in factor analysis with equilibrium constraints (FAEC), 87, 100
Eriochrome Blue Black B (EBBB), 110
Errors, in quantitative analysis by Raman spectroscopy, 110–116
Ethanol, Raman determination of, 109, 110, 111, 130
Exponential-quadratic functions, for baseline, 80
Extended X-ray absorption fine-structure analysis (EXAFS), surface analysis by, 328
Eye lens, Raman studies on, 402

Factor analysis:
  in bandfitting programs, 83–87
  basic principles of, 91–103
  data matrix **I**, 91–92
  eigenvalue, 92–93
  eigenvalue matrix **E**, 93
  eigenvalue matrix **Q**, 93
  error function, 95
  factor indicator function (IND), 94
  number of factors, 93–96
  to obtain spectra and component concentrations, 96–99
  pure wavenumber constraint, 98
  reproduced data matrix **I**′, 94
  second moment matrix, 92–93
  taking equilibria into account, 99–103
Factor analysis with equilibrium constraints (FAEC), 85, 87, 99–103
  comparison with GAUSS-*Z*, 89–90
Fano effect, 146, 167
Fermentation mixture, ethanol in, 110, 130
Fermi resonance, overtone in, in Raman spectrum of perchlorite ion, 65
Ferric chloride ion, Raman spectrum of, 66, 79
Fiber-optic illumination and collection systems, 27–28, 283–285
Fibers, Raman spectroscopy of, 233–235
Filters, 24–26, 40
Fischer–Tropsch catalysis, Raman studies of, 378–380
Fluorescence:
  in anti-Stokes Raman spectroscopy, 316
  in biological systems, 399, 400–401
  extraction of Raman signal from background of, 316–317
  in organic systems, 312–317
  as problem in Raman spectroscopy, 3–4, 31, 37, 107, 117, 124–127, 223, 276, 312–317, 331
  -quenching agents, 315, 332–333
  in Raman spectroscopy of catalysts, 339
  reduction of, 313–316, 332–333
Formation curve, of mononuclear complexes, 88
Fourier domain, Raman data transformation into, 116
Fourier transform-Raman spectroscopy, *see* FT-Raman spectroscopy
Free carrier effects:
  in polarizability theory of Raman scattering, 142–146
  in Raman scattering, 145

Frequency calibration, for gas-phase Raman instrumentation, 430–432
Frequency exclusion, in data treatment, 117
FT-gas chromatography, gas-phase applications, 429–430
FT-infrared spectroscopy, of catalysts, 327, 329
FT-Raman spectroscopy, 1, 32–36, 41–42
  advantages of, 35–36, 332–333
  collection optics, 25
  components of, 32–35
  diode lasers for, 39
  fluorescence reduction in, 127
  of polymers, 250–251
  problems in, 37
Functional groups of oganic compounds, Raman and IR intensities for, 10–11
Fungicides, Raman determination of, 109

GaAlAs, surface films, 213
GaAlAs alloys, 180, 200
  carrier concentration of, 206
  potential fluctuations of, 200–202
GaAs, 180, 183
  annealed, 189
  dispersion relations for, 153
  polish-induced strain, 193
  radiation-damaged, 193
  silicon fluoride implanted, 192
  silicon-implanted, 189, 192
  surface preparation in, 198
GaAsSb alloys, 205
GaAs wafers, 211
Gadolinium, microcrystalline, 158
Gadolinium nitrate complexes, Raman spectra of, 89
GaInAs, 213
GaInAs/InP alloys, 203
GaSb:
  polish-induced strain in, 198
  surface preparation in, 198
Gas chromatography, of catalyst reaction products, 331
Gas lasers, 2
Gas-mixture analyses, by gas-phase Raman spectroscopy, 444–445
Gas-phase Raman spectroscopy, 425–451
  analytical applications of, 443–446
  combustion diagnostics using, 441–442
  experimental aspects of, 426–434
  frequency calibration in, 430–432
  instrumentation for, 426–434
    applications, 434–446
    calibration, 430–434
    theoretical aspects, 435–437
  intensity response calibration in, 432–434
  light sources of, 426–427
  methodology in, 437–447
  monochromator and detector for, 429–430
  remote sensing by, 443–444
  sample cells for, 427–428
  species abundance measurements by, 437–438
  temperature measurements in, 438–440
  transfer optics in, 428–429
GAUSS-*Z* program, 87–90
Geochemistry:
  Raman microscopy use in, 294
  Raman spectroscopy use in, 59
Geologic samples, fluid inclusions in, 109
Geology, Raman spectroscopy use in, 59
Germanium:
  detectors, 35, 41
  microcrystalline, 158
  structural and crystalline effects in, 163–167
Glan-laser prism polarizer, 56–57
Glass, laser-crystallized silicon on, 173–174
Glass sample containers, use in Raman spectroscopy, 279–282
γ-Globulin, Raman studies on, 410
Glutathione, Raman spectroscopy of, 415–417
Group IV semiconductors, 157–180
  microcrystalline, 158
  structural and crystalline effects in, 163–167
Group theory, 63

Hadamard transform techniques, in Raman spectroscopy, 37
Headspace, in sealed vials, 109
Heme group, reaction with carbon monoxide, Raman studies on, 410
Hemoglobin, Raman spectroscopy of:
  of active site, 410
  of interaction site, 409–410
He-Ne lasers, 2, 22
Heterojunctions, in semiconductor alloys, 208–209
HgCdTe alloys, 207–208
  oxide films on, 214

High-resolution electron energy-loss spectroscopy, surface analysis by, 328
Holographic filters, 40, 41
Hot band, in Raman spectrum of perchlorite ion, 65
Hydrodesulfurization catalysts, Raman spectroscopy of, 339–342
Hydroformylation catalysis, Raman spectroscopy of, 380–381
Hydrogen, detection by gas-phase Raman spectroscopy, 445
Hydrolysis, counteractants against, 80
Hyper-Raman effect, 15
Hypochlorite ion, Raman spectrum of, 65

InAs, 184
  polish-induced strain in, 198
  surface preparation in, 198
Indium chloride ion, Raman spectrum of, 68
Inelastic electron tunneling spectroscopy (IETS), of catalysts, 327
Infrared dispersion formula, 143
Infrared spectroscopy, *see* IR spectroscopy
InGaAs detectors, 35
Inhibitory transmitter, γ-aminobutyric acid as, 413
Inorganic species:
  identification in solution, 60–61
  Raman spectroscopy of, 59–105, 108
    data compilations of, 74
InP, 183, 184
  polish-induced strain, 193
  surface preparation in, 198
  Zn-implanted, implantation and thermal annealing of, 193
InSb, polish-induced strain in, 198
In situ Raman spectroscopy of catalysts, 329–331, 351
Instrumentation for Raman spectroscopy, 2–3, 21–43, 127, 154–157
  biological systems, 398–401
  gas-phase spectroscopy, 426–434
  tuning tools for, 46–47
Insulin, Raman studies on, 410
Intensity models, for gas-phase Raman spectroscopy, 437
Intensity response calibration, for gas-phase Raman instrumentation, 432–434
Internal standards, 79
  for quantitative analysis, 76, 123
  for Raman spectroscopy, 79, 110
  for Raman spectroscopy of organic compounds and petrochemicals, 289–290
Inverse Raman spectroscopy, 15, 17
Ion-damaged silicon, laser-annealed, 167–170
Ion interactions, Raman difference spectroscopy studies on, 90
Ion scattering spectroscopy (ISS), surface analysis by, 328
Iron oxide catalysts, Raman spectroscopy of, 374–375
Irradiation-damaged semiconductors, 193
IR spectroscopy:
  of bismuth molybdate catalysts, 362–363, 372–373
  of catalysts, 325, 327
  limit of detection in, 131
  Raman spectroscopy compared to, 2, 127-131, 275
    for quantitative analysis, 127–131
  surface analysis by, 328
Isotopic shifts, Raman difference spectroscopy studies on, 90
Isotopic tracer studies, of catalyst function, 368
Isotropic spectra, 45, 61
  in gas-phase Raman spectroscopy, 435

Jacobi method, application to factor analysis, 93

Krypton lasers, 22

Laser(s):
  -annealed ion-damaged silicon, 167–170
  beams, polarization orientation of, 56–57
  "burning out" by, fluorescence reduction by, 313–314
  as light sources, 22–24, 39–40, 282–285
    fiber optic use in, 283–285
  writing, of silicon microstructures, 180
Lattice vibrations, dispersion relations for, 152–154
Least squares fitting, in quantifying spectroscopic peaks, 118, 121–123
Lenses, 24–26, 40
Leukemia, blue particles associated with, 417

Light path geometry, of monochromators, 47–53
Light sources:
for gas-phase Raman spectroscopy, 426–427
for Raman spectroscopy, 22–23
Longitudinal optical mode, 144
Low-energy electron diffraction (LEED), surface analysis by, 328
Luminous background, interference in Raman combustion diagnostics, 440
Lydanne–Sachs–Teller relationship, 144

Magnesium acetate complexes, Raman spectra of, 90
Manganese oxide catalysts, Raman spectroscopy of, 358–359
Mathematical deconvolution, use in polymerization studies, 311–312
Matrix operations, use in least squares fitting, 121
Membranes, biological, Raman spectroscopy of, 402
Metal halides, Raman spectra of, 66
Metal-ligands, Raman frequencies of, 417
Metalloporphyrins, Raman studies on, 410
Metathesis catalysis, Raman spectroscopy of, 375–377
Methanation, Raman studies on, 381
Methanol, Raman determination of, 110, 111
Methionine, Raman spectroscopy of, 415
Methionylglycine, Raman spectroscopy of, 415
N-Methylacetamide (NMA), as internal intensity reference, 80
Methyl methacrylate polymerization, Raman spectroscopy of, 310–311
Microadhesion, of silicon, 174
Microcrystalline effects, in III–V semiconductors, 185–186
Microcrystalline geometries, of silicon and other group IV semiconductors, 157–163
Microsampling, 30–31, 41
Mineral inclusions, detection by gas-phase Raman spectroscopy, 445
Mirror:
absolute horizontal-type, 46–47
front-surface-type, 46
Molar coefficient, in quantitative analysis, 77
Molar intensity, in quantitative analysis, temperature dependence, 77, 78
Molar-intensity matrix, 83
Molar scattering coefficient, 61
Molecular dynamics, Raman spectroscopy of, 408–409
Molecular point group, molecular classification by, 62
Molecules, symmetry elements of, 62
Molybdenum oxide catalysts:
activated (sulfided)-type, 349–351
background and preparation variables for, 342–344
promoter structures for
cobalt-containing, 345–347
nickel-containing, 347–349
Raman spectroscopy of, 340, 342–352
Monochromators, 28, 40
for gas-phase Raman spectroscopy, 429–430
geometry of, 47–53
optical table for, 46
Mononuclear complexes, 87
Multichannel detection systems, 29–30
Multichannel measurements, versus scanning measurements, 123–124
Multiplex spectrograph, use in gas-phase Raman spectroscopy, 429

Nd/YAG laser, 23, 24, 32
Near-infrared lasers, fluorescence reduction by, 314–315
Nervous system, Raman studies on, transmission in, 413–414
Nickel-containing catalyst promoter structures, 347–349
Nitrate ion, Raman spectra and detection of, 70, 72–74, 109
Nitric acid, Raman spectrum of, 73
Nitrite ion, Raman spectrum of, 73
Noninvasive techniques, in Raman combustion diagnostics, 441
Nonlinear Raman effects, 15–17, 131, 426
coherent anti-Stokes Raman scattering, 15
hyper-Raman effect, 15
inverse Raman spectroscopy, 15
Raman-induced Kerr effect, 15
stimulated Raman gain spectroscopy, 15
stimulated Raman scattering, 15
Nonlinear Raman spectroscopy, 317
Nonpolar groups of organic compounds, characterization of, 286–288
Normal modes of vibration, 62

activity of, 62
degenerate, 62
dipole moment charge, 63–64
doubly degenerate, 63
reduced representation of, 62
totally symmetric, 63
triply degenerate, 63
Nuclear magnetic resonance (NMR), in studies on catalysts, 333
Nuclear wastes, oxyanions in, 109
Number of components (NC), contributing to Raman intensity, factor analysis of, 83
Nyquist criterion, abscissa correction and, 116

Ockham's razor, 83
Olefin dehydrogenation, Raman studies on, 381–382
Olefin hydrogenation, Raman spectroscopy of, 377
Optical multichannel array detection system instruments (OMAs), 28, 32, 138, 154, 331
Optic phonons, 200–205
Organic compounds and materials:
functional groups of, Raman and IR intensities for, 10–11
liquids, 90
nonpolar groups of, characterization, 286–288
polymers, 223–273
Raman spectroscopy of, 275–324
surface-enhanced Raman spectroscopy of, 304
Organic reactions and systems:
fluorescence in, 312–317
Raman spectroscopy of, 290–291
time-resolved studies on, 304–312
Organisms, Raman spectroscopy of, 401–402
Orientation, of polymer fibers, low-frequency spectra and, 242–243
Oxidation catalysis, Raman spectroscopy of, 355–375
Oxides, Raman spectroscopy of, 211–214

Peak height, in quantifying spectroscopic peaks, 118–119
Penta prism, 46
Peptides, in protein structure, Raman studies on, 410
Perchlorate ion:
as internal intensity reference, 77, 80
molecular symmetry of, 62
Raman spectrum of, 64, 65
Perchloric acid, from perchlorate ion conversion, 69
Pesticides, Raman determination of, 109
Petrochemicals, Raman spectroscopy of, 275–324
Pharmaceuticals, Raman determination of, 109, 113–114
Phenolate anion, Raman spectroscopy of, 413
Phenols, Raman determination of, 109
Phonons:
disorder-activated zone-edge type, 205–206
optic, 200–205
Picosecond pulsing and gating, in extraction of Raman signal from fluorescence background, 316–317
Plasma diagnostics, gas-phase Raman spectroscopy use in, 445–446
Plasma frequency, 182
Polarizability derivatives, in gas-phase Raman spectroscopy, 435
Polarizability tensor, 139
Polarizability theory of Raman scattering, 5, 7, 138–141
free carrier effects, 142–146
Polarization analyzer, for gas-phase Raman spectroscopy, 428
Polarization measurements, 45–57
experimental geometries for, 51
Polarization scrambler, 54–56
for gas-phase Raman spectroscopy, 428
Polished-induced strain, 193, 198–199
Pollution monitoring:
by gas-phase Raman spectroscopy, 443
of water, 108
Polychloroprene, 249
Polyethylene, 249
Poly(ethylene terephthalate):
fibers, 233–235, 237–238
orientation, conformation, and crystallinity of, 237–242
Raman spectroscopy of, 224–226, 250
structure of, 224
vibrational analysis of, 226–230

Polymerizations:
  multicomponent, Raman spectroscopy of, 311–312
  Raman spectroscopy of, 308–312
Polymers:
  Raman spectroscopy of, 223–252
    parameter selection for, 235–237
  specific heat derivation for, 244–248
  vibrational analysis of, 226–230
Polynuclear complexes, formation of, 88
Polystyrene, 249
Poly(vinyl chloride), 249
Poly(vinylidene fluoride), 254
  planar zigzag configuration of, 269–272
Power, scattered light, in gas-phase Raman spectroscopy, 434
Principal component analysis (PCA), *see* Factor analysis
Process-induced damage, of semiconductors, 193–199
Propylene oxidation, 359–360, 366, 367
Proteins:
  conformation of, Raman spectroscopy of, 410–412
  disulfide bonds in, 402
Protonation in biomolecules, Raman studies on, 414–415
PtSi interface, 174
Pure wavenumber constraint, 98

q-dependence, of plasma frequency, 182
Quadrupole spectrum, in gas-phase Raman spectroscopy, 435
Quantitative analysis by Raman spectroscopy, 107–135
  abscissa errors in, 115–116
  absorption correction in, 114
  data treatment in, 116–123
  emission versus absorption in, 129–130
  factor analysis in, 83–87
  historical development of, 108–109
  of inorganic species, 76–90
  internal standards for, 289–290
  multichannel versus scanning measurements, 123–124
  ordinate errors in, 110–114
  of organic compounds and petrochemicals, 288–291
  sensitivity in, 128, 130–131
  using bandfitting programs, 80–83
  variations affecting, 109–123

Raman, C. V., 1
Raman difference spectroscopy, 90
Raman effect, 1–19
  nonlinear, 15–17
Raman-induced Kerr effect, 15
Raman intensity(ies):
  formula for, 60
  for functional groups of organic compounds, 10–11
  generalizations for, 10
  matrix, 83
  temperature factor in, 60
Raman microprobe, 138, 154, 168
  fluorescence minimization by, 333
  use in catalyst studies, 332, 343, 352
Raman microspectroscopy, 291–294, 332
Raman MOLE, 343
Raman scattering:
  first-order, 140
  polarizability theory of, 141
  quantum mechanical treatment of, 8–9
  second-order, 141–142
  theory of, 138–154
  two-phonon-type, 209–211
Raman selection rules, *see* Selection rules
Raman spectrometer(s). *See also* Instrumentation for Raman spectroscopy
  collection optics, 26–28
  components of, 21, 22–30, 39–42
  computer use with, 30
  detection systems, 29–30, 40–41
  lenses and filters, 24–26, 40
  light sources for, 22–23
  microsampling, 30–31, 41
  monochromators, 28, 40
  sample handling in, 30–31
Raman spectroscopy. *See also* Fluorescence
  advantages of, 3, 276
  analytical applications of, 3
  analytical techniques used with, 331–332
  of binary zincblende semiconductors, 180–199
  of biological systems, 13, 397–423
  of catalysts, 279, 325–395
  classical description of, 4–8

of colloids, 303
correlation with species concentration, 76–80
Fano effect, 146
of fibers, 233–235
of Fischer–Tropsch catalysis, 378–380
gas-phase-type, *see* Gas-phase Raman spectroscopy
guide to literature of, 17–18
infrared absorption spectroscopy compared to, 2, 127–131
of inorganic species, 59–105
instrumentation for, *see* Instrumentation for Raman spectroscopy; Raman spectrometer(s)
interpretation in, 62–75
organic and petrochemical applications of, 275–324
of oxidation catalysis, 355–375
of poly(ethylene terephthalate), 224–226
of polymers, 223–273
uniaxially oriented, 253–274
problems in, 3. 31
quantitative analysis by, 107–135
resonance-enhanced, *see* Resonance-enhanced Raman spectroscopy
sample-handling techniques in, 277–285
samples for, 3
selection rules for, 9–10
of semiconductors, 137–221
sensitivity to conformational change, 230–231
signal enhancement in, 295–304
of silicon semiconductors, 157–180
in situ, of catalysts, 330
solvent behavior in, 128
standardization of measurements in, 57
surface-enhanced, 301–304
thermal sample degradation in, 31
of tungsten oxide-based catalysts, 340–342, 352–359
of zeolites, 334–339
Raman tensor, 141
Rank annihilation factor analysis, 84
Rayleigh scattering, 7, 140
Reactive ion etching, of silicon, 170–172
Reduced function, 61
Reduced representation, of normal modes of vibration, 62
Reference bands, for gas-phase Raman spectroscopy, 437
Remote sensing, by gas-phase Raman spectroscopy, 443–444
Resonance-enhanced Raman spectroscopy, 109
applications to organic and petrochemical systems, 295–301
of catalysts, 327
Resonance Raman effect, 13–14, 409
Retardation, 55
Retinal isomerization, Raman studies on, 412–413
Rhodopsin, Raman spectroscopy of, 412
Rochon polarizer, 47
Rotational conformers, gas-phase Raman spectroscopy of, 443–444
Rotational Raman spectroscopy, temperature measurements in, 439
Rule of Mutual Exclusion, 9–10

Sample containers and cells, 53–54
for catalysts, 330–331
for gas-phase Raman spectroscopy, 427–428
glass, 279–282
Samples:
containers for, *see* Sample containers and cells
irradiation of, 427
of organic and petrochemical compounds, handling techniques for, 277–285
purification of, fluorescence reduction by, 313
rotating system for, 90
Scanning measurements, versus multichannel measurements, 123–124
Scattering cross section, in gas-phase Raman spectroscopy, 434
Scattering geometry, of uniaxially oriented polymers, 254–256
Secondary ion mass spectroscopy (SIMS), surface analysis by, 328
Selection rules, 62, 140–141, 146–152
atomic displacement effects, 147–150
contrasting IR and Raman spectra, 9–12
lifting of degeneracy in, 72

linear-q and surface electric field-induced effects 150–152
mutual exclusion principle, 64
relaxation of, 141
Semiconductors:
alloy-type
heterojunctions, 208–209
Raman spectroscopy of, 200–211
ion implantation and annealing of, 186–193
process-induced damage of, 193–199
Raman spectroscopy of, 137–221
a-Si:F:H alloys, 166
SiGe alloys, structural and crystalline effects in, 163
Signal enhancement, in Raman spectroscopy, 295–304
Silicon:
alloys, 166
boron-implanted, 169
dispersion relations for, 153
evaporated thin films of, 170
implanted amorphized type, 168
on insulator (SOI), 161
ion-damaged and laser annealed, 167–170
/metal interface, 174–179
microadhesion of, 174
microcrystalline, 157–163, 166
microstructures, laser writing of, 180
polycrystalline film, 169
rapid thermal annealing of, 169
reactive ion etching of, 170–172
semiconductors, 157–180
structural and crystalline effects in, 163–167
surfaces, strain effects in, 172–174
Silicon dioxide, silicon-rich, 167
Silver, surface enhancement on, for Raman spectroscopy, 109
Similar chemical species, gas-phase Raman spectroscopy of, 443
Single-channel detection systems, 29
Solid samples, Raman spectroscopy of, 113, 129–130, 137–214
Solute–solute interaction, 62
Solute–solvent interaction, 62
Solvents:
comparison for Raman and IR measurements, 128
spectra of, substraction, 90
Space-charge layer, in GaAs, 183
Spatial Correlation Model (SCM), 203, 204
for ion implantation, 186
Spatial resolution, Raman microscopy in, 291–294
Species abundance, Raman measurements of, 437–438
Specific heat of polymers, derivation of, 244–248
Spectral isolation factor analysis (SPI), 84
Spectral slit width compensation, in intensity response calibration, 433
Spontaneous Raman spectroscopy, combustion diagnostics using, 441–442
Standardization, of Raman spectroscopic measurements, 57
Statistically nonzero eigenvalues, in factor analysis, 84
Stepwise formation constants, of mononuclear complexes, 87
Stimulated Raman gain spectroscopy, 15, 17
Stokes Raman scattering, 5, 7, 8, 9, 140
Strain effects, Raman measurement of, 172–174
Strontium nitrite complexes, Raman spectra of, 89
Styrene polymerization, Raman spectroscopy of, 290–291, 308–310
Subtraction of spectra, in abscissa correction, 116
Sulfamethoxazole, Raman measurement of, 113, 115, 130–132
Sulfanilamide, Raman determination of, 113, 130–132
Sulfate ion, symmetry-lowering effect of, 69
Sulfided molybdenum-containing catalysts, 349–351
Sulfur oxyanions, Raman determination of, 109
Support materials for catalysts, lack of Raman scattering by, 327, 329
Surface-electron fields, 184
Surface-enhanced Raman spectroscopy (SERS), 85, 301–304
of catalysts, 333–334, 327
of colloids, 303
of metal-island films, 303–304
of organic materials, 304
Surface extended X-ray absorption fine-structure (SEXAFS) technique, for catalyst analysis, 327

Surface films, Raman spectroscopy of, 211–214
Swamping, of Raman signal by fluorescence, 125
Symmetry elements:
of molecules, 62, 253
of uniaxially oriented polymers, 257–268
Symmetry-forbidden transverse optical mode scattering, 186
Syngas processes, Raman spectroscopy of, 377–381

Tellurium, precipitates of, on CdTe surfaces, 214
Temperature measurements, in gas-phase Raman spectroscopy, 438–440
Temperature probe, of silicon and group IV semiconductors, 180
Tetrahedron, molecular symmetry based on, 62–64
Thermal desorption spectroscopy (TDS), 328
Thermal sample degradation, 31
Time-resolved Raman studies, on organic reactions, 304–312
Tissues, Raman spectroscopy of, 401–402
Titanium, on silicon, 178
Titanium sapphire laser, 39–40
Trace spectrum, in gas-phase Raman spectroscopy, 435
advantages of, 435–436
Transfer optics, for gas-phase Raman spectroscopy, 428–429
Transition metals, in zeolite catalysts, 338–339
Transverse optical mode, 144
Trifluoromethanesulfonate ion (TFMS), as internal intensity reference, 79
t-RNA, Raman spectroscopy of, 402–408
Tungsten oxide catalysts, Raman spectroscopy of, 340–342, 352–354
Tungsten silicide, 179
Tuning tools, 46–47
Two-phonon Raman scattering, of semiconductor alloys, 209–211
Tyrosine phenolate, association with rhodopsin, 413

Ultraviolet lasers, fluorescence reduction by, 314–315, 399
Ultraviolet photoelectron spectroscopy (UPS), surface analysis by, 328
Ultraviolet Raman spectroscopy, 37
Ultraviolet spectroscopy, 2
Uniaxially oriented polymers:
Raman spectroscopy of, 253–274
scattering geometry of, 254–256
symmetry considerations of, 257–268

Vanadium oxide catalysts, Raman spectroscopy of, 355–358
Vibrational analyses, of polymers, 249
Vibrational Raman spectroscopy, temperature measurements in, 439
Vibrations, low-frequency type, Raman spectroscopy of, 288
Vision, Raman studies on, 412–413

Water:
Raman spectrum of, 110, 111
as solvent for Raman studies, 281–282, 329
Water-vapor profiles, using gas-phase Raman spectroscopy, 443

X-ray absorption near-edge spectroscopy (XANES), of catalysts, 353
X-ray spectroscopy (XPS), surface analysis by, 328

Zeolites, Raman spectroscopy of, 334–339
Zincblende materials and semiconductors:
atomic displacement effects in, 147–150
carrier concentration and space/charge layer effects in, 181–185
Raman spectroscopy of, 180–199
two-mode behavior, 180
Zinc bromide ion, Raman spectroscopy of, 66–68, 79, 87
Zinc chloride ion, Raman spectrum of, 68

(*continued from front*)

Vol. 63. **Applied Electron Spectroscopy for Chemical Analysis.** Edited by Hassan Windawi and Floyd Ho

Vol. 64. **Analytical Aspects of Environmental Chemistry.** Edited by David F. S. Natusch and Philip K. Hopke

Vol. 65. **The Interpretation of Analytical Chemical Data by the Use of Cluster Analysis.** By D. Luc Massart and Leonard Kaufman

Vol. 66. **Solid Phase Biochemistry: Analytical and Synthetic Aspects.** Edited by William H. Scouten

Vol. 67. **An Introduction to Photoelectron Spectroscopy.** By Pradip K. Ghosh

Vol. 68. **Room Temperature Phosphorimetry for Chemical Analysis.** By Tuan Vo-Dinh

Vol. 69. **Potentiometry and Potentiometric Titrations.** By E. P. Serjeant

Vol. 70. **Design and Application of Process Analyzer Systems.** By Paul E. Mix

Vol. 71. **Analysis of Organic and Biological Surfaces.** Edited by Patrick Echlin

Vol. 72. **Small Bore Liquid Chromatography Columns: Their Properties and Uses.** Edited by Raymond P. W. Scott

Vol. 73. **Modern Methods of Particle Size Analysis.** Edited by Howard G. Barth

Vol. 74. **Auger Electron Spectroscopy.** By Michael Thompson, M. D. Baker, Alec Christie, and J. F. Tyson

Vol. 75. **Spot Test Analysis: Clinical, Environmental, Forensic and Geochemical Applications.** By Ervin Jungreis

Vol. 76. **Receptor Modeling in Environmental Chemistry.** By Philip K. Hopke

Vol. 77. **Molecular Luminescence Spectroscopy: Methods and Applications** (*in two parts*). Edited by Stephen G. Schulman

Vol. 78. **Inorganic Chromatographic Analysis.** Edited by John C. MacDonald

Vol. 79. **Analytical Solution Calorimetry.** Edited by J. K. Grime

Vol. 80. **Selected Methods of Trace Metal Analysis: Biological and Environmental Samples.** By Jon C. VanLoon

Vol. 81. **The Analysis of Extraterrestrial Materials.** By Isidore Adler

Vol. 82. **Chemometrics.** By Muhammad A. Sharaf, Deborah L. Illman, and Bruce R. Kowalski

Vol. 83. **Fourier Transform Infrared Spectrometry.** By Peter R. Griffiths and James A. de Haseth

Vol. 84. **Trace Analysis: Spectroscopic Methods for Molecules.** Edited by Gary Christian and James B. Callis

Vol. 85. **Ultratrace Analysis of Pharmaceuticals and Other Compounds of Interest.** Edited by S. Ahuja

Vol. 86. **Secondary Ion Mass Spectrometry: Basic Concepts, Instrumental Aspects, Applications and Trends.** By A. Benninghoven, F. G. Rüdenauer, and H. W. Werner

Vol. 87. **Analytical Applications of Lasers.** Edited by Edward H. Piepmeier

Vol. 88. **Applied Geochemical Analysis.** by C. O. Ingamells and F. F. Pitard

Vol. 89. **Detectors for Liquid Chromatography.** Edited by Edward S. Yeung

Vol. 90. **Inductively Coupled Plasma Emission Spectroscopy: Part I: Methodology, Instrumentation, and Performance; Part II: Applications and Fundamentals.** Edited by J. M. Boumans

Vol. 91. **Applications of New Mass Spectrometry Techniques in Pesticide Chemistry.** Edited by Joseph Rosen